Plant Cell Death Processes

Plant Cell Death Processes

Edited by
Larry D. Noodén
Department of Biology
University of Michigan

ELSEVIER
ACADEMIC PRESS

AMSTERDAM • BOSTON • HEIDELBERG • LONDON
NEW YORK • OXFORD • PARIS • SAN DIEGO
SAN FRANCISCO • SINGAPORE • SYDNEY • TOKYO
Academic Press is an imprint of Elsevier

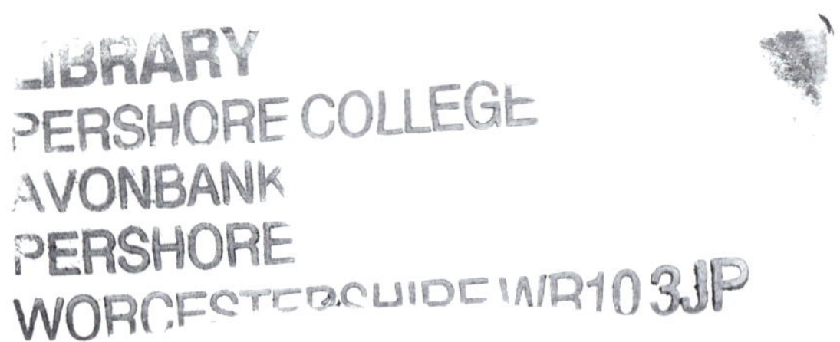

This book is printed on acid-free paper. ∞

Academic Press
An imprint of Elsevier Science
525 B Street, Suite 1900, San Diego, California 92101-4495, USA
http://www.academicpress.com

Academic Press
84 Theobald's Road, London WC1X 8RR, UK
http://www.academicpress.com

Library of Congress Control Number: 2003051842

International Standard Book Number: 0-12-520915-0

PRINTED IN THE UNITED STATES OF AMERICA
03 04 05 06 07 08 9 8 7 6 5 4 3 2 1

CONTENTS

10. Leaf Senescence and Nitrogen Metabolism 157

11. Photosynthesis and Chloroplast Breakdown 169

CONTRIBUTORS

Numbers in parentheses indicate the pages on which the authors' contributions begin.

Richard M. Amasino (91), Department of Biochemistry, University of Wisconsin, Madison, WI, 53706-1544

Tia-Lynn Ashman (349), Department of Biological Sciences; University of Pittsburgh; Pittsburgh, PA

Douglas G. Bielenberg (295), Department of Horticulture, Poole Agricultural Center, Clemson University, Clemson, SC 29634

Lars Olof Björn (285), Lund University; Department of Cell and Organism Biology; SE-223 62 Lund; Sweden

Malcolm C. Drew (19), Department of Horticultural Sciences; Texas A & M University; College Station, TX 77843

Urs Feller (107), Institute of Plant Physiology; University of Bern; Altenbergrain 21; CH-3013 Bern; Switzerland

Jean T. Greenberg (203), Department of Molecular Genetics and Cell Biology; University of Chicago; Chicago, IL 60637

Juan J. Guiamét (227), Instituto de Fisiologia Vegetal; Universidad Nacional de La Plata; c.c.327 1900—La Plata; Argentina

S. Hortensteiner (189), Department of Plant Biology; University of Zurich; Zollikerstrasse 107; CH-8008 Zurich, Switzerland

Klaus Humbeck (169), Institut für Pflanzenphysiologie; Martin-Luther-Universität Halle Wittenberg; Weinbergweg 10, D-06099 Halle, Germany

Donald A. Hunter (307), Department of Environmental Horticulture; University of California, Davis; Davis, CA 95616

Isaac John (227), Biology Department; University of Michigan; Ann Arbor, MI 48109-1048

Michelle L. Jones (51), Department of Horticulture and Crop Science; 1680 Madison Ave; The Ohio State University; OARDC; Wooster, OH 44691-4096

Kihachiro Kikuzawa (363), Center for Ecological Research and Faculty of Agriculture; Kyoto University; Kyoto; 606-8502; Japan

Keith T. Killingbeck (215), Department of Biological Sciences; University of Rhode Island; Kingston, RI 02991

Karin Krupinska (169), Botanisches Institut der Christian-Albrechts-Universität zu Kiel; Olshausenstraße-40; D-24098 Kiel; Germany

Eric Lam (37), Biotech Center; Foran Hall; Cook College; 59 Dudley Rd; Rutgers University; New Brunswick, NY 08903-0231

Nathan E. Lange (307), Department of Environmental Horticulture; University of California, Davis; Davis, CA 95616

Jennifer M. Mach (203), Department of Molecular Genetics and Cell Biology; University of Chicago; Chicago, IL 60637

Tadahiko Mae (157), Graduate School of Agricultural Science; Tohoku University; 1-1 Tsutsumidori-Amamiyamachi, Aoba-ku; Sendai 981-8555, Japan

Philippe Matile (189), Department of Plant Biology; University of Zurich; Zollikerstrasse 107; CH-8008 Zurich, Switzerland

Jennifer D. Miller (295), Department of Plant Pathology and Environmental Resources Research Institute; The Pennsylvania State University; University Park, PN 16802

Page W. Morgan (19), Department of Soil and Crop Sciences; Texas A & M University; POB 2474; College Station, Texas 77843

Stephen C. Morris (319), Sydney Postharvest Laboratory; Food Science Australia (CSIRO Afisc); North Ryde, NSW 2113; Australia

Yoo-Sun Noh (91), Department of Biochemistry; University of Wisconsin; 433 Babcock Drive; Madison, WI 53706-1544

Larry D. Noodén (1, 73, 227, 375), Biology Department; University of Michigan; Ann Arbor, MI 48109-1048

Eva J. Pell (295), Department of Plant Pathology and Environmental Resources Research Institute; The Pennsylvania State University; University Park, PN 16802

Barry J. Pogson (319), Division of Biochemistry and Molecular Biology; Australian National University; Canberra, ACT 0200, Australia

Dominique Pontier (37), Biotech Center; Foran Hall; Cook College; 59 Dudley Rd; Rutgers University; New Brunswick, NY 08903

Olga del Pozo (37), Biotech Center; Foran Hall; Cook College; 59 Dudley Rd; Rutgers University; New Brunswick, NY 08903

Betania F. Quirino (91), Pós-grad em Ciências Genômicas e Biotechnologia, Universidade Católica de Brasilia SGAN Quadra 916, Cap. 70 790-160; Brasilia; Brazil

Michael S. Reid (307), Department of Environmental Horticulture; University of California, Davis; Davis, CA 95616

Deborah Ann Roach (331), Department of Biology; University of Virginia; Charlottesville, VA 22904-4328

Michael J. Schneider (375), Department of Natural Sciences; University of Michigan, Dearborn; Dearborn, MI 48128-1491

Esa Tyystjärvi (271), Plant Physiology and Molecular Biology; University of Turku; FIN-20014; Turku; Finland

Claus Wasternack (143), Institute for Plant Biochemistry; PO Box 110432; D-06018 Halle (Saale); Germany

FOREWORD—AGING AND DEATH

The collective phenomena of aging, senescence and death are ubiquitous characteristics of the biological world. For centuries, humankind has considered these phenomena as unfair aberrations. Even in the recent past, senescence and aging have been assumed to be inevitable bad destinies. Dying organisms have been thought to be like dustbins of catastrophic biological mistakes (Kirkwood, 1979), or to be failing due to the accumulation of egregious errors (Comfort, 1956). It has been supposed that evolution was unable to correct these lethal mishaps because they so often occurred after an organism had passed beyond the reproductive state (Weismann, 1891).

From ontogenetic and evolutionary points of view, we now see that aging, senescence and death may actually provide important services. In fact, they have essential virtues in terms of cell and organ differentiation, ecological adaptation, and evolutionary change. Far from being egregious errors, aging, senescence and death can each be relatively orderly processes, regulated by distinctive sets of internal regulatory genetic programs.

Far from being restricted to old cells or tissues, senescence occurs even in some embryonic or meristematic cells. In plants, cells of xylem die and lose their contents before they can even begin to function in sap flow. Small groups of cells in plant meristems go through programmed cell death in the process that leads to differentiation of leaves, flowers and other plant tissues and organs (e.g., Sparks and Postlethwaite, 1967). In animals, the embryonic differentiation of bones, the differentiation of fingers and toes, and the shedding of unused organs such as the tadpole tail, also involve the programmed cell death of groups of cells, whether in an embryonic or a differentiated state. The continual ablation of cells from the surface of the skin or gut is a functional exercise of programmed senescence, i.e., programmed cell death (Goldstein, 1998).

In more ecological terms, localized cell death participates in adaptations against attack by disease organisms or parasites. This important type of adaptation provides for resistance to some disease organisms in plants (Kombrinck and Somssich, 1995) as well as in animals (Goldstein, 1998). The death of cells surrounding an infection site or a wound in many instances constitutes a strategy providing resistance to disease or parasitism (see Chapter 3).

One may ask whether the death of organisms could be of positive value. A few decades ago, it was assumed that death of individuals must necessarily have only negative value, and the longer that life can be extended, the greater the biological merit. But organismal death can play an intrinsic role in ecological adaptation to seasons. For example, annual plants in many instances die as a seasonal adaptation—witness the early death of spring flowers such as tulips, the synchronized midseason death of entire fields of wheat, and the late summer death of fields of maize (Leopold, 1961). Programmed organismal death is also found in some animals as illustrated by the death of entire populations of salmon after spawning, and the death of mayflies in a day or two after breeding (Clark and Rockstein, 1964). The suggestion that organismal death is beyond the reach of evolutionary change is belied by the findings that the life-span of certain insects and fish can change in response to predation pressure or even to seasonal selection of insect eggs (Rose, 1984; Reznick, 1997).

Beneficial values of programmed cell death are illustrated again in the fact that when the genetic factors that impose cell death are damaged, the result can be unhindered cell proliferation, resulting in some instances in cancer (Clark, 1999).

The death of whole organisms plays a major role in evolution. Of course organisms with the shortest life-spans such as annual weeds will be most capable of rapid evolutionary adaptation. Conversely, species with very long life-spans would be expected to be ponderously slow in making evolutionary adaptations (Leopold, 1975). In fact the ubiquitous feature of death in biological systems can be assumed to be essential for the evolutionary changes that provide the basis for the origin of species. We may assume that without organismal death, evolutionary change would be essentially prevented.

Knowing that all biological entities have a limited life span may be of little comfort to you or to me as we enter old age. Yet we can at least realize that in our senescence, aging and death, we are paying obeisance to a transcendent law of nature: the inexorable requirement for turnover.

A. Carl Leopold

References

Clark, W.R. (1999). *A Means to an End; the biological basis of aging and death.* Oxford University Press, New York.

Clark, A.M., and Rockstein, M. (1964). Aging in insects. In *The Physiology of Insects* (M. Rockstein, Ed.), Vol 1, pp. 227–281, Academic Press, New York.

Comfort, A. (1956). *The Biology of Senescence*. Rinehart, New York.

Goldstein, P. (1998). Cell death in us and others. *Science* **281**, 1283.

Kirkwood, T.B.L. (1977). Evolution of ageing. *Nature* **270**, 301–304.

Kirkwood, T.B.L. (1979). Senescence and the selfish gene. *New Scientist* **82**, 1040–1042.

Kombrinck, E., and Somssich, I.E. (1995). Defense responses of plants to pathogens. *Advances in Botanical Research* **21**, 1–34.

Leopold, A.C. (1961). Senescence in plant development. *Science* **134**, 1727–1732.

Leopold, A.C. (1975). Aging, senescence and turnover in plants. *BioScience* **25**, 659–662.

Reznick, D. (1997). Life history evolution in guppies as a model for studying the evolutionary biology of aging. *Experimental Gerontology* **32**, 245–288.

Rose, M.R. (1984). Laboratory evolution of postponed senescence in Drosophila melanogaster. *Evolution* **38**, 1004–1010.

Sparks, P.D., and Postlethwaite, S.N. (1967). Comparative morphogenesis of the dimorphic leaves of Cyamopsis tetragonaloba. *American Journal of Botany* **54**, 281–285.

Weissmann, A. (1891). *Essays upon Heredity and Kindred Biological Problems*. Oxford University Press, London.

PREFACE

Setting aside definitions for the moment, this book will deal with the developmental processes now widely recognized as senescence and/or programmed cell death (PCD) among plant biologists. The meaning of these terms will be discussed further in Chapter 1.

This volume will take up where the previous volume (*Senescence and Aging in Plants*, Noodén and Leopold, 1988) left off. That volume began reorganization of this field from an organ orientation, e.g., leaves, flowers, etc. (e.g., Thimann, 1980) to a process orientation. This volume will continue that trend; however, it will not attempt to update all the subjects covered in 1988. Since 1988, a great deal has been learned, especially in the area of apoptosis, a form of programmed cell death. In addition, several important ecological and evolutionary facets are covered here for the first time. All of these chapters aim to be more readable summaries of their fields rather than highly detailed compilations of all the facts on the subject. Hopefully, they will provide useful summary for specialists as well as easy access for non-specialists and facilitate the exchange of ideas between related but disparate fields. Readers should be aware of the general summary chapter on plant senescence and programmed cell death by Dangl, Dietrich and Thomas (2000).

I have included several chapters relating to the evolutionary and ecological aspects of senescence and programmed cell death, because this information provides valuable perspectives on the controls and functions of senescence and programmed cell death (Noodén and Guiamét, 1996). For example, the ecological issues that surround leaf longevity have great implications for crop yields and post harvest durability and this information and will ultimately provide targets for crop improvement. In addition, environmental factors play important roles in controlling senescence, and those roles need to be understood better. This interdisciplinary approach will help in the understanding of the mechanisms as well and facilitate practical application of the mechanistic information.

Of particular interest is the history of non-communication, lack of dialogue, between the physiological–molecular biology communities and the ecology–evolution communities, not to mention the animal gerontology communities, which are even more remote. One important result has been separate terminologies (sometimes even different concepts of the same term) and, of course, lack of integration of information. This volume aims to foster the integration of information from these diverse research communities.

Chapter 1 will discuss much of the terminology underlying the subjects of programmed cell death, senescence and aging; however, a different view is presented in Chapter 23. A clear terminology seems essential not only for effective communication but also to provide a focus for conceptual development. Ultimately, we will need to understand the mechanisms of these processes in order to define them well, but for now, they must remain mainly theoretical ideas. Interestingly, senescence has long been viewed as an active process by many communities of plant biologists, so programmed cell death provides a better descriptor for what is going on, and this terminology may also be less burdened with confusing connotations from other very widespread usages. Perhaps, this book is a good place to start viewing senescence as programmed cell death.

Larry D. Noodén

References

Dangl, J.L., Dietrich, R.A., and Thomas, H. (2000). Cell death and senescence. In *Biochemistry and Molecular Biology of Plants* (B.B. Buchanan, W. Gruissem, and R.L. Jones, Eds.), pp. 1044–1100. American Society of Plant Physiologists, Rockville, MD.

Noodén, L.D., and Guiamét, J.J. (1996). Genetic control of senescence and aging in plants. In *Handbook of the Biology of Aging* (4th ed.) (E.L. Schneider and J.W. Rowe, Eds.), pp. 94–118. Academic Press, San Diego, CA.

Noodén, L.D., and Leopold, A.C. (Eds.) (1988). *Senescence and Aging in Plants.* Academic Press, San Diego, CA.

Thimann, K.V. (Ed.) (1980). *Senescence in Plants.* CRC Press, Boca Baton, FL.

1

Introduction

Larry D. Noodén

I. What this Book Covers

This book will deal with the developmental processes now widely recognized as senescence and/or programmed cell death (PCD) among plant biologists. Some authors use PCD to refer to death of a small number of cells and senescence for organs and whole plants; however, senescence as conceived in the plant biology literature is in fact programmed cell death. Traditionally, senescence has been viewed broadly including processes involving only a few cells (Leopold, 1961), but not everyone shares that tradition. In many respects, programmed cell death is a better description of what is actually happening. The meanings of these terms will be discussed further in the next section; however, this chapter and most others will refer to senescence and will portray it broadly.

Senescence and programmed cell death play a very wide variety of roles in the life cycle of plants and other organisms (Leopold, 1961). Extensive listings of examples can be found in Barlow (1982), Noodén (1988a), Havel and Durzan (1996), Gray and Johal (1998), Dangl *et al.* (2000) and in this book. Of particular interest is the more in-depth coverage of cell death as a developmental process in maize (Buckner *et al.*, 2000), differentiation of certain cells (Gray and Johal, 1998; Chapter 2) and the hypersensitive response (Chapters 3 and 13). Endogenously controlled death is expressed at the cell, tissue, organ or even whole plant levels. In all cases, these deaths seem to play integral roles in the life cycle of an organism, even though those roles may not always be completely understood yet. Individual cells may be selectively targeted during plant embryo development, but PCD may occur in a whole tissue such as the tapetum in anthers or in whole organs such as flower petals or even in whole plants as in monocarpic senescence (Chapter 15). In some cases such as xylem vessel formation (Kuriyama and Fukuda, 2001) and lysigenous aerenchyma formation (Chapter 2), the protoplast is completely cleared out, but more often, a semblance of a cell is left at the end of the senescence process and the mostly emptied organ may be discarded by abscission (Butler and Simon, 1971; Biswal and Biswal, 1988; Noodén, 1988a).

This book examines not only the primary or central processes of programmed cell death/senescence but also some related processes such as chlorophyll degradation, disassembly of the photosynthetic apparatus and metabolism of the released nitrogen that may not be the proximal cause of death.

In addition to the biochemical and cellular perspectives of senescence, the broader implications of whole plant senescence, top senescence, autumnal senescence of trees and postharvest physiology are considered. Although senescence is an endogenous process, it can be induced by environmental factors, particularly those that cause stress, and these factors need to be examined in relation to senescence.

II. The Processes—Senescence, Aging, Programmed Cell Death, Apoptosis, etc.—Evolving Concepts

These terms reflect concepts, and they need to be articulated as clearly as we can, albeit imperfectly, with our current knowledge in order to foster the development of this field. They refer to processes that may be similar overall, and different terms may even refer to identical processes. Not surprisingly, different professional groups sometimes use the same terms differently. Before getting into these conceptual problems further, it is important to recognize how these ideas developed, i.e., where they came from. First, it must be noted that senescence/aging studies on plants and animals developed almost totally independently with little or no interchange until relatively recently (Noodén and Thompson, 1985; Finch, 1990). Likewise, these terms are viewed very differently by groups studying organisms or organs with a pronounced and rapid degeneration as in leaves, flower petals or monocarpic plants as compared to those studying the gradual decline apparent in polycarpic plants and most animals.

What has particularly captured the interest of plant physiologists and plant biologists in general has been the dramatic degeneration of leaves and whole plants (Molisch, 1938; Leopold, 1961; Thomas and Stoddart, 1980) and the ability of hormones, particularly cytokinin (Van Staden *et al.*, 1988) to retard that process, now widely termed senescence. Likewise, the highly visible senescence of flower parts is controlled by ethylene in many cases (Mattoo and Aharoni, 1988). These processes do progress with age, but they are not simply age-dependent. Senescence is not simply a passive accrual of "wear and tear". Senescence is often subject to correlative controls (i.e., it is controlled by other plant parts), and hormones often mediate these controls (Leopold and Noodén, 1984). Thus, these processes are clearly endogenous, not some gradual decline leading to increased vulnerability to acute life-terminating external forces such as drought, cold, herbivores or disease. These observations have led to formulation of the concept of senescence as an endogenous (i.e., developmental) process of degeneration leading to death (Leopold, 1961; Noodén and Leopold, 1978). In this sense, senescence is clearly also programmed cell death. Presently, it is easier (and maybe better) to define these terms as general concepts with a minimum of specifics until those are known with greater certainty.

By contrast, animal biologists and gerontologists focus mainly on the animals such as humans that show a gradual decline as opposed to those that die quickly after reproduction (Finch, 1990; Arking, 1991). Similarly, ecologists have tended to look at population dynamics with less emphasis on the sudden post reproductive death in monocarpic plants (see Watkinson, 1992; Chapter 23). While internal controls are recognized as possible factors in this gradual decline, external factors are clearly implicated in the final slippage toward death or the termination of life. Several scientific communities view this as senescence; however, the problem is complicated by the widespread use of aging to designate the same processes, especially in animals. Long ago, Medawar (1957) articulated the problems inherent in applying a standard dictionary term like aging, which represents the gamut of time-related changes, to biological processes occurring in living organisms. Finch (1990) also specifically avoids the use of aging to designate these processes in animals. The terminology differences between the animal and plant research communities are understandable, because endogenous controls of longevity are simply not very evident in the animals or organism populations studied. While it is true that longevity is a species characteristic and thereby can be presumed to be under genetic control, the process of decline toward death for most animals is a slow loss of vigor with time, and "wear and tear" (aging) has been the most evident explanation. Ecologists generally have been more concerned with understanding the results and less with the physiological mechanism underlying the process. In this view, senescence consists of decreasing fecundity and decreasing survival, i.e., an increase in mortality with age. Nonetheless, these declines may be attributed to internal factors (Watkinson, 1992). The extent to which the various declines with age are similar or different is unknown.

All these different approaches are valuable, but they have taken place in separate scientific communities and have produced different concepts of senescence. This book is aimed primarily at the experimental plant biology and agricultural communities and will use senescence primarily in the physiological sense. Chapter 23 will elaborate on a more demographic perspective and will bridge the understanding between the very different views in separate plant biology communities.

What is aging compared to senescence? The usage of aging is even more varied than senescence. No attempt will be made here to reconcile the different views of these concepts in the gerontology community. Indeed, they are evolving (Finch, 1990; Arking, 1991). There is a well-established, widespread, dictionary-based view of aging as passive, time-dependent changes in biological and non-biological entities. Consistent with this, in the experimental plant biology and the agricultural communities, aging is usually viewed as a passive accrual of lesions due mainly to external forces over time, e.g., "wear and tear" (Noodén and Leopold, 1978), although endogenous factors such as active oxygen species could also play a role in some tissues (Noodén and Guiamét, 1996). These endogenous factors may blur the line between aging and senescence. The clearest example of aging appears to be the loss of viability in seeds (Priestley, 1986; Roberts, 1988), but, no doubt, aging-type (passive) degeneration occurs in many places, and it must contribute to the decline of leaf function under some circumstances. For example, oxidative lesions generated during normal photosynthesis may accumulate over time in leaves (Munné-Bosch and Alegre, 2002). As the biochemical specifics of aging emerge, it should be possible to identify these in other tissues and assess their role in the life cycle and competitive ability of an organism. Aging is a very commonly used term to indicate simple changes with time, and it seems inappropriate to apply it to the complex developmental processes such as senescence and programmed cell death that are not simply determined by age/time.

The idea of programmed cell death and variations thereof have been around a long time in both the animal and plant literatures (Leopold, 1961; Kerr *et al.*, 1972; Vaux, 2002). Cell death is important in development, and it is turning out to have diverse manifestations in a wide range of organisms. Much of the investigation of cell death has been done on animal cells by medical pathologists, but they tend to deal with cell death caused by traumas and disease. Because these ideas evolved in different professional groups, the terminology and conceptual problems described above for senescence and aging also apply there. While these considerations may at first glance seem out of place in a chapter on plant cell death, it must be remembered that these are likely to be universal eukaryotic cell processes, and therefore, much of this information will apply to plant cells.

Until recently, most accepted the idea that cell death could be accomplished by two general types of processes, one (necrosis, oncosis, necrotic cell death, toxicological cell death, accidental cell death, traumatic cell death) due primarily to acute exogenous factors such trauma, stress, toxins, disease, etc. and the other (apoptosis, programmed cell death, physiological cell death, cell suicide, necrobiosis, shrinkage necrosis, autoschizis) due primarily to endogenous factors. Now, however, it is becoming apparent that each of these groupings may include several different processes.

Overwhelming stress can cause a type of cell death often termed necrosis, whereas sublethal stress can induce programmed cell death. The pathologists and some others object to the term necrosis, mainly because it has a long history of a different use, at least in the pathology literature (Majno and Joris, 1996; Levin *et al.*, 1999). Since necrosis was adapted primarily to designate cells that swell before death, it was proposed to call this "oncosis" after the Greek word "onko" for swelling, e.g., cell and organelle swelling, membrane rupture. The swelling is probably due to changes in ion concentrations and in osmotic equilibrium as the subcellular compartments break down (Levin, 1995; Trump and Berezesky, 1998). Necrosis then would refer to all types of cell death. Although concern has been expressed about confusing oncosis with malignancy; the term oncosis seems like a good idea. Certainly, the name oncosis describes the processes better. Oncosis differs from programmed cell death in other ways, e.g., ATP is required for PCD but not for oncosis (Nicotera *et al.*, 1999) and the phosphatidyl serine does not move to the outer face of the plasma membrane during oncosis (O'Brien *et al.*, 1997; Schlegel and Williamson, 2001).

Resolving the other set of terms, apoptosis, programmed cell death and variations thereof is a more complex matter, partly due to the enormous momentum these terms already have in the basic research literature (Lockshin, 1999). Ultimately, clarification will require knowing exactly what these active, endogenous cell destruction processes are. Nonetheless, a reasonable starting point has been outlined by Clarke (1990) who describes three cytologically different cell death processes: (1) apoptotic, (2) lysosomal/autophagic and (3) non-lysosomal vesiculate (Table 1-1, Fig. 1-1) with many examples of animal cells cited. These are outlined in Table 1-2 and Fig. 1-1. Probably, there are more endogenous pathways than the three listed in Table 1-1 and many variations of those. Although key features appear to distinguish these patterns, they seem more like syndromes defined by characteristics that may vary a bit. Even apoptosis and necrosis show some significant variations (Hunot and Flavell, 2001; Kumar and Vaux, 2002). For example, in apoptosis, the mitochondria may or may not play a central role or different caspases may be involved.

Table 1-1. Key Characteristics of Three Different Types of Developmental (Programmed) Cell Death Processes

I. Apoptosis	II. Autophagic degeneration	III. Non-lysosomal vesiculate degeneration
Condensation of the nucleus, chromatin and cytoplasm	Numerous autophagic vesicles form	Lysosomes not prominent, but the cytoplasm may become vacuolate
Cell membrane becomes convoluted		

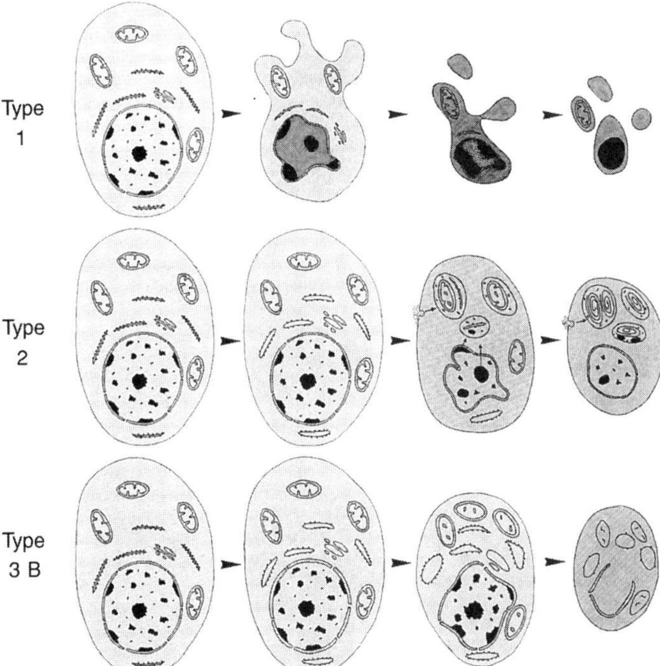

Figure 1-1. Schematic representations of the three commonest types of cell death (types 1, 2 and 3B). The drawings emphasize principally the most constant and best documented features of each type. Thus, some inconstant features, such as the swelling of mitochondria in types 2 and 3 have been omitted. However, the loss of nuclear DNA fragments to autolysosomes in type 2 cell death is shown despite the fact that this has so far been demonstrated in only one case. The different shades of gray indicate what is generally seen in ultramicrographs of lead-stained thin sections. This figure will be best understood if studied in relation to Table 1-2. From Clarke (1990).

The autophagic pathway may be fairly common in plants judging from the membrane whorls often observed within vesicles inside senescing plant cells (Noodén, 1988a); however, apoptosis is important in plants. Section IV will examine some of the key features of apoptosis found in plant cells, but these must be compared to the more complete picture available for animal cells. In any case, the pathway that is now commonly called apoptosis is by far the best characterized in plants, and it seems to be very important in normal development.

Table 1-2. Summary of the Three Main Types of Cell Death[a]

	Various designations	Nucleus	Cell membrane	Cytoplasm	Heterophagic elimination
Type 1	Apoptosis; shrinkage necrosis; precocious pycnosis; nuclear type of cell death	Nuclear condensation, clumping of chromatin leading to pronounced pycnosis	Convoluted, forming blebs	Loss of ribosomes from RER and from polysomes; cytoplasm reduced in volume becoming electron-dense	Prominent and important
Type 2	Autophagic cell death	Pycnosis in some cases. Parts of nucleus may bleb or segregate	Endocytosis at least in some cases; blebbing can occur	Abundant autophagic vacuoles; ER and mitochondria sometimes dilated; Golgi often enlarged	Occasional and late
Type 3A	Non-lysosomal disintegration	Late vacuolization, then disintegration	Breaks	General disintegration; dilation of organelles, forming "empty" spaces that fuse with each other and with the extracellular space	No
Type 3B	Cytoplasmic type	Late increase in granularity of chromatin	Rounding up of cell	Dilation of ER, nuclear envelope, Golgi and sometimes mitochondria forming "empty" spaces	Yes

[a]From Clarke (1990).

III. Apoptosis in Animals

Apoptosis is derived from Greek words meaning "falling away" or "discarding" (Kerr *et al.*, 1972). Although there are some variations in the apoptosis processes in animal cells, the sequence of cytological and biochemical events listed in Table 1-3 characterizes this process. It must be noted at the outset that not all cells undergoing what seems to be apoptosis show all these symptoms (Schulze-Osthoff *et al.*, 1994; Leist and Jäättela, 2001), so apoptosis may not be a single pathway, but a braided stream with some ability to flow through different channels. In addition, several different cell death programs may run in parallel and share some components (Ameisen, 2002). Nonetheless, apoptosis is important for the normal development and health of animals (Meier *et al.*, 2000). The role of apoptosis in determining the longevity of an animal is not clear (Lockshin and Zakeri, 1996; Warner *et al.*, 1997).

Once one starts to examine the molecular details of apoptosis and its controls, one soon becomes overwhelmed with the number of protein–protein interactions and the terminology. The literature is an alphabet soup of acronyms, and often, they differ for the same or related genes. It seems inappropriate to try to cover all these details here. Moreover, this picture and the acronyms are changing very rapidly as new information is added. Nonetheless, these are the standard to which plant studies must be compared. Apoptosis seems to consist of three phases: (1) initiation, (2) execution and (3) cleanup of the cell remains (e.g., autophagic elimination).

Table 1-3. The Main Features of Apoptosis in Animal Cells

Cytological	Biochemical
Cytoplasmic shrinkage, vacuolization and fragmentation	Cascade of caspase activation
Nuclear chromatin coalesces into clumps along the outer margin	Endolytic cleavage of many cell proteins, exs., lamins (nuclear membrane structural proteins) and poly ADPribose polymerase, (PARP a DNA repair enzyme)
Nuclear condensation and fragmentation	DNA cleavage into discrete nucleosomal fragments
Membrane blebbing → vesicles (apoptotic bodies) which may include nuclear fragments	Exposure of phosphatidyl serine on the inner surface of the plasma membrane
Phagocytosis of cell fragments	Binding and docking of the phagocyctes

Apoptosis can be initiated by a wide range of factors ranging from internal programming to external factors such as chemicals and environmental stress (Raff, 1992). This triggers a signaling pathway that appears to involve Ca^{+2} flux and protein phosphorylation. Interaction of ligands with membrane-spanning death receptors recruits certain intracellular proteins to the death domain of the receptor and initiates a caspase cascade.

Caspases are cysteine proteases that cut specifically after an aspartate. There are several different types of caspases, and they are present in the cytosol as inactive proenzymes that are activated after the death signal is triggered. They are recruited into cell-death signaling complexes through protein–protein interactions that lead to self-cleavage, oligomerization and activation. Although the release of cytochrome c from mitochondria may trigger the caspase cascade, there is an alternative pathway not involving the mitochondria (Finkel, 2001; Hunot and Flavell, 2001). Along the way, some antagonistic protein–protein interactions seem to be important in regulating apoptosis, e.g., the BCL-2 family mostly inhibits apoptosis by inhibiting caspase activation and preventing changes in the permeability of the mitochondrial membrane. However, one member of this family, BAX, promotes apoptosis, and this illustrates the complexity of these protein–protein interactions.

Apoptosis is sometimes called death by a thousand cuts due to the prominent role of the caspases, which cleave proteins internally. Nonetheless, several lines of evidence (e.g., caspase inhibitors and genetic knockouts of caspases) suggest that apoptosis can proceed in the absence of those specific caspases (Borner and Monney, 1999; Nicotera *et al.*, 1999; Joza *et al.*, 2001; Kumar and Vaux, 2002). Does this type of death represent a different pathway, perhaps one of the others cited above (Section II) or is it just a different channel in the braid of channels collectively called apoptosis or does it involve some other type of caspase that simply escaped the inhibitory treatments?

Very likely, the endolytic cleavage of the caspase substrates directly accounts for many of the characteristic changes occurring during apoptosis (Cohen, 1997). For example, cutting of the lamins in the nuclear envelope may result in the blebbing of the nucleus, and inactivation of the DNA-repair enzyme PARP [poly(ADP-ribose) polymerase] may contribute to the formation of DNA fragments; however, activation of an endonuclease seems necessary to cleave the DNA (Ameisen, 2002).

The nicking of DNA and cleavage into multiples of nucleosomal fragments are considered hallmarks of the apoptosis process. They are often measured respectively by TUNEL (terminal deoxynucleotidyl transferase-mediated dUTP nick end labeling) staining of the nuclei to detect 3′ breaks in the DNA and gel electrophoresis of the DNA to form a ladder-like pattern (internucleosomal fragments) on the gel. Certainly, these are good markers, but they are not absolute; many exceptions occur (Danon *et al.*, 2000; Joza *et al.*, 2001) and false positives may occur (Wang *et al.*, 1996). Furthermore, if the DNA is cleaved after the nucleosomal proteins are removed or if the DNA fragments are rapidly degraded, the "laddered" DNA fragments will not accumulate. In addition, a TUNEL-positive reaction may also occur in oncosis (Zakeri, 1998). What this adds up to is that there are no absolute criteria for defining/detecting apoptosis (certainly not DNA fragmentation), and it appears to be more of a syndrome. Again, this may merely be a manifestation of the braided channel nature of the process. These considerations are important in determining whether or not apoptosis occurs in plants.

IV. Apoptosis in Plants

It is widely accepted that programmed cell death occurs in plant cells, but is any of this apoptosis in the classical sense? There certainly are numerous cases where plant cells die as part of normal development, and they show some of the structural changes characteristic of apoptosis (O'Brien *et al.*, 1997; Gray and Johal, 1998; Buckner *et al.*, 2000; Dangl *et al.*, 2000; Danon *et al.*, 2000; Lam *et al.*, 2000; Jones, 2000; Hoeberichts and Woltering, 2002). These cytological and biochemical changes include nuclear condensation, aggregation of chromatin at the nuclear margins, cell shrinkage, formation of apoptotic bodies, migration of phosphatidyl serine to the outer surface of the plasma membrane (annexin binding) and cleavage of DNA into fragments corresponding to nucleosomes (laddering), TUNEL staining and DAPI staining. Perhaps, the greatest weight has been placed on the DNA cleavage; however, as noted above (Section III), these are not invariant components of apoptosis and therefore may not be definitive. Since plant cells are constrained and supported by cell walls, cell fragmentation and phagocytosis may not occur. Nonetheless, annexin binding indicates that the migration of phosphatidyl serine to the outer surface of the plasma membrane which triggers phagocytosis in animals, also occurs in plants (O'Brien *et al.*, 1997). In the end, however, the apoptosis pathway in plants will probably have to be defined in terms of the apoptotic molecular machinery (proteins) that participate. The data on plants are fragmentary, because they are mostly derived by comparison in bits and pieces to apoptosis in animal systems where most (but not all) of the basic mechanism studies are being carried out. Interestingly, much of this evidence comes from heterologous systems, where some of the animal apoptosis genes have been engineered into plants or vice versa.

The evolution of the apoptotic machinery and the degree to which it has been conserved in phylogeny will be discussed below (Section VII). With the genomes of several multicellular eukaryotes complete or nearly complete, it has been possible to search the corresponding databases for genes/proteins showing some sequence similarities to the genes/proteins believed to be important in apoptosis in animals (Aravind *et al.*, 1999), but the results depend on how the search is done, especially at low homologies. Initially, several important similarities were found in this way, and these include a family of apoptotic ATPases. Subsequently, plants have been shown to carry homologues of Apoptosis inducing factor (AIF) (Lorenzo *et al.*, 1999), Bax Inhibitor 1 (Sanchez *et al.*, 2000), PIRIN, which stabilizes quaternary complexes of apoptotic proteins (Orzáez *et al.*, 2001) and *dad-1* (defender against death, see below). In addition, plants have some apoptotic protein domains such as the nucleotide-binding domain NB-ARC of kinase 1a (van der Biezen and Jones, 1998) and a death domain (Jun *et al.*, 2002); however, it is also clear that the apoptotic machinery has not been conserved to the degree expected.

The caspases are of particular interest, and the classic vertebrate caspases apparently do not occur in plants (Uren *et al.*, 2000). There are two families of caspase-like proteins, metacaspases and paracaspases. The metacaspases occur in plants, fungi and some protozoa, while the paracaspases are found in animals and slime molds. The role of these metacaspases in plant cell death is not known; however, there is evidence that cysteine endoproteases do participate in PCD in plants (Solomon *et al.*, 1999; Xu and Chye, 1999; Lam and del Pozo, 2000; Hoeberichts and Woltering, 2002). One line of evidence for caspases comes from the ability of caspase inhibitors to block PCD. For example, the inhibitors Ac-DEVD-CHO or Ac-YVAD-CMK block the hypersensitive cell death triggered by the pathogenic bacteria (Lam and del Pozo, 2000). Caspase inhibitors can also suppress chemically induced cytological changes characteristic of apoptosis in cultured tomato cells (de Jong *et al.*, 2000). In addition, p35, a viral protein which is a fairly specific inhibitor of caspase, blocks apoptosis in plant cells (see Hoeberichts and Woltering, 2002). Yet another line of evidence comes from the cleavage of PARP into fragments characteristic of caspase action during apoptosis (see Hoeberichts and Woltering, 2002).

The release of cytC and other factors from the mitochondria may play a key role in plant cell death (Jones, 2000; Balk and Leaver, 2001; Orzáez *et al.*, 2001; Yu *et al.*, 2002b); however, some (Xu and Hanson, 2000) suggest that cytC release is not necessary. It may be that plants like animals have alternative cell death pathways that can bypass the mitochondria.

The search for animal apoptosis genes in plants appears to have emphasized *dad-1*, but this is likely a historical accident of convenience. *dad-1* codes for an essential subunit of oligosaccharyltransferase, which is not so easy to relate to apoptosis as the caspases; however, it has been shown to complex with MCL-1, a regulator of apoptosis and a member of the BCL-2 family (Makishima *et al.*, 2000), thereby providing a glimpse of why it is important. This also shows why it is so difficult to judge the importance of the many apoptosis-related factors when our knowledge of the process is so incomplete. Supporting its role in plant apoptosis, a *dad-1* homologue from Arabidopsis suppresses apoptosis in a hamster cell line (Gallois *et al.*, 1997).

There has been some uncertainty about the occurrence of BCL-2 proteins in plants; however, the evidence for BAX inhibitors seems clear. BAX is a death-promoting member of the BCL-2 family of proteins. BAX inhibitor-1 (At BI-1) has been found in Arabidopsis (Sanchez *et al.*, 2002). Furthermore, BAX from animals produces a cell death (LaComme and Santa Cruz, 1999 and vice versa (Yu *et al.*, 2002b)). In addition, an Arabidopsis and rice homologue (*AtBI-1*) of BAX Inhibitor-1 (BI-1) suppresses BAX-induced cell death in yeast (Kawai *et al.*, 1999; Sanchez *et al.*, 2000). Numerous heterologous engineered systems indicate that other BCL-2 proteins are active in plants. For example, when *bcl-*X_L from mammalian cells is engineered into tobacco with a promoter that overexpresses it, *bcl-*X_L inhibits cell death induced by UV-B irradiation or the herbicide paraquat (Mitsuhara *et al.*, 1999). Immunological procedures have also detected Bcl-2 proteins in plants (see Sanchez *et al.*, 2000).

Yet another approach has been to show that selective inhibitors such as 3-aminobenzamide, which inhibits the poly(ADP-ribose) polymerase or PARP, an enzyme implicated in apoptosis also inhibits xylem cell differentiation (Hawkins and Phillips, 1983).

Clearly, a process very similar to, but maybe not identical to, animal apoptosis exists in plants; however, its mechanism is still being worked out even in animals where a great deal more is known about the process.

If apoptosis and senescence (as defined here) are both programmed and if both of them involve cell death, then what is the relationship? This requires a further look at senescence.

V. The Senescence Syndrome

Although it is possible to define senescence conceptually as a developmental process, i.e., an endogenous degeneration process leading to death, it still is not possible to describe it in precise biochemical terms (Section II). In other words, the central or primary pathway of senescence is not known, and some of the changes that accompany senescence are likely to be peripheral or secondary as opposed to causal. For example, the glyoxylate cycle enzymes and metallothionein that increase during senescence (Buchanan-Wollaston, 1997) probably do not cause senescence. Thus, it may be appropriate to view this collection of changes as the senescence syndrome. In addition, the breakdown of the chloroplasts certainly is an important aspect of the syndrome; however, it may not directly cause the death of leaves and therefore may also be peripheral (Noodén, 1988a). This poses a dilemma because chlorophyll is the most conspicuous symptom of leaf senescence, and therefore, chloroplast-related parameters (e.g., chlorophyll and total leaf nitrogen which is mostly chloroplastic) are most often used to measure senescence. In the absence of clearly defined primary senescence parameters, it seems okay to use these chloroplast parameters as senescence markers so long as this possible limitation is recognized (Noodén, 1988a). It should be noted that these problems can be bypassed by measuring longevity and death; after all, that is the bottom line of the senescence process.

Even though the primary pathway of senescence is unknown, it seems worthwhile to think about the stages of senescence. Overall, these stages seem to be: (1) initiation, (2) degeneration (progression) and (3) terminal (Noodén *et al.*, 1997). The senescence initiators are environmental, disease and developmental, and all of these may be mediated by hormones. The degeneration phase presumably involves a decrease in cell maintenance functions and an increase in key dismantling enzymes.

The changes in ultrastructure during senescence have been described in some detail, but mostly for senescing leaves (Butler and Simon, 1971; Halevy and Mayak, 1981; Biswal and Biswal, 1988; Gepstein, 1988; Noodén, 1988a), and they provide some clues (but not definitive answers) to what is going on. Some of the earliest structural changes indicative of senescence occur in the chloroplast, i.e., changes in the grana and also formation of lipid droplets (plastoglobuli). Polysomes and ribosomes in general decrease fairly early reflecting a decrease in protein synthesis (Brady, 1988). On the other hand, the nuclei and mitochondria generally show little structural change until later. Visible disintegration of the plasma and vacuolar membranes seem to be terminal (late) events. Other changes occur, but they seem to vary somewhat among tissues and situations and are likely to be peripheral. At this point, it is difficult to say what these ultrastructural changes mean in the senescence process except that the nuclei and mitochondria may perform important functions that require their integrity until fairly late. The loss of integrity of the plasma (and probably also the vacuolar) membranes would mark death, the end of homeostasis.

It is a bit difficult to compare the changes in photosynthetic and non-photosynthetic tissues such as flower petals; however, there seems to be some general similarity but also differences. Just as differentiated cells differ in their structure, they will also differ in their disassembly. Even though nuclear breakdown usually occurs quite late in senescing tissues, there are, however, some cases where nuclear disintegration and vacuolar collapse, most notably xylem cells, occur quite early (Esau *et al.*, 1963; Fukuda, 1996; Groover *et al.*, 1997; Kuriyama and Fukuda, 2001). Furthermore, Pontier *et al.* (1999) showed that the plant disease hypersensitive response in leaves differs from leaf senescence in terms of a number of molecular markers. The visible appearance of the hypersensitive site also differs from ordinary leaf senescence. He *et al.* (2001) have clearly demonstrated both similarities and differences in gene expression during senescence of different tissues. Thus, senescence or PCD may differ somewhat among different tissues or different processes; however, it remains to be determined if these are differences in the primary pathway. Indeed, there are also indications that senescence may also differ within the same organ depending on the circumstances. For example, the chloroplasts may decrease mainly in number, presumably getting digested in the vacuole, or the number may remain the same with each shrinking in size (Gepstein, 1988). All these observations raise the possibility that there is more than one senescence process (Noodén, 1988a), or if senescence is braided channel, the different channels may look quite different.

Another set of differences in senescence occurs at the whole plant level. Whole plant or monocarpic senescence clearly depends heavily on correlative controls (sometimes referred to in the animal literature as social controls) which are influences of one part on another, and these controls may differ among different taxonomic groups, probably due to the polyphyletic nature of monocarpic senescence (Chapter 15). Since hormones often mediate these correlative controls (Leopold and Noodén, 1984), there may likewise be differences in the hormonal controls of senescence in different groups. Another distinctive feature of whole plant senescence is the importance of regeneration of organs or even ramets. Regeneration allows an escape from death, but once regeneration stops, the organism is locked onto a pathway to death.

The relationship between senescence and apoptosis is an important issue. Xylem cell differentiation does seem to exemplify apoptosis in plant cells, but there are other cases, generally involving small numbers of cells or even organs, that also seem to be apoptosis (Gray and Johal, 1998). The occurrence of cytological and biochemical changes that resemble apoptosis provides evidence for this idea. This could mean that some tissues follow a death process more like apoptosis, while others, especially whole organs, may follow a pathway more like the classical senescence. This question remains to be resolved; however, an apoptosis-like process occurs in the terminal stages of senescence (Noodén *et al.*, 1997). Indeed, a wide range of structural nuclear changes that are hallmarks of apoptosis have been reported to occur late in the senescence of leaves (Yen and Yang, 1998; Simeonova *et al.*, 2000) and flowers (Orzáez and Granell, 1997a; Xu and Hanson, 2000). Moreover, expression of *dad-1* seems to be associated with flower (Orzáez and Granell, 1997a) and pod wall (Orzáez and Granell, 1997b) senescence and barley scutella death (Lindholm *et al.*, 2000). Nonetheless, considerable study is needed to determine whether or not apoptosis is an integral (terminal) component of organ senescence. Apoptosis may be particularly important in the terminal phase of senescence (Noodén *et al.*, 1997; Delorme *et al.*, 2000).

While it is true that many metabolic activities decline, senescence is not achieved through a process of down-regulation (Chapters 4 and 5). In fact, the expression of many genes increases, particularly those connected with salvaging the nutrients invested in the senescing structures. At this point, it is not possible to distinguish those genes functioning in the primary senescence process from those in the secondary or peripheral processes. The secondary or peripheral processes are likely to vary considerably from one cell type to another.

Understandably, much emphasis has been placed on the enzymes participating in the breakdown of senescing cells, particularly the proteases, and these have been assumed to be part of the primary senescence process, i.e., the enzymes that kill the cells. Thus, reclamation or salvaging is seen as the central process of senescence, perhaps even synonymous with senescence. Nowhere is the equation of reclamation with senescence more firmly held than in monocarpic senescence, where there usually is a transfer of nutrients from the vegetative parts to the developing fruits (Chapter 15). The salvaging processes certainly are interesting and important, but they may not kill the cells. There are many disconnections between salvaging and senescence (Noodén, 1988b). Nitrogen is relatively mobile within a plant, and it is often scarce in the environment, so it is redistributed from senescing cells with a high efficiency. Most of the nitrogen in leaves is invested in the chloroplasts, and this nitrogen is reclaimed as the chloroplasts are broken down. As explained above in this section and elsewhere (Noodén, 1988a), the absence (and presumably, the removal) of the chloroplasts does not kill the cell. Furthermore, in petal cells, most of the breakdown and redistribution of cellular components occurs as the petals are growing and differentiating (Schumacher and Matthael, 1955). The mature floral structures then bide their time (sometimes for a long time) until their function is complete and they are signaled to senesce (Mayak and Halevy, 1980). Conversely, senescence may occur in leaves without the nutrient removal (Wood *et al.*, 1986; Chapter 15). In addition, the hypersensitive response in plant disease appears to involve a rapid collapse of the plant cells around an infection site with little or no reclamation (Goodman and Novacky, 1994). Thus, it does not seem appropriate to define senescence (or programmed cell death) in terms of salvaging.

Studies on gene expression (Chapter 4) and genetic controls (Chapter 5) will ultimately help to determine exactly what the central pathway of senescence is, to determine whether or not senescence in different tissues differs in basic, primary ways, to identify key promoters and to show how apoptosis may or may not fit into senescence. Thus, it is of great importance that up-regulation of gene expression during senescence be accurately identified. With this in mind, it is necessary to understand a problem that underlies the techniques used to identify up-regulation in terms of both mRNAs and proteins; that has to do with normalizing the loads on the electrophoretic gels used to analyze mRNA and protein changes during senescence (Noodén *et al.*, 1997). Because it is convenient and it is easier to get the loads right, the samples in each lane are often normalized on the basis of protein or RNA. During senescence, however, RNAs (primarily rRNA) and proteins undergo a massive decrease (Brady, 1988). This means that if the samples taken from senescing tissues at various stages are loaded with equal amounts of RNA (primarily rRNA), the gel (e.g., Northern blot) will show whether or not the mRNA rises or falls relative to total RNA (rRNA). Since the total RNA decreases during senescence, the Northern blot will show only whether the mRNA is falling more or less rapidly than total RNA, not the changes in its amount in the cell (Noodén *et al.*, 1997). Thus, the mRNA in question could be rising relative to the total RNA even while the amount per cell is decreasing. In contrast to the RNA, fresh weight remains fairly constant from organ maturity into late senescence and is easy to measure. Figure 1-2 shows how different the results look when psbA mRNA (for PSII core protein D1) is expressed on an rRNA basis versus a fresh weight basis, and of course, the interpretations would be very different. Others have likewise noted these differences for a variety of genes (Lohman *et al.*, 1994; Crafts-Brandner *et al.*, 1996). It is important to note that the same problem exists for protein gels normalized on the basis of total protein rather than fresh weight. Thus, genes whose expression is actually decreasing might seem to increase when represented on an RNA or protein basis. Not only can these data confuse the picture of which genes control senescence, but they will make it difficult to determine which promoters and other genetic control elements regulate senescence.

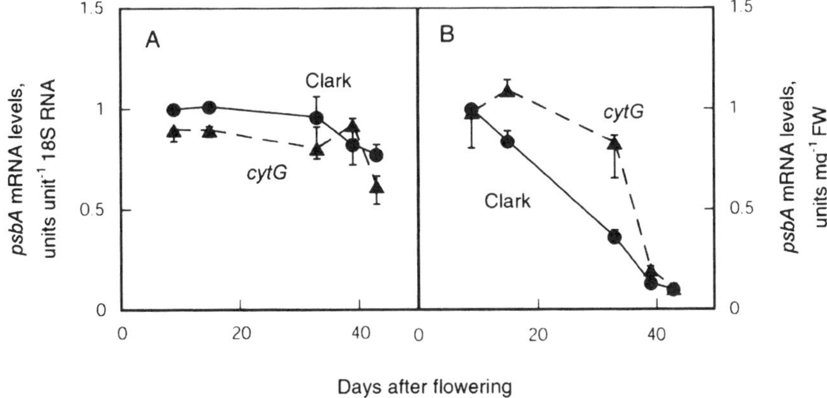

Figure 1-2. Changes in *psbA* (for PSII D1 protein) mRNA during senescence in leaves of wild-type (Clark) and the stay-green mutant (*cytG*) soybean normalized on the basis of 18s RNA (A) and fresh weight (B). From Guiamét *et al.* (2002).

It is still interesting to know the rates of decrease of different mRNAs relative to rRNA and ribosomes; however, that is a different issue.

VI. Hormonal Controls

Hormones are important controls (signals) of senescence, and that has been summarized earlier (Noodén and Leopold, 1988). Although a very wide range of hormones seem to promote and retard senescence of various plant organs and tissues, ethylene and cytokinin are particularly important. What is missing for most of these hormones is a wholistic picture of how they work in signaling at the whole plant level (a systems perspective). The hormones may interact in various ways and share some signaling components to form signaling networks (He *et al.*, 2001). Because cytokinins retard leaf and apparently also flower senescence, efforts have been made to genetically engineer cytokinin synthesis capabilities into plants (Chapter 6), and most of that progress has been possible as a result of the discovery (and availability) of the SAG12 promoter. The questions of which cytokinins are active and how they operate in controlling senescence have been addressed from a systems approach only in soybean monocarpic senescence (Noodén and Letham, 1993). The very extensive advances in explaining the earlier observations on ethylene control of senescence in terms of molecular mechanisms are summarized in Chapter 8. In addition, quite a large literature has emerged on methyljasmonate, and that is covered in Chapter 9. Since both salicylic acid and the brassinosteroids show some promise of involvement in regulating senescence, they need to be at least mentioned. Although sugars do not qualify as hormones, growing evidence does indicate that they do serve as signals that promote senescence under some circumstances (Chapter 15). In addition, other chemical signals such as nitric oxide may promote PCD (Pedroso *et al.*, 2000)

Salicylic acid mediates many plant stress responses including disease (Chapters 3, 13 and 20). It may also be required for normal senescence (Morris *et al.*, 2000).

The brassinosteroids appear to be involved in or required for several senescence processes, e.g., leaf senescence (Fujioka *et al.*, 1997) and xylem differentiation (Yamamoto *et al.*, 1997). Are brassinosteroids senescence-promoters or are they simply required for senescence processes? Exogenous applications indicate that they do, in fact, promote senescence (Clouse and Sasse, 1998).

VII. Evolution

Even though the molecular details of the senescence processes are still being elucidated, we can say something about the evolution of these processes, but that differs depending on the level of the processes, i.e., whole plant, organ or cell. As outlined by Carl Leopold in the Foreword to this volume, they play an essential role in turnover from the cell to the organism levels. The senescence processes seem to be used adaptively in a variety of ways and places, so there certainly is evolutionary plasticity in when and where they operate.

At the whole plant level, it is clear that monocarpic senescence has evolved independently in many different taxa (Noodén, 1988b). Generally, the evolution of monocarpic senescence is intertwined with the evolution of annual and biennial life cycles from perennials; however, there are some monocarpic perennials, even monocarpic trees. As a result of these diverse evolutionary histories, it seems that different groups have arrived at some different ways of carrying out monocarpic senescence. One of the more visible differences has to do with correlative controls, i.e., whether or not the developing reproductive structures control monocarpic senescence.

The extent to which organ senescence differs between different organs or between the same organs in different species is not clear (see Section II above). However, even the equivalent organs in different species may have different evolutionary and adaptive histories which may result in some physiological differences. For example, the senescence of flower petals may or may not be controlled by ethylene depending on the species (Mattoo and Aharoni, 1988).

Evolution at the cellular level (programmed cell death) has generated a lot of interest recently, and that has been tied to gene evolution. The cytological symptoms (Section IV above) certainly suggest there are many shared features between animal and plant apoptosis, which indicates a high degree of conservatism; however, that is less clear at the molecular level (Aravind *et al.*, 1999; Ameisen, 2002). While some homologous apoptosis-related genes certainly are shared between animal and plant lines, there are also some differences.

It seems likely that we will learn a lot about the evolution of senescence and related processes from studies on their molecular biology in the near future. This will also shed light on their adaptive value. Of particular interest are the genetic control elements, and eventually, they will also help to define senescence. Presumably, the senescence genes are derived from other processes, and it will be valuable to see how different genes or enzymes have been recruited into the senescence process.

References

Ameisen, J.C. (2002). On the origin, evolution, and nature of programmed cell death: a timeline of four billion years. *Cell Death and Differentiation* **9**, 367–393.

Aravind, L., Dixit, V.M., and Koonin, E.V. (1999). The domains of death: Evolution of the apoptosis machinery. *Trends in Biochemical Sciences* **24**, 47–53.

Arking, R. (1991). *Biology of Aging: Observations and Principles*. Prentice Hall, Englewood Cliffs, NJ.

Balk, J., and Leaver, C.J. (2001). The PET1-CMS mitochondrial mutation in sunflower is associated with premature programmed cell death and cytochrome C release. *Plant Cell* **13**, 1803–1818.

Barlow, P.W. (1982). Cell death—An integral part of plant development. In *Growth Regulators in Plant Senescence* (M.B. Jackson, B. Grant, and I.A. Mackenzie, Eds.), pp. 27–45. Monograph No. 8, British Plant Growth Regulator Group, Wantage, UK.

Biswal, U.C., and Biswal, B. (1988). Ultrastructural modifications and biochemical changes during senescence of chloroplasts. *International Review of Cytology* **113**, 271–321.

Borner, C., and Monney, L. (1999). Apoptosis without caspases: An inefficient molecular guillotine? *Cell Death and Differentiation* **6**, 497–507.

Brady, C.J. (1988). Nucleic acid and protein synthesis. In *Senescence and Aging in Plants* (L.D. Noodén, and A.C. Leopold, Eds.), pp. 147–179. Academic Press, San Diego, CA.

Buchanan-Wollaston, V. (1997). The molecular biology of leaf senescence. *Journal of Experimental Botany* **48**, 181–199.

Buckner, B., Johal, G.S., and Janick-Buckner, D. (2000). Cell death in maize. *Physiologia Plantarum* **108**, 231–239.

Butler, R.D., and Simon, E.W. (1971). Ultrastructural aspects of senescence in plants. *Advances in Gerontology Research* **3**, 73–129.

Clarke, P.G.H. (1990). Developmental cell-death—Morphological diversity and multiple mechanisms. *Anatomy and Embryology* **181**, 195–213.

Clouse, S.D., and Sasse, J.M. (1998). Brassinosteroids: Essential regulators of plant growth and development. *Annual Review of Plant Physiology and Plant Molecular Biology* **49**, 427–451.

Cohen, G.M. (1997). Caspases: The executioners of apoptosis. *Biochemical Journal* **326**, 1–16.

Crafts-Brandner, S.J., Klein, R.R., Klein, P., Hoelzer, R., and Feller, U. (1996). Coordination of protein and mRNA abundances of stromal enzymes and mRNA abundances of the clp protease subunits during senescence of *Phaseolus vulgaris* (L.) leaves. *Planta* **200**, 312–318.

Dangl, J.L., Dietrich, R.A., and Thomas, H. (2000). Cell death and senescence. In *Biochemistry and Molecular Biology of Plants* (B.B. Buchanan, W. Gruissem, R.L. Jones, Eds.), pp. 1044–1100. American Society of Plant Physiologists, Rockville, MD.

Danon, A., Delorme, V., Mailhac, N., and Gallois, P. (2000). Plant programmed cell death: A common way to die. *Plant Physiology and Biochemistry* **38**, 647–655.

de Jong, A.J., Hoeberichts, F.A., Yakimova, E.T., Maximova, E., and Woltering, E.J. (2000). Chemical-induced apoptotic cell death in tomato cells: Involvement of caspase-like proteases. *Planta* **211**, 656–662.

Delorme, V.G.R., McCabe, P.F., Kim, D.J., and Leaver, C.J. (2000). A matrix metalloproteinase gene is expressed at the boundary of senescence and programmed cell death in cucumber. *Plant Physiology* **123**, 917–927.

Esau, K., Cheadle, V.I., and Risley, E.B. (1963). A view of ultrastructure of Cucurbita xylem. *Botanical Gazette* **124**, 311–316.

Finch, C.E. (1990). *Longevity, Senescence and the Genome*. University of Chicago Press, Chicago, IL.

Finkel, E. (2001). The mitochondrion: Is it central to apoptosis? *Science* **292**, 624–626.

Fujioka, S., Li, J., Choi, Y.H., Seto, H., Takatsuto, S., Noguchi, T., Watanabe, T., Kuriyama, H., Yokota, T., Chory, J., and Sakurai, A. (1997). The Arabidopsis deetiolated2 mutant is blocked early in brassinosteroid biosynthesis. *Plant Cell* **9**, 1951–1962.

Fukuda, H. (1996). Xylogenesis: Initiation, progression, and cell death. *Annual Review of Plant Physiology and Plant Molecular Biology* **47**, 299–325.

Gallois, P., Makishima, T., Hecht, V., Despres, B., Laudie, M., Nishimoto, T., and Cooke, R. (1997). An Arabidopsis thaliana cDNA complementing a hamster apoptosis suppressor mutant. *Plant Journal* **11**, 1325–1331.

Gepstein, S. (1988). Photosynthesis. In *Senescence and Aging in Plants* (L.D. Noodén and A.C. Leopold, Eds.), pp. 85–109. Academic Press, San Diego, CA.

Goodman, R.N., and Novacky, R.N. (1994). *The Hypersensitive Reaction in Plants to Pathogens: A Resistance Phenomenon*. APS Press, St. Paul, MN.

Gray, J.H., and Johal, G.S. (1998). Programmed cell death in plants. In *Arabidopsis* (M. Anderson and J. Roberts, Eds.), pp. 360–394. Sheffield Academic Press, Sheffield, UK.

Groover, A., Dewitt, N., Heidel, A., and Jones, A. (1997). Programmed cell death of plant tracheary elements differentiating in vitro. *Protoplasma* **196**, 197–211.

Guiamét, J.J., Tyystjärvi, E., Tyystjärvi, T., John, I., Kairavuo, M., Pichersky, E., and Noodén, L.D. (2002). Photoinhibition and loss of photosystem II reaction center proteins during senescence of soybean leaves. Enhancement of photoinhibition by the 'stay-green' mutation *cytG*. *Physiologia Plantarum* **115**, 468–478.

Halevy, A.H., and Mayak, S. (1981). Senescence and postharvest physiology of cut flowers—Part 2. *Horticultural Reviews* **3**, 59–143.

Havel, L., and Durzan, D.J. (1996). Apoptosis in plants. *Botanica Acta* **109**, 268–277.

Hawkins, S.W., and Phillips, R. (1983). 3-Aminobenzamide inhibits tracheary element differentiation but not cell-division in cultured explants of Helianthus tuberosus. *Plant Science Letters* **32**, 221–224.

He, Y.H., Tang, W.N., Swain, J.D., Green, A.L., Jack, T.P., and Gan, S.S. (2001). Networking senescence-regulating pathways by using Arabidopsis enhancer trap lines. *Plant Physiology* **126**, 707–716.

Hoeberichts, F.A., and Woltering, E.J. (2002). Multiple mediators of plant programmed cell death: Interplay of conserved cell death mechanisms and plant-specific regulators. *Bioessays* **25**, 47–57.

Hunot, S., and Flavell, R.A. (2001). Death of a monopoly. *Science* **292**, 865–866.

Jones, A. (2000). Does the plant mitochondrion integrate cellular stress and regulate programmed cell death? *Trends in Plant Science* **5**, 225–230.

Joza, N., Susin, S.A., Daugas, E., Stanford, W.L., Cho, S.K., Li, C.Y.J., Sasaki, T., Elia, A.J., Cheng, H.Y.M., Ravagnan, L., Ferri, K.F., Zamzami, N., Wakeham, A., Hakem, R., Yoshida, H., Kong. Y.Y., Mak, T.W., Zuniga-Pflucker, J.C., Kroemer, G., and Penninger, J.M. (2001). Essential role of the mitochondrial apoptosis-inducing factor in programmed cell death. *Nature* **410**, 549–554.

Jun, J.H., Ha, C.M., and Nam, H.G. (2002). Involvement of the VEP1 gene in vascular strand development in *Arabidopsis thaliana*. *Plant and Cell Physiology* **43**, 323–330.

Kawai, M., Pan, L., Reed, J.C., and Uchimiya, H. (1999). Evolutionally conserved plant homologue of the Bax inhibitor-1 (Bi-1) gene capable of suppressing Bax-induced cell death in yeast. *FEBS Letters* **464**, 143–147.

Kerr, J.F.R., Wyllie, A.H., and Currie, A.R. (1972). Apoptosis: A basic biological phenomenon with wide ranging implications in tissue kinetics. *British Journal of Cancer* **26**, 239–247.

Kumar, S., and Vaux, D.L. (2002). A Cinderella caspase takes center stage. *Science* **297**, 1290–1291.

Kuriyama, H., and Fukuda, H. (2001). Regulation of tracheary element differentiation. *Journal of Plant Growth Regulation* **20**, 35–51.

LaComme, C., and Santa Cruz, S. (1999). Bax-induced cell death in tobacco is similar to the hypersensitive response. *Proceedings of the National Academy of Sciences of the United States of America* **96**, 7956–7961.

Lam, E., and Del Pozo, O. (2000). Caspase-like protease involvement in the control of plant cell death. *Plant Molecular Biology* **44**, 417–428.

Lam, E., Fukuda, H., and Greenberg, J. (2000). *Programmed Cell Death in Higher Plants.* Kluwer, Dordrecht.

Leist, M., and Jäättela, M. (2001). Four deaths and a funeral: From caspases to alternative mechanisms. *Nature Reviews Molecular Cell Biology* **2**, 589–598.

Leopold, A.C. (1961). Senescence in plant development. *Science* **134**, 1727–1732.

Leopold, A.C., and Noodén, L.D. (1984). Hormonal regulatory systems in plants. In *Encyclopedia of Plant Physiology, New Series*. Vol. 10. *Hormonal Regulation of Development II* (T. Scott, Ed.), pp. 4–22. Springer Verlag, Berlin.

Levin, S. (1995). A toxicologic pathologist's view of apoptosis or I used to call it necrobiosis, but now I'm singing the apoptosis blues. *Toxicologic Pathology* **23**, 533–539.

Levin, S., Bucci, T.J., Cohen, S.M., Fix, A.S., Hardisty, J.F., LeGrand, E.K., Maronpot, R.R., and Trump, B.F. (1999). The nomenclature of cell death: recommendations of an ad hoc Committee of the Society of Toxicologic Pathologists [see comments]. *Toxicologic Pathology* **27**, 484–490.

Lindholm, P., Kuittinen, T., Sorri, O., Guo, D.Y., Merits, A., Tormakangas, K., and Runeberg-Roos, P. (2000). Glycosylation of phytepsin and expression of *dad1*, *dad2* and *ost1* during onset of cell death in germinating barley scutella. *Mechanisms of Development* **93**, 169–173.

Lockshin, R.A. (1999). Commentary: The utility of apoptosis terminology. *Toxicologic Pathology* **27**, 492–493.

Lockshin, R.A., and Zakeri, Z. (1996). The biology of cell death and its relationship to aging. In *Cellular Aging and Cell Death* (N.J. Holbrook, G.R. Martin, and R.A. Lockshin, Eds.), pp. 167–180. Wiley-Liss, New York.

Lohman, K.N., Gan, S., John, M.C., and Amasino, R.M. (1994). Molecular analysis of natural leaf senescence in *Arabidopsis thaliana*. *Physiologia Plantarum* **92**, 322–328.

Lorenzo, H.K., Susin, S.A., Penninger, J., and Kroemer, G. (1999). Apoptosis inducing factor (AIF): a phylogenetically old, caspase-independent effector of cell death. *Cell Death and Differentiation* **6**, 516–524.

Majno, G., and Joris, I. (1995). Apoptosis, oncosis, and necrosis. An overview of cell death. *American Journal of Pathology* **146**, 3–15.

Makishima, T., Yoshimi, M., Komiyama, S., Hara, N., and Nishimoto, T. (2000). A subunit of the mammalian oligosaccharyltransferase, dad1, interacts with MCL-1, one of the BLC-2 protein family. *Journal of Biochemistry* **128**, 399–405.

Mattoo, A.K., and Aharoni, N. (1988). Ethylene and plant senescence. In *Senescence and Aging in Plants* (L.D. Noodén and A.C. Leopold, Eds.), pp. 241–280. Academic Press, San Diego, CA.

Mayak, S., and Halevy, A.H. (1980). Flower senescence. In *Senescence in Plants* (K.V. Thimann, Ed.), pp. 131–156. CRC Press, Boca Raton, FL.

Medawar, P.B. (1957). *The Uniqueness of the Individual*. Methuen, London.

Meier, P., Finch, A., and Evan, G. (2000). Apoptosis in development. *Nature* **407**, 796–801.

Mitsuhara, I., Malik, K.A., Miura, M., and Ohashi, Y. (1999). Animal cell-death suppressors Bcl-X(L) and Ced-9 inhibit cell death in tobacco plants. *Current Biology* **9**, 775–778.

Molisch, H. (1938). *The Longevity of Plants* (*Die Lebensdauer der Pflanze*). Science Press, Lancaster, PA.

Morris, K., Mackerness, S.A.H., Page, T., John, C.F., Murphy, A.M., Carr, J.P., and Buchanan-Wollaston, V. (2000). Salicylic acid has a role in regulating gene expression during leaf senescence. *Plant Journal* **23**, 677–685.

Munné-Bosch, S., and Alegre, L. (2002). Plant aging increases oxidative stress in chloroplasts. *Planta* **214**, 608–615.

Nicotera, P., Leist, M., and Ferrando-May, E. (1999). Apoptosis and necrosis: Different execution of the same death. *Biochemical Society Symposium* 69–73.

Nicotera, P., Leist, M., Single, B., and Volbracht, C. (1999). Execution of apoptosis: Converging or diverging pathways? *Biological Chemistry* **380**, 1035–1040.

Noodén, L.D. (1988a). The phenomena of senescence and aging. In *Senescence and Aging in Plants* (L.D. Noodén, and A.C. Leopold, Eds.), pp. 1–50. Academic Press, San Diego, CA.

Noodén, L.D. (1988b). Whole plant senescence. In *Senescence and Aging in Plants* (L.D. Noodén, and A.C. Leopold, Eds.), pp. 391–439. Academic Press, San Diego, CA.

Noodén, L.D., and Guiamét, J.J. (1996). Genetic control of senescence and aging in plants. In *Handbook of the Biology of Aging* (4th ed.) (E.L. Schneider and J.W. Rowe, Eds.). Academic Press, San Diego.

Noodén, L.D., and Leopold, A.C. (1978). Phytohormones and the endogenous regulation of senescence and abscission. In *Phytohormones and Related Compounds: A Comprehensive Treatise* (D.S. Letham, P.B. Goodwin, and T.J.V. Higgins, Eds.), Vol. II, pp. 329–369. Elsevier/North-Holland Biomedical Press, Amsterdam.

Noodén, L.D., and Leopold, A.C. (1988). *Senescence and Aging in Plants*. Academic Press, San Diego, CA.

Noodén, L.D., and Letham, D.S. (1993). Cytokinin metabolism and signaling in the soybean plant. *Australian Journal of Plant Physiology* **20**, 639–653.

Noodén, L.D., and Thompson, J.E. (1985). Aging and senescence in plants. In *Handbook of the Biology of Aging* (2nd ed.) (C.E. Finch, and E.L. Schneider, Eds.). Van Nostrand Reinhold, New York.

Noodén, L.D., Guiamét, J.J., and John, I. (1997). Senescence mechanisms. *Physiologia Plantarum* **101**, 746–753.

O'Brien, I.E., Reutelingsperger, C.P., and Holdaway, K.M. (1997). Annexin-V and TUNEL use in monitoring the progression of apoptosis in plants. *Cytometry* **29**, 28–33.

Orzáez, D., and Granell, A. (1997a). The plant homologue of the defender against apoptotic death gene is down-regulated during senescence of flower petals. *FEBS Letters* **404**, 275–278.

Orzáez, D., and Granell, A. (1997b). DNA fragmentation is regulated by ethylene during carpel senescence in Pisum sativum. *Plant Journal* **11**, 137–144.

Orzáez, D., de Jong, A.J., and Woltering, E.J. (2001). A tomato homologue of the human protein PIRIN is induced during programmed cell death. *Plant Molecular Biology* **46**, 459–468.

Pedroso, M.C., Magalhaes, J.R., and Durzan, D. (2000). Nitric oxide induces cell death in *Taxus* cells. *Plant Science* **157**, 173–180.

Pontier, D., Gan, S., Amasino, R.M., Roby, D., and Lam, E. (1999). Markers for hypersensitive response and senescence show distinct patterns of expression. *Plant Molecular Biology* **39**, 1243–1255.

Priestley, D.A. (1986). *Seed Aging*. Cornell University Press, Ithaca, NY.

Raff, M.C. (1992). Social controls on cell survival and cell death. *Nature* **356**, 397–400.

Roberts, E.H. (1988). Seed aging: The genome and its expression. In *Senescence and Aging in Plants* (L.D. Noodén, and A.C. Leopold, Eds.), pp. 465–498. Academic Press, San Diego, CA.

Sanchez, P., de Torres Zabala, M., and Grant, M. (2000). Atbi-1, a plant homologue of Bax inhibitor-1, suppresses Bax-induced cell death in yeast and is rapidly upregulated during wounding and pathogen challenge. *Plant Journal* **21**, 393–399.

Schlegel, R.A., and Williamson, P. (2001). Phosphatidylserine, a death knell. *Cell Death and Differentiation* **8**, 551–563.

Schulze-Osthoff, K., Walczak, H., Droge, W., and Krammer, P.H. (1994). Cell nucleus and DNA fragmentation are not required for apoptosis. *Journal of Cell Biology* **127**, 15–20.

Schumacher, W., and Matthael, H. (1955). Über den Zusammenhang zwischen Streckung Wachstum und Eiweisssynthese. *Planta* **45**, 213–216.

Simeonova, E., Sikora, A., Charzynska, M., and Mostowska, A. (2000). Aspects of programmed cell death during leaf senescence of mono- and dicotyledonous plants. *Protoplasma* **214**, 93–101.

Solomon, M., Belenghi, B., Delledonne, M., Menachem, E., and Levine, A. (1999). The involvement of cysteine proteases and protease inhibitor genes in the regulation of programmed cell death in plants. *Plant Cell* **11**, 431–443.

Thomas, H., and Stoddart, J.L. (1980). Leaf senescence. *Annual Review of Plant Physiology* **31**, 83–111.

Trump, B.F., and Berezesky, I.K. (1998). The reactions of cell to lethal injury: Oncosis and necrosis—The role of calcium. In *When Cells Die: A Comprehensive Evaluation of Apoptosis and Programmed Cell Death* (R.A. Lockshin, Z. Zakeri, and J.L. Tilly, Eds.), pp. 57–96. Wiley-Liss, New York.

Uren, A.G., O'Rourke, K., Aravind, L.A., Pisabarro, M.T., Seshagiri, S., Koonin, E.V., and Dixit, V.M. (2000). Identification of paracaspases and metacaspases: two ancient families of caspase-like proteins, one of which plays a key role in MALT lymphoma. *Molecular Cell* **6**, 961–967.

van der Biezen, E.A., and Jones, J.D.G. (1998). The NB-ARC domain: a novel signalling motif shared by plant resistance gene products and regulators of cell death in animals. *Current Biology* **8**, R226–R227.

Van Staden J., Cook, E.L., and Noodén, L.D. (1988). Cytokinins and senescence. In *Senescence and Aging in Plants* (L.D. Noodén and A.C. Leopold, Eds.), pp. 281–328. Academic Press, San Diego, CA.

Vaux, D.L. (2002). Apoptosis timeline. *Cell Death and Differentiation* **9**, 349–354.

Wang, H., Li, J., Bostock, R.M., and Gilchrist, D.G. (1996). Apoptosis: A functional paradigm for programmed plant cell death induced by a host-selective phytotoxin and invoked during development. *Plant Cell* **8**, 375–391.

Warner, H.R., Hodes, R.I., and Pocinki, K. (1997). What does cell death have to do with aging? *Journal of the American Geriatrics Society* **45**, 1140–1146.

Watkinson, A. (1992). Plant senescence. *Trends in Ecology and Evolution* **7**, 417–420.

Wood, L.J., Murray, B.J., Okatan, Y., and Noodén, L.D. (1986). Effect of petiole phloem disruption on starch and mineral distribution in senescing soybean leaves. *American Journal of Botany* **73**, 1377–1383.

Xu, F.-X., and Chye, M.L. (1999). Expression of cysteine proteinase during developmental events associated with programmed cell death in brinjal. *Plant Journal* **17**, 321–327.

Xu, Y., and Hanson, M.R. (2000). Programmed cell death during pollination-induced petal senescence in petunia. *Plant Physiology* **122**, 1323–1333.

Yamamoto, R., Demura, T., and Fukuda, H. (1997). Brassinosteroids induce entry into the final stage of tracheary element differentiation in cultured zinnia cells. *Plant and Cell Physiology* **38**, 980–983.

Yen, C.H., and Yang, C.H. (1998). Evidence for programmed cell death during leaf senescence in plants. *Plant and Cell Physiology* **39**, 922–927.

Yu, L.-H., Kawai-Yamada, M., Naito, M., Watanabe, K., Reed, J.C., and Uchimiya, H. (2002a). Induction of mammalian cell death by a plant Bax inhibitor. *FEBS Letters* **512**, 308–312.

Yu, X.-H., Perdue, T.D., Heimer, Y.M., and Jones, A.M. (2002b). Mitochondrial involvement in tracheary element programmed cell death. *Cell Death and Differentiation* **9**, 189–198.

Zakeri, Z. (1998). The study of cell death by the use of cellular and developmental models. In *When Cells Die: A Comprehensive Evaluation of Apoptosis and Programmed Cell Death* (R.A. Lockshin, Z. Zakeri, and J.L. Tilly, Eds.), pp. 97–129. Wiley-Liss, New York.

2

Plant Cell Death and Cell Differentiation

Page W. Morgan and Malcolm C. Drew

I. Introduction

"The concept of programmed cell death (PCD) came from plants" (Jones, 2001). This rather surprising assertion is based on early studies on plant reactions to fungal invasion (Allen, 1923) and a long standing interest in senescence by plant scientists (Molisch, 1938; Leopold, 1961). Nevertheless interest in the phenomenon was promoted greatly by animal studies focused on specific changes in cellular structure (Lockshin and Williams, 1965; Kerr *et al.*, 1972). Later Barlow (1982) reasoned regarding plants that if cells die at a predictable time and location, if the death has some beneficial effect on tissue differentiation and is encoded in the hereditary material of the species, it represents a specific process of ontogenesis. This definition excluded necrotic cell death due to accidental or random injury such as exposure to some toxins or a lethal temperature. As history has revealed, cell death in many plant processes is programmed and meets many or all of Barlow's criteria. The study of PCD has become an important area of plant biology, and this chapter will develop that theme. Owing to space limitations and the desire to consider recent findings, we have relied on reviews often and have not cited individually the papers in Barlow (1982).

There is insufficient information at present to conclude that animal and plant cells regulate PCD through a common, basic mechanism (Chapter 1). Nevertheless, some of the biochemical steps that regulate and execute PCD in animal cells have recently been identified in plant cells in culture. Early in the apoptosis of mammalian cells, phosphatidyl serine is exposed at the outer face of the plasma membrane and is detectable by the binding of annexin V (Martin *et al.*, 1995). Similarly, in protoplasts from tobacco cells, an early stage in PCD induced by camptothecin or salicylic acid was the binding of annexin V (O'Brien *et al.*, 1997, 1998).

Plant Cell Death Processes

Other indicators of PCD in animal cells are chromatin condensation and nDNA fragmentation, detected either by the occurrence of internucleosomal lengths of DNA ("DNA ladders") on agarose gels, or *in situ* by the terminal deoxynucleotidyl transferase-mediated dUTP nick end labeling (TUNEL) of exposed DNA 3′ hydroxyl groups (Gavrielli *et al.*, 1992). Cultured plant cells undergo chromatin condensation and test positive for TUNEL when exposed to agents that induce PCD (Wang *et al.*, 1996a; O'Brien *et al.*, 1997, 1998), although at high concentration some agents (e.g. H_2O_2) caused isolated cells or protoplasts to show symptoms of necrosis (oncosis) rather than PCD. Caspases, cysteine proteases that cleave polypeptides after Asp residues, are almost always required for apoptosis in mammalian cells (Salvesen and Dixit, 1997; Thornberry and Lazebnik, 1998), but a role for caspases in plant PCD is unclear (Lam and del Pozo, 2000; Beers *et al.*, 2000). Much of the evidence for participation of caspase-like proteases in PCD in plant cells relies on the activity of peptidase-specific inhibitors. Induction of PCD in cultured soybean cells by oxidative stress quickly activated cysteine proteases (Solomon *et al.*, 1999), and both cysteine protease activity and PCD were blocked by ectopic expression of the gene for the cysteine protease inhibitor, cystatin. Other examples of circumstantial evidence for the participation of caspase-like proteases in plant PCD are cited by Lam and del Pozo (2000).

Apoptosis in animal cells often involves fragmentation into membrane-bound apoptotic bodies that are quickly engulfed by surrounding macrophages or phagocytes, but no such mechanism for dealing with dead cells is available to plants. Mature plant cells, unlike animal cells, usually have a vacuole that constitutes 80–95% of the cell volume; additionally, the protoplast normally is enclosed by a cell wall that would prevent expulsion of apoptotic bodies were they to occur. It is to be expected, therefore, that some of the anatomical and biochemical steps in PCD will be unique to plant cells.

It is implicit in the concept of PCD that in most cases, at a given time, certain cells die and others do not. Why do only certain cells die? This question immediately prompts the thought that there is some selectivity in differentiation between the surviving and the suicidal cells. This chapter examines examples of PCD which seem to depend on cell differentiation as one means of targeting. The most direct possible relationship of PCD with cell differentiation would be that a cell differentiates and during that process dies. An obvious example is differentiation of xylem tracheary elements as a vital part of vascular development. A second relationship between cell death and differentiation would be that cells differentiate, perform an intermediate function for a time, and later they die. There are a number of examples of this pattern of development associated with reproduction including embryo and pollen development and aleurone cells. A third case involves cells that appear identical; yet, at a specific time and place or condition, a subset of them dies. An example would be the thin-walled parenchyma cells (Esau, 1965), including those in the root cortex, which can undergo PCD to form aerenchyma and similar internal spaces. Cell death occurs in a distinctive pattern and some of the cortical cells survive. Thus, in addition to discernable differentiation, there may be changes associated with cell lineage, position or packing within a tissue which mark or potentiate specific cells to die. Regardless, this third example is common enough to be important to plant development and will be included in this chapter.

II. The Scope of PCD in Plants

Although the widespread occurrence of PCD in plants was recognized in earlier literature (Barlow, 1982; Noodén, 1988), the importance has not been recognized broadly in biology. Ameisen (1998) has stated, "the existence of PCD in plants has long remained neglected", and he offered that evidence for its existence has not yet revealed much about the extent of its role, the nature of the mechanisms involved, or the nature of its genetic regulation. Thus, there is a real need to emphasize the scope and significance of PCD in plants.

It is now well documented that PCD plays important roles throughout plant development (Table 2-1). Not only does it occur throughout the plant life cycle, but in many parts of the plant and even whole plants. This point is now illustrated within single species

Table 2-1. Programmed Cell Death in the Life History of Plants

Stage/description	Type of regulation[a]	References
Vegetative development		
Seed		
aleurone cells	1	Wang *et al.*, 1996c; Kuo *et al.*, 1996
cotyledons	1	Barlow, 1982
scutellum	1	Lindholm *et al.*, 2000
Shoots		
holes in leaves	1	Kaplan, 1984
shoot aerenchyma	1, 2	Drew, 1997
leaf aerenchyma	1	Barlow, 1982
xylogenesis	1, 2	Fukuda, 1996; Groover and Jones, 1999
xylem ray cells	1	Barlow, 1982
secretary ducts	1	Barlow, 1982
pith cell death	1	Pappelis and Katsanos, 1969
trichomes	1	Greenberg, 1996
senescence	1, 2	Noodén, 1988
abscission zone cells	1, 2	González-Carranza *et al.*, 1998
cork/bark/wound healing	1, 2	Barlow, 1982
Roots		
root hairs	1	Noodén, 1988
xylogenesis	1, 2	Fukuda, 1996
root cap cells	1	Wang *et al.*, 1996a
cortical parenchyma	1	Barlow, 1982
aerenchyma	1, 2	Drew, 1997
lateral roots	2	Barlow, 1982
salt stress	2	Katsuhara, 1997
Reproductive development		
Floral organ abortion	1	Grant *et al.*, 1994a; Dellaporta and Calderon-Urrea, 1993
Megagametogenesis		
megaspore abortion	1	Bell, 1996; Christensen *et al.*, 1998
nucellus	1	Barlow, 1982
synergid cell	1	Barlow, 1982
Microgametogenesis	1	Goldberg *et al.*, 1993
Embryogenesis	1	Olson and Cass, 1981
antipodal cells	1	Young *et al.*, 1997
endosperm cells	1	Yeung and Meinke, 1993;
suspensor		Schwartz *et al.*, 1997
Somatic embryogenesis	1	Havel and Durzan, 1996

Continued

Table 2-1. Continued

Stage/description	Type of regulation[a]	References
Apomictic embryogenesis	1	Barlow, 1982
loculus wall cells	1	Goldberg *et al.*, 1993; Wang *et al.*, 1999
filament cells	1	Goldberg *et al.*, 1993; Wang *et al.*, 1999
tapetum cells	1	Goldberg *et al.*, 1993; Wang *et al.*, 1999
Pollen incompatibility	1	Bell, 1995
Fertilization (stigma, style and pollen cells)	1	Barlow, 1982; Wang *et al.*, 1996b
Flower petal senescence	1	Chapter 21
Post reproductive development		
Seed development		
endosperm cells	1	Knowles and Phillips, 1988; Young *et al.*, 1997
integument, palisade and epidermal cells	1	Barlow, 1982; Greenberg, 1996
pod, carpel senescence	1	Oráez and Granell, 1997
Monocarpic senescence	1	Noodén, 1988; Noodén *et al.*, 1997; Chapter 1

[a] 1, developmental regulation; 2, environmental or stress regulation.

(maize, Buckner *et al.*, 1998; arabidopsis, Gray and Johal, 1998). While participating in development, PCD also can be environmentally induced thereby contributing both to form and acclimation. PCD occurs in annuals and perennials. PCD is known in all families of angiosperms, and it also occurs in many species of green algae (Hay, 1997), thus spanning the range from simple to complex plants. Studies of PCD usually focus on the cell or cells that die, but the process can involve intricate cell to cell interactions such as incompatibility during host/pathogen interactions and pollination/fertilization processes. The scale of PCD can vary from the death of one or a few cells in some reproductive processes (Barlow, 1982) to the massive death of organs or whole plants during autumnal leaf senescence or monocarpic senescence (Chapter 1).

The survey of PCD in Table 1 illustrates its scope based on evidence ranging from "predictable cell death at specific places and times" (Barlow, 1982) to more definitive biochemical and molecular evidence. The following sections will deal with some of the better studied examples.

III. Prereproductive Cell Death

A. Aleurone and Scutellum

At the beginning of the life cycle of plants propagated by seed, cells die in the process of germination. For example, aleurone cells in cereal grain seeds facilitate germination; however, they also die in the process. Their death fits the general definition of PCD, and attention has been turned to the death process. GA specifically accelerated cell death in isolated aleurone cells and protoplasts, and the GA-signaling inhibitor LY83583 prevented DNA degradation and cell death (Kuo *et al.*, 1996; Bethke *et al.*, 1999). ABA delayed cell death and overcame the hastening effect of GA. During germination aleurone cells expressed characteristics of apoptosis including positive TUNEL staining and DNA cleavage into internucleosomal fragments (Wang *et al.*, 1996c). Subsequently these findings were shown as likely to be technical artifacts due to nucleases in the aleurone cells and in the enzymes used to prepare protoplasts (Fath *et al.*, 1999). Other symptoms of PCD in protoplasts included increased vacuolation, abrupt loss of plasma membrane integrity and then rapid shrinkage of the remainder of the protoplast (Bethke *et al.*, 1999). While DNA hydrolysis began before death, it did not result in accumulation of uniform, low molecular weight fragments.

Some of the proteases and nucleases induced during germination may be specifically involved in PCD of barley aleurone cells. GA promotes PCD and up regulates proteases and nucleases while ABA prevents cell death and inhibits synthesis of these enzymes (Jones and Jacobsen, 1991; Kuo *et al.*, 1996; Wang *et al.*, 1996c; Fath *et al.*, 1999). Some of the proteases are secreted into the endosperm for germination-linked hydrolysis. Other proteases accumulate in vacuoles in the aleurone cells (Bethke *et al.*, 1996) and are presumed to be active in the death of these cells. Secretion of α-amylase from aleurone cells begins 4 to 6 h after GA treatment, while nuclease secretion begins 24 to 36 h after GA treatment (Brown and Ho, 1986). This timing is consistent with involvement of the nuclease(s) in cell death; three nucleases are present (Fath *et al.*, 1999).

Receptors for GA and ABA appear to be on the plasma membrane, as assessed mainly by α-amylase induction, but there is also evidence for a cytoplasmic receptor for the ABA-regulated *Em* gene (Gilroy, 1996). GA was shown to stimulate transport of Ca^{2+} into ER isolated from aleurone cells (Bush *et al.*, 1989); GA-enhanced synthesis of α-amylase had previously been shown to require millimolar concentrations of calcium (Bush *et al.*, 1986). In contrast, ABA decreased Ca^{2+} concentrations in the cytoplasm. Okadaic acid blocked Ca^{2+} changes, gene expression and cell death induced by GA (Kuo *et al.*, 1996), and it had a similar but less pronounced effect on ABA-induced responses. This suggests that serine/threonine protein phosphatases, which okadaic acid inhibits, are essential for the action of GA on aleurone cells, including its promotion of PCD. Protein kinase, G-protein and lipase inhibitors were ineffective on the wheat aleurone layers studied. As noted earlier, ABA inhibited PCD in barley aleurone, and this ABA effect was *promoted* by okadaic acid (Wang *et al.*, 1996c). Further, ABA induced a rapid and transient rise in mitogen-activated protein (MAP) kinase activity in barley aleurone protoplasts (Knetsch *et al.*, 1996). ABA also increased the enzyme activity of phospholipase D and increased its product, phosphatidic acid (PPA). Application of PPA caused ABA-like inhibition of α-amylase, and inhibition of phospholipase D also inhibited ABA-inducible responses. Taken together, the evidence is consistent with existence of independent signal transduction pathways for GA and ABA which eventually meet at a common point related to α-amylase production and, presumably, PCD.

Reactive oxygen species (ROS) have been added to the proposed signal transduction pathway for PCD in aleurone cells (Fath *et al.*, 2001). Hydrogen peroxide promoted PCD, and in GA-promoted PCD hydrogen peroxide scavenging enzymes and mRNA for catalase 2 were down regulated. ABA, on the other hand, maintained RSO scavenging enzyme activity and delayed cell death.

The scutellum of the barley embryo also dies during germination. An animal cell death suppressor gene named *dad* (defender against apoptotic cell death) (Gallois *et al.*, 1997; Tanaka *et al.*, 1997) has been studied in scutellum (Lindholm *et al.*, 2000). Mammalian *dad1* encodes a subunit of an enzyme for N-linked glycosylation located on the ER (Kelleher and Gilmore, 1997). During PCD of scutellum cells, expression of *dad1* declined before DNA fragmentation occurred while *dad2* and *ost1*, which encodes another subunit of the transferase complex, did not decline. The complex glycosylates a subunit of phytepsin, a vacuolar proteinase, and the failure of this step may be involved in the onset of PCD. *Dad* has also been studied in PCD of root cortical cells (see later).

B. Xylem

Xylem is a composite tissue, comprising the tracheary elements (TEs) that die in order to conduct water and solutes, together with xylem parenchyma cells, fiber cells and ray cells that remain alive after their differentiation. TEs are made up of tracheids and vessel members, both of which have no protoplast when mature. Differentiation of TEs involves formation of secondary wall thickenings followed immediately by cell death and autolysis (Fukuda, 1996; Mittler, 1998). Alterations in the structure of the tonoplast were closely identified long ago with the breakdown of the TE protoplast (Wodzicki and Brown, 1973), although early studies were done on tissue samples from intact plants making reconstruction of the sequence of events difficult. Progress in the field of xylem differentiation has been greatly advanced by the refinement of an *in vitro* culture system in which isolated, mesophyll cells from leaves of *Zinnia elegans* can be induced by IAA and cytokinins to transdifferentiate into TEs and pass through the full sequence of maturation, including secondary wall deposition, PCD and autolysis (Fukuda and Komamine, 1980). Vacuolar collapse, preceded by a sudden change in tonoplast permeability (Obara *et al.*, 2001), appears to be the irreversible step in PCD, for cytoplasmic streaming stops at once, followed by degradation of the nucleus in 10–20 min and the total degradation of all cellular contents in 6–8 hours (Fukuda, 1997; Groover *et al.*, 1997; Obara *et al.*, 2001). Leakage of vacuolar sap into the cytoplasm could promote cell death through the action of hydrolytic enzymes and by cytoplasmic acidosis; cytoplasmic acidosis is well established as a determinant of death in anoxic cells (Roberts *et al.*, 1984).

Differentiation and death of TEs involves nDNA degradation, demonstrated by the TUNEL reaction in intact plants (Mittler and Lam, 1995) and in the zinnia system in culture (Groover *et al.*, 1997). It occurs close to the completion of the secondary wall, coinciding approximately with the appearance of several DNases and RNases (Fukuda, 1997). Genes that are expressed preferentially at various stages in the differentiation of TEs have received detailed analysis (Ye and Varner, 1993; Aoyagi *et al.*, 1998). While some encode enzymes involved in secondary cell wall synthesis, or the hydrolytic enzymes that accumulate in the vacuolar sap, the role of many others is unclear.

Activity of cysteine proteases increases at or shortly before the start of autolysis in zinnia cells (Minami and Fukada, 1995; Ye and Varner, 1996; Funk *et al.*, 2002), and a vacuolar aspartate proteinase (phytepsin) is highly expressed in intact barley roots (Runeberg-Roos and Saarma, 1998), but it is uncertain whether these might be part of a death signaling system or part of the autolytic process. However, inhibitors of cysteine proteases can block the terminal differentiation of TEs in culture, suggesting perhaps a role in the signaling of death (Fukuda, 1997). Serine proteases are prominent in the late stages of differentiation in zinnia (Ye and Varner, 1996; Beers and Freeman, 1997; Groover and Jones, 1999). A secreted, 40 kD serine protease has been closely implicated in the triggering of vacuolar rupture and PCD; exposure of cultured zinnia cells to soybean trypsin inhibitor blocked both the enzyme's activity and cell death (Groover and Jones, 1999).

Calmodulin antagonists or Ca^{2+} chelators block TE differentiation in zinnia (Roberts and Haigler, 1990). Conversely, agents that increase intracellular Ca^{2+} concentration, such as the Ca^{2+}-ionophore A23187, accelerated DNA fragmentation (TUNEL positive) and cell death (Groover and Jones, 1999). Groover and Jones (1999) propose that the secreted Ser protease in the extracellular matrix activates Ca^{2+} influx at the plasma membrane, the rise in Ca^{2+} acting to signal tonoplast rupture.

The ubiquitin–proteasome pathway of proteolysis has been implicated, both in the induction of TE differentiation and in the later stages leading to PCD. Inability to generate polyubiquitin chains in tobacco expressing a mutant ubiquitin gene prevented normal differentiation of the xylem (Bachmair *et al.*, 1990). Addition of proteasome inhibitors in the zinnia system at the time of initiation of TE differentiation blocked further development (Woffenden *et al.*, 1998). Addition of inhibitors after the commitment to differentiation delayed the formation of secondary cell walls and cell death. Many of the cells treated with one of the protease inhibitors failed to complete autolysis, although the nDNA appeared degraded. However, the inhibition of autolysis was attributed to inhibition of cysteine proteases rather than of proteasomes. Thus, the importance of the ubiquitin–proteasome pathway in signaling PCD in differentiation of TEs remains uncertain.

C. Aerenchyma

Death and autolysis of cells can be induced in roots of many monocot and dicot species in response to a wide range of environmental factors: these include hypoxia (Justin and Armstrong, 1991; Drew, 1997), mechanical impedance (He *et al.*, 1996a), transient starvation of inorganic nutrients (Drew *et al.*, 1989) and desiccation (Drew, 1979; Shone and Flood, 1983). Death occurs in primary roots usually among the cortical cells in the zone where the cells have just completed their expansion, usually a few cm behind the root tip (Campbell and Drew, 1983). The function of aerenchyma is experimentally established only for hypoxia, where the oxygen status of roots growing in O_2-impoverished surroundings is improved (Drew *et al.*, 1985). However, circumstantial evidence suggests that aerenchyma aid in root penetration of dense soil layers thereby increasing drought tolerance (Drew *et al.*, 2000). The case for interpreting lysigenous aerenchyma formation as an example of PCD is supported by the inducibility of aerenchyma by the hormone, ethylene; by the precise, predictable pattern of localized cell death within the root cortex; and by the detection of TUNEL positive nuclei early in the induction of cell death (Gunawardena *et al.*, 2001).

Signs of nuclear degradation in cells during root cortical death (RCD) beginning from the outside and progressing towards the stele have been detected using acridine orange staining (Henry and Deacon, 1981). Loss of nuclear staining by acridine orange correlates with other criteria of cell death, including loss of esterase activity and with loss of ability to plasmolyze (Lascaris and Deacon, 1991). RCD apparently occurs without the intervention of external environmental factors, although deficiency of inorganic nutrients can accelerate RCD (Elliott *et al.*, 1993). It is unclear whether aerenchyma formation and RCD are the same phenomenon; however, in one study the cell death patterns for the two phenomena did not overlap (Deacon *et al.*, 1986).

PCD in the cortex of roots of maize is triggered by ethylene. With hypoxia or mechanical impedance, the activity of enzymes involved in the biosynthetic pathway is increased resulting in increased ethylene production in the roots (Drew *et al.*, 1979; Atwell *et al.*, 1988; He *et al.*, 1996a). Such PCD can be initiated by application of low concentrations of ethylene under normoxic conditions, while inhibitors of ethylene biosynthesis or action block aerenchyma formation in hypoxic or ethylene-treated roots, respectively (Drew *et al.*, 1981; Jackson *et al.*, 1985). Evidence of a signal transduction pathway leading to PCD comes from studies using agonists or antagonists of potential intermediates (He *et al.*, 1996b).

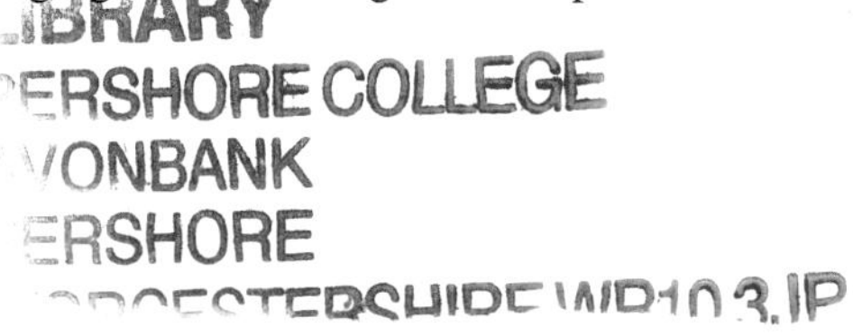

PCD was accelerated by activation of G-proteins, blocked by inhibition of protein kinases, and induced by inhibition of protein phosphatases (thereby keeping proteins in a phosphorylated form). Reagents that increased the concentration of intracellular Ca^{2+} accelerated PCD, and conversely, reagents that lowered cytosolic Ca^{2+} inhibited death. Direct measurements of phospholipase C activity and inositol 1,4,5, trisphosphate (He, Morgan and Drew, unpublished) also indicate the participation of phosphoinositides in signaling PCD.

In rice, PCD leading to aerenchyma formation is constitutive rather than inducible, although it is possible that ethylene still has a role (Justin and Armstrong, 1991). Death in rice seminal roots begins in isolated cells in the mid cortex, and it is first detected by a more acid cytoplasmic pH (stained with neutral red) and an inability to exclude Evans blue (Kawai *et al.*, 1998). The authors suggest that PCD is involved in the initiation of death, but they speculate that the subsequent spreading radially from cell to cell might be necrosis (oncosis). The use of the TUNEL reaction to identify individual cells might resolve this question.

Aerenchyma formation has many features in common with PCD in TEs. Aerenchyma formation requires the dissolution of dead cells, suggesting that an array of hydrolytic enzymes would be involved. At the transmission electron microscopy (TEM) level, the initial sign of death is the collapse inwards of isolated cells in the mid cortex suggesting loss of tonoplast integrity (Campbell and Drew, 1983), as in TE differentiation. This would allow access of vacuolar enzymes to the cytoplasm and cytoplasmic acidosis. Cell dissolution is probably complete for individual cells within 24 h of the start of cell death. Cellulase activity increases in the apical zone before any microscopic signs of cell abnormality (He *et al.*, 1994), and increased cellulase activity is always tightly coupled to cell death (He *et al.*, 1996b). Furthermore, the requirement for an increase in cytosolic Ca^{2+} to initiate death in cortical cells of roots resembles that in TEs. Whether the common features signify a similar underlying mechanism remains to be determined.

Among the relatively few genes that are experimentally linked with PCD in plants, several are expressed in roots. Mutations in the root necrosis (Rn) gene in soybean cause a progressive browning of the root after germination, and death of cells beginning in the inner cortex (Kosslak *et al.*, 1997). Maize root cortical cell delineating gene is expressed rapidly in response to mechanical impedance which induces PCD in cortical cells (Huang *et al.*, 1998), and its expression is up regulated in genetic tumors in tobacco (Fujita *et al.*, 1994). An animal cell death suppressor gene (see IIIA. Aleurone and Scutellum) homologue *Dad1* (defender against apoptotic cell death) is expressed in arabidopsis, rice and maize roots (Gallois *et al.*, 1997; Tanaka *et al.*, 1997; Finkelstein, Morgan and Drew, unpublished). The arabidopsis gene was shown to be functional in mutant animal cells lacking the cell death suppressor (Gallois *et al.*, 1997), and its expression decreases in PCD of scutellum cells (Lindholm *et al.*, 2000). Other genes in plants have been identified that have been shown to be functional as cell death suppressors (Dietrich *et al.*, 1997; Gray *et al.*, 1997). The existence of genes encoding death receptors, effectors, adaptors and regulators of animal apoptosis (Vaux and Korsmeyer, 1999) should encourage further genetic studies with plant systems.

D. Abscission Zones

The deliberate shedding of plant parts—the abscission process—results when specific cells exercise a genetic program to separate (Sexton and Roberts, 1982; Osborne, 1989). Attention has focused on the mechanism of detachment rather than the fate of the abscission zone cells. Extensive evidence indicates that zone cells synthesize and secrete hydrolases including proteases and nucleases; some of these enzymes act to separate and destroy cell walls (Sexton and Roberts, 1982; Osborne, 1989). It has been shown that any band of cells in a petiole can differentiate into abscission zone cells (McManus *et al*., 1998). The process is a true transdifferentiation which does not require cell division. Cell death certainly occurs during abscission. Many cells in the zone disintegrate or detach during the separation process. Cells on the separation surface of the organ being removed die, along with all others in the organ. On the proximal side of the abscission zone, cells enlarge and eventually form scar or wound tissue. This tissue is composed of dead cells, presumably some of the very zone cells which enlarged and aided the separation process. The abscission process shares features with PCD in response to pathogens; they include ethylene-induction of pathogenesis-related proteins and involvement of reactive oxygen species (González-Carranza *et al*., 1998). In abscission, H_2O_2 has not been implicated, but induction of peroxidases and accumulation of a metallothionen-like protein specifically in the zone cells could indicate the involvement of reactive oxygen species. A matrix metalloproteinase gene, expressed late in senescence but before appearance of DNA laddering, has been identified in cucumber cotyledons and suggested to have a role in breakdown of the extracellular matrix and PCD (Delorme *et al*., 2000). Despite the evidence available, it remains to be determined if the cells in the zone separate and die as the result of detachment or if cells enter a programmed death process during which they separate and fragment so that abscission occurs. These questions highlight the need to identify more of the genes involved in initiating PCD.

IV. Reproductive Cell Death

A. Floral Organs

Before gametes are produced by plants, reproductive structures must be differentiated. PCD plays a role in these processes. Most plants produce perfect or hermaphroditic flowers, but a few produce flowers with only pistils or only anthers. The flowers are initially perfect, but development of either pistils or anthers is arrested and PCD follows resulting in a single sex, mature flower (Irish and Nelson 1989; Dellaporta and Calderon-Urrea, 1993).

The initial bisexuality of both male and female flowers of monoecious *Zea mays* was noted early and described in detail later (Bonnett, 1948). More recently scanning electron microscopy (SEM) and TEM studies revealed that the initially bisexual tassel flowers abort gynoecium tissue in succession in both flowers of each spikelet leaving only anthers to develop (Cheng *et al*., 1983). Cells in the aborting tissue become increasingly vacuolated and the content of ribosomes and other organelles decreases as degeneration progresses. The process initiates in and spreads from the subepidermal region (Calderon-Urrea and Dellaporta, 1999). Slightly later, abortion of anthers begins in ear flowers, occurring first in the upper and later in the lower flower of each paired spikelet (Cheng *et al*., 1983). In maize the cellular phenotype of anther abortion was similar to that occurring in the gynoecium cells (Cheng *et al*., 1983; Cheng and Pareddy, 1994), but it varies somewhat in several other grasses (Le Roux and Kellogg, 1999).

Silene latifolia, white campion, illustrates regulation of PCD by sex chromosomes (Grant *et al*., 1994a). Both male and female plants begin reproductive development with perfect flowers (Grant *et al*., 1994b), but the gynoecium primordium emerges as a small, undifferentiated rod on male plants. Two developmental stages later in female plants, stamens first stop developing and later they degenerate. The timing of failure of female versus male flower parts plus genetic evidence indicate that the mechanism suppressing development of gynoecium is independent of the one leading to cell death in anthers (Grant *et al.*, 1994b). Genes are present on the Y chromosome which determine male development by converting the female default program to male (Grant *et al.*, 1994a). Presumably the Y chromosome suppresses genes on the X chromosome responsible for PCD of anthers on female plants.

Sex-linked gene loci rather than sex chromosomes activate PDC to achieve single-sex flowers on monoecious maize or dioecious *Ecballium elaterium* or *Spinacia oleracea* plants. In maize recessive mutations of several genes result in dwarf plants with perfect or male flowers in the normally female ear (Irish and Nelson, 1989). This condition is called "anther ear", a name applied to one of the mutants. Biochemical evidence has shown that gibberellin biosynthesis is blocked at different steps in these mutants (d_1, d_2, d_3, d_4, and an_1) and application of gibberellin rescues the normal phenotype (Phinney, 1984). *Anther ear 1* has been cloned (Bensen *et al.*, 1995); its sequence is homologous to a plant cyclase and biochemical evidence indicates its product synthesizes *ent*-kaurene. The d_3 gene has been shown to encode a cytochrome p450 enzyme involved in gibberellin biosynthesis (Winkler and Helentjaris, 1995). Thus some GA down stream from *ent*-kaurene, perhaps GA_1, is needed to cause PCD of anthers in ear florets. Other evidence also links high or low levels of GA with female or male development, respectively (Irish and Nelson, 1989).

The sexual development of the normally male tassel of maize is influenced by *tasselseed* genes known to occur in five loci (*Ts1*, *Ts2*, *Ts4*, *Ts5*, *Ts6*) (DeLong *et al.*, 1993; Dellaporta and Calderon-Urrea, 1993). Two genes, *ts1* and *ts2*, cause production of female flowers in the tassel when they are homozygously recessive. In wild type tassels, both stamens and a carpel are initiated in each flower, but the carpel is aborted (Irish and Nelson, 1993). In tassel flowers of *ts1* and *ts2* mutants, the carpel does not abort, but the stamens do. Thus these genes appear to control a switch for execution of PCD in either the carpel or stamens. *Ts5* is a dominant mutation which causes a base-to-tip gradient of pistillate to staminate flowers in the tassel (Irish and Nelson, 1989). This gene affects the abortion of gynoecium, and thus it also controls PCD in tassel floret development. Mutants for *Ts4* (recessive) and *Ts6* (dominant) block normal PCD of the gynoecium in tassel flowers (Dellaporta and Calderon-Urrea, 1993).

The gene encoding *ts2* has been cloned and sequenced (DeLong *et al.*, 1993). *Ts2* mRNA was expressed subepidermally in the gynoecium of tassel flowers shortly before abortion. Restoration of *Ts2* action and male development occurred when the transposon used to induce the *ts2* allele was excised. The predicted amino acid sequence of the *Ts2* protein is similar to bacterial hydroxysteroid dehydrogenases. Such an enzyme could be active in the synthesis of gibberellins or steroids (DeLong *et al.*, 1993). GAs are known to regulate floral development based on the dwarf mutants mentioned above, and brassinosteroids have been implicated in flowering based on the phenotypes of deficient mutants (Li *et al.*, 1996). Since the *Ts2* message is expressed in the gynoecium of tassel flowers and the *Ts2* gene product may synthesize a bioactive terpenoid, the *Ts2* product may trigger PCD in pistils in tassel florets. The mutant gene (*ts2*) presumably fails to produce a functional product and, therefore, the female development program is not switched off. *Ts2* also functions in ears to suppress development of the second floret in each spikelet; thus, in both tassel and ear *Ts2* causes PCD in female organs. A cell death mechanism similar to that in maize appears to be present in a wild relative, *Tripsacum dactyloides* (Li *et al.*, 1997).

B. Gametophytes

During megagametogenesis four haploid nuclei are produced. Commonly, three of the four cells resulting from meiosis die, specifically those at the micropylar end of the ovule (Barlow, 1982). Megaspore abortion is proposed to be a consequence of selective apoptosis (Bell, 1996). Gametospore degeneration in the fern *Marsilea vestita* exhibits characteristics of apoptosis (Bell, 1996). Mutations interpreted to cause an over expression of PCD have been identified in arabidopsis; four mutants show aborted female gametospores at the one nucleate stage while gametospores of five others are affected later in development (Christensen *et al.*, 1998). Tobacco plants transgenic for an antisense copy of a pistil-specific aminocyclopropane carboxylate oxidase failed to complete megasporogenesis, which was restored by the ethylene reagent ethephon (De Martinis and Mariani, 1999). During diploid parthenogenesis and early somatic embryogenesis in Norway spruce, specific nuclei die (Havel and Durzan, 1996). Nuclear elimination occurred in the ventral canal nucleus of egg-equivalents and in differentiating suspensors of early embryos. Degeneration was observed microscopically and by positive TUNEL reaction while nuclei of surviving cells fluoresced strongly with a reagent for non-fragmented DNA. Further study implicated endonucleases and ubiquitin-mediated turnover of proteins in PCD in somatic embryogenesis (Durzan, 1996).

In the anther, microgametogenesis occurs along with PCD of several cell types (Koltunow *et al.*, 1990; Goldberg *et al.*, 1993). Death of tapetal cells was recognized by Barlow (1982) to be essential to development of pollen. Cell shrinkage, polarization of the cytoplasm, vacuolation, thinning of the cell walls and cell rupture characterize the process (Chapman, 1987; Bedinger, 1992). A transgene with a tapetal-specific promoter attached to a bacterial RNase resulted in male sterility in transformed plants (Goldberg *et al.*, 1993). Both the presence of the tapetum and its degeneration via PCD appear essential for normal pollen development. Other tissues also degenerate during dehiscence of the anther including the stomium, connective and circular cell cluster (Koltunow *et al.*, 1990; Goldberg *et al.*, 1993; Beals and Goldberg, 1997). A thiol endopeptidase gene is expressed exclusively in these tissues prior to their destruction (Koltunow *et al.*, 1990). The death of these cells leads to splitting of the anther and is followed by senescence of the remaining cells, presumably also an expression of PCD. Pretreatment of anthers by hyper-osmotic shock aids in the development of plants from immature pollen in culture (androgenesis). During pretreatment, cleavage of DNA into fragments and positive TUNEL reactions identified apoptosis in loculus wall cells (adjacent to the tapetum), tapetum and filament (connective) (Wang *et al.*, 1999). TEM showed condensed chromatin in nuclei of loculus wall cells. Treatment of anthers with osmotic shock or ABA enhanced PCD in the tissues noted but enhanced viability and lowered apoptosis-linked characteristics in microspores. While not involving *in vivo* behavior, this study suggests that death of anther cells is PCD.

C. Pollination

Cell death occurs during pollination. Hair cells on the stigmatic surface, cells in the style in the pathway of the pollen tube, and cells in the nucellus die shortly after pollen germination or before flower opening (Barlow, 1982; Cheung, 1996). Recently, PCD was implicated in the death of style cells in the growth track of the pollen tube (Wang *et al.*, 1996b). The deteriorating cells exhibited mRNA poly (A) tail shortening. Another example of PCD during fertilization is exhibited by the synergid cell, one of two, which ruptures shortly before the ovule is penetrated by the pollen tube (Barlow, 1982). Eventually the pollen tubes and the other synergid degenerate too. The location of the first synergid to degenerate appears not to be random (Wu and Cheung, 2000), raising the possibility that its rich calcium stores provide a chemoattractant gradient to guide the pollen tube (Russell, 1996).

Incompatibility reactions are often described as inhibiting germination or growth of pollen tubes (Newbigin *et al.*, 1993). However, Bell (1995) cited the observation of de Nettancourt *et al.* that in *Lycopersicum* the vegetative nucellus in the incompatible tube appears to become pycnotic, a feature of nuclei in cells undergoing PCD, as a reason to propose that pollen-incompatibility is another example of PCD. One type of incompatibility involves production of RNases along the pathway of the pollen tube in the style (McClure *et al.*, 1990). If the RNase-attacked pollen tubes eventually die, and there is evidence of RNA degradation and the blurring of structural features in these cells (McClure *et al.*, 1990; Kao and McCubbin, 1996; Gietmann, 1999), then at least some forms of self-infertility may be a complex form of PCD involving both the genetics of the pollen cells and recognition by the transmitting tissue of the flower (Mittler, 1998). Pollination stimulates trimming of poly (A) tails of mRNAs and reduction or increase of message and product levels (Wang *et al.*, 1996b). One of the trimmed mRNAs encodes a pollen tube growth promoting protein (Wu *et al.*, 1995). Treatment with ethylene, ACC or phosphatase inhibitor okadaic acid induces poly (A) tail trimming (Wang *et al.*, 1996b). Further, the timing and tissue specificity of ACC oxidase expression during pistil development are paralleled by development of the self-incompatibility response (Tang *et al.*, 1994). Preliminary evidence links a xyloglucan endotransglycosylase, arabinogalactan proteins, chitinases and calcium in PCD in pollen tubes and/or style cells (Wu and Cheung, 2000).

D. Embryogenesis

After fertilization, synergid, antipodal, endosperm, coleorhiza and nucellar pillar cells die (Barlow, 1982; Murgia *et al.*, 1993). These events are all assumed to represent PCD. The suspensor, which develops at the base of the embryo in angiosperms, is another example of developmental PCD (Barlow, 1982; Raghavan, 1986). The cells of the suspensor are metabolically active, they promote embryo survival *in vitro*, and they synthesize GAs and other hormones; nevertheless, they degenerate in the later stages of embryo development (Yeung and Meinke, 1993; Schwartz *et al.*, 1997).

During seed maturation, cell death occurs in the endosperm (DeMason, 1997). Endosperm cell death in maize seeds begins at about 16 days after pollination (DAP) near the center of the endosperm, progresses centrifugally toward the top of the kernel and is complete at 40 DAP (Young *et al.*, 1997). DNA fragmentation can be observed in extracts beginning around day 28 and expression intensifies until day 44. DNA fragmentation is not seen in the embryos from the same seed. The endosperm mutation called *shrunken 2* (*sh2*), which results in 2-fold elevation of kernel sugar levels, also results in more extensive and rapid DNA fragmentation. Ethylene production rates per kernel exhibited two peaks, one around 16 to 20 DAP and the other around 32 to 40 DAP. Levels of the ethylene precursor ACC generally followed ethylene synthesis patterns. Ethylene treatments accelerated DNA fragmentation and death in the endosperm, and they reduced germination of seed. When ethylene synthesis was reduced to 20% of the control rate with AVG, DNA fragmentation was reduced but cell death was not prevented in *sh2* kernels. The data suggest that ethylene may act as a signal in PCD in maize endosperm cells, and more recent evidence indicates that ABA and reactive oxygen species may also be involved (Young and Gallie, 2000). PCD in wheat endosperm cells exhibits common and unique features with the similar program in maize (Young and Gallie, 1999).

Endosperm PCD in castor bean occurs during germination and is associated with DNA fragmentation and accumulation of small organelles termed ricinosomes (Schmid *et al.*, 1999). The ricinosomes contain a cysteine aminopeptidase which is released into the cytoplasm and then activated by cleavage. Homologous proteases occur in other species and tissues.

V. Conclusions

There is evidence that PCD occurs in a wide range of cell types during differentiation and that PCD plays an important role during plant development and survival. It is too soon to conclude whether or not PCD in plant cell differentiation faithfully parallels apoptosis in animal cells, but there are sufficient indications to suggest that the possibility of commonality deserves further attention. It is impressive that the process is conserved in all life forms. Whether PCD in plant cells follows many divergent and independent pathways or has a common basis, is far from clear at present. Cytosolic Ca^{2+} is one second-messenger that is implicated in many different cell types during PCD, but the specific proteins that are activated by Ca^{2+} thereby leading to PCD are unknown. Also too little is known about the genes encoding regulators of cell death in plant cells. Improved understanding of PCD, and its manipulation for plant improvement, are likely goals for this area of plant research in the future.

References

Allen, R.J. (1923). A cytological study of infection of baart and kanred wheats by *Puccinia grominis tritici. J. Agric. Res.* **23**, 131–152.

Ameisen, J.C. (1998). The evolutionary origin and role of programmed cell death in single-celled organisms. In *When Cells Die* (R.A. Lockshin, Z. Zakeri and J.L. Tilly, Eds.), pp. 3–56. Wiley–Liss, New York.

Aoyagi, S., Sugiyama, M., and Fukuda, H. (1998). BENI and ZENI cDNAs encoding S1-type Dnases that are associated with programmed cell death in plants. *FEBS* **429**, 134–138.

Atwell, B.J., Drew, M.C., and Jackson, M.B. (1988). The influence of oxygen deficiency on ethylene synthesis, 1-aminocyclopropane-1-carboxylic acid levels and aerenchyma formation in roots of *Zea mays* L. *Physiol. Plant.* **72**, 15–22.

Bachmair, A., Becker, F., Masterson, R.V., and Schell, J. (1990). Perturbation of the ubiquitin system causes leaf curling, vascular tissue alteration and necrotic lesions in a higher plant. *EMBO J.* **9**, 4543–4549.

Barlow, P.W. (1982). Cell death—an integral part of plant development. In *Growth Regulators in Plant Senescence* (M.B. Jackson, B. Grout and I.A. Mackenzie, Eds.), pp. 27–45. Monograph 8, British Plant Growth Regulator Group, Wantage, UK.

Beals, T.P., and Goldberg, R.B. (1997). A novel cell ablation strategy blocks tobacco anther dehiscence. *Plant Cell* **9**, 1527–1545.

Bedinger, P. (1992). The remarkable biology of pollen. *Plant Cell* **4**, 879–887.

Beers, E.P., and Freeman, T.B. (1997). Proteinase activity during tracheary element differentiation in zinnia mesophyll cultures. *Plant Physiol.* **113**, 873–880.

Beers, E.P., Woffenden, B.J., and Zhao, C. (2000). Plant proteolytic enzymes: possible roles during programmed cell death. *Plant Mol. Biol.* **44**, 399–415.

Bell, P.R. (1995). Incompatibility in flowering plants: adaptation of an ancient response. *Plant Cell* **7**, 5–16.

Bell, R.P. (1996). Megaspore abortion: A consequence of selective apoptosis? *Int. J. Plant Sci.* **157**, 1–7.

Bensen, R.J., Johal, G.S., Crane, V.C., Tossberg, J.T., Schnable, P.S., Meeley, R.B., and Briggs, S.P. (1995). Cloning and characterization of the maize *An1* gene. *Plant Cell* **7**, 75–84.

Bethke, P.C., Hillmer, S., and Jones, R.L. (1996). Isolation of intact protein storage vacuoles from barley aleurone. *Plant Physiol.* **110**, 521–529.

Bethke, P.C., Lonsdale, J.E., Fath, A., and Jones, R.L. (1999). Hormonally regulated programmed cell death in barley aleurone cells. *Plant Cell* **11**, 1033–1045.

Bonnett, O.T. (1948). Ear and tassel development in maize. *Ann. Mo. Bot. Garden* **35**, 269–288.

Brown, P.H., and Ho, T.-H.D. (1986). Barley aleurone layers secrete a nuclease in response to gibberellic acid. *Plant Physiol.* **82**, 801–806.

Buckner, B., Janick-Buckner, D., Gray, J., and Johal, G.S. (1998). Cell-death mechanisms in maize. *Trends Plant Sci.* **3**, 218–223.

Bush, D.S., Biswas, A.K., and Jones, R.L. (1989). Gibberellic-acid-stimulated Ca^{2+} accumulation in endoplasmic reticulum of barley aleurone: Ca^{2+} transport and steady-state levels. *Planta* **178**, 411–420.

Bush, D.S., Cornejo, M.-J., Huang, C.-N., and Jones, R.L. (1986). Ca^{2+} stimulated secretion of α-amylase during development in barley aleurone protoplasts. *Plant Physiol.* **82**, 566–574.

Calderon-Urrea, A., and Dellaporta, S.L. (1999). Cell death and cell protection genes determine the fate of pistils in maize. *Development* **126**, 435–441.

Campbell, R., and Drew, M.C. (1983). Election microscopy of gas space (aerenchyma) formation in adventitious roots of *Zea mays* L. subjected to oxygen shortage. *Planta* **157**, 350–357.

Chapman, G.P. (1987). The tapetum. *Int. Rev. Cytol.* **107**, 111–125.

Cheng, P.C., Grayson, R.I., and Walden, D.B. (1983). Organ initiation and the development of unisexual flowers in the tassel and ear of *Zea mays. Amer. J. Bot.* **70**, 450–462.

Cheng, P.-C., and Pareddy, D.R. (1994). Morphology and development of the tassel and ear. In *The Maize Handbook* (M. Freeling and V. Walbot, Eds.), pp. 37–47. Springer.

Cheung, A.Y. (1996). The pollen tube growth pathway: its molecular and biochemical contributions and responses to pollination. *Sex. Plant Reprod.* **9**, 330–336.

Christensen, C.A., Subramanian, S., and Drews, G.H. (1998). Identification of gametophytic mutations affecting female gametophyte development in *Arabidopsis. Dev. Biol.* **202**, 136–151.

Deacon, J.W., Drew, M.C., and Darling, A. (1986). Progressive cortical senescence and formation of lysigenous gas space (aerenchyma) distinguished by nuclear staining in adventitious roots of *Zea mays. Ann. Bot.* **58**, 719–727.

Dellaporta, S.L., and Calderon-Urrea, A. (1993). Sex determination in flowering plants. *Plant Cell* **5**, 1241–1251.

Delorme, V.G.R., McCabe P.F., Kim, D.-J., and Leaver, C.J. (2000). A matrix metalloproteinase gene is expressed at the boundary of senescence and programmed cell death in cucumber. *Plant Physiol.* **123**, 917–927.

DeLong, A., Calderon-Urrea, A., and Dellaporta, S.L. (1993). Sex determination gene TASSELSEED2 of maize encodes a short-chain alcohol dehydrogenase required for stage-specific floral organ abortion. *Cell* **74**, 757–768.

De Martinis, D., and Mariani, C. (1999). Silencing gene expression of the ethylene-forming enzyme results in a reversible inhibition of ovule development in transgenic tobacco plants. *Plant Cell* **11**, 1061–1071.

DeMason, D.A. (1997). Endosperm structure and development. In *Cellular and Molecular Biology of Plant Seed Development* (B.A. Larkins and I.K. Vasil, Eds.), pp. 73–115. Kluwer, Dordrecht.

Dietrich, R.A., Richberg, M.H., Schmidt, R., Dean, C., and Dangl, J.L. (1997). A novel zinc finger protein is encoded by the arabidopsis *LSD1* gene and functions as a negative regulator of plant cell death. *Cell* **88**, 685–694.

Drew, M.C. (1979). Root development and activities. In *Arid-land Ecosystems: Structure, Functioning and Management* (R.A. Perry and D.W. Goodall, Eds.), pp. 573–606. Cambridge University Press, Cambridge, UK.

Drew, M. (1997). Oxygen deficiency and root metabolism. *Ann Rev. Plant Physiol. Plant Mol. Biol.* **48**, 223–250.

Drew, M.C., He, C.J., and Morgan, P.W. (1989). Decreased ethylene biosynthesis, and induction of aerenchyma, by nitrogen- or phosphate-starvation in adventitious roots of *Zea mays* L. *Plant Physiol.* **91**, 266–271.

Drew, M.C., He, C.J., and Morgan, P.W. (2000). Programmed cell death and aerenchyma formation in roots. *Trends Plant Sci.* **5**, 123–127.

Drew, M.C., Jackson, M.B., and Giffard, S. (1979). Ethylene-promoted adventitious rooting and development of cortical air space (aerenchyma) in roots may be adaptive responses to flooding in *Zea mays* L. *Planta* **147**, 83–88.

Drew, M.C., Jackson, M.B., Giffard, S.C., and Campbell, R. (1981). Inhibition by silver ions of gas space (aerenchyma) formation in adventitious roots of *Zea mays* L. subjected to exogenous ethylene or to oxygen deficiency. *Planta* **153**, 217–224.

Drew, M.C., Saglio, P., and Pradet, A. (1985). Higher adenylate energy charge and ATP/ADP ratios in aerenchymatous roots of *Zea mays* in anaerobic media as a consequence of improved internal oxygen transport. *Planta* **165**, 51–58.

Durzan, D.J. (1996). Protein ubiquination in diploid parthenogenesis and early embryos of Norway spruce. *Int. J. Plant Sci.* **157**, 17–26.

Elliott, G.A., Robson, A.D., and Abbott, L.K. (1993). Effects of phosphate and nitrogen application on death of the root cortex in spring wheat. *New Phytol.* **123**, 375–382.

Esau, K. (1965). *Plant Anatomy* (2nd ed.). John Wiley & Sons, New York.

Fath, A., Bethke, P.C., and Jones, R.L. (1999). Barley aleurone cell death is not apoptotic: characterization of nuclease activities and DNA degradation. *Plant J.* **20**, 305–315.

Fath, A., Bethke, P.C., and Jones, R.L. (2001). Enzymes that scavenge reactive oxygen species are down-regulated prior to gibberellic acid-induced programmed cell death in barley aleurone. *Plant Physiol.* **126**, 156–166.

Fujita, T., Kouchi, H., Ichikawa, T., and Syone, K. (1994). Cloning of cDNAs for genes that are specifically or preferentially expressed during the development of tobacco genetic tumors. *Plant J.* **5**, 645–654.

Fukuda, H. (1996). Xylogenesis: initiation, progression, and cell death. *Annu. Rev. Plant Physiol. Plant Mol. Biol.* **47**, 299–325.

Fukuda, H. (1997). Tracheary element differentiation. *Plant Cell* **9**, 1147–1156.

Fukuda, H., and Komamine, A. (1980). Establishment of an experimental system for study of tracheary element differentiation from single cells isolated from the mesophyll of *Zinnia elegons. Plant Physiol.* **65**, 57–60.

Funk, V., Kositsup, B., Zhao, C., and Beers, E.P. (2002). The arabidopsis xylem peptidase XCP1 is a tracheary element vacuolar protein that may be a papain ortholog. *Plant Physiol.* **128**, 84–94.

Gallois, P., Makishima, M., Hecht, V., Despres, B., Laudié, M., Nishimoto, T., and Cooke, R. (1997). An *Arabidopsis thaliana* cDNA complementing a hamster apoptosis suppressor mutant. *Plant J.* **11**, 1325–1331.

Gavrielli, Y., Sherman, Y., and Ben-Sasson, S.A. (1992). Identification of programmed cell death in situ via specific labeling of nuclear DNA fragmentation. *J. Cell Biol.* **119**, 493–501.

Gietmann, A. (1999). Cell death of self-incompatible pollen tubes: Necrosis or aoiotisus? In *Fertilization in Higher Plants: Molecular and Cytological Aspects* (M. Cresti, G. Cai and A. Moscatelli, Eds.), pp. 113–137, Springer-Verlag, Berlin.

Gilroy, S. (1996). Signal transduction in barley aleurone protoplasts is calcium-dependent and calcium independent. *Plant Cell* **8**, 2193–2209.

Goldberg, R.B., Beals, T.P., and Sanders, P.M. (1993). Anther development: basic principles and practical applications. *Plant Cell* **5**, 1217–1229.

González-Carranza, Z.H., Lozoya-Gloria, E., and Roberts, J.A. (1998). Recent developments in abscission: shedding light on the shedding process. *Trends Plant Sci.* **3**, 10–14.

Grant, S., Houben, A., Vyskot, B., Siroky, J., Pan, W.-H., Macos, J., and Saedler, H. (1994a). Genetics of sex determination in flowering plants. *Dev. Genet.* **15**, 214–230.

Grant, S.R., Hunkirchen, B., and Saedler, H. (1994b). Developmental differences between male and female flowers in the dioecious plant *Silene latifolia. Plant J.* **6**, 471–480.

Gray, J., Close, P.S., Briggs, S.P., and Johal, G.S. (1997). A novel suppressor of cell death in plants encoded by the *Lls1* gene of maize. *Cell* **89**, 25–31.

Gray, J., and Johal, G.S. (1998). Programmed cell death in plants. In *Arabidopsis* (M. Anderson and J.A. Roberts, Eds.), pp. 360–388. Sheffield Academic Press, Sheffield, UK.

Greenberg, J.T. (1996). Programmed cell death: A way of life for plants. *Proc. Natl. Acad. Sci. USA* **93**, 12094–12097.

Groover, A., DeWitt, N., Heidel, A., and Jones, A. (1997). Programmed cell death of plant tracheary elements differentiating in vitro. *Protoplasma* **196**, 197–211.

Groover, A., and Jones, A.M. (1999). Tracheary element differentiation uses a novel mechanism coordinating programmed cell death and secondary cell wall synthesis. *Plant Physiol.* **119**, 375–384.

Gunawardena, A.H., Pearce, D.M., Jackson, M.B., Hawes, C.R., and Evans, D.E. (2001). Characterization of programmed cell death during aerenchyma formation induced by ethylene or hypoxia in roots of maize (*Zea mays* L.). *Planta* **212**, 205–214.

Havel, L., and Durzan, D.J. (1996). Apoptosis during diploid parthenogenesis and early somatic embryogenesis of Norway spruce. *Int. J. Plant Sci.* **157**, 8–16.

Hay, M. (1997). Synchronous spawning: When timing is everything. *Science* **275**, 1080–1081.

He, C.J., Drew, M.C., and Morgan, P.W. (1994). Induction of enzymes associated with lysigenous aerenchyma formation in roots of *Zea mays* during hypoxia or nitrogen-starvation. *Plant Physiol.* **105**, 861–865.

He, C.J., Finlayson, S.A., Drew, M.C., Jordan, W.R., and Morgan, P.W. (1996a). Ethylene biosynthesis during aerenchyma formation in roots of *Zea mays* subjected to mechanical impedance and hypoxia. *Plant Physiol.* **112**, 1679–1685.

He, C.J., Morgan, P.W., and Drew, M.C. (1996b). Transduction of an ethylene signal is required for cell death and lysis in the root cortex of maize during aerenchyma formation induced by hypoxia. *Plant Physiol.* **112**, 463–472.

Henry, C.M., and Deacon, J.W. (1981). Natural (non-pathogenic) death of the cortex of wheat and barley seminal roots, as evidenced by nuclear staining with acridine orange. *Plant Soil Sci.* **60**, 255–274.

Huang, Y.-F., Jordan, W.R., Wing, R.A., and Morgan, P.W. (1998). Gene expression induced by physical impedance in maize roots. *Plant Mol. Biol.* **37**, 921–930.

Irish, E.E., and Nelson, T. (1989). Sex determination in monecious and dioecious plants. *Plant Cell* **1**, 737–744.

Irish, E., and Nelson, T. (1993). Development of tassel seed 2 inflorescences in maize. *Am. J. Bot.* **80**, 292–299.

Jackson, M.B., Fenning, T.M., Drew, M.C., and Saker, L.R. (1985). Stimulation of ethylene production and gas-space (aerenchyma) formation in adventitious roots of *Zea mays* L. by small partial pressures of oxygen. *Planta* **165**, 486–492.

Jones, A.M. (2001). Programmed cell death in development and defense. *Plant Physiol.* **125**, 94–97.

Jones, R.L., and Jacobsen, J.V. (1991). Regulation of synthesis and transport of proteins in cereal aleurone. *Int. Rev. Cytol.* **126**, 49–88.

Justin, S.H.F.W., and Armstrong, W. (1991). Evidence for the involvement of ethene in aerenchyma formation in adventitious roots of rice (*Oryza sativa* L.). *New Phytol.* **118**, 49–62.

Kao, T.H., and McCubbin, A.G. (1996). How flowering plants discriminate between self and non-self pollen to prevent inbreeding. *Proc. Natl. Acad. Sci. USA* **93**, 12059–12065.

Kaplan, D.R. (1984). Alternative modes of organogenesis in higher plants. In *Contemporary Problems in Plant Anatomy* (R.A. White and W.C. Dickerson, Eds.), pp. 261–300. Academic Press, New York.

Katsuhara, M. (1997). Apoptosis-like cell death in barley roots under salt stress. *Plant Cell Physiol.* **38**, 1091–1093.

Kawai, M., Samarajeewa, P.K., Barrero, R.A., Nishiguchi, M., and Uchimiya, H. (1998). Cellular dissection of the degradation pattern of cortical cell death during aerenchyma formation of rice roots. *Planta* **204**, 277–287.

Kelleher, D.J., and Gilmore, R. (1997). DAD1, the defender against apoptotic cell death, is a subunit of the mammalian oligosaccharyltransferase. *Proc. Natl. Acad. Sci. USA* **94**, 4994–4999.

Kerr, J.F.R., Wyllie, A.H., and Currie, A.R. (1972). Apoptosis: a basic biological phenomenon with side-ranging implications in tissue kinetics. *Br. J. Cancer* **26**, 239–257.

Knetsch, M.L.W., Wang, M., Snaar-Jagalska, B.E., and Heimovaara-Dijkstra, S. (1996). Abscisic acid induces-mitogen-activated protein kinase activation in barley aleurone protoplasts. *Plant Cell* **8**, 1061–1067.

Knowles, R.V., and Phillips, R.L. (1988). Endosperm development in maize. *Int. Rev. Cytol.* **112**, 97–136.

Koltunow, A.M., Truettner, J., Cox, K.H., Walbroth, M., and Goldberg, R.B. (1990). Different temporal and spatial gene expression patterns occur during anther development. *Plant Cell* **2**, 1201–1224.

Kosslak, R.M., Chamberlin, M.A., Palmer, R.G., and Bowen, B.A. (1997). Programmed cell death in the root cortex of soybean root necrosis mutants. *Plant J.* **11**, 729–745.

Kuo, A., Cappelluti, S., Cervantes-Cervantes, M., Rodriguez, M., and Bush, D.S. (1996). Okadaic acid, a protein phosphatase inhibitor, blocks calcium changes, gene expression, and cell death induced by gibberellin in wheat aleurone cells. *Plant Cell* **8**, 259–269.

Lam, E., and del Pozo, O. (2000). Caspase-like protease involvement in the control of plant cell death. *Plant Mol. Biol.* **44**, 417–428.

Lascaris, D., and Deacon, J.W. (1991). Comparison of methods to assess senescence of the cortex of wheat and tomato roots. *Soil Biol. Biochem.* **23**, 979–986.

Leopold, A.C. (1961). Senescence in plant development. *Science* **134**, 1727–1732.

Le Roux, L.G., and Kellogg, E.A. (1999). Floral development and the formation of unisexual spikelets in Andropogoneae (Poaceae). *Am. J. Bot.* **86**, 354–366.

Li, D., Blakey, C.A., Dewald, C., and Dellaporta, S.L. (1997). Evidence for a common sex determination mechanism for pistil abortion in maize and its wild relative Tripsacum. *Proc. Natl. Acad. Sci. USA* **94**, 4217–4222.

Li, J., Nagpal, P., Vitart, V., McMorris, T.C., and Chory, J. (1996). A role for brassinosteroids in light-dependent development of Arabidopsis. *Science* **272**, 398–401.

Lindholm, P., Kuittinen, T., Sorri, O., Guo, D., Merits A., Törmäkangas, K., and Rueberg-Roos, P. (2000). Glycosylation of phytepsin and expression of *dad1*, *dad2*, and *ost1* during onset of cell death in germinating barley scutella. *Mech. Dev.* **93**, 169–173.

Lockshin, R.A., and Williams, C.M. (1965). Programmed cell death. I. Cytology of the degeneration of the intersegmental muscles of the Pernyi silkmoth. *J. Insect Physiol.* **11**, 123–133.

Martin, S.J., Reutelingsperger, C.P.M., McGahon, A.J., Rader, J.A., van Schie, R.C.A.A., LaFace, D.M., and Green, D.R. (1995). Early redistribution of plasma membrane phosphatidylserine is a general feature of apoptosis regardless of the initiating stimulus: inhibition by overexpression of Bcl-2 and Abl. *J. Exp. Med.* **182**, 1545–1556.

McClure, B.A., Gray, J.E., Anderson, M.A., and Clarke, A.E. (1990). Self-incompatibility in *Nicotiana alata* involves degradation of pollen rRNA. *Nature* **347**, 757–760.

McManus, M.T., Thompson, D.S., Merriman, C., Lyne, L., and Osborne, D.J. (1998). Transdifferentiation of mature cortical cells to functional abscission cells in bean. *Plant Physiol.* **116**, 891–899.

Minami, A., and Fukuda, H. (1995). Transient and specific expression of a cysteine endopeptidase during autolysis in differentiating tracheary elements from zinnia mesophyll cells. *Plant Cell Physiol.* **36**, 1599–1606.

Mittler, R. (1998). Cell death in plants. In *When Cells Die* (R.A. Lockshin, Z. Zakeri and J.L. Tilly, Eds.), pp. 147–174. Wiley–Liss, New York.

Mittler, R., and Lam, E. (1995). *In situ* detection of nDNA fragmentation during the differentiation of tracheary elements in higher plants. *Plant Physiol.* **108**, 489–493.

Molisch, H. (1938). *The Longevity of Plants* (H. Fullington, Transl.). Science Press, Lancaster, PA.

Murgia, M., Huang, B.-Q., Tucker, S.C., and Musgrave, M.E. (1993). Embryo sac lacking antipodal cells in *Arabidopsis thaliania* (Brassicaceae). *Am. J. Bot.* **80**, 824–838.

Newbigin, E., Anderson, M.A., and Clarke, A.E. (1993). Gametophytic self-incompatibility systems. *Plant Cell* **5**, 1315–1324.

Noodén, L.D. (1988). The phenomena of senescence and aging. In *Senescence and Aging in Plants* (L.D. Noodén and A.C. Leopold, Eds.), pp. 1–50, Academic Press, San Diego.

Noodén, L.D., Guiamet, J.J., and John, I. (1997). Senescence mechanisms. *Physiol. Plant.* **101**, 746–753.

Obara, K., Kuriyama, H., and Fukuda, H. (2001). Direct evidence of active and rapid nuclear degradation triggered by vacuole rupture during programmed cell death in zinnia. *Plant Physiol.* **125**, 615–626.

O'Brien, I.E.W., Baguley, B.C., Murray, B.G., Morris, B.A.M., and Ferguson, I.B. (1998). Early stages of the apoptotic pathway in plant cells are reversible. *Plant J.* **13**, 803–814.

O'Brien, I.E.W., Reutelingsperger, C.P.M., and Holdaway, K.M. (1997). Annexin-V and TUNEL use in monitoring the progression of apoptosis in plants. *Cytometry* **29**, 28–33.

Olson, A.R., and Cass, D.D. (1981). Changes in megagametophyte structure in *Papover nudicaule* L. (Papaveraceae) following *in vitro* placental pollination. *Am. J. Bot.* **68**, 1333–1341.

Oráez, D., and Granell, A. (1997). DNA fragmentation is regulated by ethylene during carpel senescence in *Pisum sativum. Plant J.* **11**, 137–144.

Osborne, D.J. (1989). Abscission. *Crit. Rev. Plant Sciences* **8**, 103–129.

Pappelis, A.J., and Katsanos, R.A. (1969). Ear removal and cell death rate in corn stalk tissue. *Phytopath.* **59**, 129–131.

Phinney, B.O. (1984). Gibberellin A_1, dwarfism and the control of shoot elongation in higher plants. In *The Biosynthesis and Metabolism of Plant Hormones* (A. Crozier and J.R. Hillman, Eds.), pp. 17–41. Cambridge University Press, Cambridge, UK.

Raghavan, V. (1986). *Embryogenesis in Angiosperms.* Cambridge University Press, Cambridge, UK.

Roberts, A.W., and Haigler, C.H. (1990). Tracheary-element differentiation in suspension-cultured cells of *Zinnia* requires uptakes of extracellular Ca^{2+}. *Planta* **180**, 502–509.

Roberts, J.K.M., Callis, J., Jardetzky, O., Walbot, V., and Freeling, M. (1984). Cytoplasmic acidosis as a determinant of flooding intolerance in plants. *Proc. Natl. Acad. Sci. USA* **81**, 6029–6033.

Runeberg-Roos, P., and Saarma, M. (1998). Phytepsin, a barley vacuolar aspartic proteinase, is highly expressed during autolysis of developing tracheary elements and sieve cells. *Plant J.* **15**, 139–145.

Russell, S.D. (1996). Attraction and transport of male gametes for fertilization. *Sex. Plant Reprod.* **9**, 337–342.

Salvesen, G.S., and Dixit, V.M. (1997). Caspases: intracellular signaling by proteolysis. *Cell* **91**, 443–446.

Schmid, M., Simpson, D., and Geitl, C. (1999). Programmed cell death in castor bean endosperm is associated with the accumulation and release of a cysteine endopeptidase from ricinosomes. *Proc. Natl. Acad. Sci. USA* **96**, 14159–14164.

Schwartz, B.W., Vernon, D.M., and Meinke, D.W. (1997). Development of the suspensor: differentiation, communication and programmed cell death during plant embryogenesis. In *Cellular and Molecular Biology of Plant Seed Development* (B.A. Larkins and I.K. Vasil, Eds.), pp. 53–72. Kluwer, Dordrecht.

Sexton, R., and Roberts, J.A. (1982). Cell biology of abscission. *Ann. Rev. Plant Physiol.* **33**, 133–162.

Shone, M.G.T., and Flood, A.V. (1983). Effects of periods of localized water stress on subsequent nutrient uptake by bailey roots and their adaptation by osmotic adjustment. *New Phytol.* **94**, 561–572.

Solomon, M., Belenghi, B., Delledonne, M., Menachem, E., and Levine, A. (1999). The involvement of cysteine proteases and protease inhibitor genes in regulation of programmed cell death in plants. *Plant Cell* **11**, 431–443.

Tanaka, Y., Makishima, T., Sasabe, M., Ichinose, Y., Shiraishi, T., Nishimoto, T., and Yamada, T. (1997). DADI, a putative cell death gene in rice. *Plant Cell Physiol.* **38**, 383–397.

Tang, X., Gomes, A.M.T.R., Bhatia, A., and Woodson, W.R. (1994). Pistil-specific and ethylene-regulated expression of 1-aminocyclopropane-1-carboxylate oxidase genes in petunia flowers. *Plant Cell* **6**, 1227–1239.

Thornberry, N.A., and Lazebnik, Y. (1998). Caspases: enemies within. *Science* **281**, 1312–1316.

Vaux, D.L., and Korsmeyer, S.J. (1999). Cell death in development. *Cell* **96**, 245–254.

Wang, H., Li, J., Bostock, R.M., and Gilchrist, D.G. (1996a). Apoptosis: A functional paradigm for programmed plant cell death induced by a host-selective phytotoxin and invoked during development. *Plant Cell* **8**, 375–391.

Wang, H., Wu, H.-M., and Cheung, A.Y. (1996b). Pollination induces mRNA poly (A) tail shortening and cell deterioration in flower transmitting tissue. *Plant J.* **9**, 715–727.

Wang, M., Hoekstra, S., van Bergen, S., Lamers, G.E.M., Oppedijk, B.J., van der Heijden, M.W., de Priester, W., and Schilperoort, R.A. (1999). Apoptosis in developing anthers and the role of ABA in this process during androgenesis in *Hordeum vulgare* L. *Plant Mol. Biol.* **39**, 489–501.

Wang, M., Oppedijk, B.J., Lu, X., Van Duijn, B., and Schilperoort, R.A. (1996c). Apoptosis in barley aleurone during germination and its inhibition by abscisic acid. *Plant Mol. Biol.* **32**, 1125–1134.

Winkler, R.G., and Helentjaris, T. (1995). The maize dwarf 3 gene encodes a cytochrome p450-mediated early step in gibberellin biosynthesis. *Plant Cell* **7**, 1307–1317.

Wodzicki, T.J., and Brown, C.L. (1973). Organization and breakdown of the protoplast during maturation of pine tracheids. *Amer. J. Bot.* **60**, 631–640.

Woffenden, B.J., Freeman, T.B., and Beers, E.P. (1998). Proteasome inhibitors prevent tracheary element differentiation in zinnia mesophyll cell cultures. *Plant Physiol.* **118**, 419–430.

Wu, H.-M., and Cheung, A.Y. (2000). Programmed cell death in plant reproduction. *Plant Mol. Biol.* **44**, 267–281.

Wu, H.-M., Wang, H., and Cheung, A.Y. (1995). A pollen tube growth stimulatory glycoprotein is deglycosylated by pollen tubes and displays a glycosylation gradient in the flower. *Cell* **82**, 393–403.

Ye, Z.H., and Varner, J.E. (1993). Gene expression patterns associated with in vitro tracheary element formation in isolated single mesophyll cells of *Zinnia elegans. Plant Physiol.* **103**, 805–813.

Ye, Z.H., and Varner, J.E. (1996). Induction of cysteine and serine proteases during xylogenesis in *Zinnia elegans. Plant Mol. Biol.* **30**, 1233–1246.

Yeung, E.C., and Meinke, D.W. (1993). Embryogenesis in angiosperms: development of the suspensor. *Plant Cell* **5**, 1371–1381.

Young, T.E., and Gallie, D.R. (1999). Analysis of programmed cell death during wheat endosperm reveals differences in endosperm development between cereals. *Plant Mol. Biol.* **39**, 915–926.

Young, T.E., and Gallie, D.R. (2000). Programmed cell death during endosperm development. *Plant Mol. Biol.* **44**, 283–301.

Young, T.E., Gallie, D.R., and De Mason, D.A. (1997). Ethylene-mediated programmed cell death during maize endosperm development of wild-type and shrunken 2 genotypes. *Plant Physiol.* **115**, 737–751.

3

Cell Death in Plant Disease: Mechanisms and Molecular Markers

Dominique Pontier, Olga del Pozo
and Eric Lam

I. Introduction

Programmed cell death (PCD) is defined as a self-destruction process triggered by external or internal factors. It is mediated through an active genetic program which has been shown to play a crucial role in development and survival in diverse organisms (see references in Chapter 1). In plants, the interest in PCD has seen a rapid increase in recent years, with its importance at many stages of the life cycle and in response to different stimuli being recognized. In particular, cell death is a predominant feature of plant–pathogen interactions since both incompatible (resistance) as well as compatible (disease) interactions often lead to the appearance of macroscopic necrosis (Mittler and Lam, 1996; Greenberg, 1997).

The most studied phenomenon is the hypersensitive response (HR), characterized by rapid and localized cell death at the inoculation site and often associated with resistance responses (Pontier *et al.*, 1998a; Heath, 1999). Mechanistic understanding concerning disease-related cell deaths is still lacking, but recent data have emerged to show that in some cases at least, they can be considered to be PCD events as well (Wang *et al.*, 1996; Gilchrist, 1998; Navarre and Wolpert, 1999).

II. Role of Cell Death during Plant–Pathogen Interactions

The HR is characterized by both the appearance of localized cell death and the activation of defense mechanisms which result in pathogen arrest. It is triggered by specific recognition events that typically involve a specific avirulence gene from the pathogen and the corresponding resistance gene in the host (Morel and Dangl, 1997). The importance of cell death *per se* in the resistance response is unclear since its establishment is concomitant with the activation of the other defense mechanisms (such as accumulation of pathogenesis-related (PR) proteins). It is thus difficult to quantify the contribution of cell death to pathogen resistance. On the one hand, rapid elimination of infected cells may be responsible for protecting the neighboring cells from further invasion. On the other hand, recent experimental data indicate that cell death can be uncoupled from resistance and thus may not be the primary barrier against pathogen proliferation. For example, pathogen resistance may not be abolished when cell death is inhibited by either high humidity or low oxygen (Hammond-Kosack *et al.*, 1996; Mittler *et al.*, 1996). Recently, the *Rx*-mediated "extreme resistance" against potato virus X has been demonstrated to be HR cell death independent (Bendahmane *et al.*, 1999). In addition, the *dnd1* mutant of Arabidopsis appears to be defective in HR cell death but remains resistant to avirulent bacterial pathogen (Yu *et al.*, 1998). These observations suggest that cell death and pathogen arrest can be uncoupled during plant–pathogen interactions.

III. Structural and Biochemical Changes Accompanying Cell Death during Plant Disease

Structural and cytological changes have been studied in different pathosystems involving viruses, bacteria or fungi (Brown and Mansfield, 1988; Mittler *et al.*, 1997a; Heath, 1999). In contrast to apoptosis in animal systems (Jacobson *et al.*, 1997), no consensus has emerged regarding the sequence of cellular events that ultimately leads to PCD of plant cells. It appears that the particular order in cytological and nuclear changes, as well as their resulting structural manifestations may vary depending on both the pathogen and the host (Mittler and Lam, 1996; Morel and Dangl, 1997). Differences have even been observed between two resistant cultivars of cowpea infected by the same cowpea rust fungus (Skalamera and Heath, 1998). However, some common features between cell death events in plants have been revealed from comparative analyses. For example, one of the early features of cell death during HR is the disruption of the cytoskeleton. Changes in microtubule organization have been observed during cowpea rust fungus–cowpea incompatible interactions and cytochalasin E, an inhibitor of microfilament polymerization, partially arrests HR cell death (Skalamera and Heath, 1998). Cessation of cytoplasmic streaming is observed during the onset of HR, as it is the case also with cell death during xylem differentiation (Meyer and Heath, 1987; Groover and Jones, 1999).

Nuclear condensation, which is one of the hallmarks of animal apoptosis, is also observed in numerous cases of plant PCD. In plants, the nuclear material aggregates into discrete patches and is distributed throughout the nucleus instead of localized to the periphery as observed in animal cells (Mittler *et al.*, 1997a). Recent flow cytometry studies with tobacco protoplasts suggest that nuclear DNA condensation could be activated before the first irreversible step of PCD after death induction by chemical treatments (O'Brian *et al.*, 1998). Examples of DNA fragmentation have been reported during HR and disease-related cell deaths as well as tracheary element (TE) differentiation, senescence or endosperm development (Mittler and Lam, 1995a; Wang *et al.*, 1996; see also Chapters 1 and 2 and references therein). Although DNA laddering has been detected during infection by incompatible fungi (Ryerson and Heath, 1996) or disease-related cell death caused by the toxin victorin (Navarre and Wolpert, 1999), nucleosomal DNA fragments were not observed during TMV-induced HR or hormone-induced TE differentiation. Lastly, a characteristic of plant PCD is the participation of the prominent vacuole, an organelle not found in animals. Cell death execution during TE differentiation is manifested by the collapse of the large vacuole (Groover and Jones, 1999). During HR cell death, a rapid cellular disruption of the tonoplast and the plasma membrane has been observed at the very end of the process, leading to the collapse of the cell (Freytag *et al.*, 1994). The release of vacuolar hydrolytic enzymes likely plays an important role in the degradation of the cellular contents. Rapid and irreversible plasma membrane damage is one of the characteristics of plant cell death (Bestwick *et al.*, 1995), in contrast to animal cell apoptosis where it remains intact while the cell is fragmented into apoptotic bodies which are then engulfed by neighbouring cells.

IV. Definition of Steps Involved in the Signaling Process of Cell Death Induction during Plant–Pathogen Interactions

A large body of studies using biochemical analysis of elicited or infected cell cultures as well as plant tissues has revealed major components of the signaling cascade such as ion fluxes, production of reactive oxygen species (ROS), phosphorylation cascades and protease action (Yang *et al.*, 1997; Richberg *et al.*, 1998). Interestingly, many of these elements have been found to participate in TE differentiation signaling as well (Groover and Jones, 1999).

One of the earliest biochemical events subsequent to pathogen recognition is the alterations of ion fluxes characterized in particular by an uptake of H^+ and Ca^{2+}. Pharmacological studies as well as pH measurements have revealed a rapid proton uptake in cells infected with an avirulent bacterial pathogen or treated with cell death elicitors such as cryptogein or harpin (Glazener *et al.*, 1996; Viard *et al.*, 1994). More recent studies using transgenic tobacco plants expressing the bacterial proton pump, bacterioopsin (bO), suggested that proton fluxes might have an important role in the induction of cell death. Such plants exhibit HR-like spontaneous lesions and mutations in the *bO* gene that affect the proton channeling capacity abolish this phenotype. These data indicate that the ectopic expression of the bO protein is likely to induce cell death by modifying the proton fluxes rather than by an indirect effect (Mittler and Lam, 1995b; Pontier *et al.*, 2002). The important role of calcium for HR cell death signaling has been suggested by the use of calcium channel blockers, calcium chelators or ionophores (Atkinson *et al.*, 1990; Tavernier *et al.*, 1995; Levine *et al.*, 1996; Mittler *et al.*, 1997b). A recent study by Heo and collaborators (1999) indicates that this requirement of calcium in cell death signaling may involve specific isoforms of calmodulin. Transgenic plants expressing two soybean calmodulin-encoding genes show spontaneous cell death lesions as well as an induction of PR proteins on older mature leaves. However, unlike HR cell death, the induction of PR genes is salicylic acid independent. Furthermore, genes encoding these calmodulin isoforms are transcriptionally activated during plant–pathogen interactions. Thus, they may be more important in the latter part of the cell death process instead of acting as early signal sensors.

Reactive oxygen species (ROS) constitute another likely signal in the cascade leading to cell death during plant–pathogen interactions (Lamb and Dixon, 1997; Van Camp *et al.*, 1998; Mach and Greenberg, Chapter 13). The direct action of ROS in mediating HR cell death has been suggested by the suppression of PCD in the absence of oxygen while PR production is not affected (Mittler *et al.*, 1996). Furthermore, ectopic expression of glucose oxidase in transgenic plants (Kazan *et al.*, 1998) as well as suppression of endogenous catalase (Chamnongpol *et al.*, 1998) trigger HR-like cell death. However, the nature and the source of the ROS involved is still under debate and might vary according to the pathosystem. As shown by the use of the inhibitor DPI, the involvement of ROS in HR cell death is likely to implicate an NADPH oxidase analogous to the well-characterized enzymatic system in neutrophils (Levine *et al.*, 1994; Lamb and Dixon, 1997). The involvement of an NADPH oxidase in ROS production during the HR also gained support from a study in which the small GTP-binding protein Rac was overexpressed in transgenic rice plants and cell suspensions (Kawasaki *et al.*, 1999). Together with the fact that Rac is known to regulate mammalian NADPH oxidase, gain-of-function and loss-of-function mutant experiments described in this work support the involvement of NADPH oxidase and its regulator Rac in ROS production observed during HR cell death. More recently, loss-of-function Arabidopsis mutants in some of the genes encoding this enzyme have been characterized and their involvement in HR cell death is confirmed (Torres *et al.*, 2002).

Nitric oxide (NO) is an important second messenger in animals and has recently been reported to play a role in mediating HR cell death (Van Camp *et al.*, 1998). NO is an inhibitor of the respiratory cytochrome c oxidase in plant mitochondria and can generate cytotoxic compounds by reacting with ROS. The generation of NO has been shown to enhance H_2O_2-mediated cell death and defense mechanisms (Delledone *et al.*, 1998; Durner *et al.*, 1998). The other molecule implicated in the potentiation of HR cell death is salicylic acid, a key mediator of disease resistance (Ryals *et al.*, 1996; Yang *et al.*, 1997). Using soybean cell suspension cultures, Shirasu and collaborators (1997) demonstrated that SA regulates a signal amplification loop leading to cell death. Furthermore, TMV-induced HR cell death is delayed in transgenic tobacco plants expressing a bacterial salicylate hydroxylase (NahG) that converts SA to catechol, leading to spreading cell death patches instead of discrete lesions (Mur *et al.*, 1997; Chivasa and Carr, 1998). Also, transgenic expression of the *NahG* gene suppresses the appearance of spontaneous cell death in the Arabidopsis disease lesion mimic mutants *lsd6* and *lsd7* (Weymann *et al.*, 1995). Similarly, SA seems to be required for spontaneous cell death in two other Arabidopsis mutants, *ssi1* and *acd6* (Rate *et al.*, 1999; Shah *et al.*, 1999). Altogether, these data lead to the following model: the oxidative burst would trigger the production of NO and SA which could in turn amplify the cell death signal. Thus, the role of NO and SA in HR cell death could be rather complex. Since the HR lesions have a definite size, a negative signal must exist to stop the amplification loop.

SA could be involved in both potentiation of the cell death signal as well as negatively regulating the spreading lesions during TMV-induced HR.

The alternative oxidase (AOX) is another component that has recently been proposed to play an important role in protecting surrounding tissue from oxidative stress and cell death (Murphy *et al.*, 1999). Its role in modulating ROS production from the mitochondria has been demonstrated by analyzing AOX suppressed tobacco plants generated using antisense technology. In such plants, inhibition of the mitochondria electron transport leads to a massive production of ROS with concomitant activation of cell death and defense genes (Maxwell *et al.*, 1999). Furthermore, induction of AOX has been observed in Arabidopsis leaves infected with avirulent as well as virulent pathogens and is potentiated by SA in the HR-inducing interactions (Simons *et al.*, 1999). A role for AOX involvement during disease resistance has been suggested by the effect of cyanide treatment, which activates AOX expression, in abolishing the spreading phenotype of TMV-induced lesions in NahG plants (Chivasa and Carr, 1998). This effect is reversed by treatment with SHAM, an inhibitor of AOX as well as other ROS protecting enzymes. Thus, a working model could be that AOX is a component of the negative control of cell death at the periphery of HR lesions.

An important finding in the work of Simons *et al.* (1999) is the demonstration that AOX induction requires the ethylene signaling pathway. In contrast to SA, which appears to be required only for the rapid induction of AOX expression during the HR, dominant mutations in *Etr1* abolished AOX activation by either avirulent or virulent bacterial pathogens. Since ethylene synthesis is activated during plant–pathogen interactions (de Laat and van Loon, 1983), these observations suggest that AOX expression could be controlled by a combination of the SA and ethylene signaling pathways. In contrast, the induction of defense-related genes such as those encoding PR proteins during the HR does not require ethylene (Lawton *et al.*, 1994). Although ethylene may not be involved in the induction of systemic acquired resistance or defense markers *per se*, its role in the activation of senescence-like chlorosis around the infection site undergoing an HR has been speculated to participate in pathogen restriction as a secondary barrier during heavy infection (Pontier *et al.*, 1999). The role of ethylene in disease symptoms related to cell death induced by biotrophic and necrotrophic pathogens that do not induce the HR or systemic response has also been shown. In the work of Lund *et al.* (1998), the ethylene-insensitive tomato mutant *Never ripe* was found to show a reduction in disease symptoms after inoculations with virulent bacterial and fungal pathogens. The phenotype is manifested as a dramatic reduction of cell death lesions due to systemic spread of the virulent pathogens without a significant effect on their proliferation within the host. This increased "tolerance" to pathogen infection suggested that suppressing ethylene sensitivity may reduce cell death caused by virulent pathogens. This conclusion is supported by the observation that cell death and disease symptoms induced by the fungal toxin victorin from the oat pathogen *Cochliobolus victoria* are also suppressed by inhibitors of the ethylene pathway (Navarre and Wolpert, 1999).

V. Molecular Components for Cell Death Control during Plant–Pathogen Interactions

In animal apoptosis, a central switch involving three classes of proteins is now accepted as the key regulator of cell death in diverse species (Raff, 1998). The enzymatic component of this tripartite switch is a specialized class of cysteine proteases called caspases, the name of which derives from their specificity of cleaving after an asparatic acid at the P1 position of their substrate recognition sites. When activated, these proteases begin a proteolytic cascade that irreversibly commits the cell to undergo cell death. The activity of caspases is under stringent control by paired pro-apoptotic and pro-survival signals through the positive regulator called CED4/APAF-1 and the negative regulator BCL2 and its related proteins (BLPs). CED4/APAF-1 proteins apparently activate caspases through protein–protein interactions while BLPs can regulate caspases through multiple mechanisms (Raff, 1998). In mammals, a large family of BLPs has been shown to control PCD and is comprised of genes showing pro-apoptotic (i.e. Bax, Bak and Bid) as well as pro-survival (i.e. Bcl2 and Bcl-x_L) activities. No BLP homologue has been detected in plants yet. However, two recent studies involving the ectopic expression of BLPs in plants suggest the existence of similar regulatory mechanisms in the two kingdoms. In transgenic tobacco plants expressing Bcl-x_L or Ced-9, HR cell death as well as cell death induced by paraquat treatment or UV-B irradiation is delayed (Mitsuhara *et al.*, 1999). Lacomme and Santa-Cruz (1999) found that Bax expression activates cell death in plants and that its localization in mitochondria is required. A similar result has been observed also in yeast in which no BLP has been found in the sequenced genome. In animals, BLPs are known to activate or repress the leakage of cytochrome c from the mitochondria, a crucial step in the cell death signaling cascade, but BLPs can also regulate caspase activity directly. The analogy between the phenotypes induced by Bax expression in plants and in yeasts, and the required localization of Bax in the mitochondria in both of these cases are in favor of a crucial role for the mitochondria in Bax-induced cell death (Lam *et al.*, 1999). Besides the BLPs, caspases have been shown to play a crucial role in the final cascade leading to cell execution in animals (Raff, 1998). In plants, specific peptide inhibitors of caspases have been shown to abolish HR cell death during the incompatible interaction between tobacco and *Pseudomonas syringae* pv. *phaseolicola* (del Pozo and Lam, 1998). Furthermore, caspase activity has been detected during the early stages of TMV-induced HR cell death. These data raised for the first time the possibility that caspase-like proteins may play a regulatory role during cell death in plants. Other proteases may also be involved in HR cell death regulation as well. Cell death induced by a xylanase from *Trichoderma viride* in tobacco cells can be abolished by inhibitors of serine proteases (Yano *et al.*, 1999). In soybean cells, cysteine protease activity is activated during H_2O_2-induced cell death and the expression of cystatin, an endogenous cysteine protease inhibitor gene, inhibits both protease activity and cell death induced by ROS or by avirulent bacterial pathogens (Solomon *et al.*, 1999). Proteases with different substrate specificity may thus play an important role in cell death control during plant–pathogen interactions.

It is well established that HR is an active process, but the intrinsic program activated during plant–pathogen interactions remains largely unknown. It was thought that a genetic approach should open up new insights into components of the regulatory cascade since alterations in this genetic program would lead to aberrant activation of cell death. Indeed, many mutants showing spontaneous cell death that resembles HR or disease symptoms, the so-called lesion mimic mutants, have been isolated in several species (Dangl *et al.*, 1996; Büschges *et al.*, 1997; Buckner *et al.*, 1998). Although several of these genes have been cloned, no clear conservation of gene structure has emerged, unlike what has been found for the disease resistance genes where conserved motifs have been revealed. For example, the Arabidopsis *lsd1* gene whose mutation leads to a propagation lesion mimic phenotype encodes a novel zinc-finger protein that could act as a transcription factor, whereas the maize *lls1* gene defining a similar mutant phenotype encodes a probable dioxygenase whose substrate remains unknown (Dietrich *et al.*, 1997; Gray *et al.*, 1997). Other examples are the *mlo* gene from barley corresponding to a novel transmembrane protein and *Les22* from maize that encodes uroporphyrinogen decarboxylase, a key enzyme in heme biosynthesis (Büschges *et al.*, 1997; Hu *et al.*, 1998). This diversity suggests that perturbations in very different metabolic pathways can lead to spontaneous cell death. In order to ascertain whether similar pathways of cell death may be activated by these mutations, extragenic suppressors that identify critical loci downstream from these lesion mimic genes will be important tools (Jabs *et al.*, 1996).

VI. Global Analyses of Markers for Cell Death Induction by Plant Pathogens

One of the central questions in the study of cell death is that of the critical steps in the pathway that ultimately are responsible for the demise of the cell. For example, the key cellular substrates that mediate the cell death-inducing activities of caspases are not defined yet in animal apoptosis. Thus, it is still uncertain how many distinct pathways for cell death there are in any system. One approach to answer this question is the application of genetics and the study of critical components for various forms of cell death affected by defined mutations. In addition to the classical genetic approach, which could be time-consuming as well as being restricted by the types of mutant screens that can be designed, characterization of molecular markers for specific types of PCD in plants may also lead to useful tools that can generate new insights into the relationship between different types of cell death events in plants. They could enable the identification of factors that play multiple roles in plant development, which may be missed in more classical genetic screens if they are essential genes or if there are multiple genes with redundant functions.

A. Isolation and Characterization of HR Markers

Induction of cell death during the HR or by treatment with pathogen derived toxins such as victorin has been shown to require *de novo* protein synthesis (He *et al.*, 1994; Navarre and Wolpert, 1999). These observations suggest that cell death induction during resistance as well as disease requires the expression of new plant proteins. Several studies aimed at isolating and characterizing genes specifically activated in one particular type of pathogen-induced cell death have been carried out by differential screening of cDNA libraries (Marco *et al.*, 1990; Buchanan-Wollaston, 1994), subtractive hybridization (Gopalan *et al.*, 1996), differential display (Reuber and Ausubel, 1996) or PCR-based subtractive techniques (Birch *et al.*, 1999; Quirino *et al.*, 1999). One of them, called *HSR203J*, has been studied extensively and found to be a specific marker for HR cell death (Pontier *et al.*, 1994, 1998b). *HIN1* is another tobacco gene that has been isolated as specifically induced by harpin treatment or by avirulent bacterial pathogens (Gopalan *et al.*, 1996). Both of these genes are also rapidly induced by viral pathogens such as TMV that can induce the HR in tobacco, as well as treatment with copper which can induce HR-like phenotypes presumably through its ability to catalyze the formation of intracellular ROS (Pontier *et al.*, 1999). However, *HIN1* is induced also during senescence whereas *HSR203J* is not. This suggested that *HSR203J* may be a more specific marker for HR cell death while *HIN1* could be regulated by multiple pathways (Pontier *et al.*, 1999). Interestingly, both *HIN1* and *HSR203J* induction by avirulent bacteria are inhibited by caspase inhibitors and their kinetics of induction closely matched each other (del Pozo and Lam, 1998). These observations thus suggest that they are likely downstream from similar signals derived from the recognition of the invading pathogen by the plant. In contrast to the specificity of *HSR203J* to HR cell death, a number of genes isolated by differential screening approaches do not correspond to specific cell death markers *per se*. For example, *AIG1*, *AIG2* and *ELI3* showed an activation pattern that is dependent on the presence of a particular resistance gene/avirulence gene combination. This indicates that HR cell death pathways induced by different pathosystems may vary at least in their early steps (Reuber and Ausubel, 1996). Alternatively, these genes may have other functions in the particular host–pathogen interaction.

The protein encoded by *HSR203J* has recently been shown to encode an esterase (Baudouin *et al.*, 1997) and preliminary studies using an antisense approach indicated that it may be involved as a negative regulator of HR cell death in tobacco (Trouchet *et al.*, 2001). The deduced amino acid sequence of HIN1 showed significant homologies to NDR1, which is required by multiple types of resistance genes to activate the HR and systemic resistance pathway (Century *et al.*, 1997). Three new *hsr* genes from Arabidopsis have been deposited into the database recently (Lacomme and Roby, 1999). These genes were isolated as induced in Arabidopsis cell suspension cultures upon treatment after one hour with the avirulent bacterial pathogen *Xanthomonas campestris* strain 147. These induced genes all encoded proteins with known homologues: *hsr1* encodes a myb transcription factor while *hsr2* and *hsr3* encode the mitochondrial proteins for a voltage-dependent anion channel (VDAC) and AOX, respectively. These genes have all been implicated in plant pathogen interactions or cell death from previous studies (Yang and Klessig, 1996; Chivasa and Carr, 1998; Shimizu *et al.*, 1999).

B. Comparative Analyses of Cell Death Pathways Induced by Avirulent Pathogens and Senescence

Molecular cell death markers can constitute useful tools to compare different kinds of cell death and to address the question of the existence of common steps or crosstalks between them. For example, *HSR203J* has been shown to be activated during HR cell death induced by different pathogens, upon HR-like cell death induced by the ectopic expression of the bO proton pump and during some disease-related cell deaths (Pontier *et al.*, 1994, 1998b, 1999). In contrast, no activation of its promoter has been detected during leaf senescence or differentiation of tracheary elements (Pontier *et al.*, 1999; D. Pontier, unpublished results). Similarly, the *Arabidopsis SAG12* gene expression is highly associated with leaf senescence and is not detected after salicylic treatment or in the dying cells during the HR in tobacco (Weaver *et al.*, 1998; Pontier *et al.*, 1999; Quirino *et al.*, 1999). These two genes may thus be part of signaling steps specific for HR cell death and senescence respectively. However, common steps or signals and crosstalks seem to exist between developmental and pathogen-induced cell death. A comparative study of senescence and HR showed that *HIN1*, an HR cell death marker, is also expressed at late stages of leaf senescence (Pontier *et al.*, 1999). Furthermore, defense-related genes comprising the Arabidopsis *ELI3* gene showed a senescence-associated induction as well (Quirino *et al.*, 1999) and the *LSC54* gene encoding a metallothionine is also highly activated during both senescence and pathogen-related cell death in addition to mechanical wounding (Butt *et al.*, 1998). These observations may suggest that some common signals could be involved in senescence and pathogen-related cell death. Alternatively, these genes may contain promoter elements that allow them to respond to various signaling pathways.

We have recently carried out a detailed analysis of the temporal and spatial characteristics for *HSR203J*, *HIN1* and *SAG12* expression during the HR as well as during senescence. The results revealed that a senescence-like process may be triggered at the periphery of cells undergoing HR cell death (Pontier *et al.*, 1999). *SAG12* is indeed activated in this neighboring area where visible chlorosis developed after infection with relatively high titers of pathogens, whereas *HSR203J* is not induced in this region. Conversely, no activation of *SAG12* was detected in the dying cells where *HIN1* and *HSR203J* are highly expressed at early time points. Thus, the senescence cell death program or at least the branch leading to *SAG12* expression is not activated in cells undergoing HR cell death. This work demonstrates the need for specific molecular markers in comparing different physiological situations as well as for a better comprehension of the biology for complex events. Such information can only be obtained by using molecular markers which are specific for each of the events being studied. Cross-referencing between both events with these markers is crucial to minimize the possibility that different stimuli may be responsible for the activation of a single gene.

C. Future Approaches to Global Analysis of Cell Death-related Genes: Genomic-scale Gene Expression Monitoring to Mapping of Intersecting Signaling Pathways

The development of DNA microarray technology and gene chips promises to provide a powerful tool for isolating new marker genes and for comparing global gene expression patterns. In principle, a large number of DNA sequences can be attached onto a solid matrix such as glass slides and hybridized with two or more different RNA populations labeled with distinct fluorescent dyes (Schena *et al.,* 1995; DeRisi *et al.*, 1997). Since it is possible to survey a large number of genes at a time, more specific markers are likely to be identified. Furthermore, this method relies directly on variations in gene expression and genes expressed at low levels can be detected. This contrasts with the classical techniques based on differential screening where transcripts at very low levels would not be detected even if they are specifically induced. Thus the microarray technology allows one to examine and compare the expression patterns of thousands of genes in a rapid and global fashion, at the same time minimizing the bias in their detection due to varying levels of expression. This exhaustive genomic approach gives the ability to identify sets of transcripts that are coordinately expressed in response to a stimulus and may thus function together as a cluster (Brent, 1999; Iyer *et al.*, 1999; Fambrough *et al.*, 1999; Chu *et al.*, 1999).

This powerful technology is just beginning to be applied in Plant Biology and is likely to become an important tool for integration of the complex cell death mechanisms occurring in various plant–pathogen interactions. By the isolation of new molecular markers and the global analyses of RNA expression, fundamental questions can be addressed, such as: Is the HR induced by different pathogens operating via the same pathway of cell death? Are HR, disease and senescence cell deaths triggered by overlapping or independent signaling networks? One can expect that global expression monitoring using DNA microarrays would lead to the definition of clusters of genes based on their temporal and spatial activation pattern under one particular cell death situation. The similarities or the differences in the sets of genes altered by different cell death activating signals would provide a first indication of the degree of overlaps between the signaling pathways. An example of one possible application of the DNA microarray technology to the study of cell death pathways in plants is illustrated in Fig. 3-1. In order to dissect out the cell death specific markers from pathogen

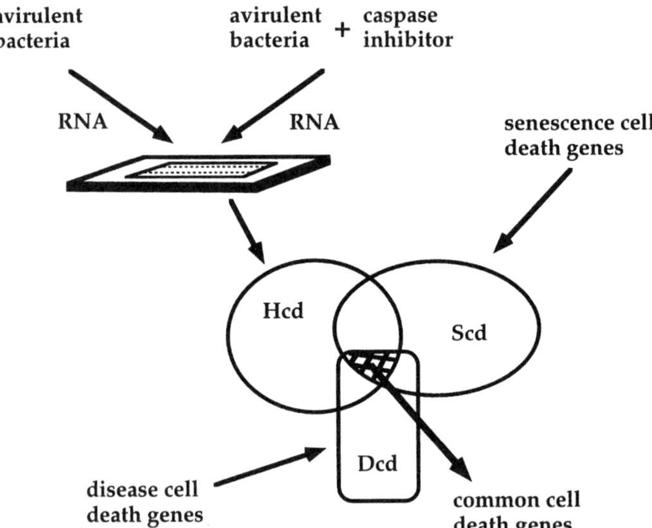

Figure 3-1. Use of DNA microarray technology to dissect cell death signaling pathways in plants. Two labeled RNA populations are hybridized to a DNA microarray and the comparison of the data leads to the identification of a set of HR cell death (Hcd) genes. Similar data can be obtained for other kinds of cell death and define disease cell death (Dcd) genes and senescence cell death (Scd) genes. Comparison of those three sets leads eventually to the identification of common cell death genes present in all sets that are similarly altered (hatched region).

resistance and other coordinately activated genes, one can choose to uncouple defense gene activation from HR cell death by the use of caspase inhibitors (del Pozo and Lam, 1998). RNA samples from plants treated with avirulent bacteria in the presence or absence of caspase inhibitor are extracted at various times and simultaneously hybridized onto a DNA microarray. Differential expression of thousands of genes under the two conditions can be measured quantitatively relative to the untreated control. This type of global gene expression analysis can reveal classes of genes specifically altered in their RNA levels during HR in the absence of inhibitor, corresponding to HR cell death genes (Hcd) that are downstream from caspase-like proteases. This same strategy can be carried out with different avirulent pathogens and the comparison of the subsets of Hcd genes identified could eventually define a class of common Hcd genes. This would indicate that the pathways triggered by different avirulent pathogens converge into a common execution cascade. The same strategy can be applied to other types of cell death such as senescence and disease-related cell death giving rise to senescence cell death (Scd) genes and disease cell death (Dcd) genes (Fig. 3-1). All those data could be integrated together in order to reveal the gene networks that are in operation during these cell death situations. The intersection between these three sets of genes would contain genes that may be critical for cell death and this result will be strong evidence for the existence of a common cell death mechanism.

This powerful technique based on DNA microarrays promises to generate lots of information that is likely to be complementary to the cytological, biochemical and molecular data concerning cell death during plant–pathogen interactions. Nevertheless, one has to keep in mind that this technology has an important limitation. Besides control at the transcriptional level, regulation by phosphorylation or degradation of proteins will likely play important roles in this process. Ultimately, we anticipate that studies as different as microscopy, genomics and biochemistry will be combined and give rise to a more complete mechanistic appreciation of cell death pathways in plants.

Acknowledgments

Sharing of preprints by Jean Greenberg and Ko Shimamoto and of unpublished data from Dominique Roby is gratefully acknowledged. Research on cell death in our laboratory is funded in part by the New Jersey Commission on Science and Technology, and a competitive grant from USDA.

References

Atkinson, M.M., Keppler, L.D., Orlandi, E.W., Baker, C.J., and Mischke, C.F. (1990). Involvement of plasma membrane calcium influx in bacterial induction of the K^+/H^+ and hypersensitive responses in tobacco. *Plant Physiol.* **92**, 215–221.

Baudouin, E., Charpenteau, M., Roby, D., Marco, Y., Ranjeva, R., and Ranty, B. (1997). Functional expression of a tobacco gene related to the serine hydrolase family. Esterase activity towards short chain dinitrophenyl acylesters. *Eur. J. Biochem.* **248**, 700–706.

Bendahmane, A., Kanyuka, K., and Baulcombe, D. (1999). The *Rx* gene from potato controls separate virus resistance and cell death responses. *Plant Cell* **11**, 781–791.

Bestwick, C.S., Bennett, M.H., and Mansfield, J.W. (1995). Hrp mutant of *Pseudomonas syringae* pv *phaseolicola* induces cell wall alterations but not membrane damage leading to the hypersensitive reaction in lettuce. *Plant Physiol.* **108**, 503–516.

Birch, P.R.J., Avrova, A.O., Duncan, J.M., Lyon, G.D., and Toth, R.L. (1999). Isolation of potato genes that are induced during an early stage of the hypersensitive response to *Phytophthora infestans. Mol. Plant-Microbe Interact.* **12**, 356–361.

Brent, R. (1999). Functional genomics: Learning to think about gene expression data. *Curr. Biol.* **9**, R338–R341.

Brown, I.R., and Mansfield, J.W. (1988). An ultrastructural study, including cytochemistry and quantitative analysis of the interactions between Pseudomonads and leaves of *Phaseolus vulgaris* L. *Physiol. Mol. Plant Pathol.* **336**, 351–376.

Buchanan-Wollaston, V. (1994). Isolation of cDNA clones for genes that are expressed during leaf senescence in *Brassica napus. Plant Physiol.* **105**, 839–846.

Buckner, B., Janick-Buckner, D., Gray, J., and Johal, G.S. (1998). Cell-death mechanisms in maize. *Trends Plant Sci.* **3**, 218–223.

Büschges, R., Hollricher, K., Panstruga, R., Simons, G., Wolter, M., Frijters, A., van Daelen, R., van der Lee, T., Diergaarde, P., Groenendijk, J., Töpsch, S., Vos, P., Salamini, F., and Schultze-Lefert, P. (1997). The barley *Mlo* gene: a novel control element of plant pathogen resistance. *Cell* **88**, 695–705.

Butt, A., Mousley, C., Morris, K., Beynon, J., Can, C., Holub, E., Greenberg, J., and Buchanan-Wollaston, V. (1998). Differential expression of a senescence-enhanced metalloprotein gene in *Arabidopsis* in response to isolates of *Peronospora parasitica* and *Pseudomonas syringae. Plant J.* **16**, 209–221.

Century, K.S., Shapiro, A.D., Repetti, P.P., Dahlbeck, D., Holub, E., and Staskawicz, B.J. (1997). *NDR1*, a pathogen-induced component required for *Arabidopsis* disease resistance. *Science* **278**, 1963–1965.

Chamnongpol, S., Willekens, H., Moeder, W., Langebartels, C., Sandermann, H., Van Montagu, M., Inze, D., and Van Camp, W. (1998). Defense activation and enhanced pathogen tolerance induced by H_2O_2 in transgenic tobacco. *Proc. Natl. Acad. Sci. USA* **95**, 5818–5823.

Chivasa, S., and Carr, J.P. (1998). Cyanide restores *N* gene-mediated resistance to tobacco mosaic-virus in transgenic tobacco plants expressing salicylic acid hydroxylase. *Plant Cell* **10**, 1489–1498.

Chu, S., DeRisi, J., Eisen, M., Mullholand, J., Botstein, D., Brown, P.O., and Herskowitz, I. (1998). The transcriptional program of sporulation in budding yeast. *Science* **282**, 699–705.

Dangl, J.L., Dietrich, R.A., and Richberg, M.H. (1996). Death don't have no mercy: cell death programs in plant-microbe interactions. *Plant Cell* **8**, 1793–1807.

Delledone, M., Xia, Y., Dixon, R.A., and Lamb, C.J. (1998). Nitric oxide functions as a signal in plant disease resistance. *Nature* **394**, 585–588.

de Laat, A.M.M., and van Loon, L.C. (1983). The relationship between stimulated ethylene production and symptom expression in virus-infected tobacco leaves. *Physiol. Plant Path.* **22**, 261–273.

del Pozo, O., and Lam, E. (1998). Caspases and programmed cell death in the hypersensitive response of plants to pathogens. *Curr. Biol.* **8**, 1129–1132.

DeRisi, J.L., Iyer, V.R., and Brown, P. (1997). Exploring the metabolic and genetic control of gene expression on a genomic scale. *Science* **278**, 680–686.

Dietrich, R.A., Richberg, M., Schmidt, R., Dean, C., and Dangl, J.L. (1997). A novel zinc finger protein is encoded by the Arabidopsis *LSD1* gene and functions as a negative regulator of plant cell death. *Cell* **88**, 685–694.

Durner, J., Wendehenne, D., and Klessig, D.F. (1998). Defense gene induction in tobacco by nitric oxide, cyclic GMP, and cyclic ADP-ribose. *Proc. Natl. Acad. Sci. USA* **95**, 10328–10333.

Fambrough, D., McClure, K., Kazlauskas, A., and Lander, E.S. (1999). Diverse signaling pathways activated by growth factor receptors induce broadly overlapping, rather than independent, sets of genes. *Cell* **97**, 727–741.

Freytag, S., Arabatzis, N., Hahlbrock, K., and Schmelzer, E. (1994). Reversible cytoplasmic rearrangements precede wall apposition, hypersensitive cell death and defense-related gene activation in potato/*Phytophthora infestans* interactions. *Planta* **194**, 123–135.

Gilchrist, D.G. (1998). Programmed cell death in plant disease: the purpose and promise of cellular suicide. *Ann. Rev. Phytopathol.* **36**, 393–414.

Glazener, J.A., Orlandi, E., and Baker, C.J. (1996). The active oxygen response of cell suspensions to incompatible bacteria is not sufficient to cause hypersensitive cell death. *Plant Physiol.* **110**, 759–763.

Gopalan, S., Wei, W., and He, S.Y. (1996). *hrp* gene-dependent induction of *hin1*: a plant gene activated rapidly by both harpins and the *avrPto* gene-mediated signal. *Plant J.* **10**, 591–600.

Gray, J., Close, P.S., Briggs, S.P., and Johal, G.S. (1997). A novel suppressor of cell death in plants encoded by the *Lls1* gene of Maize. *Cell* **89**, 25–31.

Greenberg, J. (1997). Programmed cell death in plant-pathogen interactions. *Ann. Rev. Plant Mol. Biol.* **48**, 525–545.

Groover, A., and Jones, A.M. (1999). Tracheary element differentiation uses a novel mechanism coordinating programmed cell death and secondary cell wall synthesis. *Plant Physiol.* **119**, 375–384.

Hammond-Kosack, K.E., Silverman, P., Raskin, I., and Jones, J.D.G. (1996). Race-specific elicitors of *Cladosporium fulvum* induce changes in cell morphology and the synthesis of ethylene and salicylic acid in tomato cells carrying the corresponding *Cf* disease resistance gene. *Plant Physiol.* **110**, 1381–1394.

He, S.Y., Bauer, D.W., Collmer, A., and Beer, S.V. (1994). Hypersensitive response elicited by *Erwinia amylovora* requires active plant metabolism. *Mol. Plant-Microbe Interact.* **7**, 289–292.

Heath, M. (2000). Hypersensitive response related death. *Plant Mol. Biol.*, **44**, 413–424.

Heo, W.D., Lee, S.H., Kim, M.C., Kim, J.C., Chung, W.S., Chun, H.J., Lee, K.J., Park, H.C., Choi, J.Y., and Cho, M.J. (1999). Involvement of specific calmodulin isoforms in salicylic acid-independent activation of plant disease resistance responses. *Proc. Natl. Acad. Sci. USA* **96**, 766–771.

Hu, G., Yalpani, N., Briggs, S.P., and Johal, G.S. (1998). A porphyrin pathway impairment is responsible for the phenotype of a dominant disease lesion mimic mutant of maize. *Plant Cell* **10**, 1095–1105.

Iyer, V.R., Eisen, M.B., Ross, D.T., Schuler, G., Moore, T., Lee, J.C.F., Trent, J.M., Staudt, L.M., Hudson, J., Biguski, M.S., Lashkari, D., Shalon, D., Botstein, D., and Brown, P.O. (1999). The transcriptional program in the response of human fibroblasts to serum. *Science* **283**, 83–87.

Jabs, T., Dietrich, R.A., and Dangl, J.L. (1996). Initiation of runaway cell death in an *Arabidopsis* mutant by extracellular superoxide. *Science* **273**, 1853–1856.

Jacobson, M.D., Weil, M., and Raff, M.C. (1997). Programmed cell death in animal development. *Cell* **88**, 347–354.

Kawasaki, T., Henmi, K., Ono, E., Hatakeyama, S., Iwano, M., Satoh, H., and Shimamoto, K. (1999). The small GTP-binding protein Rac is a regulator of cell death in plants. *Proc. Natl. Acad. Sci. USA* **96**, 10922–10926.

Kazan, K., Murray, F.R., Goulter, K.C., Llewellyn, D.J., and Manners, J.M. (1998). Induction of cell death in transgenic plants expressing a fungal glucose oxidase. *Mol. Plant-Microbe Interact.* **11**, 555–562.

Lacomme, C., and Roby, D. (1999). Identification of new early markers of the hypersensitive response in *Arabidopsis thaliana. FEBS Lett.* **459**, 149–153.

Lacomme, C., and Santa-Cruz, S. (1999). Bax-induced cell death in tobacco is similar to the hypersensitive response. *Proc. Natl. Acad. Sci. USA* **96**, 7956–7961.

Lam, E., del Pozo, O., and Pontier, D. (1999). BAXing in the hypersensitive response. *Trends Plant Sci.* **2**, 502–504.

Lamb, C., and Dixon, R.A. (1997). The oxidative burst in plant disease resistance. *Ann. Rev. Plant Physiol. Plant Mol. Biol.* **48**, 251–275.

Lawton, K.A., Potter, S.L., Uknes, S., and Ryals, J. (1994). Acquired resistance signal transduction in Arabidopsis is ethylene independent. *Plant Cell* **6**, 581–588.

Levine, A., Tenhaken, R., Dixon, R., and Lamb, C. (1994). H_2O_2 from the oxidative burst orchestrates the plant hypersensitive disease resistance response. *Cell* **79**, 583–593.

Levine, A., Pennell, R.I., Alvarez, M.E., Palmer, R., and Lamb, C. (1996). Calcium-mediated apoptosis in a plant hypersensitive disease resistance response. *Curr. Biol.* **6**, 427–437.

Lund, S.T., Stall, R.E., and Klee, H.J. (1998). Ethylene regulates the susceptible response to pathogen infection in plants. *Plant Cell* **10**, 371–382.

Marco, Y.J., Ragueh, F., Godiard, L., and Froissard, D. (1990). Transcriptional activation of 2 classes of genes during the hypersensitive reaction of tobacco leaves infiltrated with an incompatible isolate of the phytopathogenic bacterium *Pseudomonas solanacearum*. *Plant Mol. Biol.* **15**, 145–154.

Maxwell, D.P., Wang, Y., and McIntosh, L. (1999). The alternative oxidase lowers mitochondrial reactive oxygen production in plant cells. *Proc. Natl. Acad. Sci. USA* **96**, 8271–8276.

Meyer, S.L.F., and Heath, M.C. (1987). A comparison of the death induced by fungal invasion or toxic chemicals in cowpea epidermal cells. Responses induced by *Erysiphe cichoracearum*. *Can. J. Bot.* **66**, 624–634.

Mitsuhara, I., Malik, K.A., Miura, M., and Ohashi, Y. (1999). Animal cell-death suppressors Bcl-xL and Ced-9 inhibit cell death in tobacco plants. *Curr. Biol.* **9**, 775–778.

Mittler, R., and Lam, E. (1995a). *In situ* detection of nDNA fragmentation during the differentiation of tracheary elements in higher plants. *Plant Physiol.* **108**, 489–493.

Mittler, R., and Lam, E. (1995b). Coordinated activation of programmed cell death and defense mechanisms in transgenic tobacco plants expressing a bacterial proton pump. *Plant Cell* **7**, 29–42.

Mittler, R., and Lam, E. (1996). Sacrifice in the face of foes: pathogen-induced programmed cell death in plants. *Trends Microbiol.* **4**, 10–15.

Mittler, R., Shulaev, V., Seskar, M., and Lam, E. (1996). Inhibition of programmed cell death in tobacco plants during a pathogen-induced hypersensitive response at low oxygen pressure. *Plant Cell* **8**, 1991–2001.

Mittler, R., Simon, L., and Lam, E. (1997a). Pathogen-induced programmed cell death in tobacco. *J. Cell Science* **110**, 1333–1344.

Mittler, R., del Pozo, O., Meisel, L., and Lam, E. (1997b). Pathogen-induced programmed cell death in plants, a possible defense mechanism. *Dev. Genet.* **21**, 279–289.

Morel, J.B., and Dangl, J.L. (1997). The hypersensitive response and the induction of cell death in plants. *Cell Death Differ.* **4**, 671–683.

Mur, L.A.J., Bi, Y.M., Darby, R.M., Firek, S., and Draper, J. (1997). Compromising early salicylic acid accumulation delays the hypersensitive response and increases viral dispersal during lesion establishment in TMV-infected tobacco. *Plant J.* **12**, 1113–1126.

Murphy, A.M., Chivasa, S., Singh, D.P., and Carr, J.P. (1999). Salicylic acid-induced resistance to viruses and other pathogens: a parting of the ways? *Trends Plant Sci.* **4**, 155–160.

Navarre, D.A., and Wolpert, T.J. (1999). Victorin induction of an apoptotic/senescence-like response in oats. *Plant Cell* **11**, 237–249.

O'Brian, I.E.W., Baguley, B.C., Murray, B.G., Morris, B.A.M., and Ferguson, I.B. (1998). Early stages of the apoptotic pathway in plant cells are reversible. *Plant J.* **13**, 803–814.

Pontier, D., Godiard, L., Marco, Y., and Roby, D. (1994). *Hsr203J*, a tobacco gene whose activation is rapid, highly localized and specific for incompatible interactions. *Plant J.* **5**, 507–521.

Pontier, D., Balagué, C., and Roby, D. (1998a). The hypersensitive response. A programmed cell death associated with plant resistance. *C. R. Acad. Sci. Paris* **321**, 721–734.

Pontier, D., Tronchet, M., Rogowsky, P., Lam, E., and Roby, D. (1998b). Activation of *hsr203J*, a plant gene expressed during incompatible plant-pathogen interactions, is correlated with programmed cell death. *Mol. Plant-Microbe Interact.* **6**, 544–554.

Pontier, D., Gan, S., Amasino, R.M., Roby, D., and Lam, E. (1999). Markers for hypersensitive response and senescence show distinct patterns of expression. *Plant Mol. Biol.* **39**, 1243–1255.

Pontier, D., Mittler, R., and Lam, E. (2002). Mechanism of cell death and disease resistance induction by transgenic expression of bacterio-opsin. *Plant J.* **30**, 499–509.

Quirino, B.F., Normanly, J., and Amasino, R.M. (1999). Diverse range of gene activity during *Arabidopsis thaliana* leaf senescence includes pathogen-independent induction of defense-related genes. *Plant Mol. Biol.* **40**, 267–278.

Raff, M. (1998). Cell suicide for beginners. *Nature* **396**, 119–122.

Rate, D.N., Cuenca, J.V., Bowman, G.R., Guttman, D.S., and Greenberg, J.T. (1999). A gain-of-function Arabidopsis *acd6* mutant reveals novel regulation and function of the salicylic acid signaling pathway in controlling cell death, defenses and cell growth. *Plant Cell* **11**, 1695–1708.

Reuber, T.L., and Ausubel, F.M. (1996). Isolation of Arabidopsis genes that differentiate between resistance responses mediated by the *RPS2* and *RPM1* disease resistance genes. *Plant Cell* **8**, 241–249.

Richberg, M.H., Aviv, D.H., and Dangl, J.L. (1998). Dead cells do tell tales. *Curr. Opin. Plant Biol.* **1**, 480–485.

Ryals, J.L., Neuenschwander, U.H., Willits, M.C., Molina, A., Steiner, H.Y., and Hunt, M.D. (1996). Systemic acquired resistance. *Plant Cell* **8**, 1809–1819.

Ryerson, D.E., and Heath, M.C. (1996). Cleavage of nuclear DNA into oligonucleosomal fragments during cell death induced by fungal infection or by abiotic treatments. *Plant Cell* **8**, 393–402.

Schena, M., Shalon, D., Davis, R.W., and Brown, P.O. (1995). Quantitative monitoring of gene expression patterns with a complementary DNA microarray. *Science* **270**, 467–470.

Shah, J., Kachroo, P., and Klessig, D. (1999). The Arabidopsis *ssi1* mutation restores pathogenesis-related gene expression in *npr-1* plants and renders defensin gene expression salicylic acid dependent. *Plant Cell* **11**, 191–206.

Shimizu, S., Narita, M., and Tsujimoto, Y. (1999). Bcl-2 family proteins regulate the release of apoptogenic cytochrome c by the mitochondrial channel VDAC. *Nature* **399**, 483–487.

Shirasu, K., Nakajima, H., Rajaseskar, V.K., Dixon, R.A., and Lamb, C.J. (1997). Salicylic acid potentiates an agonist-dependent gain control that amplifies pathogen signals in the activation of defense mechanisms. *Plant Cell* **9**, 261–270.

Simons, B.H., Millenaar, F.F., Mulder, L., Van Loon, L.C., and Lambers, H. (1999). Enhanced expression and activation of the alternative oxidase during infection of Arabidopsis with *Pseudomonas syringae* pv tomato. *Plant Physiol.* **120**, 529–538.

Skalamera, D., and Heath, M.C. (1998). Changes in the cytoskeleton accompanying infection-induced nuclear movements and the hypersensitive response in plant cells invaded by rust fungi. *Plant J.* **16**, 191–200.

Solomon, M., Belenghi, B., Delledonne, M., Menachem, E., and Levine, A. (1999). The involvement of cysteine proteases and protease inhibitor genes in the regulation of programmed cell death in plants. *Plant Cell* **11**, 431–444.

Tavernier, E., Wendehenne, D., Blein, J.P., and Pugin, A. (1995). Involvement of free calcium in action of cryptogein, a proteinaceous elicitor of hypersensitive reaction in tobacco cells. *Plant Physiol.* **109**, 1025–1031.

Torres, M.A., Dangl, J.L., and Jones, J.D.G. (2002). *Arabidopsis* gp91^{phox} homologues *AtrbohD* and *AtrbohF* are required for accumulation of reactive oxygen intermediates in the plant defense response. *Proc. Natl. Acad. Sci. USA* **99**, 517–522.

Trouchet, H., Rauty, B., Marco, Y., and Roby, D. (2001). *HSR 203J* antisense suppression in tobacco accelerates development of hypersensitive cell death. *Plant J.* **27**, 115–127.

Van Camp, W., Van Montagu, M., and Inze, D. (1998). H_2O_2 and NO: redox signals in disease resistance. *Trends Plant Sci.* **3**, 330–334.

Viard, M.P., Martin, F., Pugin, A., Ricci, P., and Blein, J.P. (1994). Protein phosphorylation is induced in tobacco cells by the elicitor cryptogein. *Plant Physiol.* **104**, 1245–1249.

Wang, H., Li, J., Bostock, R.M., and Gilchrist, D.G. (1996). Apoptosis: a functional paradigm for programmed plant cell death induced by a host-selective phytotoxin and invoked during development. *Plant Cell* **8**, 375–391.

Weaver, L.M., Gan, S., Quirino, B., and Amasino, R.M. (1998). A comparison of the expression patterns of several senescence-associated genes in response to stress and hormone treatment. *Plant Mol. Biol.* **37**, 455–469.

Weymann, K., Hunt, M., Uknes, S., Neuenschwander, U., Lawton, K., Steiner, H.Y., and Ryals, J. (1995). Suppression and restoration of lesion formation in Arabidopsis *lsd* mutants. *Plant Cell* **7**, 2013–2022.

Yang, Y., and Klessig, D. (1996). Isolation and characterization of a tobacco mosaic virus-inducible *myb* oncogene homolog from tobacco. *Proc. Natl. Acad. Sci. USA* **93**, 14972–14977.

Yang, Y., Shah, J., and Klessig, D. (1997). Signal perception and transduction in plant defense responses. *Genes Dev.* **11**, 1621–1639.

Yano, A., Suzuki, K., and Shinshi, H. (1999). A signaling pathway, independent of the oxidative burst, that leads to hypersensitive cell death in cultured tobacco cells includes a serin protease. *Plant J.* **18**, 105–109.

Yu, I.C., Parker, J., and Bent, A.F. (1998). Gene-for-gene disease resistance without the hypersensitive response in *Arabidopsis dnd1* mutant. *Proc. Natl. Acad. Sci. USA* **95**, 7819–7824.

4

Changes in Gene Expression during Senescence

Michelle L. Jones

I. Introduction

Senescence represents the sequence of metabolic events occurring in the final stage of development and ultimately culminating in the programmed death of whole plants, organs, tissues, or cells. It is an actively ordered process that involves the synthesis of new RNAs and proteins and results in highly coordinated changes in metabolism and the programmed disassembly of cells. In recent years molecular biological approaches have been utilized to identify genes that may be involved in the initiation and regulation of the senescence program. The identification and characterization of these senescence-related genes has begun to provide us with an understanding of the process of senescence. An understanding of senescence, a process that limits yield, nutritional value, and marketability of many crops, will lead to ways of manipulating senescence for agricultural applications. There have been many recent reviews of gene expression during senescence (Smart, 1994; Buchanan-Wollaston, 1997; Gan and Amasino, 1997; Nam, 1997; Weaver *et al.*, 1997;

Plant Cell Death Processes

Rubinstein, 2000; Quirino *et al.*, 2000). This chapter is meant to review genes that have been identified as senescence-related, discuss their potential function during senescence, and compare the similarities and differences between the senescence of vegetative and floral tissues. Those genes identified as ripening-related will also be presented and discussed in the context of identifying the similarities between senescence and fruit ripening; however, this chapter is not meant to provide an exhaustive review of ripening-related genes.

II. Changes in Patterns of Nucleic Acids and Proteins during Senescence

While senescence is a degradative process, this degradation is not merely the result of increased rates of protein turnover and decreases in the synthesis of proteins and RNA. Although general decreases in total protein and RNA levels are observed in senescing floral and foliar tissues, specific proteins and mRNAs have been found to increase (Brady, 1988; Borochov and Woodson, 1989). Experiments with inhibitors of protein and RNA synthesis have demonstrated that senescence is a genetically programmed process that requires the selective activation of specific RNAs and proteins, and does not merely result from the inhibition of cellular metabolism by declining rates of protein and RNA synthesis. These inhibitor studies have also suggested that transcription and protein synthesis in organelles is not central to the regulation of senescence. In support of the nuclear regulation of senescence, nuclear genes have been found to encode almost all of the mRNAs found to increase during senescence (Noodén, 1988). While the later stages of ripening resemble senescence, the entire process represents more of an interaction between degradative and synthetic processes. In contrast to senescing flowers and leaves, protein levels in fruits remain constant or increase slightly during ripening (Brady, 1988). Specific mRNAs and proteins that increase during ripening have also been identified (Brady, 1988; Gray *et al.*, 1992). The focus of this chapter will be on those mRNAs and proteins that increase during senescence. These will be referred to as senescence-related (SR) genes.

III. Similarities between Senescing and Ripening Tissues

While senescence and ripening are distinct developmental events, they clearly share overlapping physiological and biochemical pathways. A primary feature in both leaves and fruit is the controlled dismantling of the photosynthetic machinery, and degradation of starch and chlorophyll (McGlasson *et al.*, 1975). A common characteristic of senescence and fruit ripening in climacteric fruits is the increase in respiration and ethylene production observed during these processes (Abeles *et al.*, 1992). Senescence of flower petals and leaves provides the plant with a mechanism by which C, N, P, and minerals from no longer needed or productive organs can be remobilized to seeds, young leaves, or storage organs before the senescent organs are shed (Thomas and Stoddart, 1980; Borochov and Woodson, 1989).

IV. Identification and Classification of Senescence-related Genes

Recent molecular studies have confirmed that the processes of senescence and ripening are accompanied by changes in gene expression. Utilizing differential screening and subtractive hybridization techniques a number of cDNAs that are up regulated during senescence have been cloned. Genes that exhibit enhanced expression during senescence have been cloned from the leaves of *Arabidopsis*, asparagus, *Brassica napus*, barley, maize, radish, and tomato (reviewed in Smart, 1994; Buchanan-Wollaston, 1997; Nam, 1997; Weaver *et al.*, 1997; Quirino *et al.*, 2000). Differential screening of senescing petal cDNA libraries and PCR-based differential display techniques have been utilized to identify genes that are up regulated during senescence of carnation and daylily flowers (Woodson *et al.*, 1993; Woodson, 1994; Valpuesta *et al.*, 1995; Guerrero *et al.*, 1998; Panavas *et al.*, 1999). Many mRNAs accompanying fruit ripening have also been identified (Davies and Grierson, 1989; Gray *et al.*, 1992). Table 4-1 includes a listing of those cDNAs identified as senescence-related whose identity has been determined by sequence homology to previously characterized genes published in DNA and protein databases. It is not the intent to provide an exhaustive review of all relevant genes but to present representative examples from multiple systems that will allow us to examine the genetic regulation of senescence from a wider perspective.

Most of the genes that have been identified as senescence-related are expressed at basal levels in non-senescing tissues (green leaves and fruits and young flowers) and increase in abundance during senescence. A smaller number of SR genes are only detectable in senescing tissues and represent senescence or ripening-specific genes. An even smaller set of genes have been identified that have high levels of expression early in development, decreased expression in young maturing tissue and increased expression at the onset of senescence. Genes that fit within this class have only been identified in vegetative tissues and represent genes that have a similar role in multiple stages of development like germination and senescence (Smart, 1994; Buchanan-Wollaston, 1997; Lohman *et al.*, 1994). More detailed classifications of the patterns of gene expression during leaf senescence can be found in reviews by Smart (1994) and Buchanan-Wollaston (1997). A classification based on patterns of gene expression is not given in Table 4-1 of this review because the patterns of expression are often different in leaves, flowers, and fruit.

Recently, concern has been voiced about the method of normalizing the samples run on RNA gels for northern blot analysis of genes in senescing tissues. This concern is based on the fact that massive degradation of RNA occurs during senescence. Therefore expression of a particular mRNA may appear to be up regulated in a senescing tissue when it is merely being maintained at steady levels or is decreasing in abundance less rapidly than the total RNA levels. In response to this concern it has been suggested that normalizing RNA loading on the basis of fresh weight, which is relatively stable until later stages of senescence, might provide a more representative pattern of gene expression during senescence (Noodén *et al.*, 1997). Incomplete extraction and variability in yield between samples following RNA extraction procedures make normalization by fresh weight difficult and have provided the main impetus for normalizations based on total RNA. In this chapter it should be noted that almost all of the reports of up-regulation of gene expression during senescence are based on northern blots normalized by equal RNA loading. This is an issue that should be kept in mind when reporting expression patterns of SR genes and will undoubtedly need to be revisited.

V. Senescence-related Genes

In the laboratory of Dr. Richard Amasino (University of Wisconsin) both differential screening of cDNA libraries made from mRNAs of naturally senescing leaf tissue and subtractive

Table 4-1. Senescence-related genes

Name	Plant	Identity/homology[a]	Up-regulation[b] LS	FS	FR	Known patterns of expression[c]	References
ERD1 (SAG15)	*Arabidopsis*	Clp protease, ClpC-like protein	Y	nd	nd	Basal L; dehydration, dark, ABA, C_2H_4	Kiyosue *et al.*, 1993; Lohman *et al.*, 1994
SAG2	*Arabidopsis*	Cysteine protease	Y	nd	nd	Basal L	Hensel *et al.*, 1993
See1	Maize	Cysteine protease	Y	nd	nd	Basal L	Smart *et al.*, 1995
LSC7	*B. napus*	Cysteine protease	Y	nd	nd		Buchanan-Wollaston, 1997
SAG12	*Arabidopsis*	Cysteine protease	Y	Y	nd	Sen specific; dark, ABA, C_2H_4	Lohman *et al.*, 1994
SENU2 (C14)	Tomato	Cysteine protease	Y	nd	N	Seed germination	Drake *et al.*, 1996
SENU3	Tomato	Cysteine protease	Y	nd	N	Seed germination	Drake *et al.*, 1996
See2	Maize	Cysteine protease	Y	nd	nd	Basal L	Smart *et al.*, 1995
SAG23	*Arabidopsis*	Cysteine protease	Y	Y	nd	Basal L, F, R, S	Quirino *et al.*, 1999
Citvac	Citrus	Cysteine protease	nd	nd	Y	C_2H_4	Alonso and Granell, 1995
LSC790	*B. napus*	Cysteine protease	Y	nd	nd	Basal L, F; seed germination	Buchanan-Wollaston and Ainsworth, 1997
DCCP1	Carnation	Cysteine protease	Y	Y	nd	Basal L, F; C_2H_4	Jones *et al.*, 1995
Sen11	Daylily	Cysteine protease	N	Y	nd	Basal L	Valpuesta *et al.*, 1995
Sen102	Daylily	Cysteine protease	N	Y	nd	Basal L	Guerrero *et al.*, 1998
Peth1	Petunia	Cysteine protease	Y	Y	nd	Basal L, F	Tournaire *et al.*, 1996
FanR93C	Strawberry	Cysteine protease	nd	nd	Y		Manning, 1998
Smcp	*S. melogena*	Cysteine protease	Y	N	Y	Anther dehiscence	Xu and Chye, 1999
LSC760	*B. napus*	Aspartic protease	Y	nd	nd	Basal L	Buchanan-Wollaston and Ainsworth, 1997
DSA4	Daylily	Aspartic protease	N	Y	nd	Basal L; ABA	Panavas *et al.*, 1999
Sen3	*Arabidopsis*	Polyubiquitin	Y	nd	nd	Dark, C_2H_4	Park *et al.*, 1998
UU116	Banana	Polyubiquitin	nd	nd	Y		Medina-Suarez *et al.*, 1997
RNS2	*Arabidopsis*	Ribonuclease	Y	Y	nd	P starvation	Taylor *et al.*, 1993
LX RNase	Tomato	Ribonuclease	Y	N	N	Basal F	Lers *et al.*, 1998
LE RNase	Tomato	Ribonuclease	Y	N	N	Basal F	Lers *et al.*, 1998
DSA6	Daylily	S1-type nuclease	N	Y	nd	ABA	Panavas *et al.*, 1999
BFN1	*Arabidopsis*	Bifunctional nuclease	Y	nd	nd	Basal F, S; sen stems	Perez-Amador *et al.*, 2000
MS	Cucumber	Malate synthase	Y	Y	nd	Germination, cotyledons	Graham *et al.*, 1992
ICL	Cucumber	Isocitrate lyase	Y	nd	nd		McLaughlin and Smith, 1994
pTIP11	Asparagus	β-galactosidase	Y	nd	nd	Postharvest	King and Davies, 1995
SR12	Carnation	β-galactosidase	N	Y	nd	C_2H_4	Woodson, 1994
PLD	Castor bean	Phospholipase D	Y	nd	nd	Detached leaves	Ryu and Wang, 1995
Rlos1	Rose	Lipoxygenase	nd	Y	nd	C_2H_4	Fukuchi-Mizutani *et al.*, 2000
lipase	Carnation	Lipase (lipolytic acyl hydrolase)	Y	Y	nd	C_2H_4	Hong *et al.*, 2000
Gln1;1 Gln 1;3	Radish	Glutamine synthetase	Y	nd	nd	Germination, dark	Kawakami and Watanabe, 1988

Continued

Table 4-1. Continued

Atgsr2	*Arabidopsis*	Glutamine synthetase	Y	nd	nd		Bernhard and Matile, 1994
GS1	Rice	Glutamine synthetase	Y	nd	nd		Kamachi *et al.*, 1992
LSC460	*B. napus*	Glutamine synthetase	Y	nd	nd		Buchanan-Wollaston and Ainsworth, 1997
pTIP12	Asparagus	Asparagine synthetase	Y	nd	nd	Postharvest	King *et al.*, 1995
SAND1	*Sandersonia*	Asparagine synthetase	Y	N	nd		Eason *et al.*, 2000
pBAN3-6	Banana	Metallothionein I	nd	nd	Y	Basal L, Fr	Clendennen and May, 1997
LSC54	*B. napus*	Metallothionein I	Y	nd	N	Basal L, F	Buchanan-Wollaston, 1994
*SAG*17 (MTI)	*Arabidopsis*	Metallothionein I	Y	nd	nd	Basal L; dehydration, dark, ABA, C_2H_4	Lohman *et al.*, 1994
LSC210	*B. napus*	Metallothionein II	Y	nd	nd		Buchanan-Wollaston and Ainsworth, 1997
PBAN 3-23	Banana	Metallothionein II	nd	nd	Y		Clendennen and May, 1997
*SAG*14 (BCB)	*Arabidopsis*	Blue copper binding protein	Y	nd	nd	Basal L; dark, C_2H_4, dehydration	Lohman *et al.*, 1994
GSTII-27	Maize	Glutathione S-transferase (GST)	Y	nd	nd	Basal R	Smart *et al.*, 1995
*SR*8	Carnation	GST	N	Y	nd	C_2H_4	Meyer *et al.*, 1991
LSC650	*B. napus*	Catalase	Y	nd	nd		Buchanan-Wollaston and Ainsworth, 1997
SEN2 (Cat3)	*Arabidopsis*	Catalase	Y	nd	nd	dark	Park *et al.*, 1998
Cat1	*N. plumbaginifolia*	Catalase	N	Y	nd	Basal L, F	Willekens *et al.*, 1994
Cat2	*N. plumbaginifolia*	Catalase	Y	N	nd	Basal L, F, S, R; dark	Willekens *et al.*, 1994
Cat3	*N. plumbaginifolia*	Catalase	Y	N	nd	Basal L, F, S, R, Se; dark	Willekens *et al.*, 1994
LSC94	*B. napus*	PR1a	Y	nd	nd	Basal R, F, Si; SA	Hanfrey *et al.*, 1996
LSC222	*B. napus*	Chitinase	Y	nd	nd		Hanfrey *et al.*, 1996
LSC212	*B. napus*	Antifungal protein	Y	nd	nd		Buchanan-Wollaston and Ainsworth, 1997
SENU4 (P14)	Tomato	P6 PR protein	Y	nd	N	Basal L, Fr	John *et al.*, 1997
pBANUU 129	Banana	β-glucosidase	nd	nd	Y	C_2H_4	Medina-Suarez *et al.*, 1997
*SR*5	Carnation	β-glucosidase	N	Y	nd	C_2H_4	Woodson, 1994
SRG2	*Arabidopsis*	β-glucosidase	Y	Y	nd	Cell cultures	Callard *et al.*, 1996
pBAN 3-28	Banana	Thaumatin-like	nd	nd	Y		Clendennen and May, 1997
pBAN 3-24	Banana	Endochitinase	nd	nd	Y		Clendennen and May, 1997
SBA 18	Avocado	Endochitinase	nd	nd	Y		Dopico *et al.*, 1993
*SAG*27	*Arabidopsis*	Nitrilase II	Y	Y	nd	Sen specific; SA	Quirino *et al.*, 1999
AtOSM34	*Arabidopsis*	Osmotin-like	Y	nd	nd	SA	Capelli *et al.*, 1997
*SAG*29	*Arabidopsis*	MtN2 homologue	Y	Y	nd	Sen specific; SA	Quirino *et al.*, 1999
*SAG*25 (ELI3-2)	*Arabidopsis*	Cinnamyl alcohol dehydrogenase	Y	Y	nd	L sen specific, basal F; SA	Quirino *et al.*, 1999

Continued

Table 4-1. Continued

Name	Plant	Identity/homology[a]	Up-regulation[b] LS	FS	FR	Known patterns of expression[c]	References
*SAG*26	*Arabidopsis*	Cinnamyl alcohol dehydrogenase	Y	Y	nd	Sen specific, SA	Quirino *et al.*, 1999
*SAG*13	*Arabidopsis*	Short chain alcohol dehydrogenase	Y	Y	nd	Sen specific, dark, ABA, C_2H_4	Lohman *et al.*, 1994
SEN1	*Arabidopsis*	Sulfide dehydrogenase	Y	nd	nd	Dark, C_2H_4, ABA	Oh *et al.*, 1996
*SAG*21	*Arabidopsis*	Member Lea family	Y	nd	nd		Weaver *et al.*, 1998
*SAG*24	*Arabidopsis*	L10 Ribosomal protein	Y	N	nd	Basal L, F, R, S	Quirino *et al.*, 1999
*SR*132	Carnation	Carboxyphosphoenol pyruvate mutase	nd	Y	nd	Basal L, F, R, S	Wang *et al.*, 1993
pTOM 75	Tomato	MIP membrane channel	Y	nd	Y	Water stress	Davies and Grierson, 1989
*pTOM*13 (LE-ACO1)	Tomato	ACC oxidase	Y	Y	Y	C_2H_4, wounded leaves	Davies and Grierson, 1989; Barry *et al.*, 1996
LE-ACO2	Tomato	ACC oxidase	N	Y	N		Barry *et al.*, 1996
LE-ACO3	Tomato	ACC oxidase	Y	Y	Y	Basal L	Barry *et al.*, 1996
PBA NUU10	Banana	ACC oxidase	nd	nd	Y		Medina-Suarez *et al.*, 1997
*SR*120	Carnation	ACC oxidase	N	Y	nd	C_2H_4	Wang and Woodson, 1991
*DCACS*1	Carnation	ACC synthase	N	Y	nd	C_2H_4	Park *et al.*, 1992
*DCACS*2	Carnation	ACC synthase	N	Y	nd	C_2H_4	ten Have and Woltering, 1997
*DCACS*3	Carnation	ACC synthase	N	Y	nd	Basal F (gynoecium); pollination, auxin	Jones and Woodson, 1999
LeETR3 (NR)	Tomato	Ethylene receptor	N	N	Y	Basal L, F, Fr, R, S, Se; C_2H_4	Lashbrook *et al.*, 1998

[a]In most cases gene function has been determined by sequence homology to previously characterized genes published in DNA and protein databases.
[b]Up-regulation of senescence associated genes is indicated in senescing leaves (LS), flowers (FS), and fruits (FR). Y indicates up-regulation of the gene was detected, N indicates it was not detected and nd indicates that gene expression was not investigated in that organ during senescence or in some instances may not be applicable. In *Arabidopsis* up-regulation during fruit ripening indicates expression is up regulated in maturing siliques.
[c]Where known additional patterns of gene expression are given. Letters indicate that basal levels of transcript were detected in the non-senescing tissue of leaves (L), flowers (F), roots (R), stems (S), siliques (Si), tubers (T), fruits (Fr), or seeds (Se).

hybridization techniques designed to enrich for senescence-enhanced clones have been utilized to identify SR genes from *Arabidopsis* (Lohman *et al.*, 1994; Gan and Amasino, 1997; Weaver *et al.*, 1997, 1998, 1999; Quirino *et al.*, 1999). To date approximately 30 SR genes, referred to as SAGs (senescence associated genes) have been identified from *Arabidopsis* (Hensel *et al.*, 1993; Lohman *et al.*, 1994; Gan and Amasino, 1997; Weaver *et al.*, 1997, 1998; Quirino *et al.*, 1999). Figure 4-1 shows the patterns of expression of a selected group of SAGs during age-related senescence of *Arabidopsis* leaves (Weaver *et al.*, 1998). Most of these genes exhibit basal levels of expression in green non-senescing leaves. Within this broad classification genes are differentially regulated, with some increasing in abundance

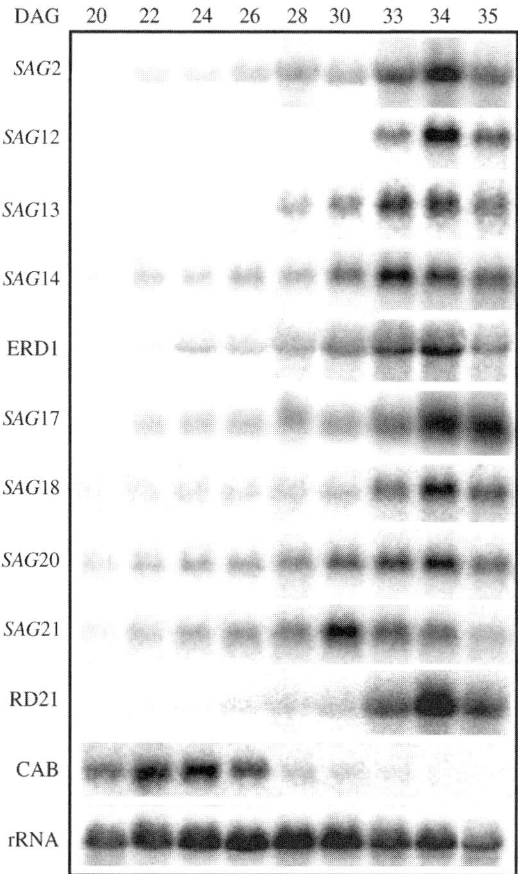

Figure 4-1. Age-dependent expression of *SAG*s in senescing *Arabidopsis* leaves. RNA was extracted from the fifth leaves and the numbers indicate days after germination (DAG). Figure reprinted with kind permission from Kluwer Academic Publishers and Dr. Richard Amasino (University of Wisconsin). Figure 1 from Weaver *et al.*, 1998.

gradually as the leaf matures and others increasing more abruptly at various stages of leaf development. Only *SAG*12 and *SAG*13 show senescence-specific expression. Among the senescence-specific genes, *SAG*13 is detected before any visible signs of leaf senescence and as such may be responsible for initiation of the senescence process, while *SAG*12 is expressed after the leaf is visibly yellowing.

Many of the genes that have been identified as senescence-related are identified from a particular plant organ and it is not known whether they are expressed in other senescing organs or during other developmental processes. Recently the expression of a number of SAGs was investigated in roots, stems, flower buds, and mature flowers of *Arabidopsis* (Quirino *et al.*, 1999; Fig. 4-2). Expression of *SAG*12, *SAG*13, *SAG*25, *SAG*26, and *SAG*29 was not detected in any non-senescing tissues but was detected in both senescing flowers and leaves, indicating a common molecular regulation of senescence in vegetative and floral tissue. Some of the SAGs show low levels of expression in multiple tissues with up-regulation in senescing leaves and flowers (*SAG*23) or up-regulation detected only in senescing leaves

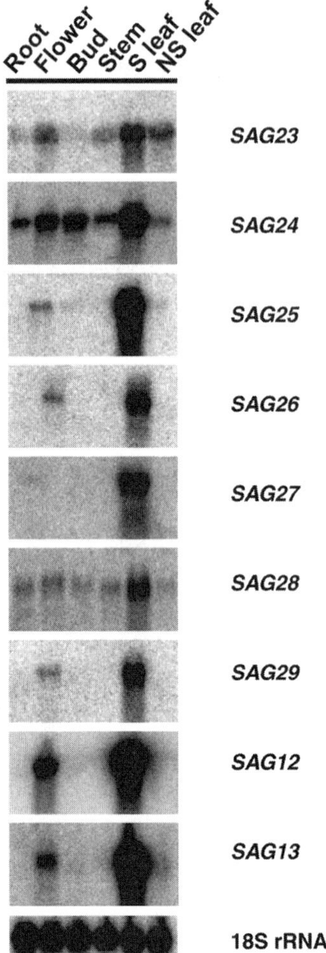

Figure 4-2. Expression of SAGs in different tissues of *Arabidopsis* plants. RNA was isolated from roots, mature flowers, flower buds, stems of 4 week old plants, fully senescent leaves (S leaf) and non-senescent leaves (N leaf) just prior to full expansion. Figure reprinted with kind permission from Kluwer Academic Publishers and Dr. Richard Amasino (University of Wisconsin). Figure 3 from Quirino *et al.*, 1999.

(*SAG*28 and *SAG*24). *SAG*27 shows strictly leaf senescence-specific expression (Quirino *et al.*, 1999).

Dr. Buchanan-Wollaston (University of London) and co-workers have also identified a number of SR genes from *Brassica napus* utilizing both differential library screening and subtractive hybridization (Buchanan-Wollaston, 1994, 1997; Hanfrey *et al.*, 1996; Buchanan-Wollaston and Ainsworth, 1997). It is not known whether any of these genes are up regulated during petal senescence, but similar to the *Arabidopsis* SAGs, they show differential patterns of expression during the development of the leaf. The homologue of the *Arabidopsis SAG*12 gene has been cloned in *B. napus* and shows similar senescence-specific expression after the onset of leaf senescence (Buchanan-Wollaston, unpublished; Noh and Amasino, 1999b).

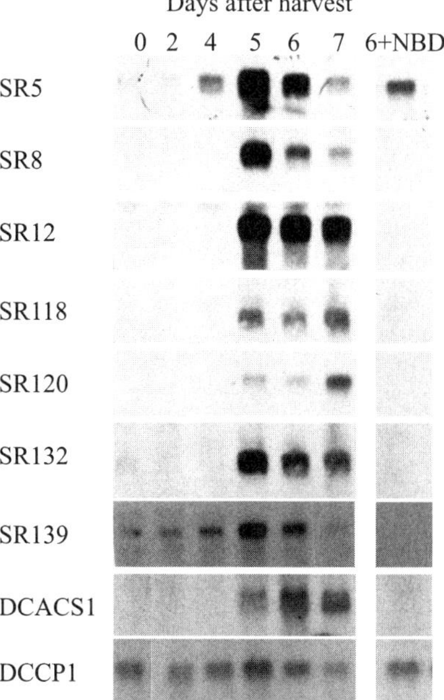

Figure 4-3. Temporal patterns of expression of SR genes in naturally senescing carnation petals. Total RNA was isolated from petals at various times after harvest and subjected to RNA gel blot analysis using the SR cDNA clones as hybridization probes. For identity of SR cDNAs see Table 4-1. RNA blots were also probed with the cDNAs for ACC synthase (DCACS3) and cysteine protease (DCCP1). Printed with permission from Dr. William R. Woodson (Purdue University).

While the majority of the research on SR gene expression has focused on vegetative tissues, there is significant evidence that petal senescence in some flowers is also a genetically programmed event that requires *de novo* protein synthesis and transcription of new genes (Borochov and Woodson, 1989; Woodson, 1987, 1994). *In vitro* translation of carnation petal mRNAs has revealed that the initiation of petal senescence is associated with increases in certain mRNAs (Lawton *et al.*, 1989; Woodson, 1994). Differential screening of a cDNA library from senescing carnation petals has identified nine cDNAs that represent unique senescence-related mRNAs. Expression patterns of seven of these genes are shown in Fig. 4-3. More recently, a cysteine protease (*DCCP*1) and three ACC synthase (*DCACS*1, *DCACS*2, and *DCACS*3) cDNAs were identified from carnation petals using RT-PCR (Park *et al.*, 1992; Jones *et al.*, 1995; ten Have and Woltering, 1997; Jones and Woodson, 1999). Only *SR*139 and *DCCP*1 transcripts are detected in preclimacteric petals. Most of the SR genes are detected in petals at 5 days after harvest, corresponding to the first detectable ethylene production from the petals. Eight of the eleven SR genes from carnation are flower specific while low levels of *SR*139, *SR*123, and *DCCP*1 are detectable in leaves (Woodson *et al.*, 1992, 1993; Jones, unpublished). Those SR genes whose putative identity has been determined are listed in Table 4-1. Identifying the function of additional flower specific SR genes will help to identify differences between the regulation of vegetative and floral senescence.

To date, tomato has been the predominant model system for studies of the regulation of fruit ripening. Several laboratories have screened cDNA libraries from ripening fruit and identified a number of ripening-related cDNAs (for a review see Gray *et al.*, 1992). The screening of expression libraries, sequencing, creation of antisense transgenic plants, and other approaches have led to the identification of some of the enzymes encoded for by these cDNAs, while the identity of many remains unknown. Virtually all of the ripening-related mRNAs in tomato have been reported to be absent or at a low level in immature, green fruit with dramatic increases detected during fruit ripening (Gray *et al.*, 1992). Expression of seven out of twelve genes identified as ripening-related was found to also increase in senescing leaves (Davies and Grierson, 1989). Of those twelve cDNAs, only *pTOM* 5, *pTOM* 6, *pTOM* 96, and *pTOM* 99 were found to be ripening-specific (Davies and Grierson, 1989). Expression in flowers was not reported. Recently a large number of ripening enhanced cDNAs have also been identified from banana (Medina-Suarez *et al.*, 1997; Clendennen and May, 1997), strawberry (Manning, 1998), and avocado (Dopico *et al.*, 1993).

A. Regulation of SR Gene Expression by Stresses

Little is known about the transcriptional regulation of the SR genes other than that their mRNAs accumulate in one or more senescing tissues. While senescence is under developmental regulation, it can be accelerated by certain environmental stresses (Noodén, 1988). A few of the genes that have recently been identified as senescence-related were previously identified based on increased transcription following exposure to various stresses including drought and darkness (Oh *et al.*, 1996; Buchanan-Wollaston and Ainsworth, 1997; Park *et al.*, 1998; Weaver *et al.*, 1998). Recent studies by Weaver *et al.* (1998) have shown that a number of the SAGs from *Arabidopsis* are induced by stresses. Exposing excised leaves to darkness was one of the strongest inducers of SAG transcript accumulation. While detaching leaves in the light did not result in leaf chlorosis, it did result in the enhanced expression of most SAGs (Weaver *et al.*, 1998). Only about half of the SAGs tested showed increased expression following dehydration treatments. In general, these responses were influenced by the developmental stage of the leaves with induction strongest in older leaves and no effect or only moderate induction detected in young leaves (Weaver *et al.*, 1998).

Physiological and biochemical changes such as decline in photosynthetic rate and chlorophyll content have been used to identify stages of senescence, but artificial methods such as detachment and darkness that rapidly induce senescence have become popular models for studying leaf senescence (Thomas and Stoddart, 1980). While the detachment of leaves in the dark accelerates senescence and serves as a means of coordinating the senescence process between leaves, its use as a model system for natural age-related senescence should be approached with caution. Following the identification of three cDNAs that showed increased expression in dark-treated leaves, experiments revealed that only one of the three showed enhanced expression during natural senescence (Becker and Apel, 1993). Only four of seven genes recently identified from *Arabidopsis* as dark-induced also showed enhanced expression in naturally senescing leaves (Oh *et al.*, 1996; Park *et al.*, 1998).

B. Regulation of SR Gene Expression by Cytokinins

Cytokinins act as senescence-retarding hormones. The aging of leaves and petals is accompanied by a decline in endogenous levels of cytokinins and the exogenous application of cytokinins has been shown to delay both leaf and flower senescence (Van Staden *et al.*, 1988). Recent experiments with detached *Arabidopsis* leaves have demonstrated that the exogenous application of cytokinins delays leaf chlorosis and inhibits the expression of most SAGs. Cytokinins inhibited the induction of all the SAGs shown in Fig. 4-1, with the strongest repression of SAG expression detected in young leaves (Weaver *et al.*, 1998).

C. Regulation of SR Gene Expression by Ethylene

The plant hormone ethylene has been implicated in the regulation of both fruit ripening and leaf and flower senescence (Abeles *et al.*, 1992). A number of SR and ripening-related genes have been found to be up regulated by the exogenous application of ethylene (Grbic and Bleeker, 1995; Davies and Grierson, 1989; Lawton *et al.*, 1990; Weaver *et al.*, 1998). All of the SAGs shown in Fig. 4-1 are induced by exogenous ethylene in *Arabidopsis* leaves (Weaver *et al.*, 1998). Treatment of preclimacteric flowers with ethylene results in the induction of all the SR genes identified from carnation (Lawton *et al.*, 1990; Jones *et al.*, 1995; Jones and Woodson, 1997, 1999). In tomato, the highest level of expression of pTOM genes in fruit was detected at the orange stage when ethylene production was highest, and enhanced expression in leaves coincided with the first visible symptoms of leaf yellowing. Treatment with exogenous ethylene resulted in increased expression of pTOM genes in fruit and leaves, providing evidence that ethylene-controlled gene expression is involved in both fruit ripening and leaf senescence (Davies and Grierson, 1989).

Never ripe (NR) tomatoes, which are insensitive to ethylene due to a mutation in the ethylene receptor, produce fruit in which ripening is inhibited, have flower petals that do not senesce, and have leaves with delayed leaf yellowing (Lanahan *et al.*, 1994). In the fruit of *NR* tomatoes ripening-related transcripts accumulate to much lower levels than in wild-type fruit (DellaPenna *et al.*, 1989). *Arabidopsis* plants with a mutated ethylene receptor, *etr1-1*, also show delayed leaf senescence but, once initiated, the process of senescence and the level of SAG expression is similar to that detected in wild-type leaves (Grbic and Bleeker, 1995). The treatment of tomato plants with the ethylene action inhibitor, silver thiosulfate, delays both fruit and leaf senescence and greatly reduces the expression of the mRNAs for *pTOM* 31, *pTOM* 36, and *pTOM* 137, and to a lesser degree *pTOM* 13, *pTOM* 66, and *pTOM* 75 in both fruit and leaves (Davies and Grierson, 1989). Treatment of carnation flowers with the ethylene action inhibitor, norbornadiene (NBD), delays the age-related accumulation of all SR genes except *SR*5 (Lawton *et al.*, 1990; Woodson *et al.*, 1992, 1993; Fig. 4-3). Treatment with NBD also reduces the basal levels of *DCCP*1 transcript in petals (Jones *et al.*, 1995; Fig. 4-3). These experiments indicate that while many SR genes are regulated by ethylene, they are also regulated by developmental or temporal cues.

The ability of plant organs to respond to exogenous ethylene appears to be developmentally regulated as the enhanced expression of SAGs in ethylene treated leaves is greatest in old leaves and not detectable or only moderately induced in young green leaves (Weaver *et al.*, 1998). Immature tomato fruits and flowers also do not respond to exogenous ethylene with ripening or petal senescence. This ethylene treatment does not induce the expression of ripening-related genes in immature green fruit or SR genes in petals from flowers in the bud stage (Lincoln *et al.*, 1987; Lawton *et al.*, 1990). While some flowers like daylily and non-climacteric fruits like strawberry are not regulated by ethylene, it is clear that ethylene plays a regulatory role in both senescence and fruit ripening through the transcriptional regulation of SR genes.

The observed differences in the timing of the response of various SR genes to external stresses and plant hormones indicate that some of the SR genes may respond directly to stress while others may be regulated by senescence that results from the stress or hormone application (Weaver *et al.*, 1998). Further characterization of the response of SR genes to various stresses will help to identify those genes that are primarily responsive to senescence and are thus key regulators of senescence. While we have discussed many genes that are up regulated during senescence and involved in the activation and coordination of senescence, the down-regulation of genes that act as repressors of senescence may play an equally important role in regulating senescence. Currently, most of the genes identified as down regulated during senescence are genes involved in photosynthesis (Hensel *et al.*, 1993; Jiang *et al.*, 1993; John *et al.*, 1997). Transcript levels for the pea homologue of the *defender against apoptotic death* (*dad*) gene, a gene known to function as a repressor of programmed cell death (PCD) in *Caenorhabditis elegans* and mammals, have been found to decrease during flower development (Orzaez and Granell, 1997). While the *dad-1* cDNA from rice can rescue temperature sensitive *dad-1* mutants of hamster from PCD, the function of the dad gene in plant senescence is still unclear (Tanaka *et al.*, 1997).

VI. Function of SR Genes in Senescence

While the enzymatic functions of some of the SR genes have been demonstrated, most have been predicted based on sequence homology (Gan and Amasino, 1997). Some of the SR genes have no sequence similarity to genes in the database. The function of the SR genes falls into four categories: (1) those involved in degradative processes including genes encoding proteases, nucleases, lipid and carbohydrate metabolizing enzymes and those involved in nutrient mobilization; (2) those that code for protective or stress response proteins, many of which have previously been identified as pathogen-responsive; (3) those that code for enzymes involved in ethylene biosynthesis and perception; and (4) those whose identity has been determined based on sequence homology but for which no know function in senescence is obvious. While identification of genes encoding hydrolytic enzymes predominates in senescing tissues, a majority of the genes identified as ripening-related code for enzymes involved in cell wall degradation and anthocyanin biosynthesis (Gray *et al.*, 1992).

A. Genes Involved in Protein Degradation

Decreases in total proteins during senescence result from increases in proteolytic enzyme activity and decreases in protein synthesis (Brady, 1988). The degradation of proteins and remobilization of amino acids to developing tissues is the predominant metabolic process during senescence. Cysteine proteases are believed to be the main proteases involved in general protein hydrolysis, and recently a number of cysteine proteases have been identified from senescing leaves, senescing flowers, and ripening fruit (Table 4-1).

Of those cysteine proteases identified from senescing tissues most share sequence homology with γ oryzain from rice, a cysteine protease that has been implicated in the mobilization of reserve proteins during seed germination. These include *SAG*2, *See*1, LSC7, *SEN*U2 and *SEN*U3 (Table 4-1). The expression patterns of these five genes are similar, with low levels of expression in young leaves and increased expression during senescence (Hensel *et al.*, 1993; Smart *et al.*, 1995; Drake *et al.*, 1996; Buchanan-Wollaston, 1997; Weaver *et al.*, 1998). Both tomato cysteine proteases, SenU2 and SenU3, and See1 from maize also show patterns of up-regulation during seed germination, indicating that these proteases may play similar roles in protein degradation during germination and leaf senescence (Drake *et al.*, 1996; Smart *et al.*, 1995). While common to germination and leaf senescence, the

*SENU*2 and *SENU*3 transcripts were not up regulated during fruit ripening (Drake *et al.*, 1996). *SAG*12, which encodes a papain-like cysteine protease, is one of the few SR genes to display senescence-specific regulation. *SAG*12 mRNAs are not detectable in roots, stems, green leaves, or young flowers, but increase in abundance in senescing petals as well as leaves (Lohman *et al.*, 1994; Quirino *et al.*, 1999). This senescence-specific expression suggests that the *SAG*12 protease might play a key role in the large-scale increases in protein degradation during senescence.

The dismantling of the chloroplast, which contains greater than 50% of the leaf's total protein, is a prominent process in leaf senescence (Thomas and Stoddart, 1980). While many SR genes have been identified as proteases only one of these has been found to be localized to the chloroplast (*Erd1*: Kiyosue *et al.*, 1993; Lohman *et al.*, 1994). Transcript levels of the clp protease have been reported to increase during leaf senescence but protein levels were found to decline, suggesting that the clp protease does not play a primary role in the programmed disassembly of the chloroplast during senescence (Weaver *et al.*, 1999). A recent study by Guiamet *et al.* (1999) has reported that the chloroplasts of senescing soybean leaves excrete plastoglobuli-containing constituents of the chloroplast. The dismantling of these chloroplast components then occurs outside the chloroplast where SR proteases are localized.

Similar to leaf senescence, protein degradation has been demonstrated to be a major part of petal senescence and the remobilization of N to the developing ovary (Nichols, 1976). A few of the SR cysteine proteases have been shown to be up regulated in both leaves and petals (*SAG*12, Quirino *et al.*, 1999; *DCCP*1, Jones *et al.*, 1995, Jones, unpublished; *Peth1*, Tournaire *et al.*, 1996). Large increases in proteolytic activity during the senescence of the ephemeral flower, daylily, have been well documented and recently this proteolytic activity was correlated with increases in the expression of two cysteine protease genes (*Sen11* and *Sen102*) during the senescence of tepals (Valpuesta *et al.*, 1995; Guerrero *et al.*, 1998; Stephenson and Rubinstein, 1998). In contrast to the cysteine proteases from carnation and petunia, transcripts are not detectable in young daylily flowers (buds) and the level of transcript does not increase in senescing leaves (Guerrero *et al.*, 1998). Both daylily cysteine proteases appear to be flower senescence-specific.

B. Genes Involved in the Breakdown of Nucleic Acids

While levels of total DNA remain relatively constant, total RNA has been shown to decrease in senescing tissues (Brady, 1988). This RNA is an important source of C, N, and P, which is recycled to metabolically active tissues during cell death and when P is limiting. This is accomplished through the activity of ribonucleases (RNases). The activity of RNases has been shown to increase substantially during senescence (Brady, 1988; Green, 1994). In *Arabidopsis*, three RNase genes were identified (*RNS*1, *RNS*2, *RNS*3; Taylor and Green, 1991). Up-regulation of *RNS*2 mRNAs was observed in both senescing petals and leaves, indicating that *RNS*2 may play an important role in the remobilization of P from senescing tissues (Taylor *et al.*, 1993). Two senescence associated RNases have also been identified from tomato leaves (Lers *et al.*, 1998). They appear to be differentially regulated and it has been proposed that the major function of *LX* RNase is the catabolism of RNA at the final stages of senescence, while *LE* RNase functions during wounding and defense responses. Neither *LE* RNase nor *LX* RNase were detected in ripening fruit and expression in flowers was only slightly up regulated during senescence (Lers *et al.*, 1998). This suggests that other RNase genes are likely responsible for RNA catabolism during petal senescence and fruit ripening in tomato. Increased RNase activity has been detected during the senescence of daylily and petunia petals (Panavas *et al.*, 1998; Xu and Hanson, 2000). A putative S1-type nuclease that is up regulated during tepal senescence and following ABA treatment, supports the involvement of RNases in the execution of floral senescence (Panavas *et al.*, 1999).

C. Genes Involved in Membrane Disassembly

The disruption of membrane integrity is a major degradative process during senescence (Thompson, 1988). A characteristic of membrane deterioration in leaves and petals is a decrease in phospholipids and an enrichment in fatty acids (Thompson, 1988). Recently a gene encoding phospholipase D, the enzyme that catalyzes the first step in the membrane lipid degradation pathway, has been shown to be up regulated during dark-induced senescence of excised castor bean leaves (Ryu and Wang, 1995). In castor bean it is suggested that differential expression of different PLD isoforms regulates membrane disassembly during growth and development. PLD1 is proposed to play a role in lipid-turnover during rapid growth, PLD2 a maintenance role throughout development, and PLD3 is thought to be responsible for membrane deterioration during senescence (Ryu and Wang, 1995). While the activities of phospholipase A, C, and D have been detected in flower petals, only PLC and PLA activities increased during petal senescence (Borochov and Woodson, 1989). The differences in the activities of various phospholipases detected during the senescence of leaves versus flowers may represent differences in the catabolism of membranes with different lipid components (Thompson, 1988). While galactolipids are the major lipid component of thylakoid membranes the main phospholipid is phosphatidylglycerol. In petals, phosphatidylcholine and phosphatidylethanolamine make up 75% of the membranes' phospholipids (Thompson, 1988; Borochov and Woodson, 1989). A senescence-induced lipase with lipolytic acyl hydrolase activity has recently been identified from carnation flowers (Hong *et al.*, 2000). The abundance of the lipase mRNA increases just as carnation petals begin to inroll and is enhanced by treatment with ethylene. Reducing the expression of a similar senescence-induced lipase in *Arabidopsis*, by expressing the gene in the antisense orientation, results in significant delays in leaf senescence and subsequent increases in yield (Thompson *et al.*, 2000).

D. Genes Involved in Remobilization of Nitrogen

Following the degradation of macromolecules, nitrogen must be exported from senescing organs to metabolically active organs. The deamination of amino acids and catabolism of nucleic acids during senescence releases ammonia, which is exported via the phloem in the form of the amino acids glutamine or asparagine (Kamachi *et al.*, 1992). The conversion of ammonia to glutamine or asparagine is catalyzed by the enzymes glutamine synthetase (GS) or asparagine synthetase (AS) respectively. In plants there are two forms of GS, cytosolic (GS1) and chloroplastic (GS2) (Kamachi *et al.*, 1992). Cytosolic GS1 may function in the mobilization of N from proteins degraded in the cytosol of senescing organs. In support of the role of GS1 in senescence, GS1 genes that are up regulated during senescence have been identified in a number of plants (Kawakami and Watanabe, 1988; Kamachi *et al.*, 1992; Bernhard and Matile, 1994; Buchanan-Wollaston and Ainsworth, 1997). In *Arabidopsis*, three isoforms of GS1 have been identified, transcription of only one of the genes (*Atgsr2*) increases during leaf senescence (natural and dark-induced) (Bernhard and Matile, 1994). Of the three GS1 genes identified from radish, only two (*Gln1;1* and *Gln1;3*) increase in abundance during both age-related and dark-induced senescence (Watanabe *et al.*, 1994). All three *Gln1* genes increase during germination indicating that the GS1 enzyme functions during both germination and senescence to remobilize N from source to sink organs (Watanabe *et al.*, 1994). It was not reported whether expression of any of the GS1 genes was up regulated during flower petal senescence. Increased expression of genes encoding AS during the senescence of asparagus ferns, asparagus spears and *Sandersonia* tepals indicates that asparagine is also utilized to export nitrogen from senescing tissues (Davies and King, 1993; King *et al.*, 1995; Eason *et al.*, 2000).

E. Genes Encoding Protective or Defense-Response Proteins

A second functional class of SR genes appears to encode proteins with a protective or stress response function, and includes a number of genes with homology to previously identified pathogenesis-related (PR) genes. These defense-related genes may serve to protect vulnerable senescing tissues from pathogen attack until the senescence program has been completed and to prevent pathogens from spreading to healthy parts of the plant. A number of these genes may also play a role in detoxification of the by-products of macromolecule and organelle catabolism within the cell and may function in maintaining cell viability until the cell's components have been salvaged.

A number of the SR genes are defense-related and have significant homology to genes that have previously been identified as pathogen-inducible. The SR gene *SENU4* from tomato was previously identified as a pathogenesis-related protein (P6) that was isolated from tomato leaves infected with *Cladosporium fulvum* (van Kan *et al.*, 1992; John *et al.*, 1997). In *B. napus* genes encoding a PR1a-like protein, a chitinase, and an antifungal protein have been identified as SR genes (Hanfrey *et al.*, 1996). In *Arabidopsis*, five of the SAGs identified are defense-related genes. These include two cinnamyl alcohol dehydrogenases, a gene with homology to nitrilase II, an osmotin-like gene reported to have antifungal activity, and a possible MtN3 homologue (Capelli *et al.*, 1997; Quirino *et al.*, 1999). The senescence-enhanced expression of these SAGs has been confirmed in pathogen-free plants (Quirino *et al.*, 1999). These SR genes may merely be induced in response to a stress similar to both pathogen invasion and senescence or they may have a fundamental role in the developmental regulation of senescence. A role for these genes in senescence is supported by the fact that six of the defense-related genes exhibit senescence-specific expression and are not detected in young leaves (*SENU4*, John *et al.*, 1997; LSC 222, Hanfrey *et al.*, 1996; SAG26, SAG29, ELI3-2, NitII, Quirino *et al.*, 1999).

During programmed senescence, mechanisms that protect the cell from free radicals, reduced oxygen species, and other toxic by-products of senescence must be in place in order for the cell to remain viable. Transcripts that encode for catalase, an enzyme that detoxifies H_2O_2, show differential regulation during senescence in *Nicotiana plumbaginifolia* (Willekens *et al.*, 1994). *Cat*1 is believed to play a principal role in scavenging photorespiratory H_2O_2. It is expressed at high levels throughout leaf development and increases dramatically in senescing petals. *Cat*2 is constitutively expressed in all tissues and is up regulated only in senescing leaves. The *cat*3 gene is suggested to function in the glyoxysomal process and is expressed at high basal levels in flowers and seeds but increases in abundance only during leaf senescence (Willekens *et al.*, 1994). A number of genes encoding metallothionein-like or heavy metal binding proteins have been identified by differentially screening senescing and ripening tissues (Buchanan-Wollaston, 1994; Lohman *et al.*, 1994; Buchanan-Wollaston and Ainsworth, 1997; Clendennen and

May, 1997). It is believed that metallothioneins may play a role in detoxification of metal ions released during the catabolism of proteins, or they may also play a salvage role by storing or mobilizing proteins to sink tissues for use in the synthesis of new proteins (Buchanan-Wollaston, 1997). In maize, transcripts corresponding to glutathionine S-transferase II (GST), were enhanced during leaf senescence (Smart *et al.*, 1995) and a flower specific GST mRNA was reported to accumulate during the senescence of carnation petals (Meyer *et al.*, 1991). It has been suggested that the endogenous function of GST during plant tissue senescence is to detoxify lipid hydroperoxides (Meyer *et al.*, 1991).

F. Genes Involved in Ethylene Biosynthesis and Perception

The biosynthesis and perception of the plant hormone ethylene are known to modulate specific components of leaf senescence, fruit ripening, and flower senescence (Borochov and Woodson, 1989; Abeles *et al.*, 1992; Gray *et al.*, 1992; Grbic and Bleeker, 1995). All three processes are also known to be accompanied by increases in the synthesis of ethylene (Abeles *et al.*, 1992), and therefore it is reasonable to assume that SR genes would include those involved in ethylene biosynthesis. Two enzymes, 1-aminocyclopropane-1-carboxylate (ACC) synthase and ACC oxidase, have been identified as catalyzing rate-limiting steps in ethylene biosynthesis (Kende, 1993). While no ACC synthase genes have specifically been isolated by differential screening of senescing or ripening tissues, three SR clones have been identified that encode ACC oxidase; these include *pTOM*13 from tomato (Hamilton *et al.*, 1991); *SR*120 from carnation petals (Wang and Woodson, 1991); and *pBANUU*10 from banana (Medina-Suarez *et al.*, 1997). While *pTOM*13 is up-regulated in leaves, flowers and fruits, *SR*120 is flower specific (Davies and Grierson, 1989; Woodson *et al.*, 1992; Barry *et al.*, 1996). Upon the identification of additional ACC synthase and ACC oxidase genes, transcriptional up-regulation has been reported during flower and leaf senescence and fruit ripening in many species (Abeles *et al.*, 1992). While the ethylene biosynthetic pathway is well established, components involved in ethylene perception and signal transduction have only recently been identified. Initial studies on the expression of genes encoding the ethylene receptor report that specific receptor genes are up regulated during fruit ripening and senescence while others appear to be constitutively expressed in multiple tissues (Payton *et al.*, 1996; Lashbrook *et al.*, 1998; Tieman and Klee, 1999).

VII. Summary

There are many common molecular events occurring among senescing tissues. The degradation and remobilization of cellular constituents is predominant during senescence and, correspondingly, the activities of hydrolytic enzymes and their mRNAs increase. A number of SR genes have also been identified that encode products with homology to PR proteins. While it is not known what the role of these proteins is in senescence, it appears that they may serve a protective role similar to their role during the defense-response. These patterns of expression indicate that throughout plant development, common molecular mechanisms are regulated by the same genes in multiple tissues. A few of these genes have been identified as having leaf, flower, or fruit senescence-specific expression. Of these genes, many encode different isoforms of the same enzyme, which may be differentially regulated within the plant organs. Similar to fruit ripening, the developmental regulation of germination and senescence also share common molecular mechanisms. This is especially evident when investigating the expression of genes involved in protein and lipid degradation and remobilization.

While differential cDNA screening, differential display and cDNA subtraction have identified a number of senescence-related genes, the expression of most genes has not been investigated in flowers, leaves, and fruits. The use of enhancer trap lines in *Arabidopsis* has resulted in the identification of over one hundred lines that have reporter gene expression in senescing but not in non-senescing tissues (He *et al.*, 2001). This technology starts to reveal the complexity of the network of senescence-regulated pathways and will allow for the identification of many additional SR genes. The identification of senescence specific promoter elements (Noh and Amasino, 1999a) and the generation of mutants and transgenic plants will help us to better understand the regulation of SR genes during senescence. DNA microarrays will allow temporal and spatial expression patterns to be determined for hundreds of genes involved in senescence. These technologies will lead to an increased understanding of the initiation and execution of senescence which will allow us to increase vase life and horticultural performance of ornamentals, increase yield in agronomic crops, and decrease postharvest losses of fruits and vegetables.

Acknowledgments

I would like to thank Drs. William Woodson and David Clark and Ms. Rebecca Riggle for reading the manuscript and Drs. William Woodson, Richard Amasino, Vicky Buchanan-Wollaston and Betania Quirino for generously supplying figures. I would also like to acknowledge the assistance of Ms. Catherine Conover.

References

Abeles, F.B., Morgan, P.W., and Saltveit, Jr., M.E. (1992). *Ethylene in Plant Biology*. Academic Press, London.

Alonso, J.M., and Granell, A. (1995). A putative vacuolar processing protease is regulated by ethylene and also during fruit ripening in *Citrus* fruit. *Plant Physiol.* **109**, 541–547.

Barry, C.S., Blume, B., Bouzayen, M., Cooper, W., Hamilton, A.J., and Grierson, D. (1996). Differential expression of the 1-aminocyclopropane-1-carboxylate oxidase gene family of tomato. *Plant Journal* **9**, 525–535.

Becker, W., and Apel, K. (1993). Differences in gene expression between natural and artificially induced leaf senescence. *Planta* **189**, 74–79.

Bernhard, W.R., and Matile, P. (1994). Differential expression of glutamine synthetase genes during the senescence of *Arabidopsis thaliana* rosette leaves. *Plant Science* **98**, 7–14.

Borochov, A., and Woodson W.R. (1989). Physiology and biochemistry of flower petal senescence. *Hortic. Rev.* **11**, 15–43.

Brady, C.J. (1988). Nucleic acid and protein synthesis. In *Senescence and Aging in Plants* (L.D. Noodén, Ed.), pp. 281–328. Academic Press, San Diego.

Buchanan-Wollaston, V. (1994). Isolation of cDNA clones for genes that are expressed during leaf senescence in *Brassica napus*. Identification of a gene encoding a senescence-specific metallothionein-like protein. *Plant Physiol.* **105**, 839–846.

Buchanan-Wollaston, V. (1997). The molecular biology of leaf senescence. *J. Exp. Bot.* **48**, 181–199.

Buchanan-Wollaston, V., and Ainsworth, C. (1997). Leaf senescence in *Brassica napus*: cloning of senescence-related genes by subtractive hybridisation. *Plant Mol. Biol.* **33**, 821–834.

Callard, D., Axelos, M., and Mazzolini, L. (1996). Novel molecular markers for late phases of the growth cycle of *Arabidopsis thaliana* cell-suspension cultures are expressed during organ senescence. *Plant Physiol.* **112**, 705–715.

Capelli, N., Diogen, T., Greppin, H., and Simon, P. (1997). Isolation and characterization of a cDNA clone encoding an osmotin-like protein from *Arabidopsis thaliana*. *Gene* **191**, 51–56.

Clendennen, S.K., and May, G.D. (1997). Differential gene expression in ripening banana fruit. *Plant Physiol.* **115**, 463–469.

Davies, K.M., and Grierson, D. (1989). Identification of cDNA clones for tomato (*Lycopersicon esculentum* Mill.) mRNAs that accumulate during fruit ripening and leaf senescence in response to ethylene. *Planta* **179**, 73–80.

Davies, K.M., and King, G.A. (1993). Isolation and characterization of a cDNA clone for a harvest-induced asparagine synthetase from *Asparagus officinalis* L. *Plant Physiol.* **102**, 1337–1340.

DellaPenna, D., Lincoln, J.E., Fischer, R.L., and Bennett, A.B. (1989). Transcriptional analysis of polygalacturonase and other ripening associated genes in Rutgers, *rin*, *nor*, and *Nr* tomato fruit. *Plant Physiol.* **90**, 1372–1377.

Dopico, B., Lowe, A.L., Wilson, I.D., Merodio, C., and Grierson, D. (1993). Cloning and characterization of avocado fruit mRNAs and their expression during ripening and low-temperature storage. *Plant Mol. Biol.* **21**, 437–449.

Drake, R., John, I., Farrell, A., Cooper, W., Schuch, W., and Grierson, D. (1996). Isolation and analysis of cDNAs encoding tomato cysteine proteases expressed during leaf senescence. *Plant Mol. Biol.* **30**, 755–767.

Eason, J.R., Johnston, J.W., de Vre, L., Sinclair, B.K., and King, G.A. (2000). Amino acid metabolism in senescing *Sandersonia aurantiaca* flowers: cloning and characterization of asparagine synthetase and glutamine synthetase cDNAs. *Australian J. Plant Physiol.* **27**, 389–396.

Fukuchi-Mizutani, M., Ishiguro, K., Nakayama, T., Utsunomiya, Y., Kusumi, T., and Ueda, T. (2000). Molecular and functional characterization of a rose lipoxygenase cDNA related to flower senescence. *Plant Science* **160**, 129–137.

Gan, S., and Amasino, R.M. (1997). Making sense of senescence. *Plant Physiol.* **113**, 313–319.

Graham, I.A., Leaver, C.J., and Smith, S.M. (1992). Induction of malate synthase gene expression in senescent and detached organs of cucumber. *Plant Cell* **4**, 349–357.

Gray, J., Picton, S., Shabbeer, J., Schuch, W., and Grierson, D. (1992). Molecular biology of fruit ripening and its manipulation with antisense genes. *Plant Mol. Biol.* **19**, 69–87.

Grbic, V., and Bleeker, A.B. (1995). Ethylene regulates the timing of leaf senescence in *Arabidopsis*. *Plant J.* **8**, 595–602.

Green, P.J. (1994). The ribonucleases of higher plants. *Annu. Rev. Plant Physiol. Plant Mol. Biol.* **45**, 421–445.

Guerrero, C., de la Calle, M., Reid, M.S., and Valpuesta, V. (1998). Analysis of the expression of two thiolprotease genes from daylily (*Hemerocallis* spp.) during flower senescence. *Plant Mol. Biol.* **36**, 565–571.

Guiamet, J.J., Pichersky, E., and Noodén, L.D. (1999). Mass exodus from senescing soybean chloroplasts. *Plant Cell Physiol.* **40**, 986–992.

Hamilton, A.J., Bouzayen, M., and Grierson, D. (1991). Identification of a tomato gene for the ethylene-forming enzyme by expression in yeast. *Proc. Natl. Acad. Sci. USA* **88**, 7434–7437.

Hanfrey, C., Fife, M., and Buchanan-Wollaston, V. (1996). Leaf senescence in *Brassica napus*: expression of genes encoding pathogenesis-related proteins. *Plant Mol. Biol.* **30**, 597–609.

He, Y., Tang, W., Swain, J.D., Green, A.L., Jack, T.P., and Gan, S. (2001). Networking senescence-regulating pathways by using *Arabidopsis* enhancer trap lines. *Plant Physiol.* **126**, 707–716.

Hensel, L.L., Grbic, V., Baumgarten, D.A., and Bleeker, A.B. (1993). Developmental and age-related processes that influence the longevity and senescence of photosynthetic tissue in *Arabidopsis*. *Plant Cell* **5**, 553–564.

Hong, Y., Wang, T.-W., Hudak, K.A., Schade, F., Froese, C.D., and Thompson, J.E. (2000). An ethylene-induced cDNA encoding a lipase expressed at the onset of senescence. *Proc. Natl. Acad. Sci. USA* **97**, 8717–8722.

Jiang, C.-Z., Rodermal, S.R., and Shibles, R.M. (1993). Photosynthesis, rubisco activity and amount and their regulation by transcription in senescing soybean leaves. *Plant Physiol.* **101**, 105–112.

John, I., Hackett, R., Cooper, W., Drake, R., Farrell, A., and Grierson, D. (1997). Cloning and characterization of tomato leaf senescence-related cDNAs. *Plant Mol. Biol.* **33**, 641–651.

Jones, M.L., Larsen, P.B., and Woodson, W.R. (1995). Ethylene-regulated expression of a carnation cysteine proteinase during flower petal senescence. *Plant Mol. Biol.* **28**, 505–512.

Jones, M.L., and Woodson, W.R. (1997). Pollination-induced ethylene in carnation. Role of stylar ethylene in corolla senescence. *Plant Physiol.* **115**, 205–212.

Jones, M.L., and Woodson, W.R. (1999). Differential expression of three members of the 1-aminocyclopropane-1-carboxylate synthase gene family in carnation. *Plant Physiol.* **119**, 755–764.

Kamachi, K., Yamaya, T., Hayakawa, T., Mae, T., and Ojima, K. (1992). Changes in cytosolic glutamine synthetase polypeptide and its mRNA in a leaf blade of rice plants during natural senescence. *Plant Physiol.* **98**, 1323–1329.

Kawakami, N., and Watanabe, A. (1988). Change in gene expression in radish cotyledons during dark-induced senescence. *Plant Cell Physiol.* **29**, 33–42.

Kende, H. (1993). Ethylene biosynthesis. *Ann. Rev. Plant Physiol.* **44**, 283–307.

King, G.A., and Davies, K.M. (1995). Cloning of a harvest-induced β-galactosidase from tips of harvested asparagus spears. *Plant Physiol.* **108**, 419–420.

King, G.A., Davies, K.M., Stewart, R.J., and Borst, W.M. (1995). Similarities in gene expression during the postharvest-induced senescence of spears and natural foliar senescence of Asparagus. *Plant Physiol.* **108**, 125–128.

Kiyosue, T., Yamaguchi-Shinozaki, K., and Shinozaki, K. (1993). Characterization of a cDNA for a dehydration-inducible gene that encodes a CLP A,B-like protein in *Arabidopsis thaliana* L. *Biochem. Biophys. Res. Commun.* **196**, 1214–1220.

Lanahan, M.B., Yen, H.-C., Giovannoni, J.J., and Klee, H.J. (1994). The *Never Ripe* mutation blocks ethylene perception in tomato. *Plant Cell* **6**, 521–530.

Lashbrook, C.C., Tieman, D.M., Klee, H.J. (1998). Differential regulation of the tomato ETR gene family throughout plant development. *Plant J.* **15**, 243–252.

Lawton, K.A., Huang, B., Goldsbrough, P.B., and Woodson, W.R. (1989). Molecular cloning and characterization of senescence-related genes from carnation flower petals. *Plant Physiol.* **90**, 690–696.

Lawton, K.A., Raghothama, K.G., Goldsbrough, P.B., and Woodson, W.R. (1990). Regulation of senescence-related gene expression in carnation flower petals by ethylene. *Plant Physiol.* **93**, 1370–1375.

Lers, A., Khalchitski, A., Lomaniec, E., Burd, S., and Green, P.J. (1998). Senescence-induced RNases in tomato. *Plant Mol. Biol.* **36**, 439–449.

Lincoln, J.E., Cordes, S., Read, E., and Fischer, R.L. (1987). Regulation of gene expression by ethylene during *Lycopersicon esculentum* (tomato) fruit development. *Proc. Natl. Acad. Sci. USA* **84**, 2793–2797.

Lohman, K.N., Gan, S., John, M.C., and Amasino, R.M. (1994). Molecular analysis of natural leaf senescence in *Arabidopsis thaliana. Physiol. Plant.* **92**, 322–328.

Manning, K. (1998). Isolation of a set of ripening-related genes from strawberry: their identification and possible relationship to fruit quality traits. *Planta* **205**, 622–631.

McGlasson, W.B., Poovaiah, B.W., and Dostal, H.C. (1975). Ethylene production and respiration in aging leaf segments and in disks of fruit tissue of normal and mutant tomatoes. *Plant Physiol.* **56**, 547–549.

McLaughlin, J.C., and Smith, S.M. (1994). Metabolic regulation of glyoxylate-cycle enzyme synthesis in detached cucumber cotyledons and protoplasts. *Planta* **195**, 22–28.

Medina-Suarez, R., Manning, K., Fletcher, J., Aked, J., Bird, C.R., and Seymour, G.B. (1997). Gene expression in the pulp of ripening bananas. *Plant Physiol.* **115**, 453–461.

Meyer, R.C., Goldsbrough, P.B., and Woodson, W.R. (1991). An ethylene-responsive flower senescence-related gene from carnation encodes a protein homologous to glutathione S-transferase. *Plant Mol. Biol.* **17**, 277–281.

Nam, H.G. (1997). The molecular genetic analysis of leaf senescence. *Curr. Opin. Biotechnol.* **8**, 200–207.

Nichols, R. (1976). Cell enlargement and sugar accumulation in the gynaecium of the glasshouse carnation (*Dianthus caryophyllus* L.) induced by ethylene. *Planta* **130**, 47–52.

Noh, Y.-S., and Amasino, R.M. (1999a). Identification of a promoter region responsible for the senescence-specific expression of SAG12. *Plant Mol. Biol.* **41**, 181–194.

Noh, Y.-S., and Amasino, R.M. (1999b). Regulation of developmental senescence is conserved between *Arabidopsis* and *Brassica napus*. *Plant Mol. Biol.* **41,** 195–206.

Noodén, L.D. (1988). The phenomena of senescence and aging. In *Senescence and Aging in Plants* (L.D. Noodén, Ed.), pp. 281–328. Academic Press, San Diego.

Noodén, L.D., Guiamet, J.J., and John, I. (1997). Senescence mechanisms. *Physiol. Plant.* **101**, 746–753.

Oh, S.A., Lee, S.Y., Chung, I.K., Lee, C.-H., and Nam, H.G. (1996). A senescence-associated gene of *Arabidopsis thaliana* is distinctively regulated during natural and artificially induced leaf senescence. *Plant Mol. Biol.* **30**, 739–754.

Orzaez, D., and Granell, A. (1997). The plant homologue of the *defender against apoptotic death* gene is down-regulated during senescence of flower petals. *FEBS Letters* **404**, 275–278.

Panavas, T., Walker, E.L., and Rubinstein, B. (1998). Possible involvement of abscisic acid in senescence of daylily petals. *J. Exp. Bot.* **49**, 1987–1997.

Panavas, T., Pikula, A., Reid, P.D., Rubinstein, B., and Walker, E.L. (1999). Identification of senescence-associated genes from daylily petals. *Plant Mol. Biol.* **40**, 237–248.

Park, K.Y., Drory, A., and Woodson, W.R. (1992). Molecular cloning of an 1-aminocyclopropane-1-carboxylate synthase from senescing carnation flower petals. *Plant Mol. Biol.* **18**, 377–386.

Park, J.-H., Oh, S.A., Kim, Y.H., Woo, H.R., and Nam, H.G. (1998). Differential expression of senescence-associated mRNAs during leaf senescence induced by different senescence-inducing factors in *Arabidopsis. Plant Mol. Biol.* **37**, 445–454.

Payton, S., Fray, R.G., Brown, S., and Grierson, D. (1996). Ethylene receptor expression is regulated during fruit ripening, flower senescence, and abscission. *Plant Mol. Biol.* **31**, 1227–1231.

Perez-Amador, M.A., Abler, M.L., De Rocher, E.J., Thompson, D.M., van Hoof, A., LeBrasseur, N.D., Lers, A., and Green, P.J. (2000). Identification of BFN1, a bifunctional nuclease induced during leaf and stem senescence in *Arabidopsis*. *Plant Physiol.* **122**, 169–179.

Quirino, B.F., Normanly, J., and Amasino, R.M. (1999). Diverse range of gene activity during *Arabidopsis thaliana* leaf senescence includes pathogen-independent induction of defense-related genes. *Plant Mol. Biol.* **40**, 267–278.

Quirino, B.F., Noh Y.-S., Himelblau, E., and Amasino, R.M. (2000). Molecular aspects of leaf senescence. *Trends Plant Sci.* **5**, 278–282.

Rubinstein, B. (2000). Regulation of cell death in flower petals. *Plant Mol. Biol.* **44**, 303–318.

Ryu, S.B., and Wang, X. (1995). Expression of phospholipase D during castor bean leaf senescence. *Plant Physiol.* **108**, 713–719.

Smart, C.M. (1994). Gene expression during leaf senescence. *New Phytol.* **126**, 419–448.

Smart, C.M., Hosken, S.E., Thomas, H., Greaves, J.A., Blair, B.G., and Schuch, W. (1995). The timing of maize leaf senescence and characterization of senescence-related cDNAs. *Plant Physiol.* **93**, 673–682.

Stephenson, P., and Rubinstein, B. (1998). Characterization of proteolytic activity during senescence in daylily. *Plant Physiol.* **104**, 463–473.

Tanaka, Y., Makishima, T., Sasabe, M., Ichinose, Y., Shiraishi, T., Nishimoto, T., and Yamada, T. (1997). *Dad-1*, a putative programmed cell death suppressor gene in rice. *Plant Cell Physiol.* **38**, 379–383.

Taylor, C.B., and Green, P.J. (1991). Genes with homology to fungal and S-gene RNases are expressed in *Arabidopsis thaliana*. *Plant Physiol.* **96**, 980–984.

Taylor, C.B., Bariola, P.A., DelCardayre, S.B., Raines, R.T., and Green, P.J. (1993). RNS2: A senescence-associated RNase of *Arabidopsis* that diverged from the S-RNases before speciation. *Proc. Natl. Acad. Sci. USA* **90**, 5118–5122.

ten Have, A., and Woltering, E.J. (1997). Ethylene biosynthetic genes are differentially expressed during carnation (*Dianthus caryophyllus* L.) flower senescence. *Plant Mol. Biol.* **34**, 89–97.

Thomas, H., and Stoddart, J.L. (1980). Leaf senescence. *Annu. Rev. Plant Physiol.* **31**, 83–111.

Thompson, J.E. (1988). The molecular basis for membrane deterioration. In *Senescence and Aging in Plants* (L.D. Noodén and A.C. Leopold, Eds.) pp. 51–83. Academic Press, San Diego.

Thompson, J.E., Taylor, C., and Wang, T.-W. (2000). Altered membrane lipase expression delays leaf senescence. *Biochem. Soc. Trans.* **28**, 775–777.

Tieman, D.M., and Klee, H.J. (1999). Differential expression of two novel members of the tomato ethylene-receptor family. *Plant Physiol.* **120**, 165–172.

Tournaire, C., Kushnir, S., Bauw, G., Inze, D., de la Serve, B.T., and Renaudin, J.-P. (1996). A thiol protease and an anionic peroxidase are induced by lowering cytokinins during callus growth in *Petunia*. *Plant Physiol.* **111**, 159–168.

Valpuesta, V., Lange, N.E., Guerrero, C., and Reid, M.S. (1995). Up-regulation of a cysteine protease accompanies the ethylene-insensitive senescence of daylily (*Hemerocallis*) flowers. *Plant Mol. Biol.* **28**, 575–582.

van Kan, J.A.L., Joosten, M., Wagemakers, C.A.M., van der Berg-Velthuis, G.C.M., and de Wit, P. (1992). Differential accumulation of mRNAs encoding extracellular and intracellular PR proteins in tomato by virulent and avirulent races of *Cladosporium fulvum*. *Plant Mol. Biol.* **20**, 513–527.

Van Staden, J., Cook, E.L., and Noodén, L.D. (1988). Cytokinins and senescence. In *Senescence and Aging in Plants* (L.D. Noodén, Ed.), pp. 281–328. Academic Press, San Diego.

Wang, H., and Woodson, W.R. (1991). A flower senescence-related mRNA from carnation shares sequence similarity with fruit ripening-related mRNAs involved in ethylene biosynthesis. *Plant Physiol.* **96**, 1000–1001.

Wang, H., Brandt, A.S., and Woodson, W.R. (1993). A flower senescence-related mRNA from carnation encodes a novel protein related to enzymes involved in phosphonate biosynthesis. *Plant Mol. Biol.* **22**, 719–724.

Watanabe, A., Hamada, K., Yokoi, H., and Watanabe, A. (1994). Biphasic and differential expression of cytosolic glutamine synthetase genes of radish during seed germination and senescence of cotyledons. *Plant Mol. Biol.* **26**, 1807–1817.

Weaver, L.M., Himelblau, E., and Amasino, R.M. (1997). Leaf senescence: gene expression and regulation. In *Genetic Engineering: Principles and Methods* (J.K. Setlow, Ed.), Vol. 19, pp. 215–234. Plenum Press, New York.

Weaver, L.M., Gan, S., Quirino, B., and Amasino, R.M. (1998). A comparison of the expression patterns of several senescence-associated genes in response to stress and hormone treatment. *Plant Mol. Biol.* **37**, 455–469.

Weaver, L.M., Froehlich, J.F., and Amasino, R.M. (1999). Chloroplast-targeted ERD1 protein declines but its mRNA increases during senescence in *Arabidopsis*. *Plant Physiol.* **119**, 1209–1216.

Willekens, H., Langebartels, C., Tire, C., Van Montagu, M., Inze, D., and Van Camp, W. (1994). Differential expression of catalase genes in *Nicotiana plumbaginifolia* (L.) *Proc. Natl. Acad. Sci. USA* **91**, 10450–10454.

Woodson, W.R. (1987). Changes in protein and mRNA populations during carnation petal senescence. *Physiol. Plant.* **71**, 495–502.

Woodson, W.R. (1994). Molecular biology of flower senescence in carnation. In *Molecular and Cellular Aspects of Plant Reproduction* (R.J. Scott and A.D. Stead, Eds.), pp. 225–267. Cambridge University Press, Cambridge, UK.

Woodson, W.R., Park, K.Y., Drory, A., Larsen, P.B., and Wang, H. (1992). Expression of ethylene biosynthetic pathway transcripts in senescing carnation flowers. *Plant Physiol.* **99**, 526–532.

Woodson, W.R., Brandt, A.S., Itzhaki, H., Maxon, J.M., Park, K.Y., and Wang, H. (1993). Regulation and function of flower senescence-related genes. *Acta Hort.* **336**, 41–46.

Xu, F.-X., and Chye, M.-L. (1999). Expression of cysteine proteinase during developmental events associated with programmed cell death in brinjal. *Plant J.* **17**, 321–327.

Xu, Y., and Hanson, M.R. (2000). Programmed cell death during pollination-induced petal senescence in Petunia. *Plant Physiol.* **122**, 1323–1333.

5

Genes that Alter Senescence

Hye Ryun Woo, Pyung Ok Lim,
Hong Gil Nam
and Larry D. Noodén

I. Introduction

Senescence is a sequence of biochemical events that comprise the final stage of development (Chapter 1). Senescence in plants is now regarded as an evolutionarily acquired and genetically programmed developmental strategy rather than as a passive and degenerative process. The genetic influences on aging are discussed elsewhere (Noodén and Guiamét, 1996). Identification of genes that alter senescence through molecular genetic approaches will be critical not just for understanding the mechanism of senescence but also for practical purposes such as improvement of plant productivity, postharvest storage, and stress tolerance.

Senescence is a developmental process that leads to death of a cell, an organ, or a whole plant and occurs at the final stage of their development (Noodén and Leopold, 1978).

Senescence in higher plants is a type of programmed cell death (PCD). It is a conspicuous and critical developmental process, albeit terminal development. Even though the biochemical and cellular events during senescence have been extensively studied, the exact biochemical definition of senescence is not clear yet. Thus, senescence may be used rather loosely to refer to what is actually a syndrome of primary and secondary changes (Noodén, 1988a).

Although apoptosis-like processes may participate in the terminal phase of leaf and petal senescence (see Chapter 1 of this volume), this chapter will not cover apoptosis-like processes *per se*. There simply are not yet clear mutations implicating them except for genetically engineered constructs (Chapter 1).

Senescence is not a random disintegration process, but a highly regulated process that involves orderly and sequential change of cellular physiology, biochemistry, and gene expression. In fact, it appears that, during senescence, activities or quantities of most cellular components including the expression of most genes are changed, increasing or decreasing (Chapter 4). Thus, it is expected that many genes participate in senescence.

Although the biochemistry of senescing structures, especially leaves and petals, has been extensively investigated, we still do not know what is the central (primary) pathway and which processes are peripheral or secondary (Chapter 1); however, mutations should help with this. Likewise, it is not known if the central processes constitute a single channel or multiple parallel intersecting channels, a braided pathway, but the latter seems likely and this is supported by the observations on mutations. Furthermore, it is not clear whether the core (primary) senescence processes are the same or different in different tissues.

II. Senescence as a Genetically Programmed Process

Interestingly, selective inhibitors of protein and mRNA synthesis give some clues about the genetic control of senescence and about what might be expected from senescence mutations. For example, enucleation and selective inhibitors of RNA synthesis and cytoplasmic protein synthesis generally block senescence (Brady, 1988; Noodén, 1988c), and this indicates that synthesis of new mRNAs and proteins (e.g., gene expression) is required for senescence. Although senescing tissues undergo an overall decrease in RNA and protein synthesis (Brady, 1988), these syntheses continue during senescence and some genes are up regulated (Chapter 4 and other chapters in this volume). Many of the known senescence mutations are recessive nuclear genes (simple Mendelian inheritance), but there are some interesting exceptions. One is *G*, which keeps the seed coat green in soybean and shows Mendelian inheritance (discussed below in Section IIIB); however, it is dominant. Yet another is the dominant monogene causing the stay-green trait in *Zea mays* (Gentinetta *et al.*, 1986).

While a majority of mutations affecting senescence show Mendelian inheritance, *cytG* in soybean shows maternal inheritance. Possibly, it is a chloroplastic gene. This is of particular interest, because senescence seems to be imposed on the chloroplast by the nucleus, and the chloroplasts may play very little role in their own execution (Noodén and Leopold, 1978).

Although senescence occurs in a concerted and ordered manner among various cellular components, some mutations clearly disconnect this coordinated sequence. For example, *cytG* (preserves primarily the LHCII chlorophyll–protein complexes but not the rest of the cell) indicates that at least some of the cellular senescence process can be uncoupled (Guiamét *et al.*, 2002).

Early senescence symptoms may result from almost any strong perturbation of homeostasis in cells, and therefore, the genes represented by most of the early senescence mutations may not directly be involved in control of senescence. Given the likelihood that many mutations that cause metabolic disruptions can also induce senescence, the alleles that promote senescence will generally not be covered here.

Recent genetic and molecular studies on senescence also provide direct evidence that senescence is a genetically programmed event. Firstly, many genetic loci that alter at least a part of senescence symptoms have been identified in *Arabidopsis* as well as in a variety of crop plants (Thomas and Stoddart, 1993; Grbić and Bleecker, 1995; Oh *et al.*, 1997). Secondly, senescence is associated with up-regulation of many genes that are involved in chlorophyll degradation, nucleic acid breakdown, protein degradation, or nitrogen and lipid remobilization (Lohman *et al.*, 1994; John *et al.*, 1997; Buchanan-Wollaston and Ainsworth, 1997). Thirdly, senescence is subject to a variety of correlative (remote internal) controls, for example, maturing pods in soybeans induce senescence of the leaves and the rest of the plant or pollination of flowers causes the petals to senesce (Noodén, 1988a).

The initiation and progression of senescence can be modulated by a variety of environmental factors such as temperature, mineral deficiency, drought conditions, and pathogen infection. It is also known that internal factors such as plant hormones, reproduction, and cellular differentiation influence senescence (Noodén and Leopold, 1988; Smart, 1994; Nam, 1997). At the whole plant level, environmental conditions that influence reproductive development also alter senescence.

Because *Arabidopsis* has allowed so many breakthroughs, it will be heavily utilized in this review. *Arabidopsis thaliana* is a monocarpic plant; that is it has one reproductive phase followed by death and it shows clear leaf as well as whole plant senescence (Napp-Zinn, 1985; Noodén and Penney, 2001). In addition, *Arabidopsis* has a short life span and amenability to molecular genetic analysis, making this plant a good model system to study senescence and to find regulatory genes by a forward genetic approach. *Arabidopsis* does, however, differ from many other monocarpic plants in that leaf longevity is not controlled by the developing reproductive structures (Hensel *et al.*, 1993; Noodén and Penney, 2001). Although different groups of plants differ a bit in terms of their correlative and hormonal controls of senescence, particularly at the whole plant level (Noodén, 1988b), evolutionary conservation of genes and functions would allow considerable extrapolation from *Arabidopsis* to other species that are less amenable to molecular genetic analyses.

While the end result of senescence is death, and death may be an important measure of the overall process, it will be necessary to break the senescence process into components for genetic analyses and many of these alterations by themselves may not be sufficient to prevent death. Mutations and genetic analyses will help to resolve all these issues. Since the exact identity of the primary senescence process(es) is not known, it is difficult organize an overall picture of genes that alter senescence. It is also uncertain whether or not senescence is the same or different in different tissues. The peripheral (secondary) processes certainly seem to differ, and the activity of senescence promoters (He *et al.*, 2001) also differs. In this chapter, we will try to give an overview of the genes that alter senescence symptoms or are potential regulatory genes of senescence (Table 5-1) and will also suggest what can or should be done for further identification of genes that alter. The coverage of gene expression and genetic constructs will mostly be left to other chapters, particularly Chapters 4 and 6.

As mentioned above, senescence in plants is not just affected by endogenous developmental factors such as reproduction and hormones but is also affected by external factors such as various environmental stresses and nutrient supply. Given the rudimentary nature of our understanding of the biochemical nature of the primary senescence process and that

Table 5-1. Genes that alter Senescence or that are Potential Regulatory Genes for Senescence Process

Genes	Molecular nature	Effects of mutations on senescence phenotype or characteristics
Genes involved in execution of senescence		
CYTG	Unknown	*cytG* delays chlorophyll and LHHCII breakdown
SID	Unknown	*sid* delays chlorophyll breakdown in the
PLDa	Phospholipase D	Retardation of ABA- and ethylene-promoted senescence, when expression is blocked
AtAPG8	Autophagy gene	Early leaf senescence in the knockout line
AtAPG9	Autophagy gene	Early leaf senescence in the knockout line
SAG101	Acyl hydrolase	Delays leaf senescence when expression is blocked
Genes affecting senescence through action on the hormonal control		
ACS	ACC synthase	Delays fruit ripening when expression is blocked
EIN2 (= ORE3)	Putative metal ion transporter	Ethylene insensitive; *ein2* delays leaf senescence
ETR1	Ethylene receptor	Ethylene insensitive; *etr* delays leaf senescence
RIN	Unknown	Reduces ethylene biosynthesis; slow softening of cell walls
DET2	Steroid 5α-reductase	*det2* delays chlorophyll loss
KN1	Homeobox protein	Delays leaf senescence when overexpressed in senescence
Genes that alter senescence in response to environmental factors		
PHYA	Phytochrome A	Delays chlorophyll and protein loss when overexpressed
PHYB	Phytochrome B	Delays leaf senescence when overexpressed
Regulatory genes and intracellular signaling		
ORE1	Unknown	*ore1* delays leaf senescence
ORE4	Plastid ribosomal protein subunit 17	*ore4* delays leaf senescence
ORE9	F-box protein	*ore9* delays leaf senescence
DLS1	Ariginyl t-RNA transferase	*dls1* delays leaf senescence
SENU5	NAC domain family	Senescence-induced
PPF-1	Membrane-associated protein	Senescence-induced
ATWRKY6	WRKY transcription factor	Senescence-induced
WRKY53	WRKY transcription factor	Senescence-induced
SARK	Receptor-like kinase	Senescence-induced

most of the genes that alter senescence are not well characterized in terms of their mechanism or mode of action, it is as yet difficult to clearly classify the genes. However, for the purpose of organizing this chapter, we have tried to group the genes that are known to alter senescence according to their mode of action wherever possible. Some of these genes alter senescence through seemingly direct effects, while others clearly act indirectly.

III. Genes Involved in Execution of Senescence

A. Overview

For the senescence program to proceed, there should be genes that execute the degeneration process. Many senescence-associated genes have been identified from a variety of plants on the basis of mRNA increases (see Chapter 4). Many of these are involved in degradation processes, including chlorophyll breakdown, and nitrogen and lipid remobilization. These individual genes are likely to affect only some senescence symptoms. It is possible that these genes may have to be combined in order to exert wider effects on senescence, as has been observed in the case of soybean GG$d_1d_1d_2d_2$ plants (see Section IIIB below). Other senescence-associated genes, however, are involved in the assimilation of the products of degradation during senescence, and they probably do not cause death. These include those coding for enzymes in the glyoxylate cycle, nitrogen metabolism and metallothionein (Buchanan-Wollaston, 1997). Although it is clear that all genes participating in senescence are not represented by mutations, many interesting mutations that seem to affect senescence have been found.

B. Chlorophyll Breakdown

Many naturally occurring so-called "stay-green" mutants have been isolated from crop plants, and a few have been at least partially characterized (Thomas and Stoddart, 1993). These may affect chlorophyll breakdown directly, but most seem to be indirect, acting on many processes. None appears to affect all of the senescence symptoms, possibly suggesting that coordinated senescence of cellular components may be uncoupled in these mutants.

A maternally inherited mutation of soybean, *cytG*, selectively preserves LHCII, resulting in higher levels of chlorophyll a/b binding proteins and chlorophyll a and especially chlorophyll b compared to wild type during leaf senescence (Guiamét *et al.*, 1991). However, *cytG* does not delay the decline in photosynthetic rate, and it causes some photoinhibition (Guiamét *et al.*, 1990, 2002). Similarly, two recessive genetic loci of soybean, d_1 and d_2, delay degradation of components of LHCII, plasma membrane degradation, and chloroplast degeneration, although neither d_1 nor d_2 works alone (Guiamét *et al.*, 1990; Guiamét and Giannibelli, 1994, 1996). However, decline of photosynthetic rate is not prevented during leaf senescence of the $d_1d_1d_2d_2$ mutant plant. When *G* (a dominant allele preventing seed coat yellowing and keeping the seed coat green) is combined with d_1 and d_2 to produce GG$d_1d_1d_2d_2$ plants, a broader spectrum of senescence symptoms is delayed including photosynthetic decline and degradation of leaf soluble proteins, plasma membrane, and chloroplast. Since G seems to preserve the seed coat and the seed coat transfers photosynthate to the seeds, G may act by maintaining the sink activity of the seeds. Without strong sinks, the photosynthetic machinery would otherwise be adjusted downward (Paul and Foyer, 2001).

The most-studied "stay-green" mutant is the recessive nuclear mutation of fescue grass (*Festuca pratensis* Huds), *sid*. The *Sid* locus encodes or regulates the gene for pheophorbide a oxygenase, an enzyme involved in the third step of chlorophyll breakdown (Vicentini *et al.*,1995). Blocking removal or synthesis of a key chloroplast component such as chlorophyll may have broad pleiotropic effects like those observed with *sid* (Von Wettstein, 1959).

This explains why the *sid* mutation maintains many thylakoid proteins, including the LHCII (Thomas and Stoddart, 1993; Vicentini *et al.*, 1995). Nonetheless, the *sid* mutation does not preserve all the thylakoid proteins and the photosynthesis declines similarly to wild type.

In contrast to *cytG* which preserves only peripheral LHCII (Guiamét *et al.*, 1991, 2002), *sid* preserves all the core and peripheral proteins of PSII during senescence. Thus, *cytG* appears to work differently from *sid*.

A non-yellowing mutation of *Phaseolus vulgaris* L. maintains chlorophyll a and b during senescence (Ronning *et al.*, 1991). Other phenotypes observed during leaf senescence in the mutant include lack of plastoglobuli in the senescing chloroplast and preservation of two bands among the soluble proteins examined by one-dimensional SDS-PAGE gels. However, the other soluble proteins appear to decline like those in wild type plants.

Clearly, the mutations that block chlorophyll breakdown are a very diverse group and many probably alter chlorophyll breakdown only indirectly. Some probably reflect deep regulatory factors. Many may also be placed in other categories below (Section VII).

C. Membrane and Lipid Breakdown

Membrane deterioration is one of the most characteristic features in plant senescence and is accompanied by progressive decline of phospholipid levels (Thompson *et al.*, 1997). Phospholipase D (PLD) plays an important role in membrane degradation by initiating the first reaction of the breakdown of phospholipids to produce phosphatidic acid. Fan *et al.* (1997) examined a role of PLD in plant senescence using transgenic plants with a reduced PLDα activity. Three distinct PLDs have been reported in *Arabidopsis*. When the level of PLDα was reduced, the plants showed delay of ABA- and ethylene-promoted senescence in detached leaves, as revealed by lower ion leakage, greater photosynthetic activity, and higher chlorophyll and phospholipid content in the transgenic leaves. Suppression of PLDα did not influence plant growth and development, including natural senescence. Thus, it was suggested that PLDα is a mediator in phytohormone-mediated senescence of detached leaves but is not involved in natural senescence. It is surprising that PLDα is not involved in natural senescence, since membrane degradation is a key feature of senescence. The results from the transgenic plants may well be expected from the expression pattern of the PLDα gene which is higher in meristems and newly divided cells than in senescing tissues. Perhaps related isozymes are involved in natural senescence. It is also expected that genes involved in phytohormone-induced senescence would also be involved in natural senescence, since phytohormones are integral regulators of plant senescence. To resolve these discrepancies, the authors suggested that PLDα itself does not initiate senescence. However, when premature senescence is induced by an increased level of phytohormones during stress conditions while a higher level of PLDα is still present, PLDα may accelerate senescence by degradation of phospholipids. It was also suggested that PLDα may be involved in signal perception or signal transduction cascades of the phytohormones and a reduced level of PLDα may delay senescence of detached leaves in the transgenic plants due to a defect in signaling of the phytohormones. In any event, the transgenic phenotype may be useful for postharvest storage and handling of vegetables or fruits.

A role of genes that are associated with lipid hydrolysis and metabolism was indicated by the study of the *Arabidopsis* acyl hydrolase gene. RNA interference of expression of this gene in transgenic plants delayed the onset of leaf senescence, whereas overexpression of the gene caused early senescence (He and Gan, 2002).

D. Other Genes Involved in Macromolecule Degradation

There are numerous reports of increased activities of enzymes involved in macromolecule degradation during senescence and these are described in other chapters. Here, we will describe only the genes known to be induced during senescence and involved in macromolecule degradation. Although it is expected that most of them are involved in a specific set of senescence execution processes, it would still be interesting to see their *in planta* functions in plant senescence.

Many lines of evidence show that protein degradation is an important component of the senescence syndrome even if it is not know how or if this kills cells. Several proteases increase during senescence (Chapters 7, 10 and 11) and several genes coding for proteases are up regulated during senescence (Chapter 4). These include genes associated with the ubiquitin pathway such as E2-type ubiquitin carrier protein genes (*UbcAt3*, *UbcAt4a*, and *UbcAt4b*) and *SEN3* in *Arabidopsis* (Genschik *et al.*, 1994; Park *et al.*, 1998). This suggests that ubiquitin-dependent degradation of proteins occurs during senescence.

Autophagy, an intracellular process for vacuolar bulk degradation of cytoplasmic components that would be needed for nutrient recycling, appears to contribute to senescence. In mutants carrying a T-DNA insertion within *Arabidopsis* autophagy genes (*AtAPG8* and *AtAPG9*), leaf senescence occurred more rapidly, implying that these proteins are required for maintaining cellular activities during leaf senescence (Doelling *et al.*, 2002; Hanaoka *et al.*, 2002).

Of course, other degradative enzymes such as RNase activity are also induced during senescence (Oh *et al.*, 1997); however, their genetic control is not well known.

IV. Genes Affecting Senescence through Action on the Hormonal Controls

A. Overview

Hormones often mediate signaling within a plant, e.g., correlative controls (Noodén and Leopold, 1978) including control of senescence (Noodén and Leopold, 1988). Most of the plant hormones are involved in various cellular processes including senescence. In fact, the effect of plant hormones, including cytokinins, ethylene, ABA, MeJA, and auxins, on senescence has been extensively studied during the past several decades for agronomic and commercial purposes. Although ethylene and cytokinins are known to have a major effect on senescence, other plant hormones also affect senescence. Thus, it is likely that the genes that alter production, perception, or signal cascades of these hormones will be included as genes that alter senescence. While it may seem that hormonal signals should be important in the initiation phase of senescence, they certainly can also act on the degradation and terminal phases.

It is also well known that reproductive development and pollination affect senescence at the whole plant or organ levels. However, the nature of the genes involved in control of senescence through these processes is mostly unknown.

B. Delay of Senescence by Cytokinin

Extensive research has identified cytokinins as the most effective senescence-retarding growth regulators. Cytokinins have been known to regulate various aspects of plant growth and development, including cell differentiation, release of lateral buds from apical dominance, and senescence. Much evidence suggests that cytokinins play an important role in senescence (see Chapter 6 for further details). Interestingly, in the transgenic plants expressing the *KNOTTED 1* (*KN1*) gene under the control of the *SAG12* promoter (*SAG*: senescence-associated gene), delayed senescence was accompanied by the increase of cytokinin content in older leaves. Thus, the effect of *KN1* on senescence may be mediated through changes in cytokinin levels (Ori *et al.*, 1999). In sorghum, a dominant gene produces a stay-green trait, apparently through promoting cytokinin formation by the roots (Ambler *et al.*, 1992; Walulu *et al.*, 1994).

A recent observation that a mutation in a putative cytokinin receptor of *Arabidopsis* delays senescence adds more intriguing support for the role of cytokinin in senescence (Kim, Lim, Woo, Hwang, and Nam, unpublished results).

C. Regulation of Senescence by Ethylene

Ethylene has long been known as an endogenous regulator of senescence, including fruit ripening and flower and leaf senescence. The involvement of ethylene in a senescence process is described in detail in Chapter 8. Many mutations acting on a wide range of steps in ethylene signaling also influence senescence, including leaf senescence, sometimes fairly selectively, sometimes as part of a pleiotropic spectrum. The importance of ethylene in controlling senescence is illustrated by *ore3*, which is an allele of *ein2* (*ethylene-insensitive2* mutation) (Oh *et al.*, 1997).

D. Change of Senescence by Brassinosteroid Biosynthesis

Besides promotion of cell elongation and division, brassinosteroids, as other plant hormones, also exhibit broad effects throughout plant development, including retardation of abscission, promotion of ethylene biosynthesis, and enhancement of stress resistance (Fujioka *et al.*, 1997). Thus, it is expected that the genes controlling biosynthesis or perception of brassinosteroids also affect senescence, although the effect of brassinosteroids on plant senescence has not been extensively examined. There is genetic evidence that brassinosteroids may be also involved in leaf senescence. The *det2* (*de-etiolated2*) mutation has a defect at an early step in brassinosteroid biosynthesis, and this was reported to confer delayed leaf senescence symptoms, i.e., delayed leaf yellowing (Chory *et al.*, 1991). It will be necessary to further examine the phenotype with other senescence markers to establish that the genes related to synthesis or perception of brassinosteroids alter senescence.

V. Genes that Alter Senescence in Response to Environmental Factors

One of the unique properties of plant growth and development is that it is highly influenced by environmental conditions. This may be an evolutionary strategy for a sessile organism to cope with environmental fluctuations. Senescence is an integrated response of plants to external environmental factors as well as to endogenous developmental signals. Thus, at least some genes that are involved in environmental response are expected to alter senescence.

A. Genes in Light Signaling

Light is perceived by a variety of photoreceptors and affects plant developmental processes throughout the life span, including germination, chloroplast development, seedling establishment, and flowering time (Quail, 1991; Furuya, 1993). Light control of senescence is covered in detail in Chapter 26. The light effects on senescence may be direct or indirect as with photoperiod effects on reproductive development (see Section VI below). Several genetic constructs also provide evidence that light can play a direct role (Jordan *et al.*, 1995; Thiele *et al.*, 1999).

B. Stress-Related Genes

Genes involved in stress responses may directly affect longevity and senescence through protection of cells and tissues from environmental stresses. Recently, expression profiles of 402 potential stress-related genes that encode known or putative transcription factors from *Arabidopsis* were monitored in various organs, at different developmental stages, and under various biotic and abiotic stresses (Chen *et al.*, 2002). Among the 43 transcription factor genes that have been reported to be induced during senescence, 28 of them also are induced by stress treatment, suggesting extensive overlap responses to these stresses. There are several reports on increasing plant tolerance to various environmental stresses through genetic manipulation, for example, expression of DREB1A from the stress-inducible rd29A promoter gives rise to drought, salt, and freezing tolerance (Kasuga *et al.*, 1999). There is no formal report on the role of stress response genes in plant senescence yet; however, it seems likely that enhanced stress resistance would also increase longevity. Preliminary evidence, albeit with oxidative stress, suggests that this is true (Kurepa *et al.*, 1998).

It is also known that many stress-related genes are induced during senescence. These include metallothionein-like genes from rice and *Arabidopsis* (Hensel *et al.*, 1993; Hsieh *et al.*, 1995) and genes involved in the oxidative stress response such as ascorbate oxidase (Callard *et al.*, 1996) and anionic peroxidase (Tournaire *et al.*, 1996). These genes may play a role in protecting the cellular functions required for progression and completion of senescence. It is likely that they affect progression of senescence at least partially. Interestingly, the *Arabidopsis* thaliana protein CEF makes yeast more tolerant to hydroperoxides, affirming the genetic control of stress resistance (Belles-Boix *et al.*, 2000).

VI. Genes Controlling Vegetative Growth (Regeneration) and Monocarpic Senescence

Monocarpic senescence is the degeneration leading to death of the whole plant at the end of the reproductive phase in monocarpic plants. As a prerequisite to or as a part of monocarpic senescence, vegetative growth ceases, so the plant can no longer regenerate its assimilatory structures. Eventually, the existing leaves and other parts senesce, and the plant dies (Noodén, 1988b). The mutations that make plants live longer do so mainly by extending their regenerative ability (Noodén and Penney, 2001). The shift to clonal growth is also an example of life-extending growth.

Whereas genes in the above categories are expressed mainly at the cell and organ level, this group is expressed more at the organ and whole plant levels. Given the important role of correlative controls in whole plant senescence, it is expected that factors that alter those controls likewise influence monocarpic senescence. In many monocarpic plants, the reproductive structures, especially the fruits, control monocarpic senescence and the longevity of the plants (Noodén, 1988b; Noodén and Penney, 2001). This enables the plant to maximize its reproductive output. The older genetics literature contains numerous examples of changes in environmental control of reproductive development with consequences for monocarpic senescence. For example, alteration of a single gene can remove the vernalization requirement in sugar beet and allow reproductive development to proceed without passing through a winter thereby changing the plant from a biennial to an annual (Whaley, 1965). Changing the photoperiod requirement for flowering and fruit development can cause similar alterations of senescence and longevity.

Other mutations seem to affect growth apical meristem activity more directly. One interesting case is the dominant gene combination *Sn Hr* which makes pea shoot apex senescence dependent on long days (Zhu and Davies, 1997). The result is cessation of vegetative growth and leaf production, locking the plant into monocarpic senescence (Noodén, 1988b; Chapter 15).

Sterility mutations may also delay leaf and whole plant senescence by preventing fruit development, especially in soybean (Noodén, 1988b). Interestingly, sterility mutants in *Arabidopsis* also prolong the life of a plant but not individual leaves (Hensel *et al.*, 1993; Noodén and Penney, 2001). In *Arabidopsis*, they act by prolonging growth and the production of new leaves which keeps the plant alive.

Other mutations such as the clavata types act more directly on the apical meristem to prolong leaf production and the life of the plant (Noodén and Penney, 2001). Thus, maintaining regenerative capacity is crucial to maintaining the life of the whole plant.

KN1 is an *Arabidopsis* gene encoding a homeobox-containing transcriptional factor and is preferentially expressed in the shoot meristem. Loss of function mutation in the *KN1* gene shows that the KN1 is essential for meristem maintenance and/or initiation. In addition, phenotypes of plants overexpressing KN1 suggested that the gene is required for maintaining cellular indeterminacy and preventing cellular differentiation (Hake *et al.*, 1995). The transgenic lines overexpressing *KN1* under the control of the *SAG12* promoter showed delayed leaf senescence. The authors suggested that KN1 is a natural negative regulator of senescence. Since the result was based on overexpression of KN1, it is important to check it with a loss of function mutation or with transgenic plants that specifically lower expression of *KN1* on senescence. The authors presented an interesting hypothesis that KN1 may act on senescence and meristems by similar mechanisms, blocking cellular differentiation or blocking developmental progression. This explanation may be in accordance with the general perception that senescence starts after maturation of an organ and KN1 may delay the maturation and thus delay senescence, whether through overproduction of cytokinin and/or through direct regulation of down stream genes involved in senescence.

In many animal tissues, the telomeres shorten as mitosis proceeds and eventually cause cell division to cease or even trigger cell death; however, this does not seem to occur in plant cells (Shippen and McKnight, 1998), which seem to be endowed with more open ended growth potential. Nonetheless, an additional protein was associated via protein–protein interaction with the telomere–protein complex during the onset of senescence (Zentgraf *et al.*, 2000).

VII. Regulatory Genes and Intracellular Signaling

In order to understand senescence fully, it will be necessary to understand the regulation of the genes that participate in senescence, and this picture is beginning to emerge. In addition, these regulatory elements will be very useful in genetic engineering applications. For example, the isolation of the promoter for *SAG*12 has been extremely useful (Chapter 6).

Given that senescence is an active process involving induction as well as down-regulation of gene expression (Section II above), it is particularly significant that transcription factors are implicated in the senescence process. The mRNAs from microarray experiments have indicated that there are at least 402 distinct transcription factor genes of which 43 are induced during senescence and the WRKY proteins seem particularly important (Chen *et al.*, 2002). WRKY proteins constitute a large family of plant-specific transcription factors in *Arabidopsis*, and all family members contain the WRKY domain, a 60-amino-acid domain with the conserved WRKYGQK motif at the N-terminal end, together with a novel zinc-finger motif (Eulgem *et al.*, 2000). Two WRKY genes, *AtWRKY6* and *WRKY53*, appear to be involved in the regulation of senescence. *WRKY53* is expressed very early in leaf senescence but decreases again at a late stage (Hinderhofer and Zentgraf, 2001). This indicates that WRKY53 might play a regulatory role in the early events of leaf senescence. Significantly, several external and internal signals involved in triggering senescence influence the expression of *AtWRKY6* (Robatzek and Somssich, 2001). AtWRKY6 may be a mediator for senescence, as its expression is not only strongly induced during leaf senescence, but is also associated with early to intermediate senescence stages.

A MADS domain-containing factor appears to relate to *Arabidopsis* flower senescence (Fang and Fernandez, 2002). A *KN1* (see above) may be involved in maintaining regenerative growth (Hake *et al.*, 1995).

The tomato *SENU5* is a senescence-up regulated gene (John *et al.*, 1997) and encodes a protein that belongs to the NAC domain family. The NAC domain family proteins include *Petunia* No Apical Meristem (NAM), *Arabidopsis* NAP (a target of the homeotic AP3/PI proteins), and GRAB (a protein interacting with Gemini virus RepA protein). Thus, this protein family appears to have regulatory roles in plant growth and differentiation (Xie *et al.*, 1999). It is conceivable that the senescence-up regulated *SENU5* gene belonging to this family may have a regulatory role in senescence.

Some senescence-related promoters have been found. For example, the mannopine synthase (mas) promoter is active in tobacco flower senescence but apparently not leaf senescence (Ursin and Shewmaker, 1993). The promoter for the senescence-related gene *SAG*12 encoding a cysteine protease in *Arabidopsis* has been isolated and characterized (Gan and Amasino, 1995; Noh and Amasino, 1999). In addition, the promoter for the gene *opr1*, a senescence-related gene that codes for the enzyme 12-oxo-phytodienoic acid-10,11-reductase, has been identified (He and Gan, 2001). Of particular interest is the report (Chen *et al.*, 2002) that the promoter regions for 23 genes that are induced during leaf senescence contain the WRKY binding site. Using enhancer-trap lines in *Arabidopsis*, 125 senescence-associated promoters have been tagged, and these have revealed that the activity of the senescence-associated promoters may differ among different tissues (He *et al.*, 2001).

Some information, albeit limited, is starting to emerge on the intracellular signaling pathway that induces senescence. A senescence-associated receptor-like kinase (*SARK*) gene is expressed exclusively during senescence in bean leaves, especially prior to the chlorophyll loss (Hajouj *et al.*, 2000). Light and cytokinin delay *SARK* gene expression as they do senescence, but darkness and ethylene advance the initial appearance of the gene. Thus, SARK may play a role in the regulation of leaf senescence.

Using *Arabidopsis* as a model system, Nam *et al.* have undertaken a systematic genetic screening to identify the regulatory genes of senescence. The *ore1*, *ore3*, and *ore9* mutations were initially isolated based on delayed yellowing of leaf during *in planta* senescence or upon dark-induced senescence of detached leaves (Oh *et al.*, 1997; Woo *et al.*, 2002). The *ore* stands for *oresara* which means "long-living" in Korean. In addition to extending

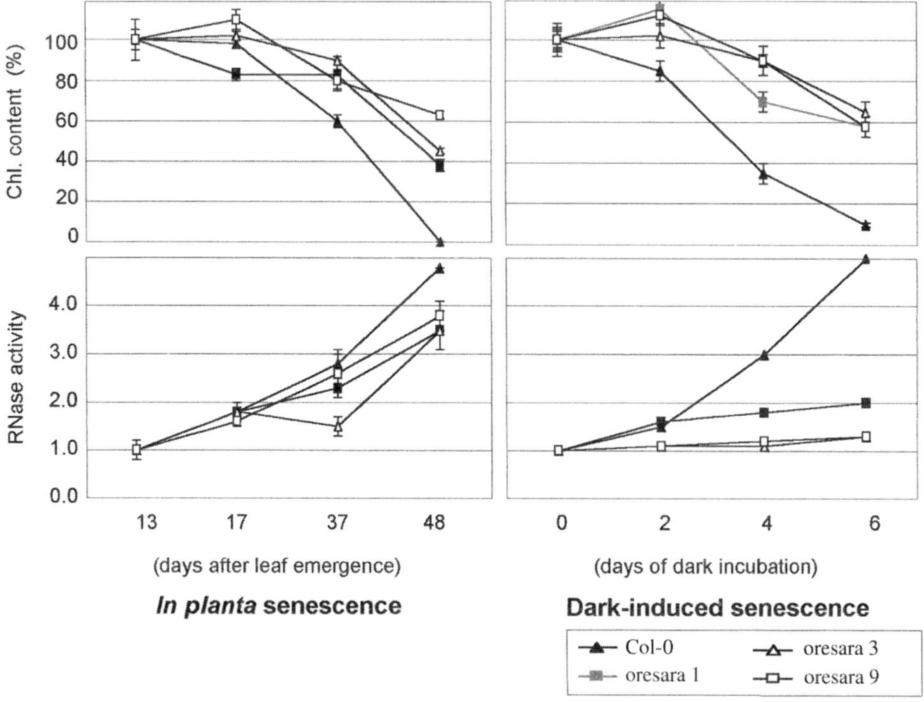

Figure 5-1. Delayed senescence symptoms of the *oresara* mutants during *in planta* and dark-induced senescence. Chlorophyll content (upper) as a marker of catabolic activity and RNase activity (lower) as an indicator of anabolic activity were examined at several developmental ages of leaves *in planta* (left) or at the given times after incubating detached leaves in darkness (right) using the third and fourth foliar leaves of wild type (Col-0), *oresara1*, *oresara3*, and *oresara9*. Data were obtained from 12 independent leaves. Shown are relative values as percentage of the initial point value. The vertical bars denote standard deviations (Oh *et al.*, 1997).

leaf longevity, the *ore* mutations delayed the decrease of anabolic activities, including chlorophyll content, the photochemical efficiency of the photosystem II, and the relative amount of the large subunit of ribulose-1,5-bisphosphate carboxylase/oxygenase (Rubisco) as well as delaying increase in catabolic activities such as RNase and peroxidase activities during *in planta* senescence and dark-induced senescence (Fig. 5-1). All of the mutations are inherited as monogenic recessive traits and comprise three complementation groups.

The existence of *ore1*, *ore3*, and *ore9* provides clear genetic evidence that leaf senescence in *Arabidopsis* is a genetically controlled event involving several monogenic genetic elements. Since these mutations affect a wide variety of senescence symptoms, these genes may be senescence regulatory genes. All of the mutations are recessive, suggesting that the genes defined by these mutations are positive regulators of senescence. The result also suggested that senescence could proceed through multiple pathways, since senescence is only delayed but not blocked in all of the mutants although there is a possibility that all of these mutations are weak alleles. Studies on expression of senescence-associated genes in the *ore* mutants will be needed to further define their role in controlling senescence. In addition, double mutant analysis will be required to elucidate the interaction among these genes in controlling senescence.

In addition to the delayed-senescence phenotype of the mutants in age and dark-induced senescence, these mutations delayed the responses to several phytohormones such as ethylene, abscisic acid (ABA), and methyl jasmonate (MeJA) (Woo, Park, and Nam, unpublished data). This suggests that ORE1, ORE3, and ORE9 may function at a common step of senescence affected by these factors. The *ORE1*, *ORE3*, and *ORE9* genes may be required for proper progression of leaf senescence induced by the phytohormones, as well as age and darkness (Fig. 5-1). In fact, *ore3*, also designated as *ein2-34*, is an allele of *ein2* (*ethylene-insensitive2*) mutations that are known to affect seedling responses to various hormones (see Section IVC). In contrast, the other mutants do not have a defect in general hormone responses but in hormone-induced senescence.

One of the most interesting glimpses of the regulatory cascade for senescence and findings to come is offered by *ore9* (Woo *et al.*, 2001). The *ore9* gene encodes a protein containing an F-box motif and 18 leucine-rich repeats. The F-box motif of ORE9 protein interacts with ASK1, which implicates ubiquitination in the senescence process. It was thus suggested that ORE9 functions to limit leaf longevity by removing, through ubiquitin-dependent proteolysis, target proteins that are required to delay the leaf senescence program in *Arabidopsis* (Woo *et al.*, 2001). This view is consistent with a report that proteolysis by the N-end rule pathway, one of the ubiquitin pathways, appears to be a mechanism involved in regulation of leaf senescence in *Arabidopsis*. The *delayed-leaf-senescence 1* (*dls1*) mutant showed a delay of leaf senescence symptoms (Yoshida *et al.*, 2002a). The mutant is due to a defect in arginyl tRNA:protein transferase (R-transferase), a component of the N-end rule proteolytic pathway that transfers arginine to the amino terminus of proteins with amino terminal glutamyl or aspartyl residues, thereby targeting the proteins for ubiquitin-dependent proteolysis. Like ORE9, DLS1 was suggested to have a role in senescence by degrading target proteins that negatively regulate leaf senescence.

Another interesting insight into the regulation of senescence comes from *ore4-1*. This mutation inhibits leaf senescence, yellowing, and death, and reduces the leaf growth rate (Woo *et al.*, 2002). Interestingly, the mutation delayed natural leaf senescence but not hormone or dark-induced senescence. The *ore4-1* mutant has a partial lesion in the chloroplast function including the function of photosystem I, which resulted from reduced expression of the plastid ribosomal protein small subunit 17 (*PRPS17*) gene. It is conceivable that the delayed leaf senescence phenotype observed in the *ore4-1* mutant is due to reduced metabolic rate, since the chloroplasts, a major energy source for plant growth through photosynthesis, are only partially functioning in the mutant. This interpretation is consistent with findings that metabolic rate is one of the key mechanisms involved in animal aging, although further evidence is needed.

The onset of leaf senescence was reported to be associated with a complex formation between ATBP1/ATBP2 and telomeric DNA (Zentgraf *et al.*, 2000). This opens up the possibility that telomeric structure could be involved in post-mitotic senescence in plants.

VIII. Conclusions

In this chapter, we have collected and reviewed the genes that alter senescence. The number of genes known to alter senescence seems limited currently. To some extent, this collection is limited by the procedures used to find the mutations. On the other hand, lack of our understanding of the senescence pathway and perhaps the general nature of senescence also might contribute. Despite these limitations, it is clear that senescence is now readily amenable to molecular and genetic analyses. It is expected that we soon should be able to obtain a deeper understanding of genes that alter senescence through the study of the genetic mutations and the putative regulatory SAGs.

The *oresara* series (Oh *et al.*, 1997; Woo *et al.*, 2002) and the promoter activity patterns found by He *et al.* (2001) suggest that the senescence process may be a braided channel rather than a single pathway. The braided channel idea very likely applies at the whole plant level (Noodén, 1988b). In this case, single mutations may not block the overall process and many of these may show only partial or weak effects on senescence. Furthermore, senescence is certainly regulated by various internal and external factors as mentioned above. Thus, it is expected that there will be many genes that alter senescence, because they are involved in integrating these factors into the senescence process. In order to identify senescence regulatory elements from the complex regulatory network, it is necessary to isolate a diverse range of novel and informative mutants. For this, more creative screening methods that consider the nature of senescence and the functional redundancy in pathways need to be utilized. It will also be important that mutants displaying subtle effects on senescence should not be ignored. Furthermore, it may be necessary to examine senescence phenotypes of double, triple, and higher order combination of mutants.

Many of the genes that alter senescence also appear to be involved in other biological processes and are not solely devoted to senescence, so distinguishing indirect and direct effects of genes on senescence can be difficult. For example, gene expression patterns show considerable overlap between senescence and other processes (He *et al.*, 2001; Robatzek and Somssich, 2001; Chen *et al.*, 2002). Many of the SAGs are also expressed at earlier stages (Buchanan-Wollaston, 1997). Thus, the number of genes specifically involved in senescence may be limited. The effect of these genes on senescence may be revealed by using senescence-specific promoters in transgenic plants, as seen in the case of the *KN1* gene (Hake *et al.*, 1995). A chemically inducible expression of a gene may also help for this purpose, as reported in the case of the acyl hydrolase gene (He and Gan, 2002). At the same time, genes that function only in senescence probably do exist. This is inferred from the presence of genes induced specifically during senescence and from identification of genetic mutations that affect senescence with a minimal effect on other developmental processes.

Genes that alter senescence could exert both positive and negative regulatory effects. Regulation of senescence may be viewed as a balancing between two antagonistic self-maintenance and senescence-promoting gene activities (Nam, 1997). In addition to up-regulation of the *SAGs*, down-regulation of the genes for self-maintenance may also be important for initiation and progression of senescence. These genes would also comprise the positive regulators of senescence. The senescence process is tightly regulated to prevent it from occurring prematurely. Thus, regulatory components to suppress senescence should also be important for controlling initiation and progression of senescence. We, therefore, envision that there will be genes that negatively regulate the senescence program. These will comprise the negative regulators of senescence. A potential negative regulator of senescence may be the HYPERSENESCENCE 1 (*HYS*1) gene, mutation of which shows an early senescence phenotype (Yoshida *et al.*, 2002b). This mutation was found as an allele of the CONSTITUTIVE *EXPRESSOR* OF PATHOGENESIS-RELATED GENES 5 (*CPR5*) gene, indicating an overlap between leaf senescence and pathogen responses. Some of the delayed senescence mutations identified from an activation tagging pool of *Arabidopsis* may also belong to the negative regulator of senescence (Lim and Nam, unpublished results).

Genes that alter senescence may be isolated through map-based cloning and from the T-DNA or transposon insertional mutant lines. A reverse genetic approach would also be possible. For example, *in planta* function of senescence-associated genes may be identified by isolating an insertional line in the gene of interest or by making transgenic plants that express the gene at a higher level or down regulate expression of the gene by an antisense or RNAi approach preferably using senescence-specific promoters. The candidate genes obviously include the genes for transcriptional factors and signal transduction components that are induced during senescence. It should also be possible to isolate transcription factors that regulate senescence-associated genes such as *SAG12* using a yeast one-hybrid system. Likewise, the yeast two-hybrid system should reveal the genes participating in the senescence signaling pathways through protein–protein interactions. Another useful approach to analyzing the senescence regulatory network would be the use of enhancer-trap lines in *Arabidopsis* (He *et al.*, 2001). Senescence may be controlled not just by differential regulation at the transcriptional level but also by regulating protein levels and activities. A proteomic approach may better show the patterns of gene expression at the protein level. Analysis of the gene expression changes associated with senescence using DNA microarray and proteomics technology will provide a more complete view of the genes that participate in senescence. We also suggest that analysis of senescence mutants with these technologies will give a clearer view on the role of the mutation in control of senescence.

Acknowledgments

The work by H.G.N. was supported by the National Research Laboratory Program of the Republic of Korea (M1-9911-00-0024). The work by P.O.L. was partially supported by the Korea Research Foundation (2001-050-D00031).

References

Ambler, J.R., Morgan, P.W., and Jordan, R.W. (1992). Amounts of zeatin and zeatin riboside in xylem sap of senescent and nonsenescent sorghum. *Crop Sci.* **32**, 411–419.

Belles-Boix, E., Babiychuk, E., Van Montagu, M., Inze, D., and Kushnir, S. (2000). CEF, a Sec24 homologue of *Arabidopsis thaliana*, enhances the survival of yeast under oxidative stress conditions. *J. Exp. Bot.* **51**, 1761–1762.

Brady, C.J. (1988). Nucleic acid and protein synthesis. In *Senescence and Aging in Plants* (L.D. Noodén and A.C. Leopold, Eds.), pp. 147–179. Academic Press, San Diego, CA.

Buchanan-Wollaston, V. (1997). The molecular biology of leaf senescence. *J. Exp. Bot.* **48**, 181–199.

Buchanan-Wollaston, V., and Ainsworth, C. (1997). Leaf senescence in *Brassica napus*: cloning of senescence related genes by subtractive hybridisation. *Plant Mol. Biol.* **33**, 821–834.

Callard, D., Axelos, M., and Mazzolini, L. (1996). Novel molecular markers for late phases of the growth cycle of *Arabidopsis thaliana* cell-suspension cultures are expressed during organ senescence. *Plant Physiol.* **112**, 705–715.

Chen, W., Provart, N.J., Glazebrook, J., Katagiri, F., Chang, H.S., Eulgem, T., Mauch, F., Luan, S., Zou, G., Whitham, S.A., Budworth, P.R., Tao, Y., Xie, Z., Chen, X., Lam, S., Kreps, J.A., Harper, J.F., Si-Ammour, A., Mauch-Mani, B., Heinlein, M., Kobayashi, K., Hohn, T., Dangl, J.L., Wang, X., and Zhu, T. (2002). Expression profile matrix of *Arabidopsis* transcription factor genes suggests their putative functions in response to environmental stresses. *Plant Cell* **14**, 559–574.

Chory, J., Nagpal, P., and Peto, C.A. (1991). Phenotypic and genetic analysis of *det2*, a new mutant that affects light-regulated seedling development in *Arabidopsis*. *Plant Cell* **3**, 445–459.

Doelling, J.H., Walker, J.M., Friedman, E.M., Thompson, A.R., and Vierstra, R.D. (2002). The APG8/12-activating enzyme APG7 is required for proper nutrient recycling and senescence in *Arabidopsis thaliana*. *J Biol. Chem.* **277**, 33105–33114.

Eulgem, T., Rushton, P.J., Robatzek, S., and Somssich, I.E. (2000). The WRKY superfamily of plant transcription factors. *Trends Plant Sci.* **5**, 199–206.

Fan, L., Zheng, S., and Wang, X. (1997). Antisense suppression of phospholipase Dα retards abscisic acid- and ethylene-promoted senescence of postharvest *Arabidopsis* leaves. *Plant Cell* **9**, 2183–2196.

Fang, S.C., and Fernandez, D.E. (2002). Effect of regulated overexpression of the MADS domain factor AGL15 on flower senescence and fruit maturation. *Plant Physiol.* **130**, 78–89.

Fujioka, S., Li, J., Choi, Y.H., Seto, H., Takatsuto, S., Noguchi, T., Watanabe, T., Kuriyama, H., Yokota, T., Chory, J., and Sakurai, A. (1997). The *Arabidopsis deetiolated2* mutant is blocked early in brassinosteroid biosynthesis. *Plant Cell* **9**, 1951–1962.

Furuya, M. (1993). Phytochromes: Their molecular species, gene families, and functions. *Annual Rev. Plant Physiol. Plant Mol. Biol.* **44**, 617–645.

Gan, S., and Amasino, R.M. (1995). Inhibition of leaf senescence by autoregulated production of cytokinin. *Science* **270**, 1986–1988.

Genschik, P., Durr, A., and Fleck, J. (1994). Differential expression of several E2-type ubiquitin carrier protein genes at different developmental stages in *Arabidopsis thaliana* and *Nicotiana sylvestris*. *Mol. Gen. Genet.* **244**, 548–556.

Gentinetta, E., Ceppi, D., Lepori, C., Perico, G., Motto, M., and Salamini, F. (1986). A major gene for delayed senescence in maize. Pattern of photosynthates accumulation and inheritance. *Plant Breeding* **97**, 193–203.

Grbić, V., and Bleecker, A.B. (1995). Ethylene regulates the timing of leaf senescence in *Arabidopsis*. *Plant J.* **8**, 595–602.

Guiamét, J.J., and Giannibelli, M.C. (1994). Inhibition of the degradation of chloroplast membranes during senescence in nuclear 'stay green' mutants in soybean. *Physiol. Plant.* **91**, 395–402.

Guiamét, J.J., and Giannibelli, M.C. (1996). Nuclear and cytoplasmic "stay-green" mutations of soybean alter the loss of leaf soluble proteins during senescence. *Physiol. Plant.* **96**, 655–661.

Guiamét, J.J., Teeri, J.A., and Noodén, L.D. (1990). Effects of nuclear and cytoplasmic genes altering chlorophyll loss on gas exchange during monocarpic senescence in soybean. *Plant Cell Physiol.* **31**, 1123–1130.

Guiamét, J.J., Schwartz, E., Pichersky, E., and Noodén, L.D. (1991). Characterization of cytoplasmic and nuclear mutations affecting chlorophyll and chlorophyll-binding proteins during senescence in soybean. *Plant Physiol.* **96**, 227–231.

Guiamét, J.J., Tyystjärvi, E., Tyystjärvi, T., John, I., Kairavuo, M., Pichersky, E., and Noodén, L.D. (2002). Photoinhibition and loss of photosystem II reaction centre proteins during senescence of soybean leaves. Enhancement of photoinhibition by the 'stay-green' mutation *cytG*. *Physiol. Plant.* **115**, 468–478.

Hajouj, T., Michelis, R., and Gepstein, S. (2000). Cloning and characterization of a receptor-like protein kinase gene associated with senescence. *Plant Physiol.* **124**, 1305–1314.

Hake, S., Char, B.R., Chuck, G., Foster, T., Long, J., and Jackson, D. (1995). Homeobox genes in the functioning of plant meristems. *Philos. Trans. R. Soc. Lond. B. Biol. Sci.* **350**, 45–51.

Hanaoka, H., Noda, T., Shirano, Y., Kato, T., Hayashi, H., Shibata, D., Tabata, S., and Ohsumi, Y. (2002). Leaf senescence and starvation-induced chlorosis are accelerated by the disruption of an *Arabidopsis* autophagy gene. *Plant Physiol.* **129**, 1181–1193.

He, Y.H., and Gan, S.S. (2001). Identical promoter elements are involved in regulation of the Opr1 gene by senescence and jasmonic acid in *Arabidopsis*. *Plant Mol. Biol.* **47**, 595–605.

He, Y.H., and Gan, S.S. (2002). A gene encoding an acyl hydrolase is involved in leaf senescence in *Arabidopsis*. *Plant Cell* **14**, 805–815.

He, Y.H., Tang, W.N., Swain, J.D., Green, A.L., Jack, T.P., and Gan, S.S. (2001). Networking senescence-regulating pathways by using *Arabidopsis* enhancer trap lines. *Plant Physiol.* **126**, 707–716.

Hensel, L.L., Grbić, V., Baumgarten, D.A., and Bleecker, A.B. (1993). Developmental and age-related processes that influence the longevity and senescence of photosynthetic tissues in *Arabidopsis*. *Plant Cell* **5**, 553–564.

Hinderhofer, K., and Zentgraf, U. (2001). Identification of a transcription factor specifically expressed at the onset of leaf senescence. *Planta* **213**, 469–473.

Hsieh, H.-M., Liu, W.-K., and Huang, P.C. (1995). A novel stress-inducible metallothionein-like gene from rice. *Plant Mol. Biol.* **28**, 381–389.

John, I., Hackett, R., Cooper, W., Drake, R., Farrell, A., and Grierson, D. (1997). Cloning and characterization of tomato leaf senescence-related cDNAs. *Plant Mol. Biol.* **33**, 641–651.

Jordan, E.T., Hatfield, P.M., Hondred, D., Talon, M., Zeevaart, J.A.D., and Vierstra, R.D. (1995). Phytochrome A overexpression in transgenic tobacco: correlation of dwarf phenotype with high concentrations of phytochrome in vascular tissue and attenuated gibberellin levels. *Plant Physiol.* **107**, 797–805.

Kasuga, M., Liu, Q., Miura, S., Yamaguchi-Shinozaki, K., and Shinozaki, K. (1999). Improving plant drought, salt, and freezing tolerance by gene transfer of a single stress-inducible transcription factor. *Nat. Biotechnol.* **17**, 287–291.

Kurepa, J., Smalle, J., Van Montagu, M., and Inze, D. (1998). Oxidative stress tolerance and longevity in *Arabidopsis*: The late-flowering mutant gigantea is tolerant to paraquat. *Plant J.* **14**, 759–764.

Lohman, K.N., Gan, S., John, M.C., and Amasino, R.M. (1994). Molecular analysis of natural leaf senescence in *Arabidopsis thaliana. Physiol. Plant.* **92**, 322–328.

Nam, H.G. (1997). Molecular genetic analysis of leaf senescence. *Curr. Opin. Biotech.* **8**, 200–207.

Napp-Zinn, K. (1985). *Arabidopsis* thaliana. In *CRC Handbook of Flowering* (A.H. Halevy, Ed.), pp. 492–503. CRC Press, Boca Raton, FL.

Noh,Y.-S., and Amasino, R.M. (1999). Identification of a promoter region responsible for the senescence-specific expression of SAG12. *Plant Mol. Biol.* **41**, 181–194.

Noodén, L.D. (1988a). The phenomena of senescence and aging. In *Senescence and Aging in Plants* (L.D. Noodén, and A.C. Leopold, Eds.), pp. 1–50. Academic Press, San Diego, CA.

Noodén, L.D. (1988b). Whole plant senescence. In *Senescence and Aging in Plants* (L.D. Noodén and A.C. Leopold, Eds.), pp. 391–439. Academic Press, San Diego, CA.

Noodén, L.D. (1988c). Postlude and prospects. In *Senescence and Aging in Plants* (L.D. Noodén and A.C. Leopold, Eds.), pp. 499–517. Academic Press, San Diego, CA.

Noodén, L.D., and Guiamét, J.J. (1996). Genetic control of senescence and aging in plants. In *Handbook of the Biology of Aging* (E.L. Schneider, and J.W. Rowe, Eds.), pp. 94–118. Academic Press, San Diego, CA.

Noodén, L.D., and Leopold, A.C. (1978). Phytohormones and the endogenous regulation of senescence and abscission. In *Phytohormones and Related Compounds*. Vol. II (D.S. Letham, T.J. Higgins and P.B. Goodwin, Eds.). Elsevier, Amsterdam.

Noodén, L.D., and Leopold, A.C. (1988). *Senescence and Aging in Plants*. Academic Press, San Diego, CA.

Noodén, L.D., and Penney, J.P. (2001). Correlative controls of senescence and plant death in *Arabidopsis thaliana* (*Brassicaceae*). *J. Exp. Bot.* **52**, 2151–2159.

Oh, S.A., Park, J.-H., Lee, G.I., Paek, K.H., Park, S.K., and Nam, H.G. (1997). Identification of three genetic loci controlling leaf senescence in *Arabidopsis thaliana. Plant J.* **12**, 527–535.

Ori, N., Juarez, M.T., Jackson, D., Yamaguchi, J., Banowetz, G.M., and Hake, S. (1999). Leaf senescence is delayed in tobacco plants expressing the maize homeobox gene knotted1 under the control of a senescence-activated promoter. *Plant Cell* **11**, 1073–1080.

Park, J.-H., Oh, S.A., Kim, Y.H., Woo, H.R., and Nam, H.G. (1998). Differential expression of senescence-associated mRNAs during leaf senescence induced by different senescence-inducing factors in *Arabidopsis. Plant Mol. Biol.* **37**, 445–454.

Paul, M.J., and Foyer, C.H. (2001). Sink regulation of photosynthesis. *J. Exp. Bot.* **52**, 1383–1400.

Quail, P.H. (1991). Phytochrome: A light-activated molecular switch that regulates plant gene expression. *Annu. Rev. Genet.* **25**, 389–409.

Robatzek, S., and Somssich, I.E. (2001). A new member of the *Arabidopsis* WRKY transcription factor family, AtWRKY6, is associated with both senescence- and defence-related processes. *Plant J.* **28**, 123–133.

Ronning, C.M., Bouwkamp, J.C., and Solomos, T. (1991). Observations on the senescence of a mutant non-yellowing genotype of *Phaseolus vulgaris* L. *J. Exp. Bot.* **42**, 235–241.

Shippen, D.E., and McKnight, T.D. (1998). Telomeres, telomerase and plant development. *Trends Plant Sci.* **3**, 126–130.

Smart, C.M. (1994). Gene expression during leaf senescence. *New Phytol.* **126**, 419–448.

Thiele, A., Herold, M., Lenk, I., Quail, P.H., and Gatz, C. (1999). Heterologous expression of Arabidopsis phytochrome B in transgenic potato influences photosynthetic performance and tuber development. *Plant Physiol.* **120**, 73–81.

Thomas, H., and Stoddart, J.L. (1993). Crops that stay green. *Ann. App. Biol.* **123**, 193–219.

Thompson, J.E., Froese, C.D., Hong, Y., Hudak, K.A., and Smith, M.D. (1997). Membrane deterioration during senescence. *Can. J. Bot.* **75**, 867–879.

Tournaire, C., Kushnir, S., Bauw, G., Inze, D., de la Serve, B.T., and Renaudin, J.-P. (1996). A thiol protease and an anionic peroxidase are induced by lowering cytokinins during callus growth in *Petunia. Plant Physiol.* **111**, 159–168.

Ursin, V.M., and Shewmaker, C.K. (1993). Demonstration of a senescence component in the regulation of the mannopine synthase promoter. *Plant Physiol.* **102**, 1033–1036.

Vicentini, F., Hörtensteiner, S., Schellenberg, M., Thomas, H., and Matile, P. (1995). Chlorophyll breakdown in senescent leaves: identification of the biochemical lesion in a stay-green genotype of Festuca pratensis Huds. *New Phytol.* **129**, 247–252.

Von Wettstein, D. (1959). In *The Photochemical Apparatus, Its Structure and Function* (Brookhaven National Laboratory, Ed.), pp. 138–159. Biology Dept, Brookhaven National Laboratory, Upton, NY.

Walulu, R.S., Rosenow, D.T., Wester, D.B., and Nguyen, H.T. (1994). Inheritance of the stay green trait in sorghum. *Crop Sci.* **34**, 970–972.

Whaley, W.G. (1965). The interaction of the genotype and environment in plant development. In *Encyclopedia of Plant Physiology* (A. Lang, Ed.), pp. 74–99. Springer-Verlag, Berlin.

Woo, H.R., Chung, K.M., Park, J.H., Oh, S.A., Ahn, T., Hong, S.H., Jang, S.K., and Nam, H.G. (2001). ORE9, an F-box protein that regulates leaf senescence in *Arabidopsis*. *Plant Cell* **13**, 1779–1790.

Woo, H.R., Goh, C.H., Park, J.H., Teyssendier, B., Kim, J.H., Park, Y.I., and Nam, H.G. (2002). Extended leaf longevity in the *ore4-1* mutant of *Arabidopsis* with a reduced expression of a plastid ribosomal protein gene. *Plant J.* **31**, 331–340.

Xie, Q., Sanz-Burgos, A.P., Guo, H., Garcia, J.A., and Gutierrez, C. (1999). GRAB proteins, novel members of the NAC domain family, isolated by their interaction with a geminivirus protein. *Plant Mol. Biol.* **39**, 647–656.

Yoshida, S., Ito, M., Callis, J., Nishida, I., and Watanabe, A. (2002a). A delayed leaf senescence mutant is defective in arginyl-tRNA:protein arginyltransferase, a component of the N-end rule pathway in *Arabidopsis*. *Plant J.* **32**, 129–137.

Yoshida, S., Ito, M., Nishida, I., and Watanabe, A. (2002b). Identification of a novel gene HYS1/CPR5 that has a repressive role in the induction of leaf senescence and pathogen-defence responses in *Arabidopsis thaliana*. *Plant J.* **29**, 427–437.

Zentgraf, U., Hinderhofer, K., and Kolb, D. (2000). Specific association of a small protein with the telomeric DNA-protein complex during the onset of leaf senescence in *Arabidopsis thaliana*. *Plant Mol. Biol.* **42**, 429–438.

Zhu, Y.X., and Davies, P.J. (1997). The control of apical bud growth and senescence by auxin and gibberellin in genetic lines of peas. *Plant Physiol.* **113**, 631–637.

6

Senescence and Genetic Engineering

Yoo-Sun Noh, Betania F. Quirino
and Richard M. Amasino

I. Introduction

Senescence is a natural stage in the development of plants. Leaf senescence yields nutrients to younger, growing parts of the plant or to storage sites such as seeds or roots (Leopold, 1961). In agriculture, however, senescence is not always beneficial. Leaf senescence is undesirable in vegetable crops in which the leaf is the part consumed. Moreover, if leaves are kept photosynthetically active for a longer period of time, this may have a positive influence on seed yield. For producers of fresh cut flowers, petal senescence reduces the commercial value of the crop.

There are several approaches to controlling senescence. One is to establish better postharvest practices. For example, senescence of apples can be delayed by storage in a controlled atmosphere with low O_2 and high CO_2 (Abeles *et al.*, 1992). Inhibitors of ethylene can be applied to flowers to prolong their vase-life (Borochov and Woodson, 1989). Another approach is to select for plants that have traits of interest in breeding programs. Genetic engineering can provide a complementary approach. The understanding of the molecular mechanisms of senescence combined with the tools of molecular biology are beginning to provide the tools to manipulate plant senescence.

It is well known that the phytohormones cytokinin and ethylene strongly affect senescence in a variety of plant species. Therefore, it is not surprising that manipulation of their synthesis or signal transduction pathways provides opportunities to alter senescence. Here we review recent progress in manipulating plant senescence through genetic engineering.

Plant Cell Death Processes

II. The Relationship of Cytokinins and Senescence

A. Manipulation of Cytokinin Levels

Cytokinins were originally discovered as factors promoting plant cell division (Miller *et al.*, 1956). Since this discovery, cytokinins have been shown to have effects on many different aspects of plant development. For example, cytokinin suppresses apical dominance and promotes the growth of lateral buds (Phillips, 1975) and viviparous leaves (Estruch *et al.*, 1991), chloroplast development (Parthier, 1979), leaf expansion (Miller *et al.*, 1956), and delays tissue senescence (Richmond and Lang, 1957; Van Staden *et al.*, 1988). The extensive literature on the role of cytokinins has been reviewed elsewhere (e.g., Binns, 1994; Mok and Mok, 1994), and this section will focus on the role of cytokinins in senescence and the genetic engineering efforts to delay senescence by modulating endogenous cytokinin levels.

The first evidence that cytokinin can delay senescence was from the study of exogenously applied cytokinin (Richmond and Lang, 1957). Subsequently an extensive literature has developed describing general antagonistic effects of applied cytokinin on senescence (reviewed in Van Staden *et al.*, 1988). The discovery of an inverse correlation between cytokinin levels and the progression of senescence (e.g., Noodén and Letham, 1993; Ambler *et al.*, 1992; Soejima *et al.*, 1995) provided further evidence that endogenously produced cytokinin has a negative effect on plant senescence. Relatively recent approaches using transgenic plants expressing the gene encoding isopentenyl transferase (IPT) from the T-DNA of *Agrobacterium tumefaciens* (Akiyoshi *et al.*, 1983; Barry *et al.*, 1984) have provided a tool to manipulate endogenous cytokinin levels and to study the role of cytokinin in plant development. The transgenic plant approach has the advantage of precisely manipulating both spatial and temporal levels of cytokinin (Klee, 1994). One of the consistent results obtained from studies using transgenic plants with altered cytokinin biosynthesis is that senescence is delayed in plant tissues in which cytokinin levels are elevated.

A variety of expression systems have been used to modulate endogenous cytokinin levels. Generally, plant tissues transformed with *IPT* under the control of constitutive promoters spontaneously produce shoots, but the transformed shoots cannot form roots as expected for cytokinin overproducing cells (Schmülling *et al.*, 1989; Binns, 1994). Therefore, when plants containing constitutively expressed *IPT* are regenerated, there has probably been selection for weak expression, particularly in roots (Gan and Amasino, 1996). Many different approaches have been undertaken to avoid this problem.

One approach is based upon inducible promoter systems to provide for conditional *IPT* expression. Heatshock promoters have often been used. Smart *et al.* (1991) reported that transgenic tobacco plants transformed with *IPT* under the control of a soybean heatshock promoter showed phenotypes associated with higher cytokinin levels such as shorter stature, lateral shoot growth, and delayed leaf senescence even without heatshocks. These phenotypes were more profound after several heatshocks of whole plants or defined areas of leaves. However, Medford *et al.* (1989) did not observe a phenotypic difference before and after heatshocks in transgenic tobacco and Arabidopsis plants transformed with *IPT* under the control of a maize heatshock promoter. With or without heatshock, the transgenic plants exhibited reduced apical dominance and a delay in leaf senescence. As expected from the phenotypic change in non-heatshocked plants, cytokinin levels in transgenic plants were higher than control plants even in the absence of heatshock, but after heatshock cytokinin levels were further elevated. Other studies using heat-inducible promoters also revealed leaky expression of *IPT* without heatshock (Schmülling *et al.*, 1989; Smigocki, 1991; Ainley *et al.*, 1993), and the low level of IPT activity driven by this leaky expression is usually enough to cause cytokinin overproduction phenotypes. The elevated cytokinin levels caused by the leaky expression may be above a threshold such that additional cytokinin produced by heatshock does not result in stronger phenotypes (Binns, 1994). A concern with heat-inducible promoter systems is that heatshock could itself affect plant responses. Other inducible promoters responding to wounding (Smigocki *et al.*, 1993) and light (Beinsberger *et al.*, 1992) have also been used to direct *IPT* expression. More recently, tetracycline- (Faiss *et al.*, 1997) and copper-inducible (McKenzie *et al.*, 1998) systems have been used to direct *IPT* expression in transgenic tobacco plants. Generally, these systems have tighter control compared to heat-inducible systems, although delayed senescence was observed from a transgenic tobacco line harboring the tetracycline-inducible system even without tetracycline treatment. In the case of transgenic tobacco plants harboring the copper-inducible system, a copper-treatment-dependent delay in leaf senescence was observed. Perhaps the most significant limitation of the inducible systems is that the delayed tissue senescence obtained from transgenic plants expressing *IPT* under the control of inducible promoters is always associated with other abnormal phenotypes resulting from cytokinin overproduction.

A second approach is based on tissue- or developmental-specific promoter systems. In theory, these systems could result in more specific overproduction of cytokinin and minimized abnormal phenotypes in undesired tissues or developmental stages. When an auxin-inducible *SAUR* promoter was used to drive *IPT* expression, numerous phenotypes commonly associated with cytokinin overproduction were observed in transgenic tobacco plants (Li *et al.*, 1992). Delayed leaf senescence was observed when these plants were grown in a growth chamber, while accelerated leaf senescence was observed when the same plants were grown in tissue-culture vessels. Different growth conditions might have resulted in different translocation patterns of cytokinins. The *SAUR* promoter is most active in elongating regions of transgenic tobacco hypocotyls and stems. Active transpiration in growth chamber-grown plants might have resulted in sufficient translocation of cytokinins to leaves resulting in delayed senescence, while lack of transpiration in culture-grown plants might account for the different phenotype (Li *et al.*, 1992). Martineau *et al.* (1994) used a fruit-specific promoter to drive the expression of *IPT* in tomato, and observed an altered fruit phenotype: islands of green pericarp tissue remaining on otherwise red ripe fruit. Cytokinin levels in these transgenic tomato fruit were 10- to 100-fold higher than in control fruit. A four-fold increase in cytokinin levels in the leaves of transgenic tomatoes was also observed despite a lack of detectable *IPT* mRNA accumulation in this tissue.

To specifically provide cytokinin to senescing leaves, a senescence-delaying system has been developed. When the expression of *IPT* was controlled by the promoter of an Arabidopsis senescence-specific gene, *SAG12*, transgenic tobacco plants containing this expression cassette exhibited a strong delay in leaf senescence (Gan and Amasino, 1995; Fig. 6-1). This senescence-delaying system is autoregulatory: at the onset of senescence the senescence-specific promoter is activated resulting in cytokinin production, and the increase in cytokinin will in turn inhibit senescence and attenuate the promoter activity. This system has three important features: spatially, cytokinin production is targeted to tissues where the promoter is active, i.e. to leaves and floral parts; temporally, cytokinin production is controlled developmentally, i.e. cytokinins are produced only during senescence; and, quantitatively, cytokinins are maintained at a minimum level necessary for the inhibition of senescence owing to autoregulation (Gan and Amasino, 1996). Transgenic tobacco plants

A

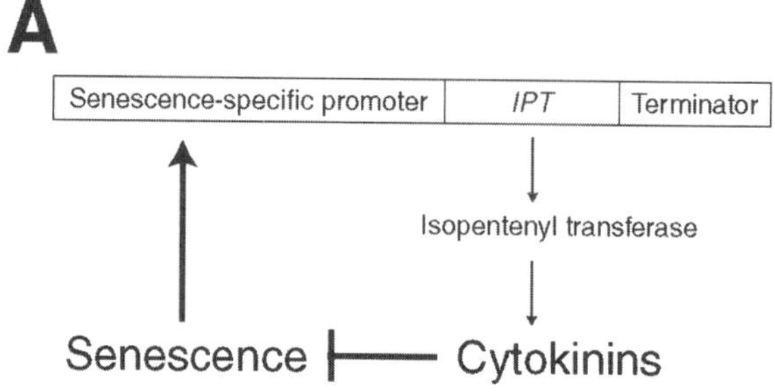

B

Figure 6-1. Autoregulated senescence-inhibition system. (A) Schematic of the system. Senescence is inhibited through the autoregulation between the senescence-specific production of cytokinins and the suppression of the senescence-specific promoter by the produced cytokinins. (B) Delayed leaf senescence in a transgenic tobacco plant containing the autoregulated senescence-inhibition system. Left, transgenic plant; right, wild-type. (Modified from Gan and Amasino, 1996.)

containing this expression system are developmentally and morphologically normal except for the significant delay in leaf senescence. These transgenic plants also show an increase in seed yield and biomass, probably because of the prolonged photosynthetic activity of senescence-retarded leaves. Because the cytokinin production is controlled genetically and developmentally in this system, no special treatments or growth conditions are required to obtain delayed senescence and increased yield.

It is notable that overexpression of certain genes other than *IPT* also causes increased cytokinin levels. Expression of a gene encoding a rice small GTP-binding protein in tobacco caused six-fold higher cytokinin levels and phenotypes associated with cytokinin overproduction (Kamada *et al.*, 1992; Sano *et al.*, 1994). Higher cytokinin levels and associated phenotypes were also observed in transgenic tobacco plants overexpressing either a tobacco gene, *NTH15* (Tamaoki *et al.*, 1997), or a rice gene, *OSH1* (Kusaba *et al.*, 1998). Proteins encoded by these genes belong to the Knotted class of homeodomain proteins. Thus this class of homeodomain proteins may regulate cytokinin synthesis. Moreover, when the maize *Knotted1* gene was expressed under the control of a senescence-specific promoter in tobacco, the resulting transgenic plants showed increased cytokinin levels in senescing leaves and delayed leaf senescence (Ori *et al.*, 1999). Therefore, several reports demonstrate that genetic manipulation of endogenous cytokinin levels can be achieved by the altered expression of certain plant genes other than *IPT*. Transgenic plant systems that have resulted in altered endogenous cytokinin levels are summarized in Table 6-1.

B. Cytokinin Signaling and Mechanism of Action

The diverse effects of cytokinin on plant development suggest that the expression of a variety of genes are affected by this hormone, and indeed many classes of genes including those involved in plant defense responses, photosynthesis, and senescence have been shown to be regulated by cytokinin. Cytokinin signaling pathways leading to this gene expression are not yet fully understood, although significant progress has been made recently.

Early studies on cytokinin regulation of gene expression were performed in cytokinin-dependent soybean cell suspension cultures. Polyribosome formation and new protein synthesis were observed after cytokinin treatment (Fosket and Tepfer, 1978) and 20 mRNAs, including some of ribosomal proteins, that exhibit rapid accumulation following cytokinin treatment were identified (Crowell *et al.*, 1990). These studies indicated that translation and transcription of many genes are affected by cytokinin treatment. However, many of the genes induced by cytokinin were also induced by auxin, indicating that their induction may be specific to growth stimulation rather than to a particular hormone (Binns, 1994).

Many genes thought to be involved in defense responses show altered expression in response to cytokinin. *β-1,3-Glucanase*, *chitinase*, and some *pathogenesis-related* (PR) genes exhibit decreased expression in response to increased cytokinin (Eichholz *et al.*, 1983; Mohnen *et al.*, 1985; Felix and Meins, 1986; Shinshi *et al.*, 1987). In other reports, *chitinase* and PR genes exhibited increased expression in cells with increased cytokinin levels (Memelink *et al.*, 1987; Simmons *et al.*, 1992; Martineau *et al.*, 1994). Sano *et al.* (1994) demonstrated that transgenic tobacco plants expressing *rgp1*, a gene encoding a Ras-related small GTP-binding protein, exhibit increased cytokinin levels, high salicylic acid levels, increased acidic PR gene expression, and a high level of resistance against tobacco mosaic virus infection.

Table 6-1. [a]Transgenic Plant Systems that have Resulted in Altered Endogenous Cytokinin Levels

Systems	Source of promoter	Transgenic species	References
IPT systems			
IPT native promoter	From pTi15955	*Nicotiana tabacum*	Schmülling *et al.*, 1989
		Solanum tuberosum	Ooms *et al.*, 1991
	From pTiC58	*N. tabacum*	Yushibov *et al.*, 1991; Beinsberger *et al.*, 1992
	From pTi15955	*N. tabacum*	Zhang *et al.*, 1995
		Lycopersicon esculentum	Groot *et al.*, 1995
Constitutive expression	CaMV 35S	*N. tabacum*	Smigocki and Owens, 1988
		N. plumbaginifolia	
		Cucumis sativa	
Transposition	CaMV 35S promoter disrupted by Ac transposon	*N. tabacum*	Estruch *et al.*, 1991, 1993
Random insertion	Unknown	*N. tabacum*	Hewelt *et al.*, 1994
Inducible			
Heat	Maize hsp70	*N. tabacum*	Medford *et al.*, 1989
		Arabidopsis thaliana	
	Drosophila hsp70	*N. tabacum*	Schmülling *et al.*, 1989
		N. plumbaginifolia	Smigocki, 1991
	Soybean HS6871	*N. tabacum*	Smart *et al.*, 1991
	Soybean Gmhsp17.5-E	*N. tabacum*	Ainley *et al.*, 1993
Wounding	Potato proteinase inhibitor II-K gene	*N. plumbaginifolia*	Smigocki *et al.*, 1993
Light	Pea rubisco small subunit gene	*N. tabacum*	Beinsberger *et al.*, 1992
Tetracycline	Tetracycline-dependent CaMV 35S	*N. tabacum*	Faiss *et al.*, 1997
Copper	Yeast copper-metallothionein regulatory system	*N. tabacum*	McKenzie *et al.*, 1998
Tissue-specific			
Elongating region	Soybean SAUR gene	*N. tabacum*	Li *et al.*, 1992
Fruit-specific	Tomato 2A11 gene	*L. esculentum*	Martineau *et al.*, 1994
Ovary-preferential	Tomato unknown gene	*L. esculentum*	Martineau *et al.*, 1994
Development-specific			
Senescence-specific	Arabidopsis *SAG12*	*N. tabacum*	Gan and Amasino, 1995
Non-IPT systems			
Homeobox protein			
NTH15	CaMV 35S	*N. tabacum*	Tamaoki *et al.*, 1997
OSH1	CaMV 35S	*N. tabacum*	Kusaba *et al.*, 1998
	NOS	*N. tabacum*	
Knotted1	Arabidopsis *SAG12*	*N. tabacum*	Ori *et al.*, 1999
Small GTP binding protein	CaMV 35S	*N. tabacum*	Sano *et al.*, 1994

[a]Updated from Gan and Amasino (1996).

Because photosynthetic activity declines during plant senescence (Gepstein, 1988; Hensel *et al.*, 1993; Jiang *et al.*, 1993), the effect of cytokinin on photosynthesis is of particular interest. Consistent with the function of cytokinin in chloroplast development, genes involved in photosynthesis are up regulated by cytokinin treatment (reviewed in Crowell and Amasino, 1994). Cytokinin is known to induce the expression of *chlorophyll a/b binding protein* (*CAB*) and *ribulose-1,5-bisphosphate* small and large subunit mRNAs in many different plant species, while the decrease of *CAB* mRNA and chlorophyll content in senescing Arabidopsis leaves is delayed by cytokinin treatment (Weaver *et al.*, 1998). Light-dependent activation of a wheat protein kinase gene, *wpk4*, was found to be dependent on cytokinin (Sano and Youssefian, 1994), suggesting a role of cytokinin in mediating light-dependent activation of regulatory genes.

During the past several years, a number genes showing increased or specific expression during senescence (senescence-associated genes or SAGs) have been isolated from a variety of plant species (Davies and Grierson, 1989; Azumi and Watanabe, 1991; Becker and Apel, 1993; Hensel *et al.*, 1993; Taylor *et al.*, 1993; Buchanan-Wollaston, 1994; Lohman *et al.*, 1994; King *et al.*, 1995; Smart *et al.*, 1995; Drake *et al.*, 1996; Hanfrey *et al.*, 1996; Oh *et al.*, 1996; Buchanan-Wollaston and Ainsworth, 1997; Park *et al.*, 1998; Weaver *et al.*, 1998; Quirino *et al.*, 1999; Noh and Amasino, 1999a,b; Chapter 4). Studies on the expression pattern of SAGs have revealed that induction of most SAG mRNAs is delayed or repressed by cytokinin treatment (Weaver *et al.*, 1998; Noh and Amasino, 1999a,b). Because the function of most SAGs during senescence is not yet clear, it is not known whether the decreased expression of SAGs is the cause or the result of the delayed senescence after cytokinin treatment. Elucidation of the function of individual SAGs during senescence will help to reveal the mechanisms of cytokinin action in delaying plant senescence.

Difficulties in obtaining mutants having altered cytokinin activity or responsiveness have hindered the application of genetics to understanding cytokinin metabolism and signal transduction. However, several recent reports have implicated two types of signaling pathway components in cytokinin action: a G-protein-coupled receptor and a two-component histidine kinase (Estelle, 1998). Reduced cytokinin sensitivity has been observed from transgenic Arabidopsis constitutively expressing antisense mRNA of *GCR1*, a G-protein-coupled receptor homologue (Plakidou-Dymock *et al.*, 1998). Kakimoto (1996) demonstrated that ectopic expression of *CKI1*, a homologue of two-component histidine kinases that consist of a sensor and a response regulator, results in cytokinin-independent cell division in cultured Arabidopsis tissues. Moreover, Brandstatter and Kieber (1998) demonstrated that *IBC6*, a member of the response regulator group, exhibits rapid cytokinin-specific induction of its mRNA. Considering the broad effects of cytokinin on plant development, downstream cytokinin signaling pathways are likely to be branched and complex. Alterations in cytokinin perception or early signaling events, especially in a tissue- or developmental-specific manner, have much potential for crop-improving efforts.

III. The Relationship of Ethylene and Senescence

As discussed in Chapter 8, the effects of ethylene in fruit-ripening and flower senescence are of particular interest in agriculture. In climacteric fruits such as apple, pear, tomato, banana, papaya, mango, and avocado, an increase in respiration stimulated by ethylene accompanies the start of the ripening process. It is thought that at the molecular level ethylene coordinates the expression of the genes that participate in ripening (Gray *et al.*,

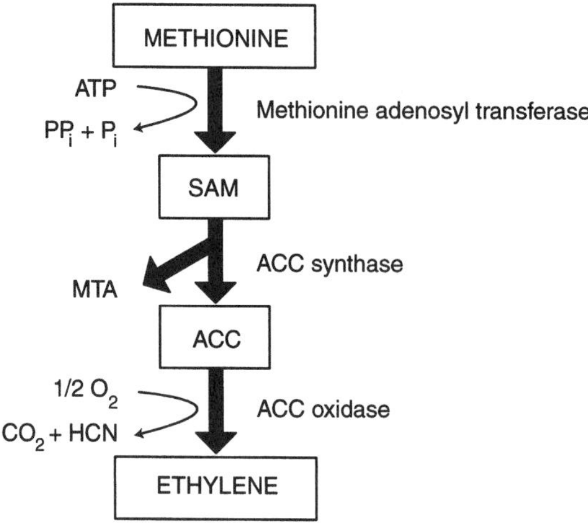

Figure 6-2. Ethylene biosynthetic pathway. SAM, S-adenosylmethionine; ACC, 1-aminocyclopropane-1-carboxylic acid; MTA, methylthioadenosine.

1992). The protein products of these genes are presumably involved in causing the changes that take place during ripening such as changes in fruit color (due to chlorophyll degradation and carotenoid pigment synthesis), development of sweetness (due to breakdown of starch into sugars or sugar translocation from other parts of the plant) and softening of the fruit (due to changes in cell wall components).

The understanding of basic aspects of ethylene biology has lagged behind the manipulation of ethylene as a horticultural practice. However, the pathway for ethylene biosynthesis is known (Fig. 6-2; also see Chapter 8) and more recently advances have been made in elucidating how ethylene signal transduction occurs. The signal transduction pathway initiates with the receptor which was identified in 1995 (Schaller and Bleecker, 1995). There are also several ethylene signaling mutants available and these are providing new insights into how the ethylene signal transduction pathway works beyond the receptor. Currently ethylene is the best understood plant hormone system at the molecular level. These recent advances in basic aspects of ethylene biology have also triggered several attempts to manipulate senescence using genetic engineering approaches.

Most attempts to block ethylene-mediated senescence have involved manipulating ethylene biosynthesis or perception. This has been accomplished by reducing ethylene synthesis by "turning off" ethylene-synthesizing enzymes or by reducing ethylene levels by depletion of ethylene precursors. More recently, a dominant negative approach using a mutant ethylene receptor has been employed.

In 1991, Oeller *et al.* reported a successful manipulation of tomato fruit ripening by turning off ACC synthase expression. ACC synthase is encoded by a gene family, two members of which are expressed during fruit ripening, *LE-ACC2* and *LE-ACC4*. The cDNA for *LE-ACC2* was placed in reverse orientation in front of the strong constitutive CaMV 35S promoter and transformed into plants. Tomato fruits from an *ACC synthase* antisense line that showed the strongest phenotype had ethylene production inhibited by 99.5%, and RNA blot analyses with gene-specific probes showed that mRNAs for both *LE-ACC2* and *LE-ACC4* were undetectable in the transgenic fruits. The red color characteristic of fruit ripening did not develop and these fruits never softened or produced a ripe aroma. Treatment with exogenous ethylene was able to reverse this phenotype leading to fruits that were indistinguishable from wild-type ripe fruits.

The turning off strategy has also been tried with the last enzyme in the pathway of ethylene synthesis, ACC oxidase (ACO). Transgenic tomato plants showing a reduction in ethylene synthesis were obtained by antisense expression of pTOM13, believed to encode a tomato *ACO* gene, driven by the CaMV 35S promoter (Hamilton *et al.*, 1990). In plants expressing antisense pTOM13 mRNA, wound-induced expression of this gene was greatly reduced and its expression was undetectable in ripening fruits. In detached fruits from certain plants containing two copies of the antisense transgene, ethylene evolution was inhibited by 97%. Although the time at which fruit ripening initiated was not changed, the transgenic fruits were less susceptible to overripening and shriveling when kept at room temperature compared to control fruits. Interestingly, Picton *et al.* (1993) reported that the time at which the fruit is detached from the vine is related to the extent that ripening is affected. The most extreme delay in fruit ripening was observed when transgenic fruits were collected at the mature-green stage (i.e. full-sized fruit with its basal part starting to show chlorophyll loss). In addition to the fruit-related phenotype, a delay in the onset of foliar senescence was observed in the transgenic plants. At 8 weeks post-germination, control plants had leaves with advanced senescence while the transgenic counterparts had no visible signs of chlorophyll loss. Wild-type lower leaves showed approximately 40% lower chlorophyll levels than the corresponding leaves of transgenic tomato plants. This delay in leaf senescence is similar to results obtained with the ethylene-insensitive Arabidopsis mutant *etr1-1*. *Etr1-1* leaves live 30% longer than wild-type indicating that ethylene plays a role in the timing of leaf senescence but that ultimately senescence occurs in the absence of ethylene perception (Grbic and Bleecker, 1995).

An antisense version of an *ACO* gene from melon has been introduced into Charentais cantaloupe which has poor storage capability (Ayub *et al.*, 1996). In one transgenic line a 99% reduction in ethylene production was observed in fruits. Interestingly, the transgenic fruits did not abscise from the plant because there was no activation of the peduncular abscission zone. At the time of commercial harvest, transgenic fruits were more than two-fold firmer than wild type. Exogenous ethylene application was able to restore yellowing of the rind, softening of flesh, and abscission zone activation in the peduncule.

Studies in *Dianthus caryophyllus* (carnation) illustrate the effects of ethylene in flower senescence (Nichols *et al.*, 1983). After pollination, the petals of carnation flowers wilt in 1 to 2 days. Petal senescence is accompanied by an increase in respiratory activity and a climacteric-like increase in ethylene production. By 3 hours after pollination there is a 10-fold increase in ethylene production by stigmas. Petals increase ethylene production later. Application of ethylene inhibitors is able to retard petal senescence. Senescing carnation petals exhibit a characteristic “inrolling” behavior which also occurs in response to exogenous ethylene. This inrolling can be delayed by inhibitors of ethylene synthesis. The gene for ACO from petunia was cloned and used to generate transgenic carnation plants that express an antisense *ACO* gene under the control of a strong constitutive promoter (Savin *et al.*, 1995). The transgenic carnations did not display the typical inrolling behavior. While the normal vase life of the cultivars used is approximately 5 days from harvest to inrolling, some of the transgenic flowers had their vase life increased up to 9 days.

Ethylene levels can also be reduced by the expression of enzymes that metabolize ethylene precursors. A *Pseudomonas* strain that is able to use ACC as its sole nitrogen source for growth was identified (Klee *et al.*, 1991). From this strain, the gene responsible for the ACC-degrading activity was cloned and found to encode ACC-deaminase, which degrades ACC to ammonia and alpha-ketobutyric acid. The *ACC-deaminase* gene was then introduced into tomato plants under the control of the CaMV 35S promoter. Transgenic plants expressing *ACC-deaminase* were phenotypically indistinguishable from wild-type plants. However, their fruits showed a significant delay in the progression of ripening and reduced ethylene synthesis (Klee, 1993).

Another study utilized the S-adenosylmethionine hydrolase (SAMase) from bacteriophage T3 (Good *et al.*, 1994). SAMase degrades SAM to methylthioadenosine and homoserine, lowering levels of the metabolic precursor to ethylene. This approach has been successful in tomato plants where SAMase expression was achieved in a tissue- and developmental stage-specific manner directed by the *E8* promoter. This regulated expression of SAMase was necessary to avoid disturbing other pathways in other tissues such as DNA methylation, polyamine and phospholipid biosynthesis that involve SAM. The *E8* promoter has been shown to be ethylene inducible and specifically activated during tomato fruit ripening (Lincoln *et al.*, 1987). Transgenic tomato fruits at the time of picking had ethylene levels similar to controls. However, their ability to produce ethylene declined at later timepoints such that ethylene synthesis was reduced by approximately 80% and the fruits took twice as long as controls to reach the final ripened stage.

An innovative way to manipulate senescence involving a dominant negative approach arose from the finding that the *ETR* gene encodes an ethylene receptor in Arabidopsis (Schaller and Bleecker, 1995). A mutated version of *ETR*, encoded by the *etr1-1* allele, confers dominant ethylene insensitivity in Arabidopsis. When *etr1-1* is heterologously expressed in petunia and tomato plants there is a significant delay of flower senescence and abscission and fruit ripening (Wilkinson *et al.*, 1997). For example, the corolla of a transgenic petunia line remained turgid for at least 8 days while wild-type flowers collapsed by the third day. Interestingly, it was found that in transgenic petunia plants the ethylene production that normally follows the pollination of petunia flowers exceeded that of wild-type. This suggests that there is a feedback mechanism to control ethylene production after pollination involving the wild-type ethylene receptors, and that this feedback was disrupted by the transgene expression. Similar to that which was observed with transgenic tomato fruits obtained by the approaches described above, detached tomato fruits from transgenic lines expressing *etr1-1* remained yellow even after 3 months whereas the wild-type control fruits turned red, softened and eventually rotted.

IV. Concluding Remarks

Senescence often limits crop yield and contributes to the postharvest loss of vegetables. Identification of antagonistic or promoting roles of cytokinin or ethylene, respectively, in plant senescence provides a framework to overcome problems associated with senescence in agriculture. Understanding the biosynthesis and signal transduction pathways of these plant hormones along with the development of techniques to introduce genes into plants have enabled progress in the genetic engineering of senescence.

Many studies have reported successful modulation of endogenous cytokinin levels which resulted in delayed tissue senescence. However, in many studies an undesirable abnormal phenotype accompanied the delayed tissue senescence because of the pleiotropic effects of cytokinin. Phenotypic effects resulting from cytokinin production were apparent even with extremely low levels of introduced *IPT* gene activity. Therefore, tightly controlled spatial and/or temporal cytokinin production systems are useful to minimize adverse effects. The autoregulated production of cytokinin using a senescence-specific promoter (Gan and Amasino, 1995) is an example of the successful use of tightly controlled cytokinin production. Although genetic modulation of cytokinin levels has proven to be effective in delaying foliar senescence, results of the efforts to delay fruit and flower senescence using this approach have not been as successful. Developing more finely tuned spatial/temporal cytokinin overproduction systems might be necessary for this application, or in many cases the antagonistic effect of cytokinin on senescence may not be as strong in fruits and flowers as in leaves.

In contrast, numerous studies have reported that blocking either synthesis or perception of ethylene can be effective in delaying fruit and flower senescence. However, only a few of these studies showed a significant delay in leaf senescence. In summary, while modulation of cytokinin levels has resulted in successful delay of foliar senescence, modulation of ethylene synthesis/perception has been useful in manipulating fruit and flower senescence. Most efforts to block synthesis or perception of ethylene have used strong constitutive promoters. Therefore, the minor effects of the altered ethylene synthesis/perception in foliar senescence compared to those of cytokinin may indicate the limitation of ethylene-related approaches in this area. The limited delay in leaf senescence observed in the Arabidopsis *etr1-1* mutant (Grbic and Bleecker, 1995) is consistent with this possibility. Although much success has been achieved in altering ripening/senescence of climacteric fruits where ethylene is involved, genetic modulation of fruit ripening/senescence of non-climacteric fruits has not been reported. Because the ripening processes of non-climacteric fruits are not well understood, a more detailed understanding of the ripening processes of these fruit species is necessary. In this regard, cytokinin-based approaches may prove useful.

To date a limited number of species have been used for the genetic modification of cytokinin and ethylene levels. It will be interesting to apply successful approaches to a broad range of species in the future. A combination of cytokinin- and ethylene-based approaches could prove very effective. Development of more finely tuned systems to regulate senescence may be achieved by the construction of spatially and temporally well-defined expression systems to control hormone levels or components of the downstream regulatory pathways of these hormones.

References

Abeles, F.B., Morgan, P. W., and Saltveit, M.E., Jr. (1992). *Ethylene in Plant Biology* (2nd ed.). Academic Press, London.

Ainley, W.M., Mcneil, K.J., Hill, J.W., Lingle, W.L., Simpson, R.B., Brenner, M.L., Nagao, R.T., and Key, J.L. (1993). Regulatable endogenous production of cytokinins up to toxic levels in transgenic plants and plant tissues. *Plant. Mol. Biol.* **22**, 13–23.

Akiyoshi, D.E., Morris, R.O., Hinz, R., Mischke, B.S., Kosuge, T., Garfinkel, D.J., Gordon, M.P., and Nester, E.W. (1983). Cytokinin/auxin balance in crown gall tumors is regulated by specific loci in the T-DNA. *Proc. Natl. Acad. Sci. USA* **80**, 407–411.

Ambler, J.R., Morgan, P.W., and Jordan, W.R. (1992). Amounts of zeatin and zeatin riboside in xylem sap of senescent and nonsenescent sorghum. *Crop Sci.* **32**, 411–419.

Ayub, R., Guis M., Amor, M.B., Gillot, L., Roustan, J., Latche, A., Bouzayen, M., and Pech, J. (1996). Expression of ACC oxidase antisense inhibits ripening of cantaloupe melon fruits. *Nature Biotechnol.* **14**, 862–866.

Azumi, Y., and Watanabe, A. (1991). Evidence for a senescence-associated gene induced by darkness. *Plant Physiol.* **95**, 577–583.

Barry, G.F., Rogers, S.G., Fraley, R.T., and Brand, L. (1984). Identification of a cloned cytokinin biosynthetic gene. *Proc. Natl. Acad. Sci. USA* **81**, 4776–4780.

Becker, W., and Apel, K. (1993). Differences in gene expression between natural and artificially induced leaf senescence. *Planta* **189**, 74–79.

Beinsberger, S.E.I., Valcke, R.L.M., Clijsters, H.M.M., De Greef, J.A., and Van Onckelen, H.A. (1992). Effects of enhanced cytokinin levels in *ipt* transgenic tobacco. In *Physiology and Biochemistry of Cytokinins in Plants* (M. Kaminek, D.W.S. Mok, and E. Zazimalova, Eds.), pp. 77–82. SPB Academic Publishing, The Hague.

Binns, A.N. (1994). Cytokinin accumulation and action, biochemical, genetic and molecular approaches. *Annu. Rev. Plant Physiol. Plant. Mol. Biol.* **45**, 173–196.

Borochov, A., and Woodson, W.R. (1989). Physiology and biochemistry of flower petal senescence. *Hort. Rev.* **11**, 14–43.

Brandstatter, I., and Kieber, J.J. (1998). Two genes with similarity to bacterial response regulators are rapidly and specifically induced by cytokinin in Arabidopsis. *Plant Cell* **10**, 1009–1019.

Buchanan-Wollaston, V. (1994). Isolation of cDNA clones for genes that are expressed during leaf senescence in *Brassica napus. Plant Physiol.* **105**, 839–846.

Buchanan-Wollaston, V., and Ainsworth, C. (1997). Leaf senescence in *Brassica napus*—cloning of senescence related genes by subtractive hybridization. *Plant. Mol. Biol.* **33**, 821–834.

Crowell, D.N., and Amasino, R.M. (1994). Cytokinins and plant gene regulation. In *Cytokinins: Chemistry, Activity and Function* (D.W.S. Mok and M.C. Mok, Eds.), pp 233–242. CRC Press, Boca Raton, FL.

Crowell, D.N., Kadlecek, A.T., John, M.C., and Amasino, R.M. (1990). Cytokinin-induced mRNAs in cultured soybean cells. *Proc. Natl. Acad. Sci. USA* **87**, 8815–8819.

Davies, K.M., and Grierson, D. (1989). Identification of cDNA clones for tomato (*Lycopersicon esculentum* Mill.) mRNAs that accumulate during fruit ripening and leaf senescence in response to ethylene. *Planta* **179**, 73–80.

Drake, R., John, I., Farrell, A., Cooper, W., Schuch, W., and Grierson, D. (1996). Isolation and analysis of cDNAs encoding tomato cysteine proteases expressed during leaf senescence. *Plant. Mol. Biol.* **30**, 755–767.

Eichholz, R., Harper, J., Felix, G., and Meins, F.J. (1983). Evidence for an abundant 33,000-dalton polypeptide regulated by cytokinins in cultured tobacco tissues. *Planta* **158**, 410–415.

Estelle, M. (1998). Cytokinin action, two receptors better than one? *Curr. Biol.* **8**, R539–R541.

Estruch, J.J., Prinsen, E., Vanonckelen, H., Schell, J., and Spena, A. (1991). Viviparous leaves produced by somatic activation of an inactive cytokinin-synthesizing gene. *Science* **254**, 1364–1367.

Estruch, J.J., Granell, A., Hansen, G., Prinsen, E., Redig, P., Van, Onckelen, H., Schwarz-Sommer, Z., Sommer, H., and Spena, A. (1993). Floral development and expression of floral homeotic genes are influenced by cytokinins. *Plant J.* **4**, 379–384.

Faiss, M., Zalubilova, J., Strnad, M., and Schmülling, T. (1997). Conditional transgenic expression of the *ipt* gene indicates a function for cytokinins in paracrine signaling in whole tobacco plants. *Plant J.* **12**, 401–415.

Felix, G., and Meins, F.J. (1986). Developmental and hormonal regulation of b-1,3-glucanase in tobacco. *Planta* **167**, 206–211.

Fosket, D.E., and Tepfer, D.A. (1978). Hormonal regulation of growth in cultured plant cells. *In Vitro* **14**, 63–75.

Gan, S., and Amasino, R.M. (1995). Inhibition of leaf senescence by autoregulated production of cytokinin. *Science* **270**, 1986–1988.

Gan, S., and Amasino, R.M. (1996). Cytokinins in plant senescence, from spray and pray to clone and play. *Bioessays* **18**, 557–565.

Gan, S., and Amasino, R.M. (1997). Making sense of senescence. *Plant Physiol.* **113**, 313–319.

Gepstein, S. (1988). Photosynthesis. In *Senescence and Aging in Plants* (L.D. Noodén and A.C. Leopold, Eds.), pp. 85–109. Academic Press, San Diego.

Good, X., Kellogg, J.A., Wagoner W., Langhoff, D., Matsumura, W., and Bestwick, R.K. (1994). Reduced ethylene synthesis by transgenic tomatoes expressing S-adenosylmethionine hydrolase. *Plant. Mol. Biol.* **26**, 781–790.

Gray, J., Picton, J., Shabbeer, J., Schuch, W., and Grierson, D. (1982). Molecular biology of fruit ripening and its manipulation with antisense genes. *Plant. Mol. Biol.* **19**, 69–87.

Grbic, V., and Bleecker, A.B. (1995). Ethylene regulates the timing of leaf senescence in *Arabidopsis. Plant J.* **8**, 595–602.

Groot, S.P.C., Bouwer, R., Busscher, M., Lindhout, P., and Dons, H.J. (1995). Increase of endogenous zeatin riboside by introduction of the *ipt* gene in wild type and the *lateral suppressor* mutant of tomato. *Plant Growth Regul.* **16**, 27–36.

Hamilton, A.J., Lycett, G.W., and Grierson, D. (1990). Antisense gene that inhibits synthesis of the hormone ethylene in transgenic plants. *Nature* **346**, 284–287.

Hanfrey, C., Fife, M., and Buchanan-Wollaston, V. (1996). Leaf senescence in *Brassica napus*—expression of genes encoding pathogenesis-related proteins. *Plant. Mol. Biol.* **30**, 597–609.

Hensel, L.L., Grbic, V., Baumgarten, D.A., and Bleecker, A.B. (1993). Developmental and age-related processes that influence the longevity and senescence of photosynthetic tissues in Arabidopsis. *Plant Cell* **5**, 553–564.

Hewelt, A., Prinsen, E., Schell, J., Van Onckelen, H., and Schmülling, T. (1994). Promoter tagging with a promoterless *ipt* gene leads to cytokinin-induced phenotypic variability in transgenic tobacco plants: implications of gene dosage effects. *Plant J.* **6**, 879–891.

Jiang, C.-Z., Rodermel, S.R., and Shibles, R.M. (1993). Photosynthesis, rubisco activity and amount, and their regulation by transcription in senescing soybean leaves. *Plant Physiol.* **101**, 105–112.

Kakimoto, T. (1996). CKI1, a histidine kinase homolog implicated in cytokinin signal transduction. *Science* **274**, 982–985.

Kamada, I., Yamaguchi, S., Youssefian, S., and Sano, H. (1992). Transgenic tobacco plants expressing rgp1, a gene encoding a ras-related GTP-binding protein from rice, show distinct morphological characteristics. *Plant J.* **2**, 799–807.

King, G.A., Davies, K.M., Stewart, R.J., and Borst, W.M. (1995). Similarities in gene-expression during post-harvest-induced senescence of spears and natural foliar senescence of asparagus. *Plant Physiol.* **108**, 125–128.

Klee, H.J. (1993). Ripening physiology of fruit from transgenic tomato (*Lycopersicon esculentum*) plants with reduced ethylene synthesis. *Plant Physiol.* **102**, 911–916.

Klee, H.J. (1994). Transgenic plants and cytokinin biology. In *Cytokinins: Chemistry, Activity and Function* (D.W.S. Mok and M.C. Mok, Eds.), pp. 289–293. CRC Press, Boca Raton, FL.

Klee, H.J., Hayford, M.B., Kretzmer, K.A., Barry, G.F., and Kishore, G.M. (1991). Control of ethylene synthesis by expression of a bacterial enzyme in transgenic tomato plants. *Plant Cell* **3**, 1187–1193.

Kusaba, S., Kano-Murakami, Y., Matsuoka, M., Tamaoki, M., Sakamoto, T., Yamaguchi, I., and Fukumoto, M. (1998). Alteration of hormone levels in transgenic tobacco plants overexpressing the rice homeobox gene *OSH1*. *Plant Physiol.* **116**, 471–476.

Leopold, A.C. (1961). Senescence in plant development. *Science* **134**, 1727–1732.

Li, Y., Hagen, G., and Guilfoyle, T.J. (1992). Altered morphology in transgenic tobacco plants that overproduce cytokinins in specific tissues and organs. *Dev. Biol.* **153**, 386–395.

Lincoln, J.E., Cordes, S., Read, E., and Fischer, R.L. (1987). Regulation of gene expression by ethylene during *Lycopersicon esculentum* (tomato) fruit development. *Proc. Natl. Acad. Sci. USA* **84**, 2793–2797.

Lohman, K., Gan, S., John, M., and Amasino, R.M. (1994). Molecular analysis of natural leaf senescence in *Arabidopsis thaliana*. *Physiol. Plant.* **92**, 322–328.

Martineau, B., Houck, C.M., Sheehy, R.E., and Hiatt, W.R. (1994). Fruit-specific expression of the *A. tumefaciens* isopentenyl transferase gene in tomato—effects on fruit ripening and defense-related gene expression in leaves. *Plant J.* **5**, 11–19.

McKenzie, M.J., Mett, V., Stewart Reynolds, P.H., and Jameson, P.E. (1998). Controlled cytokinin production in transgenic tobacco using a copper-inducible promoter. *Plant Physiol.* **116**, 969–977.

Medford, J.I., Horgan, R., El-Sawi, Z., and Klee, H.J. (1989). Alterations of endogenous cytokinins in transgenic plants using a chimeric isopentenyl transferase gene. *Plant Cell* **1**, 403–413.

Memelink, J., Hoge, J.H.C., and Schilperoort, R.A. (1987). Cytokinin stress changes the developmental regulation of several defense-related genes in tobacco. *EMBO J.* **6**, 3579–3583.

Miller, C.O., Skoog, F., Okomura, F.S., von Salta, M.H., and Strong, F.M. (1956). Isolation, structure and synthesis of kinetin, a substance promoting cell division. *J. Am. Chem. Soc.* **78**, 1375–1380.

Mohnen, D., Shinshi, H., Felix, G., and Meins, F.J. (1985). Hormonal regulation of b-1,3-glucanase messenger RNA levels in cultured tobacco tissues. *EMBO J.* **4**, 1631–1635.

Mok, D.W.S., and Mok, M.C. (Eds.) (1994). *Cytokinins: Chemistry, Activity and Function*. CRC Press, Boca Raton, FL.

Nichols, R., Bufler, G., Mor, Y., Fujino, D.W., and Reid, M.S. (1983). Changes in ethylene production and 1-aminocyclopropane-1-carboxylic acid content of pollinated carnation flowers. *J. Plant Growth Regul.* **2**, 1–8.

Noh, Y.-S., and Amasino, R.M. (1999a). Identification of a promoter region responsible for the senescence-specific expression of SAG12. *Plant Mol. Biol.* **41**, 181–194.

Noh, Y.-S., and Amasino, R.M. (1999b). Regulation of developmental senescence is conserved between *Arabidopsis* and *Brassica napus*. *Plant Mol. Biol.* **41**, 195–206.

Noodén, L.D., and Letham, D.S. (1993). Cytokinin metabolism and signaling in the soybean plant. *Aust. J. Plant Physiol.* **20**, 639–653.

Oeller, P.W., Wong, L.M., Taylor, L.P., Pike, D.A., and Theologis, A. (1991). Reversible inhibition of tomato fruit senescence by antisense RNA. *Science* **254**, 437–439.

Oh, S.A., Lee, S.Y., Chung, I.K., Lee, C.H., and Nam, H.G. (1996). A senescence-associated gene of *Arabidopsis thaliana* is distinctively regulated during natural and artificially induced leaf senescence. *Plant. Mol. Biol.* **30**, 739–754.

Ooms, G., Risiott, R., Kendall, A., Keys, A., Lawlor, D., Smith, S., Turner, J., and Young, A. (1991). Phenotypic changes in T-*cyt*-transformed potato plants are consistent with enhanced sensitivity of specific cell types to normal regulation by root-derived cytokinin. *Plant. Mol. Biol.* **17**, 727–743.

Ori, N., Juarez, M.T., Jackson, D., Yamaguchi, J., Banowetz, G.M., and Hake, S. (1999). Leaf senescence is delayed in tobacco plants expressing the maize homeobox gene *knotted1* under the control of a senescence-activated promoter. *Plant Cell* **11**, 1073–1080.

Park, J.H., Oh, S.A., Kim, Y.H., Woo, H.R., and Nam, H.G. (1998). Differential expression of senescence-associated mRNAs during leaf senescence induced by different senescence-inducing factors in Arabidopsis. *Plant. Mol. Biol.* **37**, 445–454.

Parthier, B. (1979). The role of phytohormones (cytokinins) in chloroplast development. *Biochem. Physiol. Pflanzen* **174**, 173–214.

Phillips, I.D.J. (1975). Apical dominance. *Annu. Rev. Plant Physiol.* **26**, 341–367.

Picton, S., Barton, S.L., Bouzayen, M., Hamilton, A.J., and Grierson, D. (1993). Altered fruit ripening and leaf senescence in tomatoes expressing an antisense ethylene-forming enzyme transgene. *Plant J.* **3**, 469–481.

Plakidou-Dymock, S., Dymock, D., and Hooley, R. (1998). A higher plant seven-transmembrane receptor that influences sensitivity to cytokinins. *Curr. Biol.* **8**, 315–324.

Porat, R., Borochov, A., Halevy, A.H., and O'Neill, S.D. (1994). Pollination-induced senescence of *Phalaenopsis* petals. *Plant Growth Regul.* **15**, 129–136.

Quirino, B.F., Normanly, J., and Amasino, R.M. (1999). Diverse range of gene activity during *Arabidopsis thaliana* leaf senescence includes pathogen-independent induction of defense-related genes. *Plant. Mol. Biol.* **40**, 267–278.

Richmond, A.E., and Lang, A. (1957). Effect of kinetin on protein content and survival of detached *Xanthium* leaves. *Science* **125**, 650–651.

Sano, H., and Youssefian, S. (1994). Light and nutritional regulation of transcripts encoding a wheat protein kinase homolog is mediated by cytokinins. *Proc. Natl. Acad. Sci. USA* **91**, 2582–2586.

Sano, H., Seo, S., Orudgev, E., Youssefian, S., Ishizuka, K., and Ohashi, Y. (1994). Expression of the gene for a small GTP binding protein in transgenic tobacco elevates endogenous cytokinin levels, abnormally induces salicylic acid in response to wounding, and increases resistance to tobacco mosaic virus infection. *Proc. Natl. Acad. Sci. USA* **91**, 10556–10560.

Savin, K.W., Baudinette, S.C., Graham, M.W., Michael, M.Z., Nugent, G.D., Lu, C., Chandler, S.F., and Cornish, E.C. (1995). Antisense ACC oxidase RNA delays carnation petal senescence. *Hort Sci.* **30**, 970–972.

Schaller, G.E., and Bleecker, A.B. (1995). Ethylene-binding sites generated in yeast expressing the Arabidopsis *ETR1* gene. *Science* **270**, 1809–1811.

Schmülling, T., Beinsberger, S., De Greef, J., Schell, J., and Van Onckelen, H. (1989). Construction of a heat-inducible chimeric gene to increase the cytokinin content in transgenic plant tissue. *FEBS Lett.* **249**, 401–406.

Shinshi, H., Mohnen, D., and Meins, F.J. (1987). Regulation of plant pathogenesis-related enzyme, inhibition of chitinase and chitinase mRNA accumulation in cultured tobacco tissues by auxin and cytokinin. *Proc. Natl. Acad. Sci. USA* **84**, 89–93.

Simmons, C.R., Litts, J.C., Huang, N., and Rodriguez, R.L. (1992). Structure of a rice beta-glucanase gene regulated by ethylene, cytokinin, wounding, salicylic acid and fungal elicitors. *Plant. Mol. Biol.* **18**, 33–45.

Smart, C.M., Scofield, S.R., Bevan, M.W., and Dyer, T.A. (1991). Delayed leaf senescence in tobacco plants transformed with *tmr*, a gene for cytokinin production in *Agrobacterium*. *Plant Cell* **3**, 647–656.

Smart, C.M., Hosken, S.E., Thomas, H., Greaves, J.A., Blair, B.G., and Schuch, W. (1995). The timing of maize leaf senescence and characterization of senescence-related cDNAs. *Physiol. Plant.* **93**, 673–682.

Smigocki, A.C. (1991). Cytokinin content and tissue distribution in plants transformed by a reconstructed isopentenyl transferase gene. *Plant. Mol. Biol.* **16**, 105–115.

Smigocki, A.C., and Owens, L.D. (1988). Cytokinin gene fused with a strong promoter enhances shoot organogenesis and zeatin levels in transformed plant cells. *Proc. Natl. Acad. Sci. USA* **85**, 5131–5135.

Smigocki, A., Neal, J.W., Mccanna, I., and Douglass, L. (1993). Cytokinin-mediated insect resistance in *Nicotiana* plants transformed with the *ipt* gene. *Plant. Mol. Biol.* **23**, 325–335.

Soejima, H., Sugiyama, T., and Ishihara, K. (1995). Changes in the chlorophyll contents of leaves and in levels of cytokinins in root exudates during ripening of rice cultivars Nipoonbare and Akenohoshi. *Plant Cell Physiol.* **36**, 1105–1114.

Tamaoki, M., Kusaba, S., Kano-Murakami, Y., and Matsuoka, M. (1997). Ectopic expression of a tobacco homeobox gene, *NTH15*, dramatically alters leaf morphology and hormone levels in transgenic tobacco. *Plant Cell Physiol.* **38**, 917–927.

Taylor, C.B., Bariola, P.A., Delcardayre, S.B., Raines, R.T., and Green, P.J. (1993). RNS2—a senescence-associated RNase of Arabidopsis that diverged from the S-RNases before speciation. *Proc. Natl. Acad. Sci. USA* **90**, 5118–5122.

Van Staden, J., Cook, E., and Noodén, L.D. (1988). Cytokinins and senescence. In *Senescence and Aging in Plants* (L.D. Noodén, and A.C. Leopold, Eds.), pp. 281–328. Academic Press, San Diego.

Weaver, L.M., Gan, S., Quirino, B., and Amasino, R.M. (1998). A comparison of the expression patterns of several senescence-associated genes in response to stress and hormone treatment. *Plant. Mol. Biol.* **37**, 455–469.

Wilkinson, J.Q., Lanahan, M.B., Clark, D.G., Bleecker, A.B., Chang, C., Meyerowitz, E.M., and Klee, H.J. (1997). A dominant mutant receptor from Arabidopsis confers ethylene insensitivity in heterologous plants. *Nature Biotechnol.* **15**, 444–447.

Yushibov, V.M., Il, P.C., Andrianov, V.M., and Piruzian, E.S. (1991). Phenotypically normal transgenic T-*cyt* tobacco plants as a model for the investigation of plant gene expression in response to phytohormonal stress. *Plant. Mol. Biol.* **17**, 825–836.

Zhang, R., Zhang, X., Wang, J., Letham, D.S., Mckinney, S.A., and Higgins, T.J.V. (1995). The effect of auxin on cytokinin levels and metabolism in transgenic tobacco tissue expressing an *ipt* gene. *Planta* **196**, 84–94.

7

Proteolysis

Urs Feller

I. Introduction

The cleavage of proteins and peptides in plants is catalyzed by a series of peptide hydrolases differing in their subcellular localization, in their substrate specificities or in their regulatory properties (Barrett, 1986). The hydrolysis of peptide bonds is not restricted to the catabolism of mature proteins to free amino acids, but is also relevant for the modification and maturation of proteins. The removal of signal or transit peptides from larger precursors is linked to protein synthesis and intracellular sorting. The degradation of damaged or unassembled polypeptides is important for housekeeping in a cell. Furthermore, protein turnover is a prerequisite for the adaptation of the protein pattern (e.g., enzymes, translocator proteins) to changing conditions. All the previously mentioned processes occur in differentiating or mature cells and are not necessarily related to cell death. The proteolytic enzymes involved and the regulatory mechanisms may not be identical for protein maturation and for protein remobilization.

Different physiological situations may lead to the death of a cell (Greenberg, 1996; Noodén *et al.*, 1997; Bleecker, 1998; Guarente *et al.*, 1998). The degradation of macromolecules in such cells and the export of low molecular weight compounds to sinks within the same plant represent a last contribution of these cells to other plant parts. Since proteins represent the predominant nitrogen fraction in leaves or seeds and these proteins are not transported as such in the phloem, the important role of proteolysis and its control for the redistribution of nitrogen is obvious. Under certain conditions, proteolysis may not only allow the export of nitrogen compounds after initiation of cell death, but it may also directly be involved in causing the death of a cell by degrading essential proteinaceous constituents.

II. Selective Hydrolysis of Peptide Bonds

A. Protein Maturation and Removal of Peptides from Larger Precursors

The removal of peptides from larger precursors is specific for the protein and for the peptide bond within the protein. The cotranslational removal of the signal peptide in the rough endoplasmic reticulum (Chrispeels, 1991) and the post-translational cleavage of the transit peptide from proteins imported from the cytosol into plastids or mitochondria (Eriksson *et al.*, 1996; Schatz and Dobberstein, 1996; Whelan and Glaser, 1997; Keegstra and Cline, 1999) are examples of a highly specific proteolytic process allowing the proper sorting and maturation of newly synthesized polypeptides. Processing enzymes in protein-storage vacuoles and in lytic vacuoles cleave larger precursor polypeptides also at well-defined positions (Okamoto *et al.*, 1994; Hara-Nishimura *et al.*, 1998). This type of protein cleavage is related to protein synthesis and not primarily to protein degradation or cell death.

B. Degradation of Damaged Proteins and of Unassembled Subunits

The degradation of damaged (Desimone *et al.*, 1996; Stieger and Feller, 1997), mistargeted (Halperin and Adam, 1996) or not properly assembled (Schmidt and Mishkind, 1983) proteins is specific for the polypeptide, but may be unspecific for the peptide bonds within the polypeptide. After the initial cleavage, the polypeptide may be degraded completely to free amino acids. In general, fragments produced by a first cleavage of a mature protein are rapidly degraded to amino acids or small peptides and do not accumulate in the cell.

C. Protein Turnover and Adaptation of the Metabolism to Changing Conditions

Some proteins are quite stable in a plant cell, while other proteins are turned over rapidly (Mattoo *et al.*, 1984). Protein turnover (synthesis and degradation) allows a modification of the protein pattern and as a consequence also an adaptation of the metabolism to altered conditions (Dungey and Davies, 1982). The susceptibility of proteins to proteolysis is an important intrinsic property and may be further influenced by the actual environment (Feller and Fischer, 1994). Proteins in the same compartment may differ considerably in their stability (Mitsuhashi and Feller, 1992).

D. Degradation of Mature Proteins in Relation to Nitrogen Remobilization

Mature proteins can be rapidly degraded during nitrogen mobilization from germinating seeds (Harvey and Oaks, 1974; Chrispeels and Boulter, 1975) or from cells in vegetative plant parts (Noodén *et al.*, 1997) prior to cell death. The storage proteins in seeds are located in the protein bodies (Chrispeels and Boulter, 1975). A cysteine endopeptidase is *de novo* synthesized into the endoplasmic reticulum at the onset of germination, processed and then either transported in vesicles to the protein bodies (e.g., legumes) or secreted from the aleurone layer and the scutellar epithelium into the endosperm of cereals (Callis, 1995; Müntz *et al.*, 1998). The proteins in photosynthesizing leaf cells differ considerably in their properties and in their subcellular compartmentation from storage proteins in seeds. In mesophyll cells of C3-plants, proteins (e.g., soluble enzymes, membrane proteins, regulatory proteins) are present mainly in the chloroplasts (Peoples and Dalling, 1988). During germination and during senescence, the final steps of protein remobilization coincide with the death of the cells, although the initial situation is quite different in storage tissues of seeds and in metabolically active leaf cells. The proteolytic systems involved and the regulation of proteolysis are not identical for germination and leaf senescence.

III. Proteolytic Activities in Plants

A. Classification of Peptide Hydrolases

Exopeptidases remove an amino acid, a dipeptide or a tripeptide from the N-terminus (aminopeptidases) or from the C-terminus (carboxypeptidases) of a peptide or of a protein (Barrett, 1986). The exopeptidases in higher plants were reviewed in detail by Mikola and Mikola (1986). Since no clear function (especially no regulatory function) of exopeptidases in relation to cell death has been identified so far, these types of peptide hydrolases are not considered in more detail here.

Endopeptidases hydrolyze polypeptide chains to fragments of three or more amino acid residues in length. Based on the active center, various classes of endopeptidases can be distinguished (Barrett, 1986). Serine endopeptidases (EC 3.4.21), cysteine endopeptidases (EC 3.4.22), aspartate endopeptidases (EC 3.4.23) and metal endopeptidases (EC 3.4.24) represent such classes. The various types of endopeptidases may differ in their substrate specificities, in their subcellular localization, in their physiological function and in their inhibition properties.

Cysteine endopeptidases have been detected in a series of plant species (Ryan and Walker-Simmons, 1981; Callis, 1995). These enzymes are most active in the slightly acidic pH range and were found to be localized in the vacuole (Heck *et al.*, 1981; Canut *et al.*, 1985). Cysteine endopeptidases are key enzymes for the remobilization of storage proteins during germination (Chrispeels and Boulter, 1975; Callis, 1995). High activities of such enzymes were also detected in leaves (Feller *et al.*, 1977; Peoples and Dalling, 1988). An increased level of transcription of genes encoding a cysteine proteinase during senescence has been reported by several groups (Buchanan-Wollaston, 1997; Valpuesta *et al.*, 1995). Navarre and Wolpert (1999) suggested that a calcium-activated cysteine protease causes the cleavage of the large subunit of Rubisco in oats after treatment with victorin, a host-selective fungal toxin. Caspases, the well known cell death-related proteases in animals, belong also to the cysteine proteases (Ruoslahti and Reed, 1999; Wolf and Green, 1999). Caspases represent a special family of highly conserved cysteine endopeptidases, exist often as latent zymogens and are involved in inflammation and/or in regulatory cascades during apoptosis. It remains open whether this type of "cell killer" enzyme plays an important role in senescence of plant cells. The presence of caspase-like plant protease(s) in tobacco tissues infected with the tobacco mosaic virus (hypersensitive response) was reported recently (Del Pozo and Lam, 1998). The question, to what extent caspases or caspase-like proteases are important in plant cell death or certain types of plant cell death, remains to be answered (Chapter 1).

Far less information than for cysteine endopeptidases is available for plant serine endopeptidases. A serine-type protease has been found to mediate the degradation of a light-stress protein (Adamska *et al.*, 1996). A marked increase of a vacuolar serine endopeptidase was detected in maize roots under sugar starvation (James *et al.*, 1996). As judged from activity gel assays, a 60-kDa serine endopeptidase accumulated during *in vitro* tracheary element differentiation in *Zinnia elegans* (Ye and Varner, 1996). The activity of a 70-kDa serine endopeptidase was found to be low in young leaves and high in artificially senescing parsley leaves (Jiang *et al.*, 1999). The latter reports suggest but cannot prove that serine endopeptidases play a role in plant cell death.

Metallo-endopeptidases contain zinc and were detected in plants by several groups (Ragster and Chrispeels, 1979; Barrett, 1986; McGeehan *et al.*, 1992; Bushnell *et al.*, 1993). In soybean leaves, a fraction of a monomeric metallo-proteinase with a molecular mass of only 15 to 20 kDa is partially present in the extracellular space, where the specific activity is 50-fold higher than in crude leaf extracts (Graham *et al.*, 1991). The level of this enzyme increases with leaf age. Metallo-endopeptidases were also detected in the chloroplasts (Abad *et al.*, 1989; Musgrove *et al.*, 1989; Bushnell *et al.*, 1993; Roulin and Feller, 1998). In intact chloroplasts isolated from pea leaves, the degradation of stromal proteins in darkness is strongly inhibited by EDTA (Fig. 7-1). This activity in intact chloroplasts is stimulated by divalent cations (Roulin and Feller, 1998). Considering these results, a key role of metallo-endopeptidase(s) in the degradation of stromal proteins in the intact chloroplasts appears likely.

Aspartic endopeptidases are characterized by a low pH optimum (below pH 5). This type of endopeptidase was also detected in plants, but little is known about their functions (Callis, 1995).

B. Complex Proteolytic Systems

Proteolytic activities can be integrated in larger polypeptide complexes (Lupas *et al.*, 1997). The proteolytic cores of the cytosolic proteasome (20S proteasome) and of the plastidial Clp system (ClpP) form a barrel-shaped structure with the proteolytic sides oriented to the central pore. Only polypeptides entering the central pore can be degraded by the intact structures (Lupas *et al.*, 1997). The proteolytic core may be able to degrade unfolded polypeptides in an ATP-independent manner, while the additional ATPase moieties of the two systems may serve as a funnel and may be required for unfolding proteins present in a stabilized three-dimensional structure. Complete unfolding and the absence of disulfide bonds are prerequisites for the degradation of polypeptides by the 20S proteasome. Possible interactions between complex proteolytic systems and substrate proteins are schematically shown in Fig. 7-2.

A series of reports concern the plant proteasome (Vierstra, 1993; Belknap and Garbarino, 1996; Bahrami and Gray, 1999). This complex proteolytic system is ubiquitous in eukaryotic cells (located in the cytosol and in the nucleus) and is present under two forms (the 20S proteasome and the 26S proteasome). The 20S proteasome represents the barrel-shaped proteolytic core with a narrow pore in the center and contains the Ntn hydrolases (Lupas *et al.*, 1997). The assembly of the 20S proteasome is quite well known (Chen and Hochstrasser, 1996), but the disassembly of this complex is not yet elucidated. The 26S proteasome is a very complex structure and contains 14 different subunits in the proteolytic core (20S proteasome) and about 20 different subunits in the two 19S caps (Lupas *et al.*, 1997). The 19S caps have ATPase and ubiquitin-binding activities and are attached to the end of a 20S core. The precise functions are not yet known for the various subunits of this self-compartmentalizing protease (Baumeister *et al.*, 1998). The proteasome is involved in the

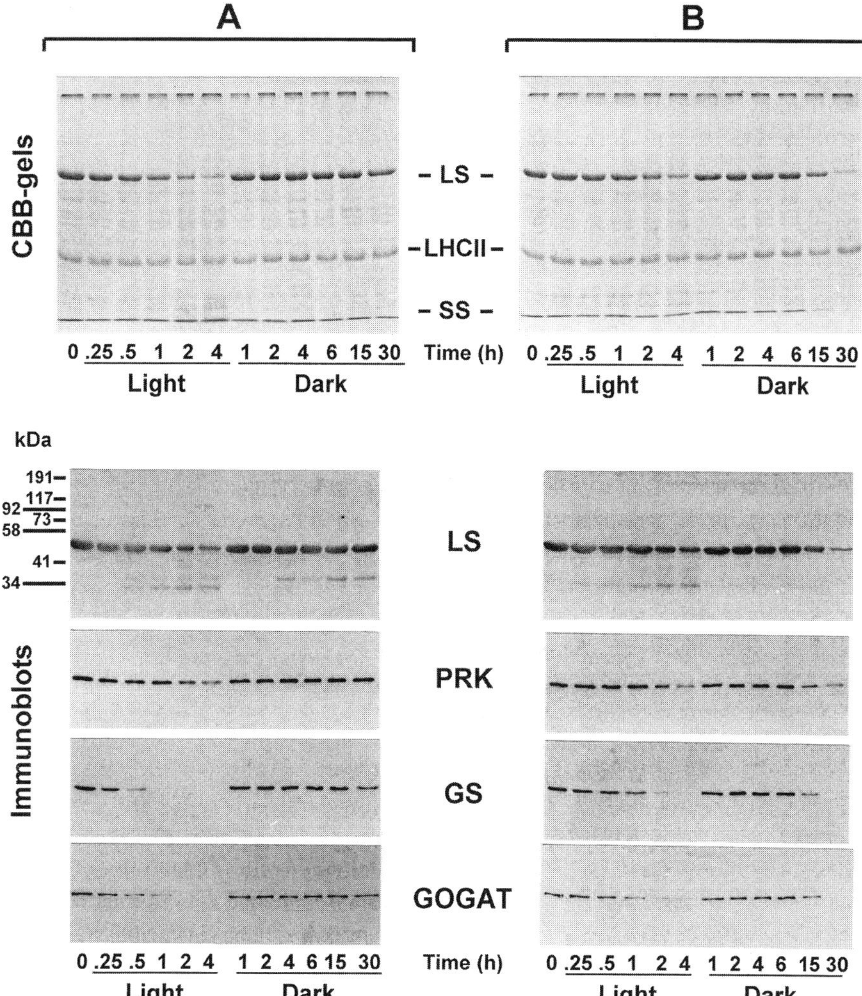

Figure 7-1. Influence of EDTA on proteolysis in intact pea (*Pisum sativum*) chloroplasts incubated at 25°C in the light (45 μmol m^{-2} s^{-1}) or in darkness. Intact chloroplasts were isolated on Percoll steps and incubated in a medium containing 1 mM DTT, 1 mM $MgCl_2$ and 2 mM EDTA (A) or in a modified medium lacking EDTA (B). Following incubation and re-isolation of the intact organelles, changes in the levels of the large (LS) and small (SS) subunit of Rubisco and of the light-harvesting chlorophyll *a*/*b*-binding protein (LHCII) were visualized by coomassie brilliant blue-staining (CBB) of the gels. The degradation of LS, phosphoribulokinase (PRK), glutamine synthetase (GS) and ferredoxin-dependent glutamate synthase (GOGAT) was detected by immunoblotting with specific antibodies. Equal quantities of chlorophyll (1 μg) were loaded on each lane. [From Roulin and Feller (1998), with permission.]

ubiquitin-dependent degradation of proteins (Vierstra, 1993). The N-terminus, the presence of a lysine residue for the covalent attachment of ubiquitin (isopeptide bond) and the three-dimensional structure of the protein must be considered as important factors influencing the susceptibility of a protein for degradation in this pathway (Vierstra, 1993; Byrd *et al.*, 1998). The ubiquitin-dependent proteolysis in the cytosol is most likely important for the

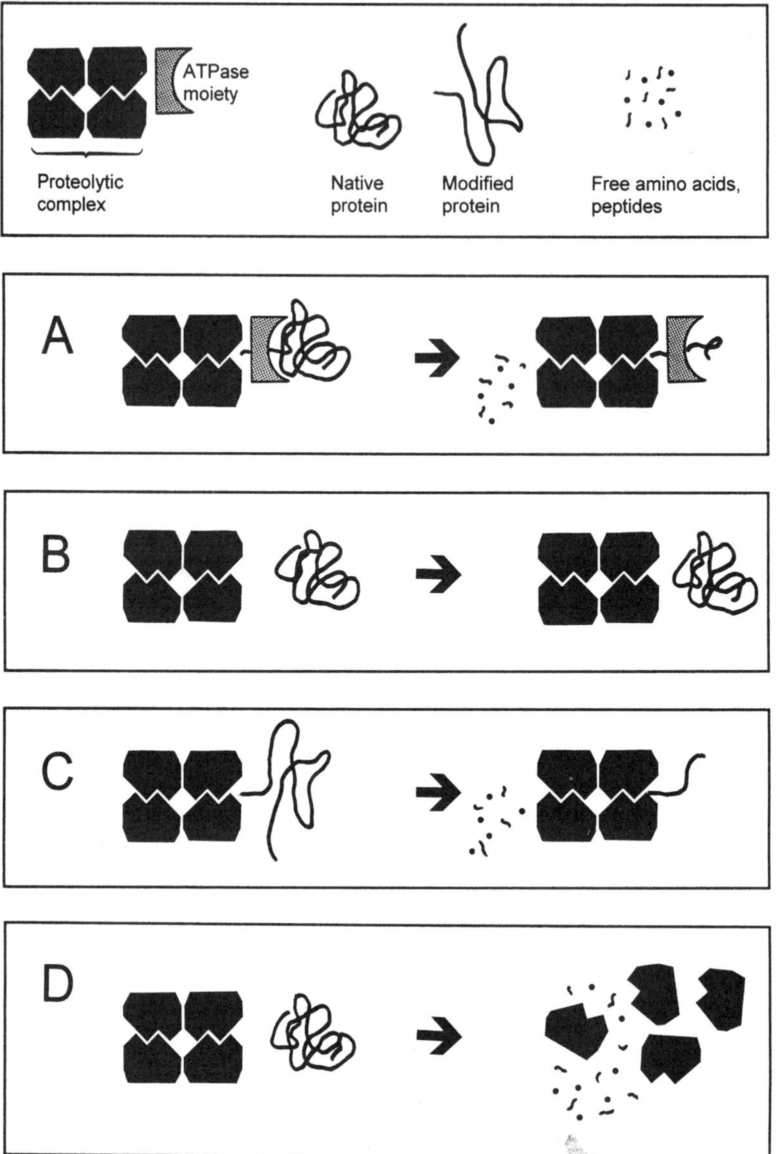

Figure 7-2. Scheme representing possible interactions between complex proteolytic systems and substrate proteins. (A) The ATPase moiety may serve as a funnel and feed the substrate protein into the proteolytic moiety. (B) Native proteins may be stable in the absence of a functional ATPase moiety. (C) Modified (unfolded) proteins may directly serve as substrates for the proteolytic moiety. (D) The separation of the subunits forming the proteolytic complex may expose catalytic sites of the peptide hydrolase(s) and allow a rather unspecific attack of proteins.

turnover of short-lived proteins, for the degradation of abnormal proteins, for the regulation of the cell cycle and for stress responses (Vierstra, 1993; Belknap and Garbarino, 1996). On the other hand, it appears unlikely that a major portion of leaf proteins is degraded during senescence via the ubiquitin/proteasome pathway (Bahrami and Gray, 1999).

During the past few years, more information has become available concerning Clp (Shanklin *et al.*, 1995; Schaller and Ryan, 1995; Crafts-Brandner *et al.*, 1996; Ostersetzer and Adam, 1996; Desimone *et al.*, 1997; Sokolenko *et al.*, 1998; Nakabayashi *et al.*, 1999). This complex proteolytic system is located in the plastids and consists of two types of subunits (ClpP and ClpC). ClpP, a 23-kDa polypeptide, is encoded in the plastome and contributes the proteolytic moiety to the complex (Shanklin *et al.*, 1995). ClpC (a polypeptide of about 90 kDa equivalent to ClpA in *E. coli*) is nuclear-encoded, synthesized as a larger precursor in the cytosol, imported into the plastids and cut to its final size by the removal of the transit peptide. ClpC bears the ATPase activity. The mRNAs (Crafts-Brandner *et al.*, 1996; Ostersetzer and Adam, 1996) and the proteins (Shanklin *et al.*, 1995) for ClpP and ClpC are present throughout the life cycle of a leaf and are not expressed in a senescence-specific manner. However, Nakabayashi *et al.* (1999) reported an increased transcription of some *clp* genes during dark-induced and natural senescence. Nuclear genes encoding ClpP (*nclpP*) have been detected recently beside the *clpP* gene on the plastome (Sokolenko *et al.*, 1998; Nakabayashi *et al.*, 1999). The exploration of the various components of the Clp system, of the genes encoding them and of their physiological functions remains a challenge for future research.

C. ATP-Dependency of Proteolysis

No ATP is required for the hydrolytic cleavage of an accessible peptide bond. Nevertheless, several ATP-dependent proteolytic systems have been identified in plant cells. Such proteolytic activities are present in various subcellular compartments (Vierstra, 1993; Shanklin *et al.*, 1995; Whelan and Glaser, 1997). In some cases the ATPase and the peptide hydrolase activities are present on different subunits (e.g., proteasome, Clp), while in other cases the two activite sites are present in the same polypeptide (e.g., in the mitochondrial protease Lon) as summarized by Lupas *et al.* (1997). Since ATP is not directly necessary for the cleavage of an accessible peptide bond, the ATP hydrolysis may be required for changes in the conformation of substrate proteins making them susceptible to proteolysis by exposing hydrolyzable peptide bonds to the active center of a peptide hydrolase. The ATPase moiety of proteolytic systems may be responsible for the substrate specificity by unfolding only those polypeptides with certain properties (e.g., destabilizing sequences, damage, partial denaturation).

D. Compartmentation

The vacuole contains various endopeptidases and carboxypeptidases (Heck *et al.*, 1981; Canut *et al.*, 1985; Barrett, 1986). The vacuole is frequently considered as the lytic compartment in plant cells. Although vacuoles contain a set of peptide hydrolases capable of hydrolyzing a series of leaf proteins including plastidial proteins, the role of this compartment in the remobilization of proteins prior to cell death is not yet clear. Several lines of evidence indicate that at least the initial steps in the degradation of plastidial proteins during senescence can occur inside the chloroplast (Feller and Fischer, 1994 and references therein). A transfer of peptides generated by plastidial protein degradation from the chloroplasts to the vacuole for final hydrolysis or the release of vacuolar enzymes into other cell compartments after membrane rupture could lead to an involvement of these peptide hydrolases in the catabolism of extravacuolar proteins and also to the remobilization of foliar proteins during senescence or other types of plant cell death (Guiamét *et al.*, 1999). A defensive function of vacuolar peptide hydrolases was also considered. Since protein bodies represent a special type of protein-storing vacuoles, proteolysis in these vacuoles during germination is very important for the mobilization of nitrogen and the function of peptide hydrolases is obvious.

Peptide hydrolases have also been identified in the extracellular space (Graham *et al.*, 1991; Groover and Jones, 1999). It was suggested that a 40-kDa extracellular protease plays an important role in the differentiation of tracheary elements (Groover and Jones, 1999), and that this protease is important at least for certain types of programmed cell death in plants.

The endoplasmic reticulum (Chrispeels, 1991), the mitochondria (Eriksson *et al.*, 1996; Schatz and Dobberstein, 1996), the peroxysomes (Distefano *et al.*, 1997) and the plastids (Reinbothe *et al.*, 1995; Shanklin, 1995; Roulin and Feller, 1998) have their own set of proteolytic activities. The proteolytic enzymes in the various compartments may be involved in the processing of precursor proteins, in the removal of damaged or not properly assembled proteins, in protein turnover and finally also in the net protein remobilization prior to cell death.

Proteolytic activities in chloroplasts are of special interest for three reasons: (1) Most of the proteins present in a mesophyll cell of a C_3-plant are in the chloroplasts (Peoples and Dalling, 1988). (2) Chloroplast functions and also chloroplast proteins are lost in an early phase of senescence, while other cell compartments are not yet affected (Makino *et al.*, 1983; Gepstein, 1988; Feller and Fischer, 1994). (3) Intact chloroplasts are able to degrade abundant proteins present inside these organelles (Mitsuhashi *et al.*, 1992; Desimone *et al.*, 1996). Light is a major factor influencing protein catabolism in isolated pea chloroplasts (Fig. 7-1). Furthermore, different degradation products of the large subunit of Rubisco accumulated in the presence and absence of light. Only proteolysis in darkness was strongly inhibited by EDTA (most likely by inhibiting a metallo-endopeptidase). Possible mechanisms for the effect of light on the degradation of stromal proteins are summarized in Fig. 7-3. Several soluble and insoluble peptide hydrolases have been detected in chloroplasts (Musgrove *et al.*, 1989; Bushnell *et al.*, 1993; Shanklin *et al.*, 1995; Lindahl *et al.*, 1996; Ostersetzer *et al.*, 1996). At least two types of proteases in the chloroplasts depend on ATP: the Clp system in the stroma with ClpP (protease moiety) and ClpC (ATPase moiety) and a membrane-bound homologue of the bacterial FtsH protease (Shanklin *et al.*, 1995; Lindahl *et al.*, 1996). The functions of these proteolytic enzymes in relation to cell death have not yet been identified.

IV. Proteolysis in Relation to Cell Death

A. Cell Death during Differentiation

Cell death is in plants also associated with differentiation processes. Some tissues are functional after cell death (e.g., xylem vessels, sklerenchyma). In this situation, the death is restricted to some well-defined cells in a growing plant part (Groover and Jones, 1999). Ye and Varner (1996) detected in activity gel assays an increase of a cysteine protease (apparent molecular mass of 20 kDa) and of a serine protease (apparent mass of 60 kDa) during xylogenesis.

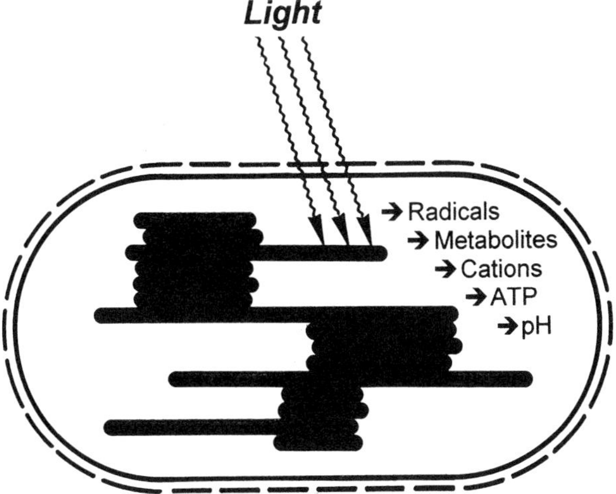

Figure 7-3. Possible mechanisms for the regulation of proteolysis in the stroma by light.

B. Senescence

Protein degradation is a prerequisite for an efficient remobilization of nitrogen from senescing plant parts (Peoples and Dalling, 1988; Huffaker, 1990; Feller and Fischer, 1994). Most of the nitrogen in photosynthesizing mesophyll cells of a C_3-plant is present in the chloroplasts and proteins represent the predominant nitrogen fraction in these organelles (Peoples and Dalling, 1988). Therefore the degradation of chloroplast proteins is a key process for the export of nitrogen from leaf cells during senescence (Feller and Fischer, 1994). Chloroplasts are disassembled in an early phase of senescence characterized by a decline in photosynthetic capacity and a loss of proteins involved in assimilatory processes (Makino *et al.*, 1983; Gepstein, 1988; Feller and Fischer, 1994). Depending on the actual environmental conditions (e.g., illumination, nutrient availability), senescence is reversible during the early phases (Wittenbach, 1977). This reversibility is progressively lost and senescence becomes irreversible after a certain time. It must be borne in mind that senescence can be initiated in the field and in the laboratory by various factors (e.g., insufficient illumination, nutrient limitation, drought). The processes during at least the early phases of senescence (including proteolysis in chloroplasts) may depend on the factor(s) causing senescence (Feller and Fischer, 1994 and references therein). Senescence is considered as a type of programmed cell death (Noodén *et al.*, 1997). Apoptosis in animal cells is also called programmed cell death, but this program differs considerably from the senescence program (Chapter 1). Senescence is in general a rather fine-tuned process controlled by a series of internal and external factors and allows the reallocation of phloem-mobile nutrients. However, other types of programmed cell death occur also in plants (Chapter 2) and may be more closely related to apoptosis in animal cells than senescence (e.g., hypersensitive reaction, xylem differentiation, aerenchyma formation). In these latter processes, the death of the cell *per se* and not a fine-tuned remobilization of nutrients is the most relevant aspect.

C. Hypersensitive Response

The hypersensitive response is induced by certain plant pathogens and leads to cell death (Chapter 3). Caspase-like proteolytic activity has been detected in tobacco leaves after infection with tobacco mosaic virus (del Pozo and Lam, 1998). This type of cell death is rapid and initiated locally by an external factor (pathogen), while cells surrounding the necrotic spots remain alive and are even protected by the rapid death of cells at the infection spot.

V. Regulation of Protein Catabolism

A. Changes in the Pattern of Proteolytic Enzymes

The rate of protein degradation depends on the activity of the peptide hydrolases present in the same compartment. An example of the initiation of a rapid protein degradation by an increased endopeptidase level is the *de novo* synthesis and maturation of a cysteine endopeptidase at the onset of germination (Mitsuhashi and Minamikawa, 1989; Okamoto *et al.*, 1994; Callis, 1995). Since the activity of the peptide hydrolases is important in this context and not the quantity of the enzyme protein, the actual conditions (e.g., pH, solutes) are important. Furthermore, peptides and proteins may serve as protease inhibitors and lower the activity of some peptide hydrolase present in the same compartment (Björk *et al.*, 1998). An up-regulation during senescence was discussed for some plant peptide hydrolases (Feller *et al.*, 1977; Peoples and Dalling, 1988; Callis, 1995; Buchanan-Wollaston, 1997; Noodén *et al.*, 1997), but it is still open to debate to what extent an up-regulation is a prerequisite for the remobilization of proteins prior to leaf cell death. Chloroplasts isolated from mature pea leaves and incubated in a suitable medium are able to degrade most of their proteins in a rather short time interval indicating that the import of new nuclear-encoded peptide hydrolases from the cytosol is not required (Fig. 7-1). The synthesis of proteolytic enzymes inside the isolated plastids appears rather unlikely, but this possibility cannot be completely ruled out. The involvement of proteases in xylem element differentiation was discussed by Ye and Varner (1996) and by Groover and Jones (1999). Their results suggest that an up-regulation of certain peptide hydrolases may be relevant for xylogenesis, a special type of plant cell death.

B. Altered Susceptibility of Substrate Proteins

The velocity of protein degradation depends equally on the pattern of peptide hydrolases and on the susceptibility of the substrate proteins (Feller and Fischer, 1994; Callis, 1995). The stability of a polypeptide in the presence of proteolytic enzymes may be altered by damage (e.g., under oxidative stress), covalent modifications (e.g., phosphorylation or dephosphorylation, ubiquitination) or other changes in the three-dimensional structure (e.g., association or dissociation of subunits, ligand binding). As shown in Fig. 7-4A, glutamine synthetase can be selectively protected in intact chloroplasts by its inhibitor methionine sulfoximine (substrate analogue). The interaction between glutamine synthetase and inhibitor affects the three-dimensional structure of the enzyme and as a consequence also its susceptibility to proteolysis. The reduction in disulfide bridges is an example of a reversible effect on protein stability (Kamber and Feller, 1998).

Isoforms of the same enzyme may not only differ in their expression pattern, their subcellular compartmentation and their catalytic properties, but they may also differ in their

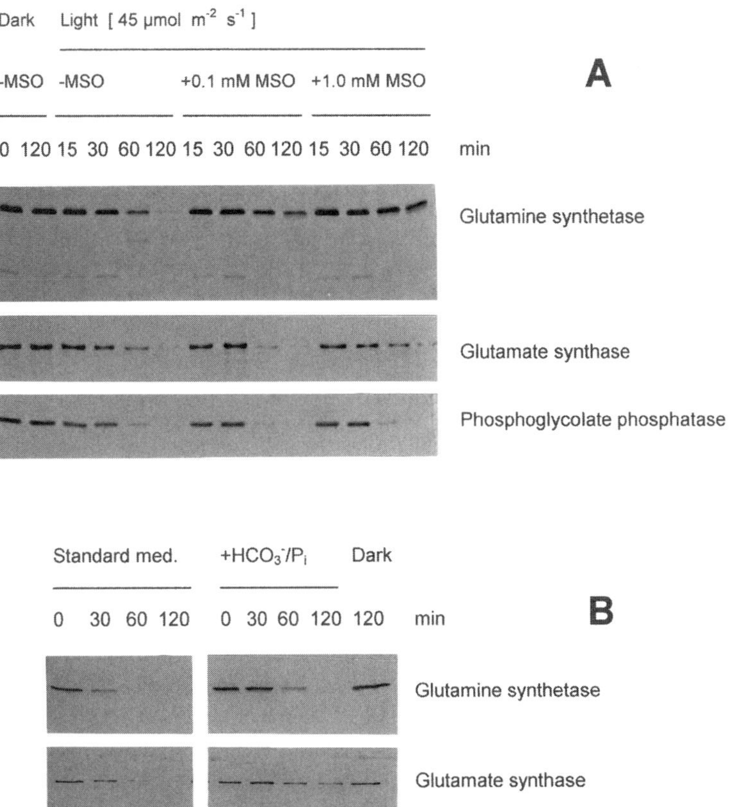

Figure 7-4. Effects of methionine sulfoximine (MSO; an inhibitor of glutamine synthetase) and of bicarbonate/phosphate on the degradation of glutamine synthetase, ferredoxin-dependent glutamate synthase and phosphoglycolate phosphatase in isolated pea (*Pisum sativum*) chloroplasts. Intact chloropasts were re-isolated after incubation at 25°C to ensure that only processes in intact chloroplasts were analyzed. Glutamine synthetase was stabilized specifically by MSO (A). As compared to the standard medium (Standard. med.), glutamine synthetase and glutamate synthase were both stabilized in the light (45 μmol m^{-2} s^{-1}) in a medium containing 5 mM $NaHCO_3$ and 0.5 mM KH_2PO_4 (B). The changes in the enzyme proteins were visualized by immunoblotting with specific antibodies. Equal quantities of chlorophyll (1 μg) were loaded on each lane. [From Thoenen and Feller (1998), with permission.]

stabilities in the presence of proteolytic enzymes (Streit and Feller, 1983; Thomas and Feller, 1993).

Oxidative stress may damage proteins and render them more susceptible to proteolytic attack (Chapter 12). In isolated chloroplasts, stromal proteins are in general more rapidly degraded in the light than in darkness (Mitsuhashi and Feller, 1992). As schematically shown in Fig. 7-3, light may influence protein catabolism in the stroma via the alkalinization of the stroma (Heldt *et al.*, 1973), via providing ATP (Stieger and Feller, 1997), via increased cation concentrations (Portis and Heldt, 1976), via altered concentrations of Calvin cycle intermediates or other metabolites (Thoenen and Feller, 1998) or via photooxidative processes (Stieger and Feller, 1997). Several lines of evidence lead to the conclusion that in isolated chloroplasts and most likely also in intact cells radicals play an important role in the degradation of stroma proteins in the light. In this context it is important to consider that dithiothreitol, a common component in chloroplast isolation media, is a potent inhibitor of two enzymes involved in radical detoxification in chloroplasts (Chen and Asada, 1992; Neubauer, 1993). Indeed, omitting dithiothreitol from chloroplast incubation media improves the stability of stromal enzymes in the light (Stieger and Feller, 1997; Roulin and Feller, 1998). Results obtained with antisense mutants strongly suggest a protective function of peroxiredoxins *in vivo* (Baier and Dietz, 1999). Peroxiredoxins reduce the toxic alkyl hydroperoxides to the corresponding alcohols and may play a role in the antioxidative defense of chloroplasts.

C. Aspects of Compartmentation

Changes in the compartmentation may allow the action of a previously inaccessible peptide hydrolase on a given substrate protein (Guiamét *et al.*, 1999). The compartmentation may be altered by membrane fusion or membrane disruption leading to a total or partial mixture of two aqueous phases. The fusion of endopeptidase-bearing vesicles with protein bodies in germinating seeds (Chrispeels *et al.*, 1976) or the release of vacuolar enzymes after tonoplast rupture (Cooke *et al.*, 1980; Yoshida and Minamikawa, 1996) are examples of this type of change. The release of vacuolar enzymes after tonoplast rupture was also proposed as a mechanism involved in tracheary element differentiation (Groover and Jones, 1999).

Another type of compartmentation was reported for self-compartmentalizing proteases (Lupas *et al.*, 1997; Baumeister *et al.*, 1998). Interactions of proteolytic enzymes or the corresponding proenzymes with the cytoskeleton or with membrane proteins may also prevent the action of a peptide hydrolase by separating the active site from the substrate proteins as suggested for caspases in mammalian cells (Ruoslahti and Reed, 1999). Little is known about the fate of self-compartmentalizing proteases under unfavorable conditions. As hypothetically shown in Fig. 7-2, the barrel-shaped structure may be destroyed and proteolytically active sites normally oriented towards the central pore may become accessible. The stimulatory effect of SDS on some proteolytic enzymes *in vitro* might depend on the dissociation of a latent complex to an active subunit or on the conformational change of a latent to an active protease as proposed recently by Yamada *et al.* (1998). The conversion of pre-existing protease(s) to more active or more accessible enzymes may be relevant for a rapid protein degradation associated with plant cell death.

D. Influence of Solutes

Solutes may interfere with proteolysis by interacting with proteolytic enzymes (rather general effect) or with some substrate proteins (rather specific effect). Cations must be considered as possible regulatory solutes for metallo-endopeptidases (Graham *et al.*, 1991; Bushnell *et al.*, 1993; Roulin and Feller, 1998). Changes in the pH of a compartment, the redox status and the availability of energy in the form of ATP may also influence the activity of a proteolytic system.

Inorganic compounds or metabolites can interact quite specifically with enzyme proteins bearing appropriate binding sites (Streit and Feller, 1982; Kurlandsky *et al.*, 1988; Chen and Spreitzer, 1991; Fröhlich *et al.*, 1994). The conformational changes caused by ligand binding may increase or decrease the accessibility of hydrolyzable peptide bonds to peptide hydrolases. The stabilization of the nuclear-encoded glutamine synthetase in intact pea plastids by a substrate analogue indicates that this enzyme is more slowly degraded by the

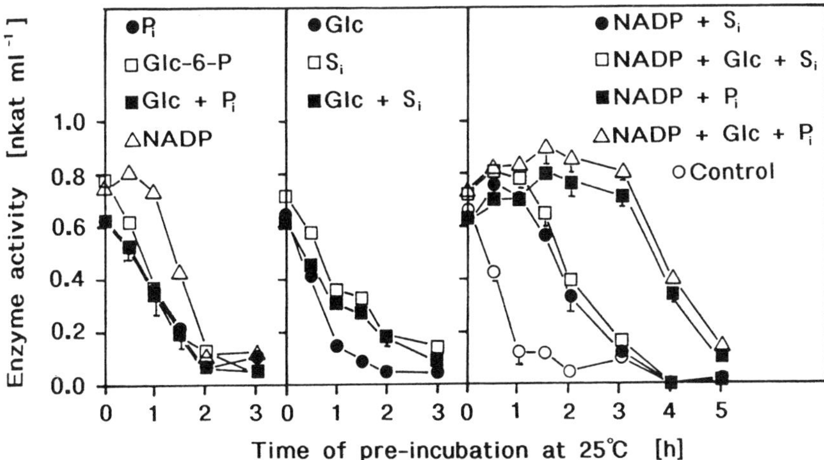

Figure 7-5. Stabilization of glucose-6-phosphate dehydrogenase by solute combinations. Extracts from ungerminated wheat caryopses (source of glucose-6-phosphate dehydrogenase) and endosperms of wheat plants germinated for 5 days (source of proteolytic activity) were pre-incubated at pH 5.4 and 25°C without additives (control) or with different effectors or effector combinations as indicated. The final concentration during pre-incubation was 1 mM for NADP and 10 mM for phosphate (Pi), sulfate (Si), glucose (Glc) and glucose-6-phosphate (Glc-6-P). Means and standard deviations for the remaining enzyme activity were computed from four pre-incubations. Standard deviations are shown when exceeding the size of the symbol (on one side only for clarity). [From Fischer *et al.* (1992), with permission.]

proteolytic system(s) present inside these organelles and that this effect is specific, since two other stromal enzymes, ferredoxin-dependent glutamate synthase and phosphoglycolate phosphatase, were not protected by this compound (Fig. 7-4A). In contrast, bicarbonate and inorganic phosphate added to the incubation medium for the isolated chloroplasts caused a stabilization of several stromal enzymes (Fig. 7-4B). This unspecific effect may be due either to interactions with the proteolytic system or to a stimulation of photosynthesis leading to a more balanced solute composition and as a consequence to a rather general stabilization of stromal proteins. These findings indicate that the actual status of the metabolism may influence protein degradation in intact organelles.

Besides with individual compounds, solute mixtures can interact in a more complex manner with proteins and as a consequence cause their stabilization or destabilization (Fig. 7-5). Ligands binding to different sites in an enzyme can act cooperatively and stabilize far more than one ligand alone. The degradation of an emzyme protein in a compartment with a given set of peptide hydrolases might be fine-tuned by altered solute concentrations caused by metabolic changes or by altered transport rates across membranes. Such effects must be considered as important regulatory mechanisms in protein turnover and may also be relevant for the regulation of protein catablism during senescence.

VI. Conclusions

Several types of proteolytic enzymes are present in various compartments of plant cells. Some of these proteolytic activities are similar to those found in animal cells (e.g., proteasome, mitochondrial peptide hydrolases), while some other enzymes are more typical for a plant cell or a plant cell compartment (e.g., Clp, some vacuolar enzymes). The functions of these enzymes in relation to plant cell death are not yet satisfactorily known. The fact that a proteolytic enzyme is able to degrade a substrate *in vitro* does not necessarily mean that this proteolytic enzyme plays a key role in the degradation of this protein *in vivo*, since a spatial separation or the actual conditions in the living cell may prevent such an interaction. Mutants and more sophisticated biochemical or biophysical techniques may offer new tools to elucidate proteolysis and its control in plants. Especially the following aspects remain to be clarified in future experiments:

(a) Are the same proteolytic enzymes involved in different types of plant cell death? Which enzymes play which role in which compartment?
(b) Is *de novo* synthesis of peptide hydrolases a prerequisite for the rapid protein catabolism prior to cell death or are pre-existing enzymes sufficient for this process? How are the activity and the accessibility of peptide hydrolases regulated *in vivo*?
(c) Is the control of proteolysis primarily due to altered susceptibilities of the substrate proteins, to changes in the compartmentation (of peptide hydrolases or of substrate proteins) or to the activity of proteolytic systems?
(d) Are the proteases just involved in reclaimation or do some of them actually cause death as certain caspases do in other systems?

References

Abad, M.S., Clark, S.E., and Lamppa, G.K. (1989). Properties of a chloroplast enzyme that cleaves the chlorophyll *a/b* binding protein precursor. Optimization of an organelle-free reaction. *Plant Physiol.* **90**, 117–124.

Adamska, I., Lindahl, M., Roobol-Boza, M., and Andersson, B. (1996). Degradation of the light-stress protein is mediated by an ATP-independent, serine-type protease under low light conditions. *Eur. J. Biochem.* **236**, 591–599.

Bahrami, A.R., and Gray, J.E. (1999). Expression of a proteasome alpha-type subunit gene during tobacco development and senescence. *Plant Mol. Biol.* **39**, 325–333.

Baier, M., and Dietz, K.-J. (1999). Protective function of chloroplast 2-cysteine peroxiredoxin in photosynthesis. Evidence from transgenic Arabidopsis. *Plant Physiol.* **119**, 1407–1414.

Barrett, A.J. (1986). The classes of proteolytic enzymes. In *Plant Proteolytic Enzymes* (M.J. Dalling, Ed.), Vol. I, pp. 1–16. CRC Press, Boca Raton, FL.

Baumeister, W., Walz, J., Zühl, F., and Seemüller, E. (1998). The proteasome: paradigm of a self-compartmentalizing protease. *Cell* **92**, 367–380.

Belknap, W.R., and Garbarino, J.E. (1996). The role of ubiquitin in plant senescence and stress responses. *Trends Plant Sci.* **1**, 331–335.

Björk, I., Nordling, K., Raub-Segall, E., Hellman, U., and Olson, S.T. (1998). Inactivation of papain by antithrombin due to autolytic digestion: a model of serpin inactivation of cysteine proteinases. *Biochem. J.* **335**, 701–709.

Bleecker, A.B. (1998). The evolutionary basis of leaf senescence: method to the madness? *Curr. Opin. Plant Biol.* **1**, 73–78.

Buchanan-Wollaston, V. (1997). The molecular biology of leaf senescence. *J. Exp. Bot.* **48**, 181–199.

Bushnell, T.P., Bushnell, D., and Jagendorf, A.T. (1993). A purified zinc protease of pea chloroplasts, EP1, degrades the large subunit of ribulose-1,5-bisphosphate carboxylase/oxygenase. *Plant Physiol.* **103**, 585–591.

Byrd, C., Turner, G.C., and Varshavsky, A. (1998). The N-end rule pathway controls the import of peptides through degradation of a transcriptional repressor. *EMBO J.* **17**, 269–277.

Callis, J. (1995). Regulation of protein degradation. *Plant Cell* **7**, 845–857.

Canut, H., Alibert, G., and Boudet, A.M. (1985). Proteases of *Melilotus alba* mesophyll protoplasts. I. Intracellular localization. *Plant Sci.* **39**, 163–169.

Chen, G.-X., and Asada, K. (1992). Inactivation of ascorbate peroxidase by thiols requires hydrogen peroxide. *Plant Cell Physiol.* **33**, 117–123.

Chen, P., and Hochstrasser, M. (1996). Autocatalytic subunit processing couples active site formation in the 20S proteasome to completion of assembly. *Cell* **86**, 961–972.

Chen, Z., and Spreitzer, R.J. (1991). Proteolysis and transition-state-analogue binding of mutant forms of ribulose-1,5-bisphosphate carboxylase/oxygenase from *Chlamydomonas reinhardtii*. *Planta* **183**, 597–603.

Chrispeels, M.J. (1991). Sorting of proteins in the secretory system. *Annu. Rev. Plant Physiol. Plant Mol. Biol.* **42**, 21–53.

Chrispeels, M.J., and Boulter, D. (1975). Control of storage protein metabolism in the cotyledons of germinating mung beans: role of endopeptidase. *Plant Physiol.* **55**, 1031–1037.

Chrispeels, M.J., Baumgartner, B., and Harris, N. (1976). Regulation of reserve protein metabolism in the cotyledons of mung bean seedlings. *Proc. Natl. Acad. Sci. USA* **73**, 3168–3172.

Cooke, R.J., Grego, S., Roberts, K., and Davies, D.D. (1980). The mechanism of deuterium oxide-induced protein degradation in *Lemna minor*. *Planta* **148**, 374–380.

Crafts-Brandner, S.J., Klein, R., Klein, P., Hölzer, R., and Feller, U. (1996). Coordination of protein and mRNA abundances of stromal enzymes and mRNA abundances of the Clp protease subunits during senescence of *Phaseolus vulgaris* (L.) leaves. *Planta* **200**, 312–318.

Del Pozo, O., and Lam, E. (1998). Caspases and programmed cell death in the hypersensitive response of plants to pathogens. *Curr. Biol.* **8**, 1129–1132.

Desimone, M., Henke, A., and Wagner, E. (1996). Oxidative stress induces partial degradation of the large subunit of ribulose-1,5-bisphosphate carboxylase/oxygenase in isolated chloroplasts of barley. *Plant Physiol.* **111**, 789–796.

Desimone, M., Weiss-Wichert, C., Wagner, E., Altenfeld, U., and Johanningmeier, U. (1997). Immunochemical studies on the Clp-protease in chloroplasts: evidence for the formation of a ClpC/P complex. *Bot. Acta* **110**, 234–239.

Distefano, S., Palma, J.M., Gomez, M., and Del Rio, L.A. (1997). Characterization of endoproteinases from plant peroxysomes. *Biochem. J.* **327**, 399–405.

Dungey, N.O., and Davies, D.D. (1982). Protein turnover in the attached leaves of non-stressed and stressed barley seedlings. *Planta* **154**, 435–440.

Eriksson, A.C., Sjöling, S., and Glaser, E. (1996). Characterization of the bifunctional mitochondrial processing peptidase (MPP)/bc_1 complex in *Spinacia oleracea*. *J. Bioenerg. Biomembr.* **28**, 285–292.

Feller, U., and Fischer, A. (1994). Nitrogen metabolism in senescing leaves. *Crit. Rev. Plant Sci.* **13**, 241–273.

Feller, U.K., Soong, T.-S.T., and Hageman, R.H. (1977). Leaf proteolytic activities and senescence during grain development of field-grown corn (*Zea mays* L.). *Plant Physiol.* **59**, 290–294.

Fischer, A., Salgó, A., Hildbrand, M., and Feller, U. (1992). Cooperative protection of glucose-6-phosphate dehydrogenase by ligands in extracts from wheat grains. *Biochem. Physiol. Pflanzen* **188**, 295–303.

Fröhlich, V., Fischer, A., Ochs, G., Wild, A., and Feller, U. (1994). Proteolytic inactivation of glutamine synthetase in extracts from wheat leaves: effect of pH, inorganic ions and metabolites. *Aust. J. Plant Physiol.* **21**, 303–310.

Gepstein, S. (1988). Photosynthesis. In *Senescence and Aging in Plants* (L.D. Noodén and A.C. Leopold, Eds.), pp. 85–109. Academic Press, San Diego.

Graham, J.S., Xiong, J., and Gillikin, J.W. (1991). Purification and developmental analysis of a metalloendoproteinase from the leaves of *Glycine max*. *Plant Physiol.* **97**, 786–792.

Greenberg, J.T. (1996). Programmed cell death: A way of life for plants. *Proc. Natl. Acad. Sci. USA* **93**, 12094–12097.

Groover, A., and Jones, A.M. (1999). Tracheary element differentiation uses a novel mechanism coordinating programmed cell death and secondary cell wall synthesis. *Plant Physiol.* **119**, 375–384.

Guarente, L., Ruvkun, G., and Amasino, R. (1998). Aging, life span, and senescence. *Proc. Natl. Acad. Sci. USA* **95**, 11034–11036.

Guiamét, J.J., Pichersky, E., and Noodén, L.D. (1999). Mass exodus from senescing soybean chloroplasts. *Plant Cell Physiol.* **40**, 986–992.

Halperin, T., and Adam, Z. (1996). Degradation of mistargeted OEE33 in the chloroplast stroma. *Plant Mol. Biol.* **30**, 925–933.

Hara-Nishimura, I., Kinoshita, T., Hiraiwa, N., and Nishimura, M. (1998). Vacuolar processing enzymes in protein-storage vacuoles and lytic vacuoles. *J. Plant Physiol.* **152**, 668–674.

Harvey, B.M.R., and Oaks, A. (1974). The hydrolysis of endosperm protein in *Zea mays*. *Plant Physiol.* **53**, 453–457.

Heck, U., Martinoia, E., and Matile, P. (1981). Subcellular localization of acid proteinase in barley mesophyll protoplasts. *Planta* **151**, 198–200.

Heldt, H.W., Werdan, K., Milovancev, M., and Geller, G. (1973). Alkalinization of the chloroplast stroma caused by light-dependent proton flux into the thylakoid space. *Biochim. Biophys. Acta* **314**, 224–241.

Huffaker, R.C. (1990). Proteolytic activity during senescence of plants. *New Phytol.* **116**, 199–231.

James, F., Brouquisse, R., Suire, C., Pradet, A., and Raymond, P. (1996). Purification and biochemical characterization of a vacuolar serine endopeptidase induced by glucose starvation in maize roots. *Biochem. J.* **320**, 283–292.

Jiang, W.B., Lers, A., Lomaniec, E., and Aharoni, N. (1999). Senescence-related serine protease in parsley. *Phytochemistry* **50**, 377–382.

Kamber, L., and Feller, U. (1998). Influence of the activation status and of ATP on phosphoribulokinase degradation. *J. Exp. Bot.* **49**, 1197–1204.

Keegstra, K., and Cline, K. (1999). Protein import and routing systems of chloroplasts. *Plant Cell* **11**, 557–570.

Kurlandsky, S.B., Hilburger, A.C., and Levy, H.R. (1988). Glucose-6-phosphate dehydrogenase from *Leuconostoc mesenteroides*: Ligand-induced conformational changes. *Arch. Biochem. Biophys.* **264**, 93–102.

Lindahl, M., Tabak, S., Cseke, L., Pichersky, E., Andersson, B., and Adam, Z. (1996). Identification, characterization, and molecular cloning of a homologue of the bacterial FtsH protease in chloroplasts of higher plants. *J. Biol. Chem.* **271**, 29329–29334.

Lupas, A., Flanagan, J.M., Tamura, T., and Baumeister, W. (1997). Self-compartmentalizing proteases. *Trends Biochem. Sci.* **22**, 399–404.

Makino, A., Mae, T., and Ohira, K. (1983). Photosynthesis and ribulose 1,5-bisphosphate carboxylase in rice leaves. Changes in photosynthesis and enzymes involved in carbon assimilation from leaf development through senescence. *Plant Physiol.* **73**, 1002–1007.

Mattoo, A.K., Hoffmann-Falk, H., Marder, J.B., and Edelman, M. (1984). Regulation of protein metabolism: coupling of photosynthetic electron transport to in vivo degradation of the rapidly metabolized 32-kilodalton protein of the chloroplast membranes. *Proc. Natl. Acad. Sci. USA* **81**, 1380–1384.

McGeehan, G, Burkhart, W., Anderegg, R., Becherer, J.D., Gillikin, J.W., and Graham, J.S. (1992). Sequencing and characterization of the soybean leaf metalloproteinase. *Plant Physiol.* **99**, 1179–1183.

Mikola, L., and Mikola, J. (1986). Occurrence and properties of different types of peptidases in higher plants. In *Plant Proteolytic Enzymes* (M.J. Dalling, Ed.), Vol. I, pp. 97–117. CRC Press, Boca Raton, FL.

Mitsuhashi, W., and Feller, U. (1992). Effects of light and external solutes on the catabolism of nuclear-encoded stromal proteins in intact chloroplasts isolated from pea leaves. *Plant Physiol.* **100**, 2100–2105.

Mitsuhashi, W., and Minamikawa, T. (1989). Synthesis and post-translational activation of sulfhydryl-endopeptidase in cotyledons of germinating *Vigna mungo* seeds. *Plant Physiol.* **89**, 274–279.

Mitsuhashi, W., Crafts-Brandner, S.J., and Feller, U. (1992). Ribulose-1,5-bis-phosphate carboxylase/oxygenase degradation in isolated pea chloroplasts incubated in the light or in the dark. *J. Plant Physiol.* **139**, 653–658.

Müntz, K., Becker, C., Pancke, J., Schlereth, A., Fischer, J., Horstmann, C., Kirkin, V., Neubohn, B., Senyuk, V., and Shutov, A. (1998). Protein degradation and nitrogen supply during germination and seedling growth of vetch (*Vicia sativa* L.). *J. Plant Physiol.* **152**, 683–691.

Musgrove, J.E., Elderfield, P.D., and Robinson, C. (1989). Endopeptidases in the stroma and thylakoids of pea chloroplasts. *Plant Physiol.* **90**, 1616–1621.

Nakabayashi, K., Masaki, I., Kiyosue, T., Shinozaki, K., and Watanabe, A. (1999). Identification of *clp* genes expressed in senescing *Arabidopsis* leaves. *Plant Cell Physiol.* **40**, 504–514.

Navarre, D.A., and Wolpert, T.J. (1999). Victorin induction of an apoptotic/senescence-like response in oats. *Plant Cell* **11**, 237–249.

Neubauer, C. (1993). Multiple effects of dithiothreitol on non-photochemical fluorescence quenching in intact chloroplasts. *Plant Physiol.* **103**, 575–583.

Noodén, L.D., Guiamét, J.J., and John, I. (1997). Senescence mechanisms. *Physiol. Plant.* **101**, 746–753.

Okamoto, T., Nakayama, H., Seta, K., Isobe, T., and Minamikawa, T. (1994). Posttranslational processing of a carboxy-terminal propeptide containing a KDEL sequence of plant vacuolar cysteine endopeptidase (SH-EP). *FEBS Lett.* **351**, 31–34.

Ostersetzer, O., and Adam, Z. (1996). Effects of light and temperature on expression of ClpC, the regulatory subunit of chloroplastic Clp protease, in pea seedlings. *Plant Mol. Biol.* **31**, 673–676.

Ostersetzer, O., Tabak, S., Yarden, O., Shapira, R., and Adam, Z. (1996). Immunological detection of proteins similar to bacterial proteases in higher plant chloroplasts. *Eur. J. Biochem.* **236**, 932–936.

Peoples, M.B., and Dalling, M.J. (1988). The interplay between proteolysis and amino acid metabolism during senescence and nitrogen reallocation. In *Senescence and Aging in Plants* (L.D. Noodén and A.C. Leopold, Eds.), pp. 181–217. Academic Press, San Diego.

Portis, A.R., and Heldt, H.W. (1976). Light-dependent changes of the Mg^{2+} dependency of CO_2 fixation in intact chloroplasts. *Biochim. Biophys. Acta* **449**, 434–446.

Ragster, L., and Chrispeels, M.J. (1979). Azocoll-digesting proteinases in soybean leaves. Characteristics and changes during leaf maturation and senescence. *Plant Physiol.* **64**, 857–862.

Reinbothe, C., Apel, K., and Reinbothe, S. (1995). A light-induced protease from barley plastids degrades NADPH, protochlorophyllide oxidoreductase complexed with chlorophyllide. *Mol. Cell. Biol.* **15**, 6206–6212.

Roulin, S., and Feller, U. (1998). Light-independent degradation of stromal proteins in intact chloroplasts isolated from *Pisum sativum* L. leaves: requirement for divalent cations. *Planta* **205**, 297–304.

Ruoslahti, E., and Reed, J. (1999). New way to activate caspases. *Nature* **397**, 479–480.

Ryan, C., and Walker-Simmons, M. (1981). Plant proteinases. In *The Biochemistry of Plants* (A. Marcus, Ed.), pp. 321–349. Academic Press, New York.

Schaller, A., and Ryan, C.A. (1995). Cloning of a tomato cDNA (GenBank L38581) encoding the proteolytic subunit of a Clp-like energy dependent protease. *Plant Physiol.* **108**, 1341.

Schatz, G., and Dobberstein, B. (1996). Common principles of protein translocation across membranes. *Science* **271**, 1519–1526.

Schmidt, G.W., and Mishkind, M.L. (1983). Rapid degradation of unassembled ribulose 1,5-bisphosphate carboxylase small subunit in chloroplasts. *Proc. Natl. Acad. Sci. USA* **80**, 2623–2636.

Shanklin, J., DeWitt, N.D., and Flanagan, J.M. (1995). The stroma of higher plant plastids contain ClpP and ClpC, functional homologs of *Escherichia coli* ClpP and ClpA: an archetypal two-component ATP-dependent protease. *Plant Cell* **7**, 1713–1722.

Sokolenko, A., Lerbs-Mache, S., Altschmied, L., and Herrmann, R.G. (1998). Clp protease complexes and their diversity in chloroplasts. *Planta* **207**, 286–295.

Stieger, P.A., and Feller, U. (1997). Requirements for the light-stimulated degradation of stromal proteins in isolated pea (*Pisum sativum* L.) chloroplasts. *J. Exp. Bot.* **48**, 1639–1645.

Streit, L., and Feller, U. (1982). Inactivation of N-assimilating enzymes and proteolytic activities in wheat leaf extracts: effects of pyridine nucleotides and of adenylates. *Experientia* **38**, 1176–1180.

Streit, L., and Feller, U. (1983). Changing activities and different resistance to proteolytic activity of two forms of glutamine synthetase in wheat leaves during senescence. *Physiol. Vég.* **21**, 103–108.

Thoenen, M., and Feller, U. (1998). Degradation of glutamine synthetase in intact chloroplasts isolated from pea (*Pisum sativum*) leaves. *Aust. J. Plant Physiol.* **25**, 279–286.

Thomas, H., and Feller, U. (1993). Leaf development in *Lolium temulentum*: differential susceptibility of transaminase isoenzymes to proteolysis. *J. Plant Physiol.* **142**, 37–42.

Valpuesta, V., Lange, N.E., Guerrero, C., and Reid, M.S. (1995). Up-regulation of a cysteine protease accompanies the ethylene-insensitive senescence of daylily (*Hemerocallis*) flowers. *Plant Mol. Biol.* **28**, 575–582.

Vierstra, R.D. (1993). Protein degradation in plants. *Annu. Rev. Plant Physiol. Plant Mol. Biol.* **44**, 385–410.

Whelan, J., and Glaser, E. (1997). Protein import into plant mitochondria. *Plant Mol. Biol.* **33**, 771–789.

Wittenbach, V.A. (1977). Induced senescence of intact wheat seedlings and its reversibility. *Plant Physiol.* **59**, 1039–1042.

Wolf, B.B., and Green, D.R. (1999). Suicidal tendencies: apoptotic cell death by caspase family proteinases. *J. Biol. Chem.* **274**, 20049–20052.

Yamada, T., Ohta, H., Masuda, T., Ikeda, M., Tomita, N., Ozawa, A., Shioi, Y., and Takamiya, K. (1998). Purification of a novel type of SDS-dependent protease in maize using a monoclonal antibody. *Plant Cell Physiol.* **39**, 106–114.

Ye, Z.-H., and Varner, J.E. (1996). Induction of cysteine and serine proteases during xylogenesis in *Zinnia elegans*. *Plant Mol. Biol.* **30**, 1233–1246.

Yoshida, T., and Minamikawa, T. (1996). Successive amino-terminal proteolysis of the large subunit of ribulose 1,5-bisphosphate carboxylase/oxygenase by vacuolar enzymes from French bean leaves. *Eur. J. Biochem.* **238**, 317–324.

8

Ethylene Signaling in Plant Cell Death

Autar K. Mattoo and Avtar K. Handa

I. Introduction

Hormonal controls singly or in combination are essential for overall control of growth, development and senescence in plants. A number of plant hormones have been implicated in these processes, namely, auxins, cytokinins, gibberellins, abscisic acid, jasmonates, and ethylene. Intracellular levels and the sensitivity of a particular cell type or tissue to hormones control plant metabolism and function. Among the plant hormones, the one most associated with promotion of senescence and cell death is ethylene. Ethylene is a simple gaseous hydrocarbon with myriad roles in plant life, namely, seed germination, diageotropism, flowering, abscission, senescence, fruit ripening, and pathogenesis (Mattoo and Aharoni, 1988; Mattoo and Suttle, 1991; Abeles *et al.*, 1992; Fluhr and Mattoo, 1996). This chapter deals with the role of ethylene in signaling in senescence and cell death.

Technically, senescence refers to all forms of programmed cell death (PCD) in plants; however, PCD is commonly used now to refer to senescence of cells or a group of cells rather than to that of a whole plant organ. To gain insight into ethylene's role in senescence, it is necessary to keep in mind other roles ethylene plays in cellular function and how other hormones and factors override the biosynthesis and action of ethylene. Several discoveries in the late-1960s and 1970s made available biochemical tools that had a major impact in understanding ethylene action and unraveling the biosynthetic pathway of ethylene. These included the discovery (Lieberman *et al.*, 1966) of methionine as a precursor of ethylene, the discovery (Owens *et al.*, 1971) of rhizobitoxine and its analogue aminoethoxyvinylglycine (AVG) as relatively specific inhibitors of ethylene biosynthesis, the use of silver salts (Beyer, 1976) and cyclic olefins (Sisler, 1977) as inhibitors of ethylene action, the breakthrough discovery (Adams and Yang, 1979; Lurssen *et al.*, 1979) of 1-aminocyclopropane-1-carboxylic acid (ACC) as the intermediate between methionine and ethylene, and finally the identification of ethylene receptors (Chang *et al.*, 1993). More recently, mutations are being used to determine the role of ethylene in senescence. Details on the steps regulated in the biosynthetic pathway of ethylene or the way ethylene receptors act can be found in several interesting reviews (Fluhr and Mattoo, 1996; Chang and Stadler, 2001; Hall *et al.*, 2001; Wang *et al.*, 2002).

II. Ethylene Biosynthesis Pathways

A. ACC Pathway

A major route of ethylene synthesis in higher plants involves the following metabolic sequence: methionine → S-adenosylmethionine (AdoMet) → ACC → ethylene. Methionine is converted to AdoMet [ATP:L-methionine S-adenosyltransferase (AdoMet synthetase, EC 2.5.1.6); Giovanelli *et al.*, 1980], AdoMet to ACC [AdoMet methylthioadenosinelyase (ACC synthase, EC 4.4.1.14); Kende, 1989] and ACC to ethylene (ACC oxidase, also called ethylene-forming enzyme; John, 1991). The enzymes catalyzing these reactions and the genes encoding them have been demonstrated in plants (Giovannoni, 2001). The genes encoding these enzymes belong to multigene families (see Fluhr and Mattoo, 1996). Generally, the rate-limiting steps in this pathway are catalyzed by ACC synthase and ACC oxidase.

ACC is only one product of ACC synthase activity, the other product produced in stoichiometric amounts is 5′-methylthioadenosine (MTA) (Adams and Yang, 1979). MTA is also a product generated from decarboxylated-AdoMet during the biosynthesis of polyamines (Schlenk, 1983; Cohen, 1998) and from AdoMet during enzymatic methylation of nucleic acids (Grefter *et al.*, 1966). Further, MTA is readily metabolized and recycled to methionine (Kushad, 1990). The recognition of MTA as a common biosynthetic product in these reactions made it apparent that the different pathways might be interlinked and developmentally regulated.

B. Non-ACC Pathway

Although ethylene production in most instances occurs via induction of ACC synthase, alternative ethylene synthesis pathways also exist during certain stresses and other situations. For example, ACC does not appear to serve as a precursor of ethylene in aquatic ferns, some aquatic angiosperms, in *Ceratocystis*-infected sweet potato root tissue, or acid-stressed Norway spruce needles (see Mattoo and White, 1991). In several of these examples, oxygen free radicals interact with fatty acids or methionine to generate ethylene. Since free radical generation is intimately associated with PCD (Chapter 13), it is possible that in some of these processes ethylene is produced via a non-ACC pathway to allow for relatively quick death of infected or damaged cells.

III. Temporal and Spatial Regulation of Ethylene Biosynthesis

Different biotic and abiotic factors regulate ethylene production in higher plants, and thus influence plant senescence and PCD (see Mattoo and White, 1991; Dangl *et al.*, 2000). Ethylene production is modulated by a number of factors: physical and chemical wounding, environmental conditions, other hormones (e.g., IAA, ABA, cytokinins and methyl jasmonate), and metabolites such as carbohydrates, orthophosphate, and polyamines (see Mattoo and Aharoni, 1988; Mattoo and White, 1991; McKeon *et al.*, 1995). The role of ethylene as a promoter of senescence and cell death involves a shift from a growth program in a plant cell to that of senescence suggesting that intricate controls exist in plants to fine tune ethylene production. The cellular redox is one of the regulatory factors that influence PCD/senescence. A change in cellular redox induced by oxidative stress promoted early senescence in some plants (Mattoo *et al.*, 1986). The change in the chloroplast redox resulted in rapid oxidation and degradation of ribulose bisphosphate carboxylase/oxygenase and membrane association of proteins (Mehta *et al.*, 1992).

IV. Ethylene Signal Transduction Pathway

In fruits—tomato, avocado and banana—ethylene is known to coordinate and complete the ripening process. Tomato and, in recent years, *Arabidopsis* have been used as plant models to unravel mechanisms surrounding ethylene action in plants. Since ripening and senescence share many common features during cell death (Gillaspy *et al.*, 1993), one favorite approach to unravel regulation has been the use of mutants as well as molecular genetics. In Arabidopsis, the mutations clearly support a role for ethylene in leaf senescence; however, mutations in the ethylene-signaling pathway indicate that ethylene promotes rather than initiates senescence (Bleecker and Patterson, 1997; Jing *et al.*, 2002). Tomato ripening single-gene mutants—*rin* (ripening inhibitor), *nor* (non-ripening), *Nr* (never ripe) (Tigchelaar *et al.*, 1978) and the "alcobaca" mutant (Leal and Tabim, 1974)—have contributed substantially to studies devoted to identification of ethylene receptors and the signal transduction pathway in tomato. The fruits of these mutants show an extended shelf life, absence of the ethylene-mediated climacteric rise in respiration, inability to fully soften, and inferior flavor and aroma (Giovannoni, 2001). In addition, leaves, petioles and abscission zone tissue of *Nr* plants exhibit greatly delayed senescence and abscission. The tomato locus *Nr* mapped together with an *Arabidopsis ETR1* gene-RFLP probe, suggesting homology (Yen *et al.*, 1995). The *NR* protein in tomato is closely related to the *AtERS1* ethylene receptor in *Arabidopsis*, indicating that pleiotropic phenotype of Nr mutation relates to ethylene insensitivity (see below; Wilkinson *et al.*, 1995; Payton *et al.*, 1996).

A large number of ethylene mutants, including *etr1*, *ein2*, *ein3*, *ein4*, *ein5*, *ein6* and *ein7*, have been isolated from *Arabidopsis*, and they have significantly contributed to our current knowledge on the regulation of ethylene biosynthesis and its perception with reference to ripening, senescence and environmental stresses (see Table 8-1, and reviews by Chang and Stadler, 2001; Hall *et al.*, 2001; Wang *et al.*, 2002). *ein5* and *ein7* may be allelic, while the other loci map separately (Romano *et al.*, 1995). The expression of *rin* and *nor* loci appears in a narrow developmental window during fruit ripening. Successful targeting of the genes via map-based cloning schemes has contributed to the understanding of the molecular nature of these lesions in *nor* and *rin* tomato mutants (Giovannoni, 2001). The *rin* locus has

Table 8-1. *Arabidopsis* Mutants Showing How Other Phytohormones Interact with the Ethylene Signaling Pathway

Mutant	Gene product	Phenotype	References
Ethylene and ABA			
abi1-1	Protein phosphatase 2C	Pleiotropic defect in ABA, response	Koorneef *et al.*, 1984; Leung *et al.*, 1994; Meyer *et al.*, 1994
era3/ein2	Membrane bound metal sensor	Ethylene sensitive allelic to *ein2*	Alonso *et al.*, 1999; Ghassemian *et al.*, 2000
ctr1	Protein kinase raf family	Reduced dormancy Constitutive triple response	Kieber *et al.*, 1993; Beaudoin *et al.*, 2000
Ethylene and sugar sensing			
gin1, aba2 *sis4, isi4* *san3, sre1*	Short chain dehydrogenase/ reductase	Growth on 6% Glu Reduced dormancy ABA-deficient, wilty Lack triple response in dark	Zhou *et al.*, 1998; Leon-Kloosterziel *et al.*, 1996; Rook *et al.*, 2001; Laby *et al.*, 2000
gin2	Hexose kinase	Growth on 6% Glu	Zhou *et al.*, 1998
gin4, *ctr1, sis1*	Protein kinase raf family	Growth on 6% Glu Reduced dormancy Constitutive triple response	Zhou *et al.*, 1998; Laby *et al.*, 2000; Kieber *et al.*, 1993; Beaudoin *et al.*, 2000
Ethylene and auxin			
aux1	Auxin amino acid permease	Auxin resistant root growth Abolish root gravicurvature Disrupts apical hook formation	Bennett *et al.*, 1996; Romano *et al.*, 1995; Marchant *et al.*, 1999
hsl1 *sur1, alf1*	N-acetyltransferase	Hookless Disrupts apical hook formation	Boerjan *et al.*, 1995; Celenza *et al.*, 1995; Lehman *et al.*, 1996
eto1		Ethylene over-producer NPA and auxin disrupt apical hook formation	Lehman *et al.*, 1996
ctr1	Protein kinase raf family	NPA and auxin disrupt apical hook formation	Lehman *et al.*, 1996
pir2/ein2		NPA resistant affected in root-elongation	Fujita and Syono, 1996
Ethylene and cytokinin			
cin5	ACC synthase	Absence of triple response in the presence of kinetin	Vogel *et al.*, 1998
ckr1/ein2		Cytokinin-resistant root growth	Su and Howell, 1992

now been shown to contain two MADS-box genes—*LeMADS-RIN* and *LeMADS-MC*—that encode transcription factors (Vrebalov *et al.*, 2002). Vrebalov and colleagues elegantly demonstrated non-hormonal regulation of ripening, upstream of ethylene in the regulatory cascade, by *LeMADS-RIN*. Seedlings of these mutants show normal sensitivity to ethylene (Lanahan *et al.*, 1994). However, in the fruit, both mutants fail to synthesize climacteric ethylene or accumulate the red carotenoid lycopene (Tigchelaar *et al.*, 1978). These mutants should aid studies for determining hormonal regulation of PCD in reproductive tissues of plants.

How do ethylene receptors regulate ethylene responses during plant growth, development and senescence? Chang *et al.* (1993) were the first to clone and characterize the gene, *AtETR1*, responsible for a dominant ethylene-insensitive mutant in *Arabidopsis* and discovered that it shared many similarities with two-component regulators in yeast and bacteria. Subsequently, the *AtETR1* gene was expressed in yeast and the recombinant protein was found to bind ethylene *in vitro* with similar affinity as that estimated from the dose-response curve for ethylene inhibition of hypocotyl growth in *Arabidopsis* seedlings (Schaller and Bleecker, 1995). The *AtETR1* ethylene receptor has three domains: a sensor, a kinase, and a receiver domain (response regulator). Ethylene binds to the N-terminal sensor domain that has three membrane-spanning helices (Schaller and Bleecker, 1995). In *Arabidopsis*, five genes make up a family of ethylene receptors (Bleecker, 1999). They all contain the three transmembrane domains required for ethylene binding, and a putative, GAF-like, cyclic nucleotide-binding domain (Bleecker, 1999). Interestingly, *AtERS1* and *AtERS2* lack a response regulator, while three of the five gene products, *AtETR2*, *AtEIN4* and *AtERS2*, do not contain the target amino acids deemed necessary for the histidine kinase activity found in *AtETR1* (Bleecker, 1999). Therefore, the role of the histidine kinase domain and response regulator in the ethylene signal transduction pathway remains to be elucidated. Wang *et al.* (2003) provide evidence that the histidine kinase domain in the ETR1 ethylene receptor is not required for ethylene signaling in *Arabidopsis*. Klee (unpublished data) has suggested that the proposed histidine kinase domain may in actuality be a serine–threonine kinase. Although the receptor proteins are structurally different, Hua and Meyerowitz (1998) proposed that at least four of them serve redundant functions in *Arabidopsis*. Analysis of the loss-of-function mutants revealed constitutive ethylene-like response, which was suggested to indicate that the ethylene response pathway is negatively regulated by the ethylene receptors in *Arabidopsis*. Two orthologues of the *Arabidopsis* ETR1 gene in tomato, eTAE1 (Zhou *et al.*, 1996a) and TFE27 (Zhou *et al.*, 1996b), renamed *LeETR1* and *LeETR2*, respectively, also possess the three domains of the *AtETR1* protein (sensor, histidine kinase and receiver domains), while *NR* (renamed *LeETR3*), like *AtERS1*, is devoid of a receiver domain. Two additional genes belonging to the tomato ethylene receptor family, *LeETR4* and *LeETR5* (Tieman and Klee, 1999), contain a sequence for a putative receiver domain but do not have the necessary domain for histidine kinase (Tieman and Klee, 1999). Models of how various gene products may interact to regulate ethylene action have been presented (Fig. 8-1; see reviews by Chang and Stadler, 2001; Hall *et al.*, 2001; Wang *et al.*, 2002) but which of these are of consequence in the various types of plant senescence is yet to be determined.

V. Ethylene Cross Talk with Other Plant Hormones

Various hormones have been shown to regulate senescence (see below). Whether similar is true of their role in plant PCD remains to be determined. Plant hormones do interact or cross talk with one another to influence plant processes. New genomics research is beginning to show that different hormones induce factors or transcription factors that are then transported into the nucleus to bind to specific regulatory sequences and control expression of a specific gene, i.e., in the 5′ region of a gene several regulatory sequences exist that have the potential to bind specific transcription factors induced by different hormones. An understanding of these interactions should shed further light onto which molecular processes are shared by PCD and senescence. It is only recently that we have begun to understand the molecular complexity involving interactions between phytohormones. *Arabidopsis* genetics and

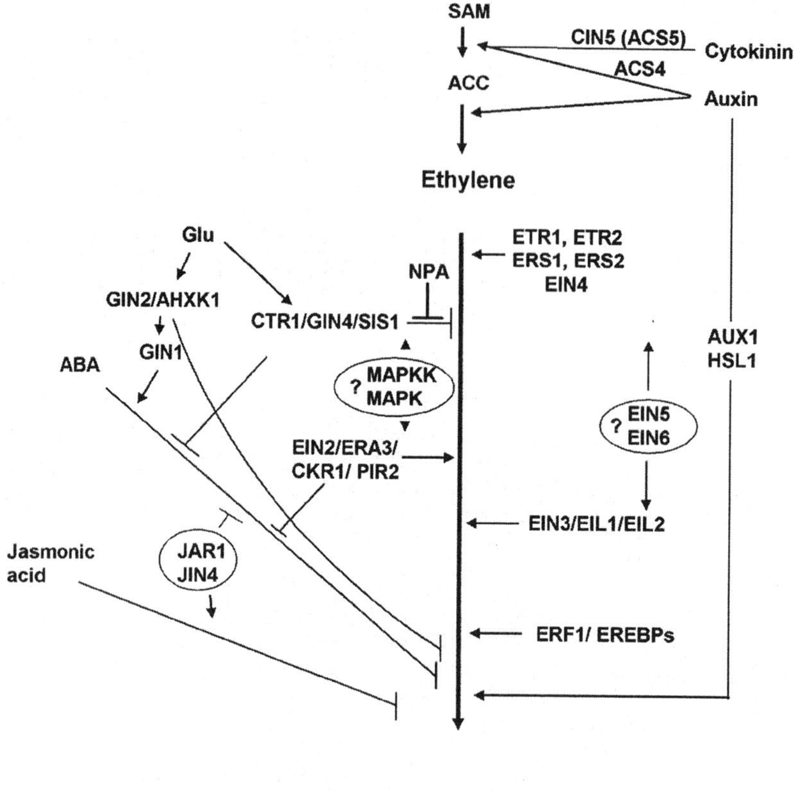

Figure 8-1. Cross-talk between ethylene and other plant hormones. Selection for altered phytohormone responses resulted in isolation of mutants that have a shared ethylene-signaling pathway. Mutants *era3*, *ckr1* and *pir2* selected for enhanced response to ABA, root growth resistance to cytokinin, and N-1-naphthylphthalamic acid (a polar auxin transport inhibitor), respectively, were alleles of ein2. Alleles of ctr1 were isolated among the mutants that showed resistance to high levels of sugar (*gin4*, *sis1*) and enhancement of *abi1-1* (ABA-resistant seed germination) mutant. *gin1* and *gin2* affect seedling response to ethylene. Mutations in an auxin amino acid permease (*aux1*) and N-acetyltransferase (*hsl1*) disrupted the apical hook formation. Both auxin and cytokinin enhance ethylene production by regulating expression of different members of the ACC synthase gene family. A mutation in ASC5 is associated with resistance to kinetin. Isolation and characterization of additional mutations impacting plant response to growth regulators would help understanding of the molecular circuitry regulating plant growth and development.

other molecular tools are helping to unravel genetic circuitries and molecular mechanisms regulating cross talk between ethylene and other plant hormones. Several laboratories have isolated *Arabidopsis* mutants that show altered response to classical phytohormones in the presence of ethylene (Table 8-1). Molecular characterization of these mutants has allowed identification of a number of genes underlying interactions between ethylene-, abscisic acid (ABA)-, auxin-, cytokinin-, sugar- and light-signaling pathways. Figure 8-1 attempts to summarize information in the literature and show emerging possibilities of cross talk between ethylene and other hormones, including convergence between the ethylene, ABA and sugar at the ethylene signal transduction pathway intermediates, CTR and EIN2.

Some of the components of hormonal signal transduction pathways act as central points where signals from different hormones merge and undergo amplification, modulation or attenuation to regulate plant growth, developmental processes, and PCD. The interactions between different signaling pathways are relatively specific, and we are just beginning to understand which hormone response loci are involved in multiple signaling pathways. Studies on additional mutants, including alleles of *ein2* in screens involving auxin transport inhibitors (Fujita and Syono, 1996) and cytokinin (Cary *et al.*, 1995) and delayed senescence (Oh *et al.*, 1997) suggest that the ethylene-signaling pathway is likely to intersect with other senescence-signaling pathways as well. In the following sections, we present examples of cross talk existence where multiple hormones control a single process, if not for anything else but to bring to light the possibility that some aspects of ethylene-regulated PCD may indeed be regulated in similar ways.

A. Ethylene, Abscisic Acid and Sugar

Genetic analysis of *Arabidopsis* mutants altered in ABA response provided the evidence that the ethylene signal transduction pathway intersects with that of ABA (Beaudoin *et al.*, 2000; Ghassemian *et al.*, 2000). Characterization of suppressor and enhancers of *abi1-1*, an *Arabidopsis* mutant showing ABA-resistant seed germination, resulted in identification of alleles of mutant *ein2* and mutant *ctr1*, respectively (Beaudoin *et al.*, 2000). At about the same time, *era3* (enhanced response to ABA3), an *Arabidopsis* mutant whose seed exhibited enhanced sensitivity of germination in response to ABA, was demonstrated to be allelic to *ein2* (Ghassemian *et al.*, 2000). Genetic characterization of ethylene-insensitive mutations showed that *ETR1*, *CTR1*, and *EIN2* inhibit ABA signaling in seeds. It was concluded that ethylene promotes seed germination by decreasing sensitivity to endogenous ABA. In contrast, ABA inhibition of root growth seems to require a functional ethylene-signaling cascade as the roots of *ein2* and *etr1-1* are resistant to both ABA and ethylene (Beaudoin *et al.*, 2000). To explain this discrepancy it has been proposed that ABA uses the ethylene-signaling pathway as a surrogate to inhibit root growth only when ethylene is not present (Ghassemian *et al.*, 2000).

An additional facet of this complexity is the growing body of evidence that suggests that the sugar-sensing mechanisms play a significant role in regulating effects of ABA and ethylene, particularly during seed germination and root growth, processes that in certain instances involve PCD (Gazzarrini and McCourt, 2001; Rolland *et al.*, 2002). Soluble sugar levels affect a diverse array of plant developmental processes, including inhibition of early seedling growth in the presence of high levels of sugar. Mutants that show either sugar insensitive [glucose insensitive (*gin*)/*sugar insensitive* (*sis*)] or oversensitive [glucose oversensitive (*glo*)/*sugar super sensitive* (*sss*)] phenotypes have been isolated and studied. Molecular characterization of *gin4* and *sis1*, two independent glucose-resistant mutants, revealed them to be allelic to *ctr1* (Gibson *et al.*, 2001; Rolland *et al.*, 2002). Further, the treatment of seedlings with the precursor of ethylene ACC in the presence of glucose resulted in a "gin" phenotype, while *eto1* and *ctr1* mutants show glucose-insensitive phenotype (Zhou *et al.*, 1998). However, dark-grown seedlings of *gin1*, unlike *ctr1* and *eto1*, do not show the triple response, indicating that these signaling pathways play a different role in seedling development. The "triple response" phenomenon—inhibition of normal root geotropic response, inhibition of root and hypocotyl elongation as well as exaggerated apical curvature, first recorded in pea (Neljubow, 1901)—was critical in identifying ethylene signal transduction mutants of Arabidopsis. Interestingly, over expression of the C-terminus of EIN2 also resulted in many constitutive ethylene responses but not the triple response (Alonso *et al.*, 1999). In contrast, the ethylene-insensitive mutants *etr1-1*, *ein2*, *ein3* and *ein6* exhibit phenotype similar to *glo* mutants (Zhou *et al.*, 1998; Rolland *et al.*, 2002). Genetic analysis of the *gin1/etr1* and *gin1/ein2* double mutants places *GIN1* downstream of the ethylene receptor (*ETR1*) and *EIN2* (Zhou *et al.*, 1998; Rolland *et al.*, 2002).

Sucrose or glucose may induce senescence via induction of ethylene production (Philosoph-Hadas *et al.*, 1985). Thus, the role of sugars in senescence/PCD may depend on other factors, such as the presence of other plant hormones. Glucose signaling has been shown to be linked to the ethylene transduction pathway (Zhou *et al.*, 1998). Induction of senescence by sugars is covered more extensively in Chapter 15.

B. Ethylene and Auxin

The complex nature of multi-hormone cross talk influencing senescence/PCD is also illustrated by the following studies on interactions between ethylene and auxin. These two hormones interact to affect root growth, root hair differentiation and elongation, apical hook formation and hypocotyl phototropism (Swarup *et al.*, 2002). Auxin induces ethylene biosynthesis in many plant species primarily by elevating the steady-state level of ACS4 mRNA, a member of the ACC synthase gene family (Beckman *et al.*, 2000; Swarup *et al.*, 2002). In *Arabidopsis*, ACS4 is up regulated by auxin—the ACS4 promoter contains four putative auxin-response elements (Abel *et al.*, 1995; Woeste *et al.*, 1999). In addition to auxin, brassinosteroids (BR) and cytokinins also regulate expression of the ACS gene family. However, there seems to be a complex interaction between ethylene, auxin, BR and light (Yi *et al.*, 1999). Light sometimes induces ethylene biosynthesis, particularly in vegetative tissues, by generating free oxygen radicals, the inducers of PCD. Such a possibility needs to be investigated. The mechanism of the inhibitory and stimulatory role of auxin and ethylene, respectively, in abscission is complex. Ethylene inhibits the movement of auxin (Riov and Goren, 1979) but may also interact directly to affect gene expression in the separation layer (Tucker *et al.*, 2002).

C. Ethylene and Cytokinins

Cytokinins (CKs) are generally considered to be plant hormones that are antisenescence in nature. However, CK is known to enhance production of ethylene in many plant species, including *Arabidopsis* (Mattoo and Suttle, 1991; Cary *et al.*, 1995). Benzyladenine (BA), a cytokinin, increased ACC synthase activity in mung bean hypocotyls (Yoshii and Imaseki, 1981). Although molecular interactions between ethylene and cytokinins have yet to be understood, Lieberman (1979) proposed that some cytokinin responses of seedlings are coupled with ethylene. Low levels of cytokinins induce ethylene production in dark-grown *Arabidopsis* seedlings and these seedlings exhibit a typical ethylene-induced triple response (Cary *et al.*, 1995). Exploiting this phenotype, various *Arabidopsis* mutants that do not show the triple response in the presence of kinetin have been isolated. These mutants, designated as cytokinin-insensitive (*cin*), do not show increased ethylene biosynthesis in the presence of kinetin. Molecular characterization of one of the complementation groups, *cin5*, showed disruption in a member of the ACC synthases, ACS5, leading to a suggestion that the CK regulates ethylene induction by post-transcriptional activation of ACS5 (Vogel *et al.*, 1998). These authors further demonstrated that the dominant ethylene-overproducing phenotype of *eto2* mutation is a result of an alteration of the carboxy terminus of ACS5 that might act as a negative regulator of ACS5 enzymatic activity. Previously, it had been shown (Li and Mattoo, 1994) that the carboxy terminal portion of a wound-inducible ACC synthase (ACS2 gene product) regulates activity and conformation of the dimeric enzyme.

Carimi *et al.* (2003) found that high concentrations of cytokinins enhanced ethylene production, inhibited cell proliferation, and induced PCD in carrot cell cultures and Arabidopsis. However, inhibitors of ethylene biosynthesis did not block these cytokinin responses. It is likely that interactions between ethylene and cytokinins are complex. Although the antisenescence activity of cytokinins has not been unambiguously demonstrated to be due to inhibition of ethylene production, it is possible that ethylene may lower levels of cytokinins *in vivo* (Sanyal and Bangerth, 1998).

D. Ethylene and Polyamines

One of the intriguing plant growth regulators is a class of nitrogenous compounds comprising diamines and higher polyamines. Like CKs, they are considered to be antisenescence factors and may inhibit PCD. Polyamines have been implicated in a variety of physiological and developmental processes in plants, including cell division, embryogenesis, root formation, floral initiation and development, fruit development and ripening, pollen tube growth and senescence, and biotic and abiotic stress responses (see Cohen, 1998; Malmberg *et al.*, 1998; Cassol and Mattoo, 2003). Both ethylene and polyamine biosynthesis share a common intermediate, SAM. *In vitro* studies have shown that polyamines inhibit ethylene biosynthesis in a variety of fruit and vegetative tissues, while ethylene suppresses the accumulation of polyamines (Apelbaum *et al.*, 1981; Li *et al.*, 1992). In short-term experiments, inhibition of ethylene biosynthesis by polyamines results in channeling of SAM into polyamine biosynthesis (Even-Chen *et al.*, 1982; Roberts *et al.*, 1984). Ethylene-mediated suppression of polyamine biosynthesis seems to occur mainly by an inhibition of SAM decarboxylase, a key enzyme in the polyamine biosynthesis. Based on these studies, a hypothesis has been developed: That a cross talk exists between the two apparently antagonistic biosynthetic pathways, namely, ethylene and polyamines, which influences specific physiological processes in plants, particularly senescence/PCD (Mehta *et al.*, 1997).

Like ethylene, the polyamine metabolic pathway is highly regulated by a multitude of developmental and environmental signals (see Cohen, 1998; Malmberg *et al.*, 1998). However, the molecular interactions between ethylene and polyamines have just begun to be understood. Mehta *et al.* (2002) expressed a yeast SAM decarboxylase gene under the control of a ripening-specific promoter E8 to evaluate the role of the SAM metabolic pool in regulating either the ethylene or the polyamine biosynthetic pathway. These studies revealed that the rates of ethylene production in the transgenic tomato fruit were consistently higher than those in the non-transgenic control fruit, suggesting that the polyamine and ethylene biosynthesis pathways can function simultaneously in a ripening tomato fruit. Further, these studies showed that genetically enhanced polyamine levels led to an increase in the carotenoid lycopene, and improved vine life (Mehta *et al.*, 2002). Lycopene is an antioxidant, and its accumulation via enhanced levels of another class of antioxidants, polyamines, raises questions that are important to studies on PCD in a ripening fruit. Interestingly, high-polyamine fruits showed pronounced hypersensitive response (HR) to pathogen attack (Mattoo and Goth, unpublished data). Characterization of genes that are differentially expressed in response to increased polyamine levels should shed light on the molecular basis of cross talk between these growth regulators.

VI. Protease Involvement and Ethylene Biosynthesis in PCD

In plant organ senescence, protease action has featured prominently for at least two reasons. One is the well-studied nitrogen redistribution that occurs between the source (leaf tissue) and the sink organs (younger leaves, root, fruit). The second is the general protein turnover that allows dynamic degradation of proteins that accumulate during growth concomitant with the synthesis of new, senescence-induced proteins (see Noodén and Leopold, 1988). In several instances, ethylene has been shown to induce senescence-related proteases. Proteases, particularly caspases/metacaspases, play a crucial role in animal PCD. In light of these observations, it is expected that proteases also regulate plant PCD.

What is known about protease action and ethylene during pathogen- and stress-induced PCD? Anderson *et al.* (1982) showed that a fungal cell wall hydrolytic enzyme, cellulysin, induces ethylene biosynthesis in tobacco leaves. Interestingly, much before plant PCD was described, these investigators showed that phenylmethanesulfonic sulfate (PMSF) and soybean trypsin inhibitor, but not pepstatin A (an inhibitor of carboxyl proteases), markedly inhibited ethylene production and ACC formation in cellulysin-treated tobacco leaf discs (Anderson *et al.*, 1982) and wound-induced increase in ACC synthase activity in tomato fruit slices (Mattoo and Anderson, 1984). These authors proposed that specific proteolytic activity *in vivo* is associated with the ethylene induction processes, possibly with the activation and/or inactivation of ACC synthase, or with the induction signal itself (Mattoo and Anderson, 1984). Such an activation mechanism was later shown to cause activation of mammalian caspases (Nicholson, 1999; Grutter, 2000). Groover and Jones (1999) also have shown that soybean trypsin inhibitor inhibits protease activity and plant PCD. Like in the case of cellulysin-induced ethylene production, studies with elicitin-induced hypersensitive response (HR) have implicated Ser proteases (Beers *et al.*, 2000). Cysteine protease activity has been implicated in cell death caused by oxidative stress of soybean cells (Solomon *et al.*, 1999).

Both ethylene and proteases have been implicated in plant PCD. Which comes first, ethylene or the protease? From the results presented above it would seem that protease induction is one of the first signaling events followed by other factors, one of which could be mediated by ethylene. Thus, plant PCD induction may require only a small threshold level of ethylene, but it would not take place unless the related protease is activated.

VII. Hormonal Regulation of Plant PCD

A well-studied PCD model in plants is camptothecin-mediated cell death in tomato cell suspensions (see De Jong *et al.*, 2002 and references therein). Camptothecin, a topo isomerase-I inhibitor and inducer of PCD in animals (Kaufmann, 1998; Simizu *et al.*, 1998), causes cell death in tomato suspension in a manner reminiscent of animal apoptosis—chromatin condensation, DNA and nuclear fragmentation. Inhibitors that inhibit generation of reactive oxygen species (ROS) (Lamb and Dixon, 1997), superoxide (O_2^-), hydroxy radicals (OH^-) and hydrogen peroxide (H_2O_2), also inhibit camptothecin-mediated cell death (De Jong *et al.*, 2002). This is quite interesting in light of the studies showing that oxidative stress induced by cupric ions generates oxygen free radicals, enhanced ethylene production and membrane fragmentation (Mattoo *et al.*, 1986). In the latter system, scavengers of hydroxy radicals inhibited ethylene production as well as senescence-related protein degradation (Mattoo *et al.*, 1986; Mehta *et al.*, 1992).

Evidence is accumulating to indicate that plant cells share features of PCD characterized in animal cells (see Chapter 1). H_2O_2 activates protein kinase cascades as well as NF-κB transcription factor, which are components of defense signaling in animals. Earlier, hydroperoxide levels were suggested to be involved in ethylene evolution and the fruit ripening process (Frenkel and Eskin, 1977). Hypoxia-induced aerenchyma formation in maize roots (He *et al.*, 1996; Gunawardena *et al.*, 2001), maize endosperm development cell death (Young and Gallie, 2000), camptothecin-induced PCD in tomato cell suspensions (De Jong *et al.*, 2002), and pea carpel senescence (Orzaez and Granell, 1997) are a few well-studied examples in plants where ethylene is directly involved in PCD.

Ozone-induced cell death in tomato leaf is preceded by a rapid increase in ethylene biosynthesis. Transcript levels for specific ACC synthase, ACC oxidase, and ethylene receptor genes are up regulated in the O_3-treated leaves within 1 to 5 h (Moeder *et al.*, 2002). These authors further produced transgenic plants containing an LE-ACO1 promoter-beta-glucuronidase fusion construct. In these plants, β-glucuronidase activity increased upon O_3 exposure and the spatial distribution of GUS resembled the pattern of extracellular H_2O_2 production. These studies show that ethylene synthesis and perception are required for ROS production and spread of cell death.

Similarly, involvement of ethylene in PCD is exemplified by studies on host-plant interactions. Host defense during pathogenesis in plants involves HR, which culminates in the death of the infected cell and, in some instances, has been shown to involve ethylene (Dangl *et al.*, 1996; Greenberg, 1996; Gilchrist, 1998; Podile and Sripriya, 2002). In recent years, ethylene's role in pathogen-mediated HR was investigated with tomato mutants that are defective in ethylene responsiveness. Never-ripe (*NR*) tomato is insensitive to ethylene and therefore its fruit do not ripen. When challenged with microbial pathogens, this mutant displays reduced disease symptoms compared to the wild type (Lund *et al.*, 1997). Similarly, mycotoxin fumonisin causes cell death in wild type tomato but less so in the *NR* mutant (Moore *et al.*, 1999). These studies show that ethylene plays a prominent role in PCD during pathogenesis. In addition to ethylene, salicylic acid has been implicated in disease-susceptible responses. Salicylic acid does not accumulate in the ethylene-insensitive plants (O'Donnell *et al.*, 2001), perhaps suggesting cross talk between ethylene and salicylic acid during disease susceptibility. The cell death lesions in HR are mimicked in plants exposed to toxic levels of ozone (O_3) and inhibitors of ethylene biosynthesis or perception prevent their development. O_3 is known to induce ethylene biosynthesis, therefore, cell death both in pathogen attack or when plants are exposed to abiotic stresses involves ethylene action (Avni *et al.*, 1994; Overmyer *et al.*, 2000).

Ethylene regulates production of O_2^- and cell death in carrot suspension cells, by activating NADPH oxidase (Chae and Lee, 2001). In other words, ethylene plays a role upstream of NADPH oxidase and other systems that produce ROS. De Jong *et al.* (2002) working with camptothecin-mediated cell death in tomato suspension cells also arrived at a similar conclusion. These authors proposed two partly overlapping cell-death pathways. One pathway involves caspases or metacaspases (De Jong *et al.*, 2000; Uren *et al.*, 2000; Elbaz *et al.*, 2002) that require low ethylene levels for activation. The other pathway is caspase independent and operates at high ethylene levels. This hypothesis is consistent with proposals that different plant responses to ethylene are mediated by independent downstream pathways that have differing thresholds for ethylene levels (Chen and Bleecker, 1995) and/or respond to hormonal cross talk (Whitelaw *et al.*, 2002).

VIII. Perspective

More and more investigations reveal common features between plant senescence and PCD; however, whole organ senescence may be regulated differently than individual cells and tissues that may warrant rapid activation of PCD to arrest life. Interestingly, Pontier *et al.* (1999) and Dangl *et al.* (2000) differentiated between PCD that takes place during senescence and the one that occurs in HR during plant–pathogen interactions. These authors describe senescence as a slow cell-death process in a tissue destined to die, involving ordered disassembly of cellular components (and structures). This disassembly is a highly regulated process, serving a beneficial need whereby the plants recover and reutilize nutrients from senescing cells by recycling them to other living, sink tissues. On the other hand, PCD in HR is quick, intensified in the cells where pathogen attacks, and is a means to kill the host tissue to prevent establishment of the pathogen (Dangl *et al.*, 2000). Thus, descriptions of PCD in plants take different shapes based on the processes involved and the need for survival of the tissue. In either situation, however, ethylene is generated and promotes PCD.

In the past decade, both ethylene biosynthesis and its perception have been genetically manipulated and transgenic plants created where processes such as fruit ripening, flower senescence and leaf senescence have been successfully modified (Wang *et al.*, 2002). Mutational and ectopic expression approaches have been used to understand factors regulating timing and progression of PCD syndrome. The use of Arabidopsis as a model system to elucidate molecular mechanisms underlying senescence in plants, including the role of ethylene, has yielded only limited information. Screening of a mutagenized population of Arabidopsis plants failed to isolate a mutant showing lack of senescence syndrome development. Among the hormone mutants of Arabidopsis, only an ethylene-insensitive mutant *etr-1* showed measurable delay in the timing of leaf senescence (Bleecker and Patterson, 1997). These results indicate that genetics of cell death/senescence in plants is complex and more creative screens are needed to identify genes regulating this process. Screening of mutants with altered expression of a reporter gene under the control of the senescence-regulated promoters would likely identify genes underlying signal transduction pathways that signal PCD.

Rapidly emerging understanding of the role of various metabolic processes and genes in regulating PCD and aging in model systems such as *Caenorhabditis elegans* (Hekimi *et al.*, 2001) and *Drosophila melanogaster* (Gorski and Marra, 2002) should greatly facilitate development and testing of hypotheses regulating PCD in animals and plants. Availability of null and enhancer trap lines should provide mutants to test the role of the candidate gene either individually or in combination. Arabidopsis and rice genomes have been sequenced and sequencing of other genomes is already making a steady progress. Synteny within and between monocot and dicot would help understanding of the molecular basis of evolution, including responses to various signals that regulate growth, development and senescence/PCD processes. Finally, the identification and characterization of the proteases (Woltering *et al.*, 2002) involved in cellulysin-mediated and wound-induced ethylene biosynthesis (Anderson *et al.*, 1982; Mattoo and Anderson, 1984) will be crucial in defining how intimately associated senescence and PCD are in vegetative versus reproductive plant tissues.

Acknowledgments

We thank Mark Tucker for a critical review of the manuscript and Ernest J. Woltering and Adi Avni for sharing their recent papers.

References

Abel, S., Nguyen, M.D., Chow, W., and Theologis, A. (1995). ACS4, a primary indoleacetic acid-responsive gene encoding 1-aminocyclopropane-1-carboxylate synthase in *Arabidopsis thaliana*. Structural characterization, expression in *Escherichia coli*, and expression characteristics in response to auxin. *J. Biol. Chem.* **270**, 19093–19099.

Abeles, F., Morgan, P., and Saltveit, M. (1992). *Ethylene in Plant Biology*. Academic Press, San Diego.

Adams, D.O., and Yang, S.F. (1979). Ethylene biosynthesis: identification of 1-minocyclopropane-1-carboxylic acid as an intermediate in the conversion of methionine to ethylene. *Proc. Natl. Acad. Sci. USA* **76**, 170–174.

Alonso, J.M., Hirayama, T., Roman, G., Nourizadeh, S., and Ecker, J.R. (1999). EIN2, a bifunctional transducer of ethylene and stress responses in *Arabidopsis*. *Science* **284**, 2148–2152.

Anderson, J.D., Mattoo, A.K., and Lieberman, M. (1982). Induction of ethylene biosynthesis in tobacco leaf discs by cell wall digesting enzymes. *Biochem. Biophys. Res. Commun.* **107**, 588–596.

Apelbaum, A., Burgoon, A.C., Anderson, J.D., Lieberman, M., Ben-Arie, R., and Mattoo, A.K. (1981). Polyamines inhibit biosynthesis of ethylene in higher plant tissue and protoplasts. *Plant Physiol.* **68**, 453–456.

Avni, A., Bailey, B.A., Mattoo, A.K., and Anderson, J.D. (1994). Induction of ethylene biosynthesis in Nicotiana tabacum by a Trichoderma viride xylanase is correlated to the accumulation of ACC synthase and ACC oxidase. *Plant Physiol.* **106**, 1049–1055.

Beaudoin, N., Serizet, C., Gosti, F., and Giraudat, J. (2000). Interactions between abscisic acid and ethylene signaling cascades. *Plant Cell* **12**, 1103–1115.

Beckman, E.P., Saibo, N.J.M., Di Cataldo, A., Regalado, A.P., Ricardo, C.P., and Rodrigues-Pousada, C. (2000). Differential expression of four genes encoding 1-aminocyclopropane-1-carboxylate synthase in *Lupinus albus* during germination and in response to indole-3-acetic acid and wounding. *Planta* **211**, 663–672.

Beers, E.P., Woffenden, B.J., and Zhao, C. (2000). Plant proteolytic enzymes: possible roles during programmed cell death. *Plant Mol. Biol.* **44**, 399–415.

Bennett, M.J., Marchant, A., Green, H.G., May, S.T., Ward, S.P., Millmer, P.A., Walker, A.R., Schulz, B., and Feldmann, K.A. (1996). *Arabidopsis AUX1* gene: a permease-like regulator of root gravitropism. *Science* **273**, 948–950.

Beyer, Jr., E.M. (1976). Silver ion: a potent antiethylene agent in cucumber and tomato. *HortScience* **11**, 195–196.

Bleecker, A.B. (1999). Ethylene perception and signaling: an evolutionary perspective. *Trends Plant Sci.* **4**, 269–274.

Bleecker, A.B., and Patterson, S.E. (1997). Last exit: senescence, abscission, and meristem arrest in *Arabidopsis*. *Plant Cell* **9**, 1169–1179.

Boerjan, W., Cervera, M.T., Delarue, M., Beeckman, T., De-Witte, W., Bellini, C., Caboche, M., Vanonckelen, H., Van Montagu, M., and Inzé, D. (1995). Superroot, a recessive mutation in *Arabidopsis*, confers auxin overproduction. *Plant Cell* **7**, 1405–1419.

Carimi, F., Zottini, M., Formentin, E., Terzi, M., and Lo Schiavo, F. (2003). Cytokinins: new apoptotic inducers in plants. *Planta* **216**, 413–421.

Cary, A.J., Liu, W., and Howell, S.H. (1995). Cytokinin action is coupled to ethylene in its effects on the inhibition of root and hypocotyl elongation in *Arabidopsis thaliana* seedlings. *Plant Physiol.* **107**, 1075–1082.

Cassol, T., and Mattoo, A.K. (2003). Do polyamines and ethylene interact to regulate plant growth, development and senescence? In *Molecular Insights in Plant Biology* (P. Nath, A.K. Mattoo, S.R. Ranade, and J.H. Weil, Eds.), pp. 121–132. Science Publishers, Inc., Enfield, NH.

Celenza, J., Grisafi, P., and Fink, G. (1995). A pathway for lateral root formation in *Arabidopsis thaliana*. *Genes Dev.* **9**, 2131–2142.

Chae, H.S., and Lee, W.S. (2001). Ethylene- and enzyme-mediated superoxide production and cell death in carrot cells grown under carbon starvation. *Plant Cell Rpts.* **20**, 256-261.

Chang, C., and Stadler, R. (2001). Ethylene hormone receptor action in Arabidopsis. *BioEssays* **23**, 619–627.

Chang, C., Kwok, S.F., Bleecker, A.B., and Meyerowitz, E.M. (1993). Arabidopsis ethylene-response gene ETR1: Similarity of product to two-component regulators. *Science* **262**, 539–544.

Chen, Q.G., and Bleecker, A.B. (1995). Analysis of ethylene signal transduction kinetics associated with seedling-growth response and chitinase induction in wild-type and mutant Arabidopsis. *Plant Physiol.* **108**, 597–607.

Cohen, S.S. (1998). *A Guide to the Polyamines*. Oxford Univ. Press, New York.

Dangl, J.L., Dietrich, R.A., and Richberg, M.H. (1996). Death don't have no mercy: Cell death programs in plant-microbe interactions. *Plant Cell* **8**, 1793–1807.

Dangl, J.L., Dietrich, R.A., and Thomas, H. (2000). Senescence and programmed cell death. In *Biochemistry and Molecular Biology of Plants* (B. Buchanan, W. Gruissem, and R. Jones, Eds.), pp. 1044–1100. American Society of Plant Physiologists, Rockville, MD.

De Jong, A.J., Hoeberichts, F.A., Yakimova, E.T., Maximova, E., and Woltering, E.J. (2000). Chemical-induced apoptotic cell death in tomato cells: involvement of caspase-like proteases. *Planta* **211**, 656–662.

De Jong, A.J., Yakimova, E.T., Kapchina, V.M., and Woltering, E. J. (2002). A critical role for ethylene in hydrogen peroxide release during programmed cell death in tomato suspension cells. *Planta* **214**, 537–545.

Elbaz, M., Avni, A., and Weil, M. (2002). Constitutive caspase-like machinery executes programmed cell death in plant cells. *Cell Death Differentiation* **9**, 726–733.

Even-Chen, Z., Mattoo, A.K., and Goren, R. (1982). Inhibition of ethylene biosynthesis by aminoethoxyvinylglycine and by polyamines shunts label from 3,4-[^{14}C] methionine into spermidine in aged orange peel discs [*Citrus sinensis*]. *Plant Physiol.* **69**, 385–388.

Fluhr, R., and Mattoo, A.K. (1996). Ethylene—biosynthesis and perception. *Crit. Rev. Plant Sci.* **15**, 479–523.

Frenkel, C., and Eskin, M. (1977). Ethylene evolution as related to changes in hydroperoxide levels in ripening tomato fruit. *HortScience* **12**, 552–553.

Fujita, H., and Syono, K. (1996). Genetic analysis of the effects of polar auxin inhibitors on root growth in *Arabidopsis thaliana*. *Plant Cell Physiol.* **37**, 1094–1101.

Gazzarrini, S., and McCourt, P. (2001). Genetic interactions between ABA, ethylene and sugar signaling pathways. *Curr. Opin. Plant Biol.* **4**, 387–391.

Ghassemian, M., Nambara, E., Cutler, S., Kawaide, H., Kamiya, Y., and McCourt, P. (2000). Regulation of abscisic acid signaling by the ethylene response pathway in *Arabidopsis*. *Plant Cell* **12**, 1117–1126.

Gibson, S.I., Laby, R.J., and Kim, D. (2001). The *sugar-insensitive1* (*sis1*) mutant of *Arabidopsis* is allelic to *ctr1*. *Biochem. Biophys. Res. Commun.* **280**, 196–203.

Gilchrist, D.G. (1998). Programmed cell death in plant disease: The purpose and promise of cellular suicide. *Annu. Rev. Phytopathol.* **36**, 393–414.

Gillaspy, G., Hilla, B.-D., and Gruissem, W. (1993). Fruit: a developmental perspective. *Plant Cell* **5**, 1439–1451.

Giovanelli, L., Mudd, S.H., and Datko, A.H. (1980). Sulfur amino acids in plants. In *Amino Acids and Derivatives. The Biochemistry of Plants: A Comprehensive Treatise* (B.J. Miflin, Ed.), Vol. 5, pp. 453–505. Academic Press, New York.

Giovannoni, J. (2001). Molecular biology of fruit maturation and ripening. *Annu. Rev. Plant Physiol. Plant Mol. Biol.* **52**, 725–749.

Gorski, S., and Marra, M. (2002). Programmed cell death takes flight: genetic and genomic approaches to gene discovery in Drosophila. *Physiological Genomics* **9**, 59–69.

Greenberg, J.T. (1996). Programmed cell death: a way of life for plants. *Proc. Natl. Acad. Sci. USA* **93**, 12094–12097.

Grefter, M., Hausmann, R., Gold, M., and Hurwitz, J. (1966). The enzymatic methylation of ribonucleic acid and deoxyribonucleic acid. *J. Biol. Chem.* **241**, 1995–2006.

Groover, A., and Jones, A.M. (1999). Tracheary element differentiation uses a novel mechanism coordinating programmed cell death and secondary cell wall synthesis. *Plant Physiol.* **119**, 375–384.

Grutter, M.G. (2000). Caspases: key players in programmed cell death. *Curr. Opin. Struct. Biol.* **10**, 649–655.

Gunawardena, A.H.L.A.N., Jackson, M.B., Hawes, C.R., and Evans, D.E. (2001). Characterization of programmed cell death during aerenchyma formation induced by ethylene of hypoxia in roots of maize (*Zea mays* L.). *Planta* **212**, 205–214.

Hall, M.A., Moshkov, I.E., Novikova, G.V., Mur, L.A.J., and Smith, A.R. (2001). Ethylene signal perception and transduction: multiple paradigms? *Biol. Rev.* **76**, 103–128.

He, C.-J., Morgan, P.W., and Drew, M.C. (1996). Transduction of an ethylene signal is required for cell death and lysis in the root cortex of maize during aerenchyma formation induced by hypoxia. *Plant Physiol.* **112**, 463–472.

Hekimi, S., Burgess, J., Bussiere, F., Meng, Y., and Benard, C. (2001). Genetics of lifespan in C-elegans: molecular diversity, physiological complexity, mechanistic simplicity. *Trends Genetics* **17**, 712–718.

Hua, J., and Meyerowitz, E.M. (1998). Ethylene responses are negatively regulated by a receptor gene family in *Arabidopsis thaliana*. *Cell* **94**, 261–271.

Jing, H.C., Sturre, M.J.G., Hille, J., and Dijkwel, P.P. (2002). *Arabidopsis* onset of leaf death mutants identify a regulatory pathway controlling leaf senescence. *Plant J.* **32**, 51–63.

John, P. (1991). How plant molecular biologists revealed a surprising relationship between two enzymes, which took an enzyme out of a membrane where it was not located, and put it into the soluble phase where it could be studied. *Plant Mol. Biol. Rptr.* **9**, 192–194.

Kaufmann, S.H. (1998). Cell death induced by topoisomerase-targeted drugs: more questions than answers. *Biochim. Biophys. Acta* **1400**, 195–211.

Kende, H. (1989). Enzymes of ethylene biosynthesis. *Plant Physiol.* **91**, 1–4.

Kieber, J.J., Rothenberg, M., Roman, G., Feldmann, K.A., and Ecker, J.R. (1993). CTR1, a negative regulator of the ethylene response pathway in *Arabidopsis*, encodes a member of the *raf* family of protein kinases. *Cell* **72**, 427–441.

Koorneef, M., Reuling, G., and Karssen, C.M. (1984). The isolation and characterization of abscisic acid-insensitive mutants of *Arabidopsis thaliana. Physiologia Plant.* **61**, 377–383.

Kushad, M.M. (1990). Recycling of 5′-deoxy-5′-methylthioadenosine in plants. In *Polyamines and Ethylene: Biochemistry, Physiology, and Interactions* (H.E. Flores, R.N. Arteca, and J.C. Shannon, Eds.), pp. 50–61. American Society of Plant Physiologists, Rockville, MD.

Laby, R.J., Kincaid, M.S., Kim, D., and Gibson, S.I. (2000). The *Arabidopsis* sugar-insensitive mutants *sis4* and *sis5* are defective in abscisic acid synthesis and response. *Plant J.* **23**, 587–596.

Lamb, C., and Dixon, R.A. (1997). The oxidative burst in plant disease resistance. *Annu. Rev. Plant Physiol. Plant Mol. Biol.* **48**, 251–275.

Lanahan, M.B., Yen, H.-C., Giovannoni, J.J., and Klee, H.J. (1994). The never ripe mutation blocks ethylene perception in tomato. *Plant Cell* **6**, 521–530.

Leal, N.R., and Tabim, M.E. (1974). Testes de conservacao natural pos colheita, alem dos 3000 dias, de frutos de alguns cultivares de tomateiro (*Lycopersicon esculentum* Mill). E. hibridos destes com ëalcobacaí. *Rev. Ceres* **21**, 310–328.

Lehman, A., Black, R., and Ecker, J.R. (1996). HOOKLESS1, an ethylene response gene, is required for differential cell elongation in the *Arabidopsis* hypocotyl. *Cell* **85**, 183–194.

Leon-Kloosterziel, K., Van De Bunt, G., Zeevaart, J., and Koornneef, M. (1996). *Arabidopsis* mutants with a reduced seed dormancy. *Plant Physiol.* **110**, 233–240.

Leung, J., Bouvier-Durand, M., Morris, P.-C., Guerrier, D., Chefdor, F., and Giraudat, J. (1994). *Arabidopsis* ABA response gene ABI1: Features of a calcium-modulated protein phosphatase. *Science* **264**, 1448–1452.

Li, N., and Mattoo, A.K. (1994). Deletion of the carboxyl-terminal region of ACC synthase, a key protein in the biosynthesis of ethylene, results in catalytically hyper-active, monomeric enzyme. *J. Biol. Chem.* **269**, 6908–6917.

Li, N., Parsons, B., Liu, D., and Mattoo, A.K. (1992). Accumulation of wound-inducible ACC synthase transcript in tomato fruit is inhibited by salicylic acid and polyamines. *Plant Mol. Biol.* **18**, 477–487.

Lieberman, M. (1979). Biosynthesis and action of ethylene. *Annu. Rev. Plant Physiol.* **30**, 533–591.

Lieberman, M., Kunishi, A., Mapson, L.W., and Wardale, D.A. (1966). Stimulation of ethylene production in apple tissue slices by methionine. *Plant Physiol.* **41**, 376–382.

Lund, S.T., Stall, R.E., and Klee, H.J. (1997). Ethylene regulates the susceptible response to pathogen infection in tomato. *Plant Cell* **10**, 371–382.

Lurssen, K., Naumann, K., and Schroder, R. (1979). 1-Aminocyclopropane-1-carboxylic acid. An intermediate of the ethylene biosynthesis in higher plants. *Z. Pflanzenphysiol.* **92**, 285–294.

Malmberg, R.L., Watson, M.B., Galloway, G.L., and Yu, W. (1998). Molecular genetic analyses of plant polyamines. *Crit. Rev. Plant Sci.* **17**, 199–224.

Marchant, A., Kargul, J., May, S.T., Muller, P., Delbarre, A., Perrot-Rechenmann, C., and Bennett, M.J. (1999). AUX1 regulates root gravitropism in *Arabidopsis* by facilitating auxin uptake within root apical tissues. *EMBO J.* **18**, 2066–2073.

Mattoo, A.K., and Aharoni, N. (1988). Ethylene and plant senescence. In *Senescence and Aging in Plants* (L.D. Noodén and A.C. Leopold, Eds.), pp. 241–280. Academic Press, San Diego.

Mattoo, A.K., and Anderson, J.D. (1984). Wound-induced increase in 1-aminocyclopropane-1-carboxylate synthase activity: Regulatory aspects and membrane association of the enzyme. In *Ethylene: Biochemical, Physiological and Applied Aspects* (Y. Fuchs and E. Chalutz, Eds.), pp. 139–147. Martinus Nijhoff/Dr W. Junk Publishers, The Hague.

Mattoo, A.K., and Suttle, J.C. (Eds.) (1991). *The Plant Hormone Ethylene*. CRC Press, Boca Raton, FL.

Mattoo, A.K., and White, W.B. (1991). Regulation of ethylene biosynthesis. In *The Plant Hormone Ethylene* (A.K. Mattoo and J.C. Suttle, Eds.), pp. 21–42. CRC Press, Boca Raton, FL.

Mattoo, A.K., Baker, J.E., and Moline, H.E. (1986). Induction by copper ions of ethylene production in *Spirodela oligorrhiza*: Evidence for a pathway independent of 1-aminocyclopropane-1-carboxylic acid. *J. Plant Physiol.* **123**, 193–202.

McKeon, T.A., Fernandez-Maculet, J.C., and Yang, S.F. (1995). Biosynthesis and metabolism of ethylene. In *Plant Hormones* (P.J. Davies, Ed.), pp. 118–139. Kluwer, Dordrecht.

Mehta, R.A., Fawcett, T.W., Porath, D., and Mattoo, A.K. (1992). Oxidative stress causes rapid membrane translocation and *in vivo* degradation of ribulose-1,5-bisphosphate carboxylase/oxygenase. *J. Biol. Chem.* **267**, 2810–2816.

Mehta, R.A., Handa, A., and Mattoo, A.K. (1997). Interactions of ethylene and polyamines in regulating fruit ripening. In *Biology and Biotechnology of the Plant Hormone Ethylene* (A.K. Kanellis, C. Chang, H. Kende and D. Grierson, Eds.), pp. 321–326. Kluwer, Dordrecht.

Mehta, R.A., Cassol, T., Li, N., Ali, N., Handa, A.K., and Mattoo, A.K. (2002). Engineered polyamine accumulation in tomato enhances phytonutrient content, juice quality and vine life. *Nature Biotechnol.* **20**, 613–618.

Meyer, K., Leube, M., and Grill, E. (1994). A protein phosphatase 2C involved in ABA signal transduction in *Arabidopsis thaliana*. *Science* **264**, 1452–1455.

Moeder, W., Barry, C.S., Tauriainen, A.A., Betz, C., Tuomainen, J., Utriainen, M., Grierson, D., Sandermann, H., Langebartels, C., and Kangasjarvi, J. (2002). Ethylene synthesis regulated by biphasic induction of 1-aminocyclopropane-1-carboxylic acid synthase and 1-aminocyclopropane-1-carboxylic acid oxidase genes is required for hydrogen peroxide accumulation and cell death in ozone-exposed tomato. *Plant Physiol.* **130**, 1918–1926.

Moore, T., Bostock, R.M., Lincoln, J.E., and Gilchrist, D.G. (1999). Molecular and genetic characterization of ethylene involvement in mycotoxin-induced plant cell death. *Physiol. Mol. Plant Pathol.* **54**, 73–85.

Neljubow, D. (1901). Ueber die horizontale Nutation der Stengel von Pisum sativum und einiger anderer. *Pflanzen Beih. Bot. Zentralbl.* **10**, 128–139.

Nicholson, D.W. (1999). Caspase structure, proteolytic substrates, and function during apoptotic cell death. *Cell Death Diff.* **6**, 1028–1042.

Noodén, L.D., and Leopold, A.C. (1988). *Senescence and Aging in Plants*. Academic Press, San Diego.

O'Donnell, P.J., Antoine, F.R., Ciardi, J., and Klee, H.J. (2001). Ethylene-dependent salicylic acid regulates an expanded cell death response to a plant pathogen. *Plant J.* **25**, 315–323.

Oh, S.A., Park, J.H., Lee, G.I., Paek, K.H., Park, S.K., and Nam, H.G. (1997). Identification of three genetic loci controlling leaf senescence in *Arabidopsis thaliana*. *Plant J.* **12**, 527–535.

Orzaez, D., and Granell, A. (1997). DNA fragmentation is regulated by ethylene during carpel senescence in *Pisum sativum*. *Plant J.* **11**, 137–144.

Overmyer, K., Tuominen, H., Kettunen, R., Betz, C., Langebartels, C., Sandermann, Jr., H., and Kangasjarvi, J. (2000). Ozone-sensitive *Arabidopsis rcd*1 mutant reveals opposite roles for ethylene and jasmonate signaling pathways in regulating superoxide-dependent cell death. *Plant Cell* **12**, 1849–1862.

Owens, L.D., Lieberman, M., and Kunishi, A. (1971). Inhibition of ethylene production by rhizobitoxin. *Plant Physiol.* **48**, 1–4.

Payton, S., Fray, R.G., Brown, S., and Grierson, D. (1996). Ethylene receptor expression is regulated during fruit ripening, flower senescence and abscission. *Plant Mol. Biol.* **31**, 1227–1231.

Philosoph-Hadas, S., Meir, S., and Aharoni, A. (1985). Carbohydrates stimulate ethylene production in tobacco leaf discs: II. Sites of stimulation in the ethylene biosynthesis pathway. *Plant Physiol.* **78**, 139–143.

Podile, A.R., and Sripriya, P. (2002). Pathogen-induced hypersensitive response (HR) as a form of programmed cell death in plants. *Annu. Rev. Plant Pathol.* **1**, 155–176.

Pontier, D., Gan, S., Amasino, R.M., Roby, D., and Lam, E. (1999). Markers for hypersensitive response and senescence show distinct patterns of expression. *Plant Mol. Biol.* **39**, 1243–1255.

Riov, J., and Goren, R. (1979). Effect of ethylene on auxin transport and metabolism in midrib sections in relation to leaf abscission of woody plants. *Plant Cell Environ.* **2**, 83–89.

Roberts, D.R., Walker, M.A., Thompson, J.E., and Dumbroff, E.B. (1984). The effects of inhibitors of polyamine and ethylene biosynthesis on senescence, ethylene production and polyamine levels in cut carnation flowers [*Dianthus caryophyllus*]. *Plant Cell Physiol.* **25**, 315–322.

Rolland, F., Moore, B., and Sheen, J. (2002). Sugar sensing and signaling in plants. *Plant Cell* **14** (suppl.), S185–S205.

Romano, C.P., Robson, P.R.H., Smith, H., Estelle, M., and Klee, H. (1995). Transgene mediated auxin overproduction in *Arabidopsis*: hypocotyl elongation phenotype and interaction with the hy6-1 hypocotyl elongation and aux1 auxin resistant mutant. *Plant Mol. Biol.* **27**, 1071–1083.

Rook, F., Corke, F., Card, R., Munz, G., Smith, C., and Bevan, M.W. (2001). Impaired sucrose-induction mutants reveal the modulation of sugar-induced starch biosynthetic gene expression by abscisic acid signalling. *Plant J.* **26**, 421–433.

Sanyal, D., and Bangerth, F. (1998). Stress induced ethylene evolution and its possible relationship to auxin-transport, cytokinin levels, and flower bud induction in shoots of apple seedlings and bearing apple trees. *Plant Growth Regul.* **24,** 127–134.

Schaller, G.E., and Bleecker, A.B. (1995). Ethylene-binding sites generated in yeast expressing the *Arabidopsis* ETR1 gene. *Science* **270**, 1809–1811.

Schlenk, F. (1983). Methylthioadenosine. *Adv. Enzymol.* **54**, 195–265.

Simizu, S., Takada, M., Umezawa, K., and Imoto, M. (1998). Requirement of caspase 3 (-like) protease-mediated hydrogen peroxide production for apoptosis induced by various anticancer drugs. *J. Biol. Chem.* **273**, 26900–26907.

Sisler, E.C. (1977). Ethylene activity of some π-acceptor compounds. *Tobacco Sci.* **17**, 68–72.

Solomon, M., Belenghi, B., Delledonne, M., Menachem, E., and Levine, A. (1999). The involvement of cysteine proteases and protease inhibitor genes in the regulation of programmed cell death in plants. *Plant Cell* **11**, 431–443.

Su, W., and Howell, S.H. (1992). A single genetic locus, *ckr1*, defines *Arabidopsis* mutants in which root growth is resistant to low concentrations of cytokinin. *Plant Physiol.* **99**, 1569–1574.

Swarup, R., Parry, G., Graham, N., Allen, T., and Bennett, M. (2002). Auxin cross-talk: integration of signalling pathways to control plant development. *Plant Mol. Biol.* **49**, 411–426.

Tieman, D.M., and Klee, H.J. (1999). Differential expression of two novel members of the tomato ethylene-receptor family. *Plant Physiol.* **120**, 165–172.

Tigchelaar, E.C., McGlasson, W.B., and Buescher, R.W. (1978). Genetic regulation of tomato fruit ripening. *HortScience* **13**, 508–513.

Tucker, M.L., Whitelaw, C.A., Lyssenko, N.N., and Nath, P. (2002). Functional analysis of regulatory elements in the gene promoter for an abscission-specific cellulase from bean and isolation, expression, and binding affinity of three TGA-type basic leucine zipper transcription factors. *Plant Physiol.* **130**, 1487–1496.

Uren, G.A., O'Rourke, K., Aravind, L., Pisabarro, T.M., Seshagiri, S., Koonin, V.E., and Dixit, M.V. (2000). Identification of paracaspases and metacaspases: two ancient families of caspase-like proteins, one of which plays a key role in MALT lymphoma. *Mol. Cell* **6**, 961–967.

Vogel, J.P., Woeste, K.W., Theologis, A., and Kieber, J.J. (1998). Recessive and dominant mutations in the ethylene biosynthetic gene *ACS5* of *Arabidopsis* confer cytokinin insensitivity and ethylene overproduction, respectively. *Proc. Natl. Acad. Sci. USA* **95**, 4766–4771.

Vrebalov, J., Ruezinsky, D., Padmanabhan, V., White, R., Medrano, D., Drake, R., Schuch, W., and Giovannoni, J. (2002). A MADS-BOX gene necessary for fruit ripening at the tomato *ripening-inhibitor* (*rin*) locus. *Science* **296**, 343–346.

Wang, K.L.-C., Li, H., and Ecker, J.R. (2002). Ethylene biosynthesis and signaling networks. *Plant Cell* **14**, S131–S151.

Wang, W., Hall, A.E., O'Malley, R., and Bleecker, A.B. (2003). Canonical histidine kinase activity of the ETR1 ethylene receptor from *Arabidopsis* is not required for signal transmission. *Proc. Natl. Acad. Sci. USA* **100**, 352–357.

Whitelaw, C.A., Lyssenko, N.N., Chen, L., Zhou, D., Mattoo, A. K., and Tucker, M. (2002). Delayed abscission and shorter internodes correlate with a reduction in the ethylene-receptor LeETR1 transcript in transgenic tomato. *Plant Physiol.* **128,** 978–987.

Wilkinson, J.Q., Lanahan, M.B., Yen, H.-C., Giovannoni, J.J., and Klee, H.J. (1995). An ethylene-inducible component of signal transduction encoded by *Never-ripe*. *Science* **270**, 1807–1809.

Woeste, K.E., Vogel, J.P., and Kieber, J.J. (1999). Factors regulating ethylene biosynthesis in etiolated *Arabidopsis thaliana* seedlings. *Physiol. Plant.* **105**, 478–484.

Woltering, E.J., van der Bent, A., and Hoeberichts, F.A. (2002). Do plant caspases exist? *Plant Physiol.* **130**, 1764–1769.

Yen, H.-C., Lee, S., Tanksley, S.D., Lanahan, M.B., Klee, H.J., and Giovannoni, J.J. (1995). The tomato never-ripe locus regulates ethylene-inducible gene expression and is linked to a homolog of the *Arabidopsis ETR*1 gene. *Plant Physiol.* **107**, 1343–1353.

Yi, H.C., Joo, S., Nam, K.H., Lee, J.S., Kang, B.G., and Kim, W. T. (1999). Auxin and brassinosteroid differentially regulate the expression of three members of the 1-aminocyclopropane-1-carboxylate synthase gene family in mung bean (*Vigna radiata* L.). *Plant Mol. Biol.* **41**, 443–454.

Yoshii, H., and Imaseki, H. (1981). Biosynthesis of auxin-induced ethylene: effects of indole-3-acetic acid, benzyladenine, and absicic acid on endogenous levels of 1-aminocyclopropane-1-carboxylic acid (ACC) and ACC synthase. *Plant Cell Physiol.* **22**, 369–379.

Young, T.E., and Gallie, D.R. (2000). Programmed cell death during endosperm development. *Plant Mol. Biol.* **44**, 283–301.

Zhou, D., Kalaitzis, P., Mattoo, A.K., and Tucker, M.L. (1996a). The mRNA for an ETR1 homologue in tomato is constitutively expressed in vegetative and reproductive tissues. *Plant Mol. Biol.* **30**, 1331–1338.

Zhou, D., Mattoo, A.K., and Tucker, M.L. (1996b). Molecular cloning of a tomato cDNA (Genbank U47279) encoding an ethylene receptor. *Plant Physiol.* **110**, 1435–1436.

Zhou, L., Jang, J.-C., Jones, T.L., and Sheen, J. (1998). Glucose and ethylene signal transduction crosstalk revealed by an *Arabidopsis* glucose-insensitive mutant. *Proc. Natl. Acad. Sci. USA* **95**, 10294–10299.

9

Jasmonates—Biosynthesis and Role in Stress Responses and Developmental Processes

Claus Wasternack

I. Introduction

Jasmonates are ubiquitously occurring plant growth regulators which are assumed to function as a "master switch" in biotic and abiotic stress responses (Wasternack and Parthier, 1997). Plants have to adapt to abiotic factors such as ultraviolet (UV) light, darkness, high-light conditions, oxidative stress, hypoxia, heat, low temperature, desiccation, water deficit, salt stress, nutrient imbalance, or to biotic factors such as pathogens or herbivores by a complex alteration of gene expression programs. Many facets of these alterations occur also during senescence, which is known to be promoted by jasmonates. The biosynthesis of jasmonates and their ubiquitous role in the reprogramming of gene expression upon environmental and developmental stimuli will be discussed.

II. Jasmonates and Related Compounds

Jasmonates are lipid-derived cyclopentanone compounds which occur in all plants tested so far, as free acid (JA), methyl ester (JAME) or amino acid conjugates such as that of

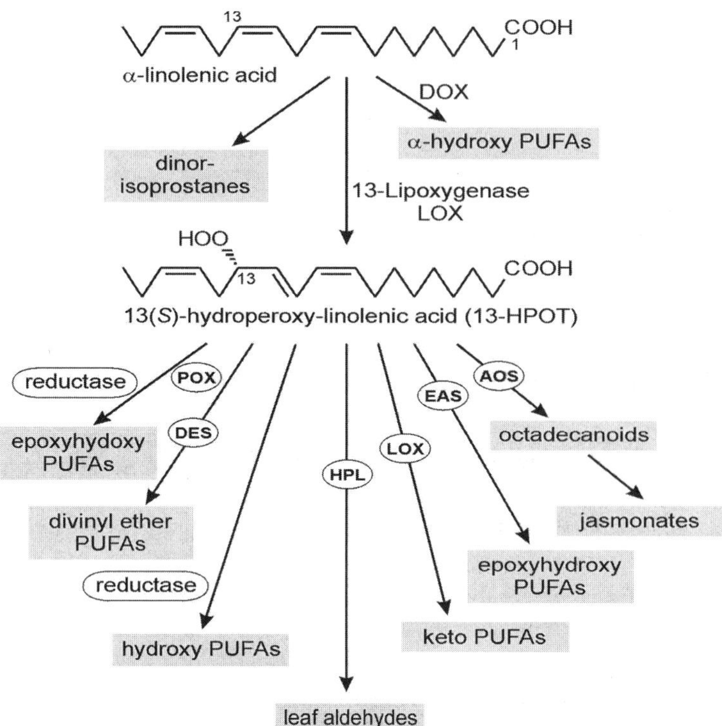

Figure 9-1. The LOX pathway. Seven different branches originate from the LOX products 13-HPOT and 13-HPOD, respectively. The AOS catalyzes the first step of the most-studied branch leading to OPDA and JA.

L-isoleucine (JA-Ile); hydroxylated and glucosylated derivatives of JA were found as minor constituents. All of these compounds are collectively named "jasmonates".

More recently, a biosynthetic precursor of JA, 12-oxo-phytodienoic acid (OPDA) as well as its methyl ester (OPDAME), could be frequently detected in plant extracts. Together with structurally related compounds, they are collectively called "octadecanoids". Octadecanoids and jasmonates originate from peroxidation of polyunsaturated fatty acids (PUFAs) by lipoxygenases (LOXs). In the so-called LOX pathway (Fig. 9-1) numerous compounds, collectively named "oxylipins", are formed from lipid peroxides.

Originally, jasmonates were found to promote senescence (Parthier, 1991). Later on, their specific effects on gene expression were found, followed by the observation that octadecanoids exert similar, but not identical biological activities. Furthermore, numerous oxylipins show biological activity such as antimicrobial effects, illustrating the overall role of the LOX pathway (Feussner and Wasternack, 2002).

III. LOX-derived Compounds and the Biosynthesis of Octadecanoids and Jasmonates

A. The LOX Pathway

Enzymatic peroxidation of lipids occurs by stereospecific dioxygenation of PUFAs such as linolenic acid (LeA) or linoleic acid (LA) catalyzed by LOXs. LeA might be released by activity of a phospholipase A_2, as shown for a wound response (Narváez-Vásquez *et al.*, 1999). Recently, JA deficiency in the *Arabidopsis* mutant *dad1* (defective in anther dehiscence 1), which is affected in a flower-specific chloroplast-located phospholipase A1, indicates the role of this enzyme in JA biosynthesis (Ishiguro *et al.*, 2001). Although unknown in function, a senescence-related cytosolic lipase of carnation petals was cloned recently (Hong *et al.*, 2000). Several forms of LOX have been shown to oxygenate LA or LeA (Brash, 1999) and many of them are chloroplast-located (Feussner *et al.*, 1995). However, only recently reverse genetic analysis revealed specific functions in development or stress-responses (Feussner and Wasternack, 2002). The *S*-configured LOX-derived hydroperoxides (HPOD or HPOT) are substrates for at least seven metabolic branches (Fig. 9-1): In one of these branches, allene oxides are formed by allene oxide synthase (AOS), and its products are octadecanoids and jasmonates. Important products of other branches are the leaf aldehydes and divinyl ether. Cloning and reverse genetic studies on lipases and the various LOX forms will show in molecular terms initial events of lipid peroxidation and its role in PCD and senescence. Furthermore, autoxidative processes in lipid peroxidation occur in leaf senescence (Berger *et al.*, 2001).

B. Octadecanoid and Jasmonate Biosynthesis

The initial step in JA biosynthesis is the synthesis of an allene oxide formed from 13-HPOT by an AOS (Fig. 9-2). AOSs have been cloned from several plants, and the AOS protein resides in the chloroplast (Maucher *et al.*, 2000), apparently in a chloroplast envelope fraction (Froehlich *et al.*, 2001). The allene oxide is highly unstable and needs to be converted by an allene oxide cyclase (AOC). The *cis*-(9*S*,13*S*)-OPDA formed is the unique precursor of the naturally occurring stereoisomer of JA and, thus, the AOC has a special regulatory role in jasmonate biosynthesis.

Recently, AOC was cloned from tomato (Ziegler *et al.*, 2000) and later from *Arabidopsis* (Stenzel *et al.*, 2002b). Wounding, osmotic stress and glucose treatment or application of jasmonates and octadecanoids lead to the expression of *AOC* in tomato leaves (Stenzel *et al.*, 2002a). Similar kinetics of transcriptional up-regulation were found for the tomato AOS (Howe *et al.*, 2000). Further conversion of OPDA takes place by an NADPH-dependent reductase (OPR). Besides the unspecific OPR1 and OPR2, only the recently cloned OPR3 of *Arabidopsis* forms exclusively the correct enantiomer *cis*-(9*S*,13*S*)-OPDA (Schaller *et al.*, 2000). OPR3 contains a peroxisomal signal sequence suggesting its peroxisomal location. Consequently, the transport has to be assumed of OPDA, from the chloroplasts into peroxisomes, where enzymes catalyzing β-oxidation are located. The final product of JA biosynthesis, (+)-*7-iso*-JA, isomerizes to the more stable (−)-JA.

A highly important issue in JA biosynthesis is its regulation. Due to the fact that all LOX, AOS, AOC and OPR3 are transcriptionally up-regulated by JA or stress, a positive feedback was assumed. However, at least within some hours, a transcriptional up-regulation for *LOX*, *AOS* and *AOC* upon treatment with jasmonates and octadecanoids was not accompanied by an endogenous rise in jasmonates measured by GC-MS quantification and isotopic dilution analysis (Kramell *et al.*, 2000). Recently, for *Arabidopsis*, indications were found that positive feedback may occur during growth (Stenzel *et al.*, 2002b). In addition, in tomato, substrate availability may control enzyme activity in JA biosynthesis (Stenzel *et al.*, 2002a). Interestingly, constitutively high JA levels, found upon overexpression of AOS in potato, did not lead necessarily to constitutive expression of JA-inducible genes,

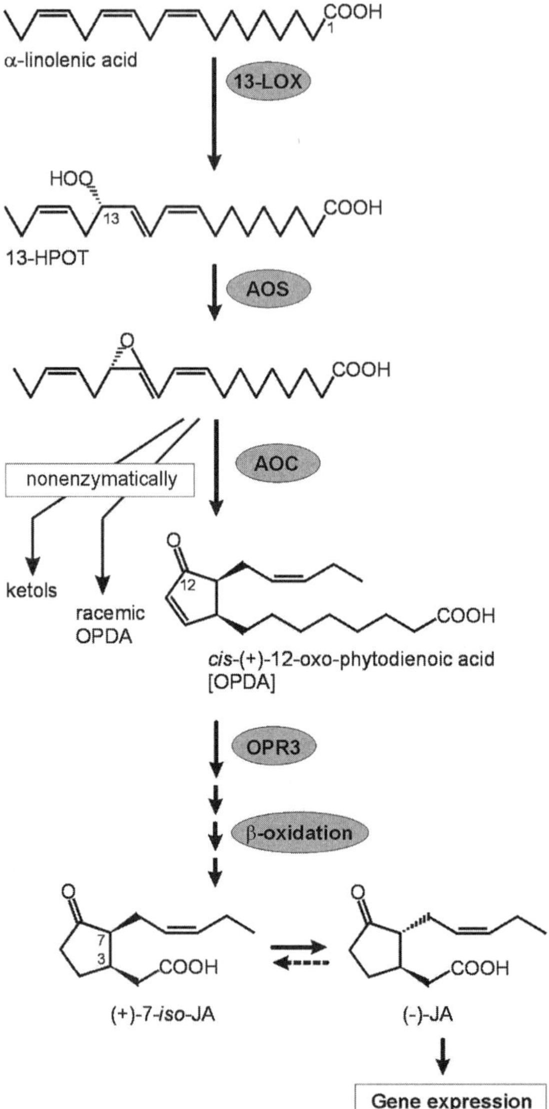

Figure 9-2. The biosynthesis of jasmonates. Peculiarities are given by instability of the allene oxide and the strict stereochemistry established by the AOC and kept by the OPR-catalyzed step.

suggesting intracellular sequestration of JA (Harms *et al.*, 1995). Recent data emphasize cell-specific AOC expression and elevation of JA levels in leaf veins (Stenzel *et al.*, 2002a).

C. Signaling Properties of Jasmonates and Octadecanoids—the Oxylipin Signature

The signaling role of jasmonates is based on only correlative data on (i) structure–activity relationships and (ii) a link between an endogenous rise upon an external stimulus and the corresponding expression of jasmonate-responsive genes. Numerous studies on structure–activity relations of substituted, deleted or stereospecifically altered jasmonic acid derivatives revealed that (i) an intact cyclopentanone ring with a (−)-enantiomeric or (+)-*7-iso*-enantiomeric structure, and (ii) a pentenyl side chain and (iii) an acetic acid side chain are necessary (Blechert *et al.*, 1999; Miersch *et al.*, 1999). More recently, it was shown that octadecanoids but not jasmonates were responsible for resistance against insect and fungal attack (Stintzi *et al.*, 2001). Furthermore, octadecanoids are the preferential signals in tendril coiling (Blechert *et al.*, 1999) and in the release of diterpenoid-derived volatiles (Koch *et al.*, 1999), whereas jasmonates were found to be preferentially active in inducing synthesis of alkaloids in cell cultures or in the release of sesquiterpene volatiles (Koch *et al.*, 1999). The OPDA derivative dinor OPDA, which is formed from the 16:3 fatty acid, appears beside OPDA (Weber *et al.*, 1997). A varying amount of different oxylipins, called the "oxylipin-signature", may function as a signal (Weber *et al.*, 1997). Such a distinct ratio of jasmonates, octadecanoids and conjugates of amino acid with JA was found recently in distinct flower organs (Hause *et al.*, 2000).

IV. Jasmonate-induced Alteration of Gene Expression

In all plants studied so far, treatment with jasmonates exerts two different changes in gene expression: (i) decrease in the expression of nuclear- and chloroplast-encoded genes mainly involved in photosynthesis (Reinbothe *et al.*, 1997), (ii) increase in the expression of specific genes. Decreased gene expression caused by jasmonates has been shown to be (i) transcriptionally regulated in the case of the small and large subunits of Rubisco and chlorophyll *a/b* binding protein (Reinbothe *et al.*, 1997; Wierstra and Kloppstech, 2000) and (ii) translationally regulated in the case of some plastid-located proteins (Reinbothe *et al.*, 1993). Both types of regulation may contribute to the senescence-promoting effect of jasmonates (see Section VI).

The genes up regulated by jasmonates code for proteins of various functions (Table 9-1). Some of these proteins were found to be encoded by SAGs indicating the overlap in plant defense response and senescence (Fig. 9-3) (cf. Sections V, VI).

V. Jasmonates and Plant Defense Reactions

Jasmonates and related compounds are signaling compounds in plant responses to abiotic and biotic stress factors. Among the abiotic factors, irradiation by UV light and high-light conditions may contribute to senescence. The role of jasmonate in UV-B-induced stress responses of plants has been shown (Conconi *et al.*, 1996). UV protection might be built up by jasmonate-mediated synthesis of anthocyanins (Franceschi and Grimes, 1991). Anthocyanins are known to exhibit protective properties against UV light and oxidative stress. Also protection against high light is related to altered gene expression induced by jasmonates. Early light-induced proteins (ELIPs) expressed during high-light conditions are down regulated by jasmonates suggesting a negative regulatory function of jasmonates in high light (Wierstra and Kloppstech, 2000, cf. Section VI).

Upon wounding, plants synthesize proteinase inhibitors (pins) which are toxic proteins to the gut of attacking insects. Plants already carrying pins are "immunized" against a subsequent insect attack. Pins seem to be formed in the following sequence: wounding → expression of the prosystemin gene in vascular bundles of a vein → processing to the

Table 9-1. Jasmonate-induced Proteins likely to be Related to Senescence belong to Many Different Functions. Examples are given in the Table

	References (only examples are given)
Enzymes of jasmonate and oxylipin biosynthesis	
LOX	
AOS	Maucher *et al.*, 2000; Howe *et al.*, 2000
AOC	Ziegler *et al.*, 2000
OPR3	Reymond *et al.*, 2000
HPL	
Proteinases and proteinase inhibitors	
Leucine amino peptidase	
Serine proteinase inhibitor (1 and 2)	Ryan, 2000
Cysteine proteinase	
Proteins in signal transduction	
-Phospholipase A_2	
MPK3	Reymond *et al.*, 2000
Systemin	Ryan, 2000
Defense-related proteins	
NPR1	
PR1	Pieterse and van Loon, 1999
Thionins	Reymond *et al.*, 2000
Defensins	
Metallothionin	
Glutathionine-S-transferase	
Storage proteins	
Vegetative storage proteins (VSPs)	
Napin	Creelman and Mullet, 1997
Protein of cell wall formation	
Hydroxyproline rich proteins (HPRPs)	Creelman and Mullet, 1997

peptide systemin → local response upon binding of systemin to its receptor → activation of a phospholipase A_2 and MAP kinase mediated by Ca^{2+} and calmodulin → release of linolenic acid → formation of JA → expression of pin2 and other defense genes such PPO or leucine amino peptidase → wound response in a systemic leaf by loading of systemin into the phloem (for reviews cf. Ryan, 2000; Wasternack and Hause, 2002). Interestingly, the initial wound response—formation of systemin within the veins—might be amplified by JA. Genes coding for prosystemin and AOC are JA-inducible, are co-expressed within the veins (Jacinto *et al.*, 1997; Hause *et al.*, 2000), and the veins elevate levels of JA upon wounding more than the intercostal regions (Stenzel *et al.*, 2002a).

In plant–pathogen interactions, numerous genes are expressed, but many of them are also expressed during senescence (cf. Section VI). Here, JA and its counterpart SA can be discussed only briefly in terms of their signaling pathways in specific responses to pathogens. Individual pathways containing salicylate (SA), ethylene and JA as signals can be distinguished in the interaction of plants with heterotrophic and pathogenic microorganisms (Pieterse and van Loon, 1999; Feys and Parker, 2000). Formation of the toxic proteins such as defensin and thionin is activated via the octadecanoid pathway in an ethylene-dependent and SA-independent manner (Thomma *et al.*, 1998; Bohlmann *et al.*, 1998). The most important pathway in plant–pathogen interactions is SA-dependent. This pathway is characterized by a local increase in SA levels which is necessary and sufficient for PR

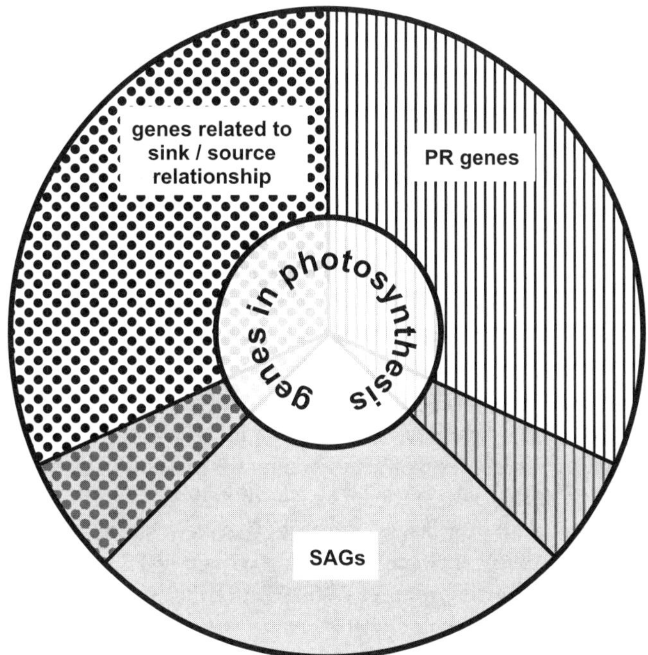

Figure 9-3. Different sets of genes exhibiting partial overlap in expression pattern are related to sink–source relationships, defense reactions and senescence. Many of them are up regulated by jasmonates and other signals. In contrast, photosynthetic genes are down regulated by jasmonates.

gene expression and systemic acquired resistance (SAR). Among the mutants affected in SA signal transduction are *pad4* and *cpr1*, both acting upstream of SA. CPR1 and the MAP-kinase MPK4 act as negative regulators, and MPK4 may perform cross-talk between SA-dependent SAR and JA-dependent defensin/thionin formation (Petersen *et al.*, 2000). Another protein, NPR1, was shown to function downstream of SA. NPR1 is a positive regulator which interacts with transcription factors of the TGA-type and with *as-1*-like promoter elements. Interestingly, NPR1 is a member of a signaling pathway in response to non-pathogenic bacteria. This pathway is clearly JA- and ethylene-dependent and leads to the so-called induced systemic resistance (ISR) (Pieterse and van Loon, 1999). Both pathways leading to SAR and ISR, respectively, can be simultaneously activated which results in an additive effect of induced protection (Van Wees *et al.*, 2000). Obviously, NPR1 seems to be a knot of cross-talk, where many environmental signals are integrated and translated into a specific gene expression pattern. Another cross-talk seems to occur by jasmonates, since the JA/ethylene-dependent SA-independent expression of thionins and defensins is also wound responsive. Mutants will facilitate analysis of the different signaling pathways. Three types of mutants are available so far to inspect the role of JA. (i) JA-deficient mutants such as *opr3* and *fad3-2fad7-2fad8*. This triple mutant is unable to form linolenic acid, the essential precursor of JA (McConn and Browse, 1996), and the *opr3* is affected in OPR3. Consequently, both mutants exhibit diminished insect defense (McConn *et al.*, 1997; Stintzi *et al.*, 2001; Vijayan *et al.*, 1998). (ii) JA-insensitive mutants such as *coi1*. (iii) Mutants carrying a constitutive JA response like the *cev* mutant (Ellis and Turner, 2001).

VI. Jasmonates in Plant Development and Senescence

At least five different developmental processes are known to be influenced by jasmonates: (i) seed germination and root growth, (ii) sink–source relationships, (iii) morphogenetic processes such as tuber formation and tendril coiling, (iv) flower and fruit development and (v) senescence. Since some sink–source relationships may play a role in controlling senescence, these and senescence will be discussed here. The other aspects are reviewed by Koda (1997), Creelman and Mullet (1997) and Wasternack and Hause (2002). Jasmonates have been shown to alter sink–source relationships, e.g., JA promotes formation of the N-rich vegetative storage proteins VSPα and VSPβ of soybean, including reallocation in pod filling. In addition to such nutrient reallocation to other parts of the plant, jasmonates cause decreases in photosynthesis and chlorophyll content, the most striking manifestations of senescence in leaves. The promotion of senescence by jasmonates is counteracted by cytokinins (Parthier, 1991); however, the underlying mechanism is unknown. In barley leaves, JA-induced down-regulation of genes coding for proteins of photosynthesis occurs in a similar fashion as during senescence. The rise of endogenous jasmonates upon stress or exogenous treatment with jasmonates correlates in time with the expression of various genes (Lehmann *et al.*, 1995), including a 23 kDa protein (JIP23), which is unknown in function, a ribosome-inactivating protein of 60 kDa (JIP60 = RIP60) and several LOX forms. JIP23 occurs during seedling development in non-green tissues, e.g., the scutellar node and the leaf base (Hause *et al.*, 1996), which also exhibit high expression of AOS and elevated levels of jasmonates (Maucher *et al.*, 2000). Interestingly, overexpression of barley JIP23 in tobacco led to down-regulation of housekeeping proteins such as Rubisco and decreased the photosynthetic capacity of these plants (Görschen *et al.*, 1997). This effect of a jasmonate-induced protein is reminiscent of the well-known effect of jasmonates themselves. In most tissues studied so far, down-regulation of the expression of photosynthetic genes is one of the most prominent effects of jasmonates, and the following scenario may occur: Elevated levels of JA in a tissue upon environmental stress or during development may within a few hours inhibit synthesis of the housekeeping proteins necessary for the photosynthetic apparatus and may induce synthesis of other proteins such as JIP23. JIP23 in turn may amplify the JA-induced down-regulation of housekeeping proteins due to decreased translation of housekeeping proteins. As a consequence, premature formation of the photosynthetic apparatus and formation of radicals occurring in photosynthesis and scavenged by chlorophyll and carotenoids is kept low in a non-green tissue such as the leaf base (Creelman and Mullet, 1997).

Also, sink-specific as well as defense-related genes were found to be regulated inversely to photosynthetic genes (Roitsch *et al.*, 2000). It will be interesting to analyze in detail the role of jasmonates in this coordinated expression and repression of genes. Recently, the coordinated up-regulation of a glucose-inducible AOC and the JA/OPDA levels in ovules of tomato flowers was shown (Hause *et al.*, 2000). This sink tissue, which is known to up regulate formation of plant defense proteins such as pin2 or PR1, may depend on jasmonates/octadecanoids to coordinate expression of genes related to sink/source relationships and plant defense (Hause *et al.*, 2000).

The capacity of jasmonates to down regulate photosynthetic genes (Creelman and Mullet, 1997) may also be one determinant in the onset of senescence. In this respect it is interesting to note that barley plants exhibit high JA levels in tissues undergoing senescence (unpublished data from our laboratory), and that ELIPs, considered to be protective against high-light conditions, are down regulated by JA (Wierstra and Kloppstech, 2000).

Recently, two cDNAs coding for the initial enzyme of chlorophyll breakdown, the chlorophyllase, were isolated from *A. thaliana* (Tsuchiya *et al.*, 1999). One of them lacks a chloroplast signal peptide, but exhibits a vacuolar sorting determinant NPIR, and the corresponding gene is expressed in response to jasmonate (Tsuchiya *et al.*, 1999). However, it is difficult to reconcile a vacuolar location of chlorophyllase with the well-known chloroplast location of the three subsequent steps of chlorophyll degradation (Matile *et al.*, 1999). Therefore, it will be interesting to analyze whether the recently described senescence-related secretion of chlorophyll into vacuoles (Guiamét *et al.*, 1999) is functionally linked to chlorophyllase activity and formation of chlorophyll catabolites.

The role of jasmonates in natural senescence and stress-induced senescence is also suggested by the fact that numerous genes isolated by screening for senescence-associated genes (SAGs) (Buchanan-Wollaston and Ainsworth, 1997; Quirino *et al.*, 2000) exhibit inducibility by jasmonate (e.g., Park *et al.*, 1998). SAG-encoded proteins function in protein degradation, nutrient reallocation, autoxidative processes, defense reactions to pathogen attack and signal transduction. Consequently, the sets of genes involved in the sink–source relationship, plant defense and senescence overlap in their expression pattern. Jasmonate usually up regulates expression of these genes as it depresses photosynthetic genes. Among the genes expressed during senescence are SAG13 and LSC54 coding for metallothionins, whereas SAG12 coding for a cysteine protease seems to be related to the onset of senescence. Interestingly, the signaling pathway leading to SAG12 expression contains the sequential action of PAD4, SA and NPR1 (Morris *et al.*, 2000). Thus, again NPR1 seems to be a point of convergence (cf. above), and senescence and plant defense overlap in terms of gene expression. Senescence is a multifactorial event regulated by internal or environmental factors. Due to the overlap of SAG expression with plant defense gene expression and gene expression related to nutrient reallocation, a network of signaling pathways has to be expected as outlined by studies on plant–pathogen interactions (Genoud and Metraux, 1999). Such a network possibly functioning in leaf senescence was recently shown by isolation of 147 enhancer trap lines of *A. thaliana* with increased senescence in response to senescence-promoting factors such as ABA, JA, ethylene, brassinosteroids, darkness and dehydration (He *et al.*, 2001). The responsiveness of all lines was found to converge via a few lines resulting in leaf senescence. Surprisingly, convergence in response to jasmonate, ethylene and ABA involves an F-box protein, called ORE9 (Woo *et al.*, 2001). This protein seems to be a positive regulator of senescence by removing through ubiquitin-dependent proteolysis those target proteins which are required to delay the leaf senescence of *A. thaliana* (Woo *et al.*, 2001). Another important key element in dissecting different signaling pathways is transcription factors. The EREBP system of ethylene response, the DREB2-DRE system, the bZIP-ABRE system and the MYC/MYB system in dehydration response as well as the DREB1-DRE system of low temperature response (Stepanova and Ecker, 2000; Shinozaki and Yamaguchi-Shinozaki, 2000) have common elements and they illustrate the impact of transcription factors on the signaling network. Interestingly, all these factors belong to a family of APETALA2-domain proteins, suggesting a modular principle of interaction at the level of transcription factors. Within the large family of WRKY transcription factors a new member, AtWRKY6, was found to be associated with senescence- and defense-related processes (Robatzek and Somssich, 2001). Within the SAG12 promoter only a senescence-specific region has been identified (Noh and Amasino, 1999), and senescence-specific transcription factors are unknown. In the case of jasmonate-induced gene expression, the first transcription factors were characterized recently (Memelink *et al.*, 2001). Also these factors are AP2-domain proteins, known to function in many stress-related responses. It will be interesting to see whether JA-specific transcription factors function in senescence. The main understanding of the regulation of senescence including the role of jasmonate will come out by analysis of transcription factors and *cis*-acting elements.

Acknowledgments

I thank Drs. B. Hause and I. Feussner for critical reading and C. Dietel for typing the manuscript.

References

Berger, S., Weichert, H., Kühn, H., Wasternack, C., and Feussner, I. (2001). Enzymatic and non-enzymatic lipid peroxidation in leaf development. *Biochimicaet Biophysica Acta* **1533**, 266–276.

Blechert, S., Bockelmann, C., Füßlein, M., v. Schrader, T., Stelmach, B., Niesel, U., and Weiler, E.W. (1999). Structure activity analyses reveal the existence of two separate groups of active octadecanoids in elicitation of the tendril-coiling response of *Bryonia dioica* Jacq. *Planta* **207**, 470–479.

Bohlmann, H., Vignutelli, A., Hilpert, B., Miersch, O., Wasternack, C., and Apel, K. (1998). Wounding and chemicals induce expression of the *Arabidopsis thaliana* gene *Thi2.1*, encoding a fungal defense thionin, via the octadecanoid pathway. *FEBS Letters* **437**, 281–286.

Brash, A.R. (1999). Lipoxygenases: Occurrence, Functions, catalysis, and acquisition of substrate. *Journal of Biological Chemistry* **274**, 23679–23682.

Buchanan-Wollaston, V., and Ainsworth, C. (1997). Leaf senescence in *Brassica napus*: cloning of senescence related genes by subtractive hybridisation. *Plant Molecular Biology* **33**, 821–834.

Conconi, A., Smerdon, M.J., Howe, G.A., and Ryan, C.A. (1996). The octadecanoid signalling pathway in plants mediates a response to ultraviolet radiation. *Nature* **383**, 826–829.

Creelman, R.A., and Mullet, J.E. (1997). Biosynthesis and action of jasmonates in plants. *Annu. Rev. Plant Physiol. Plant Molecular Biology* **48**, 355–381.

Ellis, C., and Turner, J.G. (2001). The *Arabidopsis* mutant *cev1* has constitutively active jasmonate and ethylene signal pathways and enhanced resistance to pathogens. *The Plant Cell* **13**, 1025–1033.

Feussner, I., Hause, B., Vörös, K., Parthier, B., and Wasternack, C. (1995). Jasmonate-induced lipoxygenase forms are localized in chloroplast of barley leaves (*Hordeum vulgare* cv, Salome). *The Plant Journal* **7**, 949–957.

Feussner, I., and Wasternack, C. (2002). The lipoxygenase pathway. *Annual Reviews of Plant Physiology and Plant Molecular Biology* **53**, 275–297.

Feys, B.I., and Parker, J.A. (2000). Interplay of signaling pathways in plant disease resistance. *Trends in Genetics* **16**, 449–455.

Franceschi, V.R., and Grimes, H.D. (1991). Low levels of atmospheric methyl jasmonate induce the accumulation of soybean vegetative storage proteins and anthocyanins. *Proceedings of the National Academy of Sciences of the USA* **88**, 6745–6749.

Froehlich, J.E., Itoh, A., and Howe, G.A. (2001). Tomato allene oxide synthase and fatty acid hydroperoxide lyase, two cytochrome P450s involved in oxylipin metabolism, are targeted to different membranes of chloroplast envelope. *Plant Physiology* **125**, 306–317.

Genoud, T., and Metraux, J.-P. (1999). Cross-talk in plant cell signalling: structure and function of the genetic network. *Trends in Plant Sciences* **4**, 503–507.

Görschen, E., Dunaeva, M., Reeh, I., and Wasternack, C. (1997). Overexpression of the jasmonate-inducible 23 kDa protein (JIP 23) from barley in transgenic tobacco leads to the repression of leaf proteins. *FEBS Letters* **419**, 58–62.

Guiamét, J.J., Pichersky, E., and Noodén, L.D. (1999). Mass exodus from senescing soybean chloroplasts. *Plant Cell Physiology* **40**(9), 986–992.

Harms, K., Atzorn, R., Brash, A.R., Kühn, H., Wasternack, C., Willmitzer, L., and Peña-Cortés, H. (1995). Expression of a flax allene oxide synthase cDNA leads to increased endogenous jasmonic acid (JA) levels in transgenic potato plants but not to a corresponding activation of JA-responding genes. *Plant Cell* **7**, 1645–1654.

Hause, B., Demus, U., Teichmann, C., Parthier, B., and Wasternack, C. (1996). Developmental and tissue-specific expression of JIP-23, a jasmonate-inducible protein of barley. *Plant Cell Physiology* **37**, 641–649.

Hause, B., Stenzel, I., Miersch, O., Maucher, H., Kramell, R., Ziegler, J., and Wasternack, C. (2000). Tissue-specific oxylipin signature of tomato flowers—allene oxide cyclase is highly expressed in distinct flower organs and vascular bundles. *The Plant Journal* **24**, 113–126.

He, Y., Tang, W., Swain, J.D., Green, A.L., Jack, T.P., and Gan, S. (2001). Networking senescence-regulating pathways by using *Arabidopsis* enhancer trap lines. *Plant Physiology* **126**, 707–716.

Hong, Y., Wang, T.-W., Hudak, K.A., Schade, F., Froese, C.D., and Thompson, J.E. (2000). An ethylene-induced cDNA encoding a lipase expressed at the onset of senescence. *Proceedings of the National Academy of Sciences of the USA* **97**, 8717–8722.

Howe, G.A., Lee, G.I., Itoh, A., Li, L., and DeRocher, A.E. (2000). Cytochrome P450-dependent metabolism of oxylipins in tomato. Cloning and expression of allene oxide synthase and fatty acid hydroperoxide lyase. *Plant Physiology* **123**, 711–724.

Ishiguro, S., Kawai-Oda, A., Ueda, J., Nishida, I., and Okada, K. (2001). The defective in anther dehiscence 1 gene encodes a novel phospholipase A1 catalyzing the initial step of jasmonic acid biosynthesis, which synchronyzes pollen maturation, anther dehiscence, and flower opening in *Arabidopsis*. *The Plant Cell* **13**, 191–209.

Jacinto, T., McGurl, B., Franceschi, V., Delano-Freier, J., and Ryan, C.A. (1997). Tomato prosystemin promoter confers wound-inducible, vascular bundle-specific expression of the β-glucuronidase gene in transgenic tomato plants. *Planta* **203**, 406–412.

Koch, T., Krumm, T., Jung, V., Engelberth, J., and Boland, W. (1999). Differential induction of plant volatile biosynthesis in the lima bean by early and late intermediates of the octadecanoid-signaling pathway. *Plant Physiology* **121**, 153–162.

Koda, Y. (1997). Possible involvement of jasmonates in various morphogenic events. *Physiologia Plantarum* **100**, 639–646.

Kramell, R., Miersch, O., Atzorn, R., Parthier, B., and Wasternack, C. (2000). Octadecanoid-derived alteration of gene expression and the 'oxylipin signature' in stressed barley leaves—implications for different signalling pathways. *Plant Physiology* **123**, 177–186.

Lehmann, J., Atzorn, R., Brückner, C., Reinbothe, S., Leopold, J., Wasternack, C., and Parthier, B. (1995). Accumulation of jasmonate, abscisic acid, specific transcripts and proteins in osmotically stressed barley leaf segments. *Planta* **197**, 156–162.

Matile, P., Hörtensteiner, S., and Thomas, H. (1999). Chlorophyll degradation. *Annual Reviews on Plant Physiology and Plant Molecular Biology* **50**, 67–95.

Maucher, H., Hause, B., Feussner, I., Ziegler, J., and Wasternack, C. (2000). Allene oxide synthases of barley (*Hordeum vulgare* cv. Salome)—tissue specific regulation in seedling development. *The Plant Journal* **21**, 199–213.

McConn, M., and Browse, J. (1996). The critical requirement for linolenic acid is pollen development, not photosynthesis, in an *Arabidopsis* mutant. *Plant Cell* **8**, 403–416.

McConn, M., Creelman, R.A., Bell, E., Mullet, J.E., and Browse, J. (1997). Jasmonate is essential for insect defense in *Arabidopsis*. *Proceedings of the National Academy of Sciences of the USA* **94**, 5473–5477.

Memelink, J., Verpoorte, R., and Kijne, J.W. (2001). ORCAnization of jasmonate-responsive gene expression in alkaloid metabolism. *Trends in Plant Science* **6**, 212–219.

Miersch, O., Kramell, R., Parthier, B., and Wasternack, C. (1999). Structure-activity relations of substituted, deleted or stereospecifically altered jasmonic acid in gene expression of barley leaves. *Phytochemistry* **50**, 353–361.

Morris, K., Mackerness, S.A.-H., Page, T., John, C.F., Murphy, A.M., Carr, J.P., and Buchanan-Wollaston, V. (2000). Salicylic acid has a role in regulating gene expression during leaf senescence. *The Plant Journal* **23**(5), 677–685.

Narváez-Vásquez, J., Florin-Christensen, J., and Ryan, C.A. (1999). Positional specificity of a phospholipase a activity induced by wounding, systemin, and oligosaccharide elicitors in tomato leaves. *Plant Cell* **11**, 2249–2260.

Noh, Y.-S., and Amasino, R.M. (1999). Identification of a promoter region responsible for the senescence-specific expression of SAG12. *Plant Molecular Biology* **41**, 181–194.

Park, J.-H., Oh, S.A., Kim, Y.H., Woo, H.R., and Nam, H.G. (1998). Differential expression of senescence-associated mRNAs during leaf senescence induced by different senescence-inducing factors in *Arabidopsis*. *Plant Molecular Biology* **37**, 445–454.

Parthier, B. (1991). Jasmonates, new regulators of plant growth and development: many facts and few hypothesis on their actions. *Botanica Acta* **104**, 446–454.

Petersen, M., Brodersen, P., Naested, H., Andreasson, E., Lindhart, U., Johansen, B., Nielsen, H.B., Lacy, M., Austin, M.J., Parker, J.E., Sharma, S.B., Klessig, D.F., Martienssen, R., Mattsson, O., Jensen, A.B., and Mundy, J. (2000). *Arabidopsis* MAP kinase 4 negatively regulates systemic acquired resistance. *Cell* **103**, 1111–1120.

Pieterse, C.M.J., and van Loon, B.C. (1999). Salicylic acid-independent plant defence pathways. *Trends in Plant Science* **4**, 52–58.

Quirino, B.F., Noh, Y.-S., Himelblau, E., and Amasino, R.M. (2000). Molecular aspects of leaf senescence. *Trends in Plant Science* **5**, 278–282.

Reinbothe, S., Reinbothe, C., and Parthier, B. (1993). Methyl jasmonate-regulated translation of nuclear-encoded chloroplast proteins in barley (*Hordeum vulgare* L. cv. Salome). *Journal of Biological Chemistry* **268**, 10606–10611.

Reinbothe, C., Parthier, B., and Reinbothe, S. (1997). Temporal pattern of jasmonate-induced alterations of gene expression. *Planta* **201**, 281–287.

Reymond, P., Weber, H., Damond, M., and Farmer, E.E. (2000). Differential gene expression in response to mechanical wounding and insect feeding in *Arabidopsis*. *The Plant Cell* **12**, 707–719.

Robatzek, S., and Somssich, I.E. (2001). A new member of the *Arabidopsis* WRKY transcription factor family, AWRKY6, is associated with both senescence- and defence-related processes. *The Plant Journal* **28**, 123–133.

Roitsch, T., Ehneß, R., Goetz, M., Hause, B., Hofmann, M., and Sinha, A.K. (2000). Regulation and function of extracellular invertase from higher plants in relation to assimilate partitioning, stress responses and sugar signalling. *Australian Journal of Plant Physiology* **27**, 815–825.

Ryan, C.A. (2000). The systemin signaling pathway: differential activation of plant defensive genes. *Biochimicaet Biophysica Acta* **1477**, 112–121.

Schaller, F., Biesgen, C., Müssig, C., Altmann, T., and Weiler, E.W. (2000). 12-Oxophytodienoate reductase 3 (OPR3) is the isoenzyme involved in jasmonate biosynthesis. *Planta* **210**, 979–984.

Shinozaki, K., and Yamaguchi-Shinozaki, K. (2000). Molecular responses to dehydration and low temperature: differences and cross-talk between two stress signaling pathways. *Current Opinion in Plant Biology* **3**, 217–223.

Stenzel, I., Hause, B., Maucher, M., Pitzschke, A., Miersch, O., Ziegler, J., Ryan, C.A. and Wasternack, C. (2003a). Allene oxide cyclase dependence of tomato—amplification in wound signalling. *The Plant Journal* **33**, 577–589.

Stenzel, I., Hause, B., Miersch, O., Kurz, T., Maucher, H., Weichert, H., Ziegler, J. , Feussner. I., and Wasternack, C. (2003b). Jasmonate biosynthesis and the allene oxide cyclase family of *Arabidopsis thaliana*. *Plant Molecular Biology* **51**, 895–911.

Stepanova, A.N., and Ecker, J.R. (2000). Ethylene signaling: from mutants to molecules. *Current Opinion in Plant Biology* **3**, 353–360.

Stintzi, A., and Browse, J. (2000) The *Arabidopsis* male-sterile mutant, *opr3*, lacks the 12-oxophytodienoic acid reductase required for jasmonate synthesis. *Proceedings of the National Academy of Sciences of the USA* **97**, 10625–10630.

Stintzi, A., Weber, H., Reymond, P., Browse, J., and Farmer, E.E. (2001). Plant defense in the absence of jasmonic acid: The role of cyclopentenones. *Proceedings of the National Academy of Sciences of the USA* **98**, 12837–12842.

Thomma, B.P.H.J., Eggermont, K., Penninckx, I.A.M.A., Mauch-Mani, B., Vogelsang, R., Cammue, B.P.A., and Broekaert, W.F. (1998). Separate jasmonate-dependent and salicylate-dependent defense-response pathways in *Arabidopsis* are essential for resistance to distinct microbial pathogens. *Proceedings of the National Academy of Sciences of the USA* **95**, 15107–15111.

Tsuchiya, T., Ohta, H., Okawa, K., Iwamastsu, A., Shimada, H., Masuda, T., and Takamiya, K. (1999). Cloning of chlorophyllase, the key enzyme in chlorophyll degradation: finding of a lipase motif and the induction by methyl jasmonate. *Proceedings of the National Academy of Sciences of the USA* **96**, 15362–15367.

Van Wees, S.C.M., de Swart, E.A.M., van Pelt, J.A., van Loon, L.C., and Pieterse, C.M.J. (2000). Enhancement of induced disease resistance by simultaneous activation of salicylate- and jasmonate-dependent defense pathways in *Arabidopsis thaliana*. *Proceedings of the National Academy of Sciences of the USA* **97**, 8711–8716.

Vijayan, P., Shockey, J., Leversque, C.A., Cook, R.J., and Browse, J. (1998). A role of jasmonate in pathogen defense of *Arabidopsis*. *Proceedings of the National Academy of Sciences of the USA* **95**, 7209–7214.

Wasternack, C., and Hause, B. (2002). Jasmonates and octadecanoids—Signals in plant stress responses and development. *Progress in Nucleic Acid Research* **72**, 165–221.

Wasternack, C., and Parthier, B. (1997). Jasmonate-signalled plant gene expression. *Trends in Plant Science* **2**, 302–307.

Weber, H., Vick, B.A., and Farmer, E.E. (1997). Dinor-oxo-phytodienoic acid: A new hexadecanoid signal in the jasmonate family. *Proceedings of the National Academy of Sciences of the USA* **94**, 10473–10478.

Wierstra, I., and Kloppstech, K. (2000). Differential effects of methyl jasmonate on the expression of the early light-inducible proteins and other light-regulated genes in barley. *Plant Physiology* **124**, 833–844.

Woo, H.R., Chung, K.M., Park, J.-H., Oh, S.A., Ahn, T., Hong, S.H., Jang, S.K., and Nam, H.G. (2001). ORE9, an F-box protein that regulates leaf senescence in *Arabidopsis*. *Plant Cell* **13**, 1779–1790.

Ziegler, J., Stenzel, I., Hause, B., Maucher, H., Miersch, O., Hamberg, M., Grimm, M., Ganal, M., and Wasternack, C. (2000). Molecular cloning of allene oxide cyclase: The enzyme establishing the stereochemistry of octadecanoids and jasmonates. *Journal of Biological Chemistry* **275**, 19132–19138.

10

Leaf Senescence and Nitrogen Metabolism

Tadahiko Mae

I. Introduction

Most of the nitrogen in a plant is invested in the leaves. During the course of leaf senescence, proteins, nucleic acids and other nitrogenous compounds are degraded, and their nitrogen is remobilized and translocated into the actively growing organs (Peoples and Dalling, 1988). Reutilization of the remobilized nitrogen by growing organs gives a chance for the plant to conserve this limiting resource and use it most efficiently. Leaves are composed of many different types of cells and each cell is also composed of different subcellular components. Senescence of the leaf does not start at the same time and with the same speed in all cells or in all organelles. Although all of these structures are eventually broken down, the chloroplasts contain most of the leaf nitrogen and therefore their disassembly is particularly important.

The objective of this chapter is to review characteristic features of nitrogen metabolism, especially protein metabolism in the processes of leaf senescence.

II. Leaf Senescence and Nitrogen Loss

The nitrogen content of a leaf reaches the maximum level at around the completion of leaf expansion. When mesophyll cells cease their enlargement, increase of chloroplast number and size also stops (Pyke and Leech, 1987; Baumgartner *et al.*, 1989; Sodmergen *et al.*, 1991). In mature leaves of C_3 plants, chloroplast nitrogen accounts for 70–80% of leaf nitrogen (Morita, 1980; Makino and Osmond, 1991). In the course of leaf senescence, nitrogen (protein) content in the leaf gradually decreases with a concomitant loss of the photosynthetic activity (Wittenbach, 1979; Friedrich and Huffaker, 1980; Peoples *et al.*, 1980; Makino *et al.*, 1983; Crafts-Brandner *et al.*, 1984; Ford and Shibles, 1988). Total apparent loss of chloroplast nitrogen during leaf senescence was estimated to account for 90% of the total nitrogen loss from the leaf in rice (Morita, 1980).

A. Fate of Chloroplasts and their Constituents

Two major hypotheses to explain the decline in chloroplast nitrogen or photosynthetic capability during leaf senescence have been advanced (Huffaker, 1990). It has been proposed that the decline in photosynthesis is mainly attributed to a decrease in the number of chloroplasts in the senescing leaf. Wittenbach *et al.* (1982), for example, reported a simultaneous decline in photosynthesis, chlorophyll, soluble protein, and total number of chloroplasts in senescing wheat leaves. They also presented ultrastructural evidence indicating that chloroplasts might be associated with invagination of the vacuole or may move into the vacuole during senescence. The other hypothesis is that the decline is largely attributed to events occurring within chloroplasts that did not change in number until the late stage of leaf senescence. The hypothesis fitted the ultrastructural studies by Thomson and Platt-Aloia (1987) and Inada *et al.* (1998a,b, 1999), indicating that membrane integrity and subcellular compartmentation of chloroplasts, mitochondria, peroxisomes, and vacuoles in mesophyll cells are maintained until the last stage of senescence. The latter hypothesis might be further supported by observations of non-parallel decline in chloroplast components during senescence. Preferential decline in cytochrome f content (Gepstein, 1988) and preferential retention of LHCII content (Hidema *et al.*, 1991, 1992; Mae *et al.*, 1993) were observed in leaves senesced under relatively low irradiances. The enzymes involved in the photosynthetic carbon reduction cycle did not lose activities or proteins in a synchronized manner during senescence (Makino *et al.*, 1983; Crafts-Brandner *et al.*, 1996, 1998). Stay-green mutants of *Lolium temulentum* exhibited preferential retention of Chl, P_{700}, and D1 contents until the late stage of senescence, while Rubisco was normally lost during senescence (Thomas *et al.*, 1999). Ford and Shibles (1988) reported that senescence is a two-stage process: first a decline in chloroplast function accompanied by a loss of chloroplast content, followed by a brief terminal phase when whole chloroplasts are lost as well.

III. Protein Metabolism during Leaf Senescence

The amount of a protein in a leaf is a result of its synthesis and degradation during growth. In order to understand protein metabolism in senescing leaves we need to know both synthesis and degradation of the protein throughout the life span of the leaf. Particular attention has been paid to ribulose-1,5-bisphosphate carboxylase/oxygenase (Rubisco), light-harvesting chlorophyll a/b protein complex (LHCII), and D1 protein of photosystem II because of their important roles in photosynthesis and nitrogen economy in plants.

A. Fates of DNA and RNA

In mesophyll cells, there are three sites for protein synthesis: nucleus/cytoplasm, mitochondria, and chloroplasts. Each site has its own DNA and apparatus necessary for RNA replication and protein synthesis. Chloroplasts contain a large portion of leaf protein, but not all the proteins in chloroplasts are synthesized there. A number of chloroplast proteins are encoded by nuclear DNA, synthesized as precursors in the cytoplasm, imported into the chloroplast, processed, and assembled there. It was shown that the content of nuclear DNA or mitochondrial DNA was unchanged until the last stage of leaf senescence, but the content of chloroplast DNA decreased considerably during senescence in barley (Baumgartner *et al.*, 1989) and rice (Sodmergen *et al.*, 1989, 1991; Inada *et al.*, 1998b, 1999). These data indicate that template availability might be one of the limiting factors for protein synthesis in senescing chloroplasts. Nuclear condensation and DNA fragmentation, which are indices of programmed cell death (Pennell and Lamb, 1997), were observed at the last stage of leaf senescence (Yen and Yang, 1998). Disorganization of the nucleus and degeneration of tonoplast were found at the final stage of senescence followed by complete loss of cytoplasmic components (Thomson and Platt-Aloia, 1987; Inada *et al.*, 1998a,b, 1999).

Cellular RNA is present as ribosomal (rRNA), transfer (tRNA), and messenger RNA (mRNA) components. Ribosomal RNA accounts for 85 to 90% of the total RNA (Brady, 1988). Total RNA or ribosomal RNA content in the leaf gradually decreases after full leaf expansion (Makrides and Goldthwaite, 1981; Brady, 1988; Hensel *et al.*, 1993; Lohman *et al.*, 1994; Crafts-Brandner *et al.*, 1996). For instance, a yellow leaf has about 10-fold less RNA than a green leaf (Bate *et al.*, 1991). Protein synthesis might, therefore, be limited in part by the decreased capacity for protein synthesis in senescing leaves.

B. Synthesis and Degradation of Rubisco

Rubisco is a bifunctional enzyme that catalyzes the first reaction of photosynthetic CO_2 fixation and photorespiration. Rubisco is a stromal enzyme of chloroplasts and the most abundant protein in leaves. It accounts for 12–35% of total leaf nitrogen in mature leaves of C_3 plants (Evans and Seeman, 1989). Rubisco content in a leaf is a limiting factor for the rate of CO_2 assimilation under light-saturated and ambient air conditions (Makino *et al.*, 1985). Therefore, its fate during leaf senescence directly affects the photosynthesis and nitrogen metabolism of the plant.

Rubisco content in a leaf increases during leaf expansion and reaches the maximum amount at around the time of full expansion. Thereafter, it continuously decreases during leaf senescence and reaches a non-measurable level when the leaf is completely senesced (Hall *et al.*, 1978; Wittenbach, 1979; Peoples *et al.*, 1980; Wittenbach *et al.*, 1982; Mae *et al.*, 1983; Ford and Shibles, 1988; Hensel *et al.*, 1993; Crafts-Brandner *et al.*, 1996, 1998). Total apparent loss of Rubisco nitrogen from the leaf blade of rice plants was about 40% of the total nitrogen lost from the leaf during senescence (Makino *et al.*, 1984).

Decline of Rubisco content in senescing leaves can be attributed to the decreased rate of its synthesis, or to the increased rate of its degradation, or to both. In barley (Peterson *et al.*, 1973), *Perilla*, and *Capsicum* (Brady, 1988) little or no incorporation of precursors into Rubisco protein could be measured in expanded leaves, so that little or no replacement of the protein occurs. On the other hand, in wheat (Brady, 1988) Rubisco was found to be synthesized in newly expanded leaves, and this was also the case in *Zea mays* leaves (Simpson *et al.*, 1981). Mae *et al.* (1983) and Makino *et al.* (1984) quantitatively examined the synthesis and degradation of Rubisco protein throughout the life span of a rice leaf by using ^{15}N as a tracer (Fig. 10-1). Synthesis of Rubisco peaked during leaf expansion and about 90% of Rubisco synthesized during the life span of the leaf had been formed at about a week after the completion of expansion. After the completion of leaf expansion,

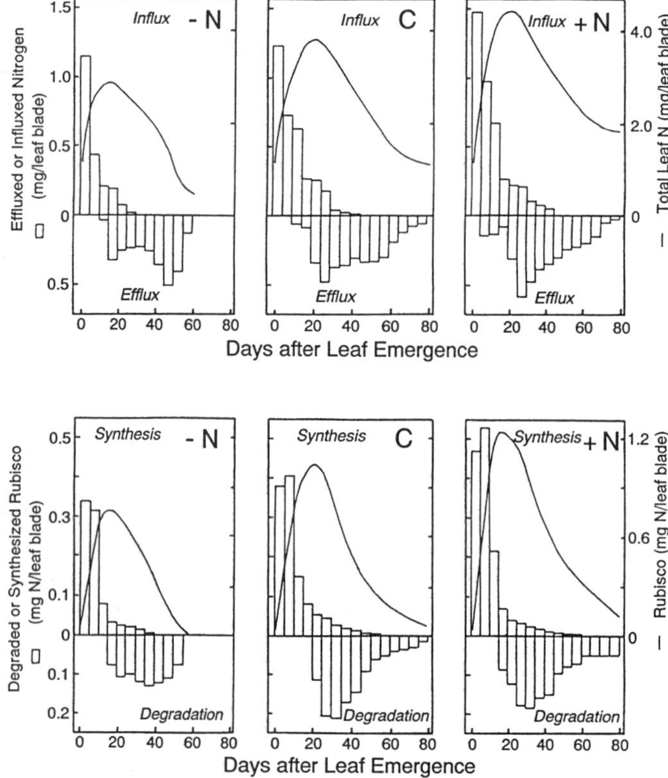

Figure 10-1. Changes in the influx and efflux of nitrogen and changes in the amount of Rubisco synthesized or degraded in the 12th leaf blades on the main stem of rice from leaf emergence through senescence. Plants were grown with the same amount of nitrogen until the emergence of the 12th leaf blades, then on 0.3 mM $(^{15}NH_4)_2SO_4$ for 3 days and then 0 [−N], 1 mM [C], or 2 mM [S] NH_4NO_3. The influx and efflux, and the amounts of Rubisco synthesized and degraded were calculated from the changes in ^{15}N content as described by Mae *et al.* (1983). The solid curve shows changes in the content of the nitrogen or the Rubisco (from Makino *et al.*, 1984).

Rubisco synthesis sharply decreased. The rate of synthesis was almost proportional to the rate of nitrogen influx into the leaf throughout the leaf's life span, indicating that the availability of the substrate might be involved as one of the factors determining the rate of Rubisco synthesis. The synthesis of Rubisco during senescence was constantly about 3% of the Rubisco amount in the leaf.

Degradation of Rubisco had already started at the time of full leaf expansion and was most active in the early stage of leaf senescence. The degradation rate was much faster than the rate of synthesis throughout the period of senescence.

C. Molecular Basis for Changes in Rubisco Synthesis during Senescence

The holoenzyme of Rubisco in higher plants consists of eight large and eight small subunits. The large subunit (*rbcL*; Mr ≈ 54,000) is coded by chloroplast DNA and translated on chloroplast ribosomes. The small subunit (*rbcS*; Mr ≈ 13,000) is coded by nuclear DNA, translated on cytoplasmic ribosomes as a precursor molecule, and processed to its mature size within the chloroplasts. The *in vivo* observed decline in synthesis of Rubisco was shown to correlate with a declining population of mRNAs for each subunit in the leaves of wheat, placing the point of control at either transcription or at mRNA stability (Brady, 1988). In leaves of soybean (Jiang *et al.*, 1993), *Arabidopsis* (Hensel *et al.*, 1993) and *Phaseolus vulgaris* (Crafts-Brandner *et al.*, 1996), the decreases in the level of Rubisco protein were correlated with the decreases in the levels of mRNAs for both subunits and in the level of total RNA during senescence. Thus, there is a close correlation between their transcription rate and mRNA abundance, an indication that transcription plays a primary role in determination of mRNA abundance (Rapp *et al.*, 1992). However, analysis of mRNA levels and transcription rates also provided evidence that plastid mRNA stability might be an important factor in determining the level of RNA (Deng and Gruissem, 1987; Mullet and Klein, 1987; Rapp *et al.*, 1992). Kim *et al.* (1993) analyzed the stability of mRNAs encoded by seven different barley chloroplast genes representing five major chloroplast functions during barley chloroplast development. Their analyses revealed that half-lives of mRNAs in barley chloroplasts range from 3 h to over 40 h. The stability of mRNA for *rbcL* increased 2.5-fold during chloroplast development in the dark, and then decreased 2-fold in chloroplasts of light-grown plants.

Transcriptional activity in plastids varies during chloroplast development. Plastid transcriptional activity and DNA copy number increased early in chloroplast development and transcriptional activity per template varied up to 5-fold during barley leaf biogenesis (Baumgartner *et al.*, 1989). Importantly, plastid transcriptional activity per cell and number of plastid DNA copies per cell were highest in the middle part of leaves when they had grown to one-third of full size. Both activity and number in the older top part of the same leaf were lower than those in the middle part. Thereafter, the transcriptional activity sharply decreased during leaf expansion and reached one-tenth of maximum in the leaf at full expansion.

Translational regulation of the genes *rbcS* and *rbcL* is also involved as one of the factors determining the levels of Rubisco in leaves. When dark-grown cotyledons of amaranth were transferred into the light, synthesis of the LSU and SSU polypeptides of Rubisco was initiated very rapidly, before any increase in the levels of their corresponding mRNAs (Berry *et al.*, 1988), indicating that the expression of *rbcS* and *rbcL* genes could be adjusted at the translational level. To address the question of whether the abundance of the SSU of Rubisco protein influences LSU protein metabolism, Rodermel *et al.* (1988) generated anti-*rbcS* transgenic tobacco plants that have reduced amounts of *rbcS* mRNAs and the SSU proteins. The LSU protein was coordinately reduced in mutant plants, but the *rbcL* mRNA level was normal. After a short-term pulse, there was less labeled LSU protein in the transgenic plants than in wild-type plants (Rodermel *et al.*, 1996), indicating that the LSU accumulation is controlled in the mutants at the translational and/or posttranslational levels.

D. Light-Harvesting Chlorophyll a/b Binding Protein of PSII

LHCII is the most abundant protein of thylakoid, constitutes approximately 33% of the total mass of membrane protein and binds approximately 50% of the total chlorophyll (Chitniss and Thornber, 1988). In addition to its central role in the light-harvesting process, LHCII is also a key component for regulating and distributing the excitation energy in response to short-term and long-term fluctuation in the light intensity and quality. The apoproteins of LHCII are encoded in the nucleus by a small multigene family. The polypeptides are translated in the cytoplasm on membrane-free polysomes as soluble precursor proteins, which are posttranslationally imported into the chloroplast. Upon entry into the chloroplast, these apoproteins are processed to their mature forms and subsequently inserted and assembled into the thylakoid membranes as pigment–protein complexes.

During greening of higher plants, the abundance of LHCII is primarily a consequence of transcriptional control of *CAB* genes (Silverthorne and Tobin, 1984). Once the leaf reaches full size, the abundance of mRNAs for LHCII genes as well as those for the small subunits of Rubisco genes sharply falls (Roberts *et al.*, 1987; Hensel *et al.*, 1993; Lohman *et al.*, 1994). The effects of growth irradiance on the levels of LHCII and other photosynthetic proteins in senescing leaves were examined in rice (Hidema *et al.*, 1991, 1992) and *Lolium temulentum* (Mae *et al.*, 1993). Rubisco protein levels decreased rapidly in both high-light and low-light leaves. The LHCII protein levels in the high-light senescing leaves decreased soon after the decreases in Rubisco level. On the other hand, LHCII levels in the low-light senescing leaves were retained at nearly as high levels as those at the time of full expansion until the late stage of senescence. However, the amounts of LHCII proteins synthesized in those senescing leaves, judged by ^{15}N incorporation, were less than 1% of the existing amount, indicating that there was little synthesis of this protein in senescing leaves. Thus, suppression of its degradation, rather than the acceleration of its synthesis could maintain the high levels of LHCII in the low-light senescing leaves. Those results, together with others by a pulse-labeling technique with isotopes and a Northern-blotting analysis (Roberts *et al.*, 1987; Bate *et al.*, 1991; Humbeck *et al.*, 1996), strongly suggest that the process of protein degradation is responsible for environmentally-induced changes during senescence. In addition, regulation of Rubisco and LHCII degradation in senescing leaves occurs independently at each protein level, rather than random degradation of the whole population. The higher ratio of light-harvesting capacity to Rubisco in the low-light leaves is likely to be an acclimation phenomenon of the senescing leaves similar to that found in expanding leaves.

E. D1 Protein

The D1 protein coded by chloroplast DNA (*psbA*) is a component of the reaction center of photosystem II (PSII) and known as a rapid turnover protein that is specifically degraded under illumination (Mattoo *et al.*, 1984). Under normal light conditions, degradation of the D1 protein is counteracted by a repair system that includes synthesis of the protein *de novo*. By contrast, under strong illumination that causes photoinhibition of PSII, the rate of degradation exceeds that of repair so that the amount of D1 protein decreases (Prasil *et al.*, 1992). D1 protein in chloroplasts is far less abundant than Rubisco protein. However, *psbA* and *rbcL* mRNAs accumulated similar levels in young leaves, and in mature leaves *psbA* mRNA was more abundant than *rbcL* mRNA (Ellis, 1981; Deng and Gruissem, 1987). The half-life of *psbA* mRNA was increased more than 2-fold in mature leaves compared with young leaves. The different stability of mRNAs could account for differences in mRNA accumulation during leaf development (Klaff and Gruissem, 1991). In bean leaves total RNA declined 10-fold, but *psbA* and *rbcL* mRNAs remained at a constant proportion of total RNA throughout senescence. By contrast, *CAB* and *rbcS* mRNAs comprised a progressively decreasing proportion of total RNA as senescence progressed. When ^{35}S-methionine was fed to leaves at late stages of senescence, 65% and 6% of total thylakoid radiolabeled protein, respectively, was incorporated into D1 and LHCII proteins, indicating preferential synthesis of D1 protein in senescing leaves (Bate *et al.*, 1991).

IV. Protein Degradation

Protein degradation in leaves during senescence has been studied for many years because of its important role in protein turnover and nitrogen recycling (Huffaker, 1990; Vierustra, 1993; Feller and Fisher, 1994; Callis, 1995). Now, it is known that protein degradation is definitely a regulated process. Protein degradation is a complex of several pathways ranging from triggering of degradation to a complete hydrolysis to amino acids. It is known that there exists a number of proteases and their homologues in senescing leaves, and such proteases are thought to be involved in some aspects of protein degradation (see Chapter 4 by Jones and Chapter 7 by Feller). Surprisingly little is known yet about their individual roles in protein degradation *in vivo*. Activation of existing proteases, change of compartmentation, or change of microenvironment around target proteins in cells or organelles would also be other ways to trigger the degradation. Phosphorylation, ubiquitination, oxidation or reduction of their amino acid residues would modulate the susceptibility of target proteins to proteolysis. Conformational change of target proteins by binding of substrates or effectors might trigger their degradation, too.

V. Remobilization of Nitrogen in Senescing Leaves

Although the contribution of remobilized nitrogen to the formation of new tissues might differ depending on the type of organs, nutritional status of nitrogen, plant species etc., reutilization of the remobilized nitrogen can be seen all the time throughout the life span of the plant. In field-grown rice, about 80% of the ear nitrogen is derived from the remobilized nitrogen from vegetative organs (Mae, 1997), and about 60% of the remobilized nitrogen is originated from the leaf blades (Mae and Ohira, 1981). The reutilization of remobilized nitrogen by the plant includes several steps: (1) degradation of the macromolecules into small molecules in source cells; (2) interconversion of the small molecules to transportable forms; (3) transport from source cells to sink cells mainly via phloem; (4) conversion of the transported compounds to available forms and their utilization in sink tissues. Here, interconversion of nitrogenous compounds during senescence, especially in preparation for transport or storage in leaf tissues, is described.

A. Amide Formation in Senescing Leaves

Analyses of phloem saps from several plant species indicate that free amino acids are the principal compounds of nitrogen transport via phloem (Pate *et al.*, 1979; Fisher and Macnicol 1986; Hayashi and Chino, 1990; Sieciechowicz *et al.*, 1988). Although the amino acid composition of phloem saps varies depending on plant species, tissues, ages, nutritional status etc., they are often rich in glutamine and asparagine. As these amides each possess two nitrogen atoms per molecule, they are thought to be efficient carriers for nitrogen transport. Glutamine accounts for 21–62% and asparagine for 12–19% of the total free amino acid nitrogen in the phloem sap of rice plants (Hayashi and Chino, 1990). These results and others indicate that the nitrogenous degradation products in senescing tissues are preferentially interconverted into these nitrogen-rich compounds.

In plants, glutamine is synthesized from ammonia and glutamate with the simultaneous hydrolysis of ATP and this reaction is catalyzed by glutamine synthetase (GS). There are two isoforms of GS in green leaves of many plants: a minor isoform located in cytosol (GS1) and a major isoform in the chloroplast stroma (GS2). Studies with mutants lacking GS2 clearly show that a major role of this isoform is the reassimilation of NH_4^+ released from photorespiration. Because the mutants were able to grow normally under non-photorespiratory conditions, GS1 in leaves could be important for normal growth and development (Lea and Ireland, 1999). It is hypothesized that GS1 might play a central role in the formation of glutamine for phloem transport in senescing leaves (Kawakami and Watanabe, 1989; Kamachi *et al.*, 1991). In rice, its activities and protein contents in senescing leaves were relatively constant throughout the senescing period, while the photosynthetic proteins such as Rubisco, ferredoxin-GOGAT, GS2, and NADH-GDH were degraded rapidly as leaf senescence proceeded (Kamachi *et al.*, 1991, 1992). Moreover, by using a tissue-print immunoblot method with specific antibodies, it was shown that GS1 was specifically located in large and small vascular bundles of all regions from mature to senescing leaf blades (Kamachi *et al.*, 1993). Further studies on leaf blades of rice indicated that GS1 was localized in companion cells of large vascular bundles, and also in vascular-parenchyma cells of both large and small vascular bundles (Sakurai *et al.*, 1996). GS1 has proved to be also specifically localized in phloem companion cells of tobacco leaves (Carvalho *et al.*, 1992; Dubois *et al.*, 1996) and the phloem of potato leaves (Pereira *et al.*, 1992). Companion cells are important in the regulation of phloem loading (Van Bel, 1993). The companion cells and metaphloem- and metaxylem-parenchyma cells are considered to be active in the transport of solutes, since they contain abundant mitochondria and endoplasmic reticulum and are interconnected by plasmodesmata (Chonan *et al.*, 1981).

Asparagine biosynthesis in plants is usually mediated by the action of glutamine-dependent asparagine synthetase (AS), which catalyzes the ATP-dependent transfer of the amide group from glutamine to aspartate, and yields glutamate and asparagine. AS activity has proved difficult to measure because of the presence of natural inhibitors and its instability *in vitro* (Ireland and Lea, 1999). There is no firm evidence to suggest that AS is located anywhere other than the cytoplasm. Far more information about asparagine synthesis has been obtained by taking a molecular approach. There are three AS genes (*ASN1*, *ASN2* and *ASN3*) in *Arabidopsis thalianta* (Lam *et al.*, 1996). The expression of one of these (*ASN1*) was repressed by light in green tissues and by supply of sucrose in the dark-grown plants. The other two genes were expressed at a lower level than *ASN1*, and appeared to be regulated in a quite different manner, being induced by light and sucrose, and repressed in the dark. Similar responses of AS genes to light and sucrose were observed in peas, tobacco, maize and soybean, indicating that the gene expression of AS is regulated by light and nitrogen/carbon ratios (Lam *et al.*, 1996). However, the relationship between gene expression of AS and asparagine synthesis in senescing leaves has not been described yet.

VI. Conclusion

Accumulating evidence indicates that decline in protein synthesis in senescing leaves can be attributed to multiple factors: slowing down of the transcription rate, limited template availability (loss of chloroplast DNA), changes of transcript stability, decreased capacity for translation, and availability of substrates and energies. However, the effects of these factors on protein synthesis may differ among individual proteins, stages of leaf senescence, and plant species. During senescence proteins are actively degraded in senescing leaves. It is now obvious that their degradation is not a simple passive decay. They are degraded in a highly ordered manner adapted to a given circumstance in the course of senescence, indicating that the degradation process is strictly controlled until death. Appearance of senescence-specific protease(s) by *de novo* synthesis or by activation, change of compartmentation of protease, change of microenvironment around target proteins degraded, and various modulations of the proteins such as phosphorylation, ubiqitination, oxidation, or reduction of amino acid residues might play roles in protein degradation during senescence. The extent of the contribution of each step to regulating the protein level might change in the course of senescence and differ among protein species and plant species. Such a multiplicity of mechanisms regulating protein synthesis and degradation would make the leaf able to senesce in a highly ordered and adapted manner for given circumstances in the process of senescence.

References

Bate, N.J., Rothstein, S.J., and Thompson, J.E. (1991). Expression of nuclear and chloroplast photosynthesis-specific genes during leaf senescence. *J. Exp. Bot.* **42**, 801–811.

Baumgartner, B.J., Rapp, J.C., and Mullet, J.E. (1989). Plastid transcription activity and DNA copy number increase early in barley chloroplast development. *Plant Physiol.* **89**, 1011–1018.

Berry, J.O., Carr, J.P., and Klessig, D.F. (1988). mRNAs encoding ribulose-1,5-bisphosphate carboxylase remain bound to polysomes but are not translated in *Amaranth* seedlings transferred to darkness. *Proc. Natl. Acad. Sci. USA* **85**, 4190–4194.

Brady, C.J. (1988). 5. Nucleic acid and protein synthesis. In *Senescence and Aging in Plants* (L.D. Noodén and A.C. Leopold, Eds.), pp. 147–179. Academic Press, San Diego.

Callis, J. (1995). Regulation of protein degradation. *Plant Cell* **7**, 845–857.

Chitnis, P.R., and Thornber, P.J. (1988). The major light-harvesting complex of Photosystem II: aspects of its molecular and cell biology. *Photosynth. Res.* 41–63.

Chonan, N., Kaneko, M., Kawahara, H., and Matsuda, T. (1981). Ultrastructure of the large vascular bundles in the leaves of rice plants. *Jpn. J. Crop Sci.* **50**, 323–331.

Crafts-Brandner, S.J., Below, F.E., Wittenbach, V.A., Harper, J.E., and Hageman, R.H. (1984). Differential senescence of maize hybrids following ear removal. II Selected leaf. *Plant Physiol.* **74**, 368–373.

Crafts-Brandner, S.H., Klein, R.R., Klein, P., Holzer, R., and Feller, U. (1996). Coordination of protein and mRNA abundances of stromal enzymes and mRNA abundances of the Clp protease subunits during senescence of *Phaseolus vulgaris* (L.) leaves. *Planta* **200**, 312–318.

Crafts-Brandner, S.J., Holzer, R., and Feller, U. (1998). Influence of nitrogen deficiency on senescence and the amounts of RNA and proteins in wheat leaves. *Physiol. Plant.* **102**, 192–200.

Devois, F., Brugiere, N., Sangwan, R.S., and Hirel, B. (1996). Localization of tobacco cytosolic glutamine synthetase enzymes and the corresponding transcripts show organ-specific and cell-specific patterns of protein synthesis and gene expression. *Plant Mol. Biol.* **31**, 803–817.

Deng, X.W., and Gruissem, W. (1987). Control of gene expression during development: the limited role of transcriptional regulation. *Cell* **49**, 379–387.

Ellis, R.J. (1981). Chloroplast proteins: Synthesis, transport and assembly. *Annu. Rev. Plant Physiol.* **52**, 111–132.

Evans, J.R., and Seeman, J.R. (1989). The allocation of protein nitrogen in the photosynthetic apparatus: cost, consequence and control. In *Photosynthesis* (W. Briggs, Ed.), pp. 183–205. Liss, New York.

Feller, U., and Fisher, A. (1994). Nitrogen metabolism in senescing leaves. *Crit. Rev. Plant. Sci.* **13**, 241–273.

Fisher, D.B., and Macnicol, P.K. (1986). Amino acid composition along the transport pathway during grain filling in wheat. *Plant Physiol.* **82**, 1019–1023.

Ford, D.M., and Shibles, R. (1988). Photosynthesis and other traits in relation to chloroplast number during soybean leaf senescence. *Plant Physiol.* **86**, 108–111.

Friedrich, J.W., and Huffaker, R.C. (1980). Photosynthesis, leaf resistance and ribulose-1,5-bisphosphate carboxylase in senescing barley leaves. *Plant Physiol.* **65**, 1103–1107.

Gepstein, S. (1988). Photosynthesis. In *Senescence and Aging in Plants* (L.D. Noodén and A.C. Leopold, Eds.), pp. 85–109. Academic Press, San Diego.

Hall, N.P., Keys, A.J., and Merret, M.J. (1978). Ribulose-1,5-bisphosphate carboxylase protein during flag leaf senescnece. *J. Exp. Bot.* **29**, 31–37.

Hayashi, H., and Chino, M. (1986). Collection of pure phloem sap from wheat and its composition. *Plant Cell Physiol.* **27**, 1387–1393.

Hayashi, H., and Chino, M. (1990). Chemical composition of phloem sap from the uppermost internode of rice plant. *Plant Cell Physiol.* **31**, 247–251.

Hensel, L.L., Grbic, V., Baumgarten, D.A., and Bleecker, A.B. (1993). Developmental and age-related processes that influence the longevity and senescence of photosynthetic tissues in *Arabidopsis*. *Plant Cell* **5**, 553–564.

Hidema, J., Makino, A., Mae, T., and Ojima, K. (1991). Photosynthetic characteristics of rice leaves aged under different irradiances from full expansion through senescence. *Plant Physiol.* **97**, 1287–1297.

Hidema, J., Makino, A., Kurita, Y., Mae, T., and Ojima, K. (1992). Changes in the levels of chlorophyll and light-harvesting chlorophyll a/b protein of PSII in rice leaves aged under different irradiances from full expansion through senescence. *Plant Cell Physiol.* **33**, 1209–1214.

Huffaker, R.C. (1990). Proteolytic activity during senescence of plants. *New Phytol.* **116**, 199–231.

Humbeck, K., Quest, S., and Krupinska, K. (1996). Functional and molecular changes in the photosynthetic apparatus during senescence of flag leaves from field-grown barley leaves. *Plant Cell Environ.* **19**, 337–344.

Inada, N., Sakai, A., Kuroiwa, H., and Kuroiwa, T. (1998a). Three dimensional analysis of the senescence program in rice (*Oryza sativa* L.) coleoptiles: investigation of tissues and cells by fluorescence microscopy. *Planta* **205**, 153–164.

Inada, N., Sakai, A., Kuroiwa, H., and Kuroiwa, T. (1998b). Three dimensional analysis of the senescence program in rice (*Oryza sativa* L.) coleoptiles: investigations by fluorescence microscopy and electron microscopy. *Planta* **206**, 585–597.

Inada, N., Sakai, A., Kuroiwa, H., and Kuroiwa, T. (1999). Senescence program in rice (*Oryza sativa* L.) leaves: Analysis of the blade of the second leaf at the tissue and cellular levels. *Protoplasma* **207**, 222–232.

Ireland, R.J., and Lea, P.J. (1999). The enzymes of glutamine, glutamate, asparagine, and aspartate metabolism. In *Plant Amino Acids. Biochemistry and Biotechnology* (B.R. Singh, Ed.). pp. 49–109. Marcel Dekker, New York.

Jiang, C., Rodermel, S.R., and Shibles, R.M. (1993). Photosynthesis, Rubisco activity and amount, and their regulation by transcription in senescing soybean leaves. *Plant Physiol.* **101**, 105–112.

Kamachi, K., Yamaya, T., Mae, T., and Ojima, K. (1991). A role for glutamine synthetase in the remobilization of leaf nitrogen during natural senescence in rice leaves. *Plant Physiol.* **96**, 411–417.

Kamachi, K., Yamaya, T., Hayakawa, T., Mae, T., and Ojima, K. (1992). Changes in cytosolic glutamine synthetase polypeptide and its mRNA in a leaf blade of rice plants during natural senescence. *Plant Physiol.* **98**, 1323–1329.

Kamachi, K., Yamaya, T., Hayakawa, T., Mae, T., and Ojima, K. (1992). Vascular bundle-specific localization of cytosolic glutamine synthetase in rice leaves. *Plant Physiol.* **99**, 1481–1486.

Kawakami, N., and Watanabe, A. (1988). Senescence-specific increase in cytosolic glutamine synthetase and its mRNA in radish cotyledons. *Plant Physiol.* **88**, 1430–1434.

Kim, M., Christopher, D.A., and Mullet, J.E. (1993). Direct evidence for selective modulation of *psbA*, *rpoA*, *rbcL* and *16S* RNA stability during barley chloroplast development. *Plant Mol. Biol.* **22**, 447–463.

Klaff, P., and Gruissem, W. (1991). Changes in chloroplast mRNA stability during leaf development. *Plant Cell* **3**, 517–529.

Lam, H.-M., Coshigano, I.C., Oliveira, I.C., Melo-Oliveira, R., and Coruzzi, G.M. (1996). The molecular-genetics of nitrogen assimilation into amino acids in higher plants. *Annu. Rev. Plant Physiol. Plant Mol. Biol.* **47**, 569–593.

Lea, P.J., and Ireland, R.J. (1999). Nitrogen metabolism in higher plants. In *Plant Amino Acids. Biochemistry and Biotechnology* (B.K. Singh, Ed.), pp. 1–47. Marcel Dekker, New York.

Lohman, K.N., Gan, S., John, M.C., and Amasino, R.M. (1994). Molecular analysis of natural leaf senescence in *Arabidopsis thaliana*. *Physiol. Plant.* **92**, 322–328.

Mae, T. (1997). Physiological nitrogen efficiency in rice: Nitrogen utilization, photosynthesis, and yield potential. *Plant Soil* **196**, 201–210.

Mae, T., and Ohira, K. (1981). The remobilization of nitrogen related to leaf growth and senescence in rice plants (*Oryza sativa* L.). *Plant Cell Physiol.* **22**, 1067–1074.

Mae, T., Makino, A., and Ohira, K. (1983). Changes in the amounts of ribulose-1,5-bisphosphate carboxylase synthesized and degraded during the life span of rice leaf (*Oryza sativa* L). *Plant Cell Physiol.* **24**, 1079–1081.

Mae, T., Thomas, H., Gay, A.P., Makino, A., and Hidema, J. (1993). Leaf development in *Lolium temulentum*: Photosynthesis and photosynthetic proteins in leaves senescing under different irradiances. *Plant Cell Physiol.* **34**, 391–399.

Makino, A., Mae, T., and Ohira, K. (1983). Photosynthesis and ribulose 1,5-bisphosphate carboxylase in rice leaves. Changes in photosynthesis and enzymes involved in carbon assimilation from leaf development through senescence. *Plant Physiol.* **73**, 1002–1007.

Makino, A., Mae, T., and Ohira, K. (1984). Relation between nitrogen and ribulose-1,5-bisphosphate carboxylase in rice leaves from emergence through senescence. *Plant Cell Physiol.* **25**, 429–437.

Makino, A., Mae, T., and Ohira, K. (1985). Photosynthesis and ribulose-1,5-bisphosphate carboxylase/oxygenase in rice leaves from emergence through senescence. Quantitative analysis by carboxylation/oxygenation and regeneration of ribulose-1,5-bisphosphate. *Planta* **166**, 414–420.

Makino, A., and Osmand, B. (1991). Effects of nitrogen nutrition on nitrogen partitioning between chloroplasts and mitochondria in pea and wheat. *Plant Physiol.* **96**, 355–362.

Makrides, S.S., and Goldthwaite, J. (1981). Biochemical changes during bean leaf growth, maturity and senescence. Content of DNA, polysomes, ribosomal RNA, protein and chlorophyll. *J. Exp. Bot.* **32**, 725–735.

Mattoo, A.K., Hoffman-Falk, H., Marder, J.B., and Edelman, M. (1984). Regulation of protein metabolism: coupling of photosynthetic electron transport to in vivo degradation of the rapidly metabolized 32 kilodalton protein of the chloroplast membrane. *Proc. Natl. Acad. Sci. USA* **81**, 1380–1384.

Miller, B.L., and Huffaker, R.C. (1982). Hydrolysis of RuBPCase endoproteinases from senescing barley leaves. *Plant Physiol.* **69**, 58–62.

Morita, K. (1980). Release of nitrogen from chloroplasts during leaf senescence in rice (*Oryza sativa* L.). *Ann. Bot.* **46**, 297–302.

Mullet, J.E., and Klein, R.R. (1987). Transcription and RNA stability are important determinants of higher plant chloroplast RNA levels. *EMBO J.* **6**, 1571–1579.

Nikolau, B.J., and Klessig, D.F. (1987). Coordinate, organ-specific and developmental regulation of ribulose-1,5-bisphosphate carboxylase gene expression in *Amaranthus hypochondriacus*. *Plant Physiol.* **85**, 167–173.

Pate, J.S., Atkins, C.A., Hamel, K., McNeil, D.L., and Layzell, D.W. (1979). Transport of organic solutes in phloem and xylem of a nodulated legume. *Plant Physiol.* **63**, 1082–1088.

Pennell, R.I., and Lamb, C. (1997). Programmed cell death in plants. *Plant Cell* **9**, 1157–1168.

Peoples, M.B., and Dalling, M.J. (1988). 6. The interplay between proteolysis and amino acid metabolism during senescence and nitrogen reallocation. In *Senescence and Aging in Plants* (L.D. Noodén and A.C. Leopold, Eds.), pp. 181–217. Academic Press, San Diego.

Peoples, M.B., Belzharz, V.C., Waters, S.P., Simpson, R.J., and Dalling, M.J. (1980). Nitrogen redistribution during grain growth in wheat (*Triticum aestivum* L.) II. Chloroplast senescence and the degradation of ribulose-1,5-bisphosphate carboxylase. *Planta* **149**, 241–251.

Pereira, S., Carvalho, H., Sunkel, C., and Salema, R. (1992). Immunocytolocalization of glutamine synthetase in mesophyll and phloem of leaves of *Solanum tuberosum* L. *Protoplasma* **167**, 66–73.

Peterson, L.W., Kleinkopf, G.E., and Huffaker, R.C. (1973). Evidence for lack of turnover of ribulose-1,5-bisphosphate carboxylase in barley leaves. *Plant Physiol.* **51**, 1042–1045.

Prasil, O., Adir, N., and Ohad, I. (1992). Dynamics of photosystem II: mechanism of photoinhibition and recovery processes. In *The Photosystems: Structure, Function, and Molecular Biology*, Topics in Photosynthesis Vol. 11 (J. Barber, Ed.), pp. 295–348. Elsevier, Amsterdam.

Pyke, K.A., and Leech, R.M. (1987). The control of chloroplast number in wheat mesophyll cells. *Planta* **170**, 416–420.

Rapp, J.C., Baumgartner, B.J., and Mullet, J. (1992). Quantitative analysis of transcription and RNA levels of 15 barley chloroplast genes: transcription rates and mRNA levels vary over 300-fold; predicted mRNA stabilities vary 30-fold. *J. Biol. Chem.* **267**, 21404–21411.

Rodermel, S.R., Abbott, M.S., and Bogorad, L. (1988). Nuclear-organelle interactions: nuclear antisense gene inhibits ribulose bisphosphate carboxylase enzyme levels in transformed tobacco plants. *Cell* **55**, 673–681.

Rodermel, S., Haley, J., Jiang, C., Tsai, C., and Bogorad, L. (1996). A mechanism for integration: Abundance of ribulose bisphosphate carboxylase small-subunit protein influences the translation of the large-subunit mRNA. *Proc. Natl. Acad. Sci. USA* **93**, 3881–3885.

Sakurai, N., Hayakawa, T., Nakamura, T., and Yamaya, T. (1996). Changes in the cellular localization of cytosolic glutamine synthetase protein in vascular bundles of rice leaves at various stages of development. *Planta* **200**, 306–311.

Sieciechowicz, K.A., Joy, K.W., and Ireland, R.J. (1988). The metabolism of asparagine in plants. *Phytochemistry*, **27**, 663–671.

Silverthorne, J., and Tobin, E.M. (1984). Demonstration of transcriptional regulation of specific genes by phytochrome action. *Proc. Natl. Acad. Sci. USA* **81**, 1112–1116.

Simpson, E., Cook, R.J., and Davies, D.D. (1981). Measurement of protein degradation in leaves of *Zea mays* using [^{3}H] acetic anhydride and tritiated water. *Plant Physiol.* **67**, 1214–1219.

Sodmergen, K.S., Kawano, S., Tano, S., and Kuroiwa, T. (1989). Preferential digestion of chloroplast nuclei (nucleoids) during senescence of the coleoptile of *Oryza sativa*. *Protoplasma* **152**, 65–68.

Sodmergen, K.S., Kawano, S., Tano, S., and Kuroiwa, T. (1991). Degradation of choroplast DNA in second leaves of rice (*Oryza sativa*) before leaf yellowing. *Protoplasma* **160**, 89–98.

Thomas, H., Morgan, W.G., Thomas, A.M., and Ougham, H.J. (1999). Expression of the stay-green character introgressed into *Lolium temulentum* Ceres from a senescence mutant of *Festuca pratensis*. *Theoret. Appl. Genetics* **99**, 92–99.

Thomson, W.W., and Plat-Aloia, K.A. (1987). Ultrastructure and senescence in plants. In *Plant Senescence and its Biochemistry and Physiology* (W.W. Thomson, E.A. Nothnagel, and R.C. Huffaker, Eds.), pp. 20–30. American Society of Plant Physiologists, Rockville, MD.

Van Bel, A.J.E. (1993). Strategies of phloem loading. *Annu. Rev. Plant Physiol. Plant Mol. Biol.* **44**, 253–281.

Vierstra, R.D. (1993). Protein degradation in plants. *Annu. Rev. Plant Physiol. Plant Mol. Biol.* **44**, 385–410.

Wittenbach, V.A. (1979). Ribulose bisphosphate carboxylase and proteolytic activity in wheat leaves from anthesis through senescence. *Plant Physiol.* **64**, 884–887.

Wittenbach, V.A., Lin, W., and Herbert, R.R. (1982). Vacuolar localization of proteases and degradation of chloroplasts in mesophyll protoplasts from senescing primary wheat leaves. *Plant Physiol.* **69**, 98–102.

Yen, C., and Yang, C. (1998). Evidence for programmed cell death during leaf senescence in plants. *Plant Cell Physiol.* **39**, 922–927.

11

Photosynthesis and Chloroplast Breakdown

Karin Krupinska and Klaus Humbeck

I. Introduction

This chapter will discuss the more general features of chloroplast breakdown, e.g. the degradation of membranes and of nucleic acids, as well as more specific aspects associated with the photosynthetic apparatus including photosynthetic performance and expression of photosynthesis-related genes. Moreover this chapter will address questions concerning the interrelation of photosynthesis and senescence. During the past ten years, classical approaches to studying photosynthesis have been complemented on the one hand by powerful non-invasive fluorescence tools and on the other hand by molecular biology techniques. The latter allow the analyses of gene expression at different levels and the generation of transgenic plants with specific modifications in the photosynthetic apparatus. Together with the collection of stay-green mutants, transgenic plants with altered leaf development may provide valuable tools for studies addressing the coordination and regulation of photosynthesis-associated changes occurring during chloroplast senescence.

II. Decline of Photosynthetic Function

A. Decrease in the Number of Chloroplasts or Gradual Degradation of Persisting Chloroplasts?

During senescence, photosynthetic capacity per unit leaf area decreases dramatically as components of the plastids are broken down. In principle, there are two alternative ways to explain the decrease in photosynthesis during leaf senescence:

(1) Whole chloroplasts are removed from the mesophyll cells by intracellular digestion of the entire organelles.

(2) Persisting plastids undergo a gradual degradation.

Ultrastructural studies showed that the number of chloroplasts is higher in mesophyll cells of younger leaves in comparison to mesophyll cells from senescing leaves of the same plant (Kura-Hotta *et al.*, 1990; Yamasaki *et al.*, 1996). This suggests the destruction of whole chloroplasts rather than a differential loss of components from persisting chloroplasts (Gepstein, 1988). Vacuolar phagocytosis of whole plastids which would underlie the decrease of the number of chloroplasts has been observed in a variety of senescing cells (Noodén *et al.*, 1997). Other investigations revealed contradictory results (Matile, 1992) indicating that the entire population of plastids does persist in senescing mesophyll cells, however, and undergoes specific changes in structure and composition leading to the decrease in photosynthetic capacity (Martinoia *et al.*, 1983; Mae *et al.*, 1984; Grover, 1993). Ultrastructural changes in senescing chloroplasts are well documented (Butler and Simon, 1971; for reviews see Biswal and Biswal, 1988; Gepstein, 1988) and comprise a reduction in the thylakoid membrane system, a loosening of the granae, a swelling of intrathylakoidal space, a shrinkage of the size and a transition from ellipsoid to circular shape (Kura-Hotta *et al.*, 1990; Pleijel *et al.*, 1997; Inada *et al.*, 1998b). It is likely that both the degradation and export of components from persisting gerontoplasts and a phagocytosis-like destruction of whole plastids may occur during senescence. In fact, in a vertical gradient of leaves of *Chenopodium album* both mechanisms could be demonstrated (Yamasaki *et al.*, 1996).

B. Photosynthetic Performance

During senescence a marked decrease of photosynthetic capacity at saturating light intensities per unit leaf area has been measured, reflecting mainly the senescence-specific degradation of Calvin cycle enzymes which is described in more detail in Section III. In contrast, only small changes occurred in the rates of photosynthesis at limiting light intensities which indicates that the quantum yield of photosynthesis remains at a high level when photosynthetic capacity has already started to decline during the early phases of senescence (Gay and Thomas, 1995; Okada and Katoh, 1998). Maximal efficiency of photosystem II photochemistry in the dark-adapted state analyzed by chlorophyll fluorescence techniques (Schreiber *et al.*, 1986) proved to be a well suited parameter for the characterization of senescence-associated changes in the photosynthetic apparatus of leaves from barley plants grown either in fields or under controlled environmental conditions in a climate chamber (Humbeck *et al.*, 1996; Miersch *et al.*, 2000). During the initial phase of senescence, photosynthetic capacity decreases while the F_v/F_m value stays high. The F_v/F_m ratio decreases sharply at a later point of time (Fig. 11-1). This coincides with a change in the organization

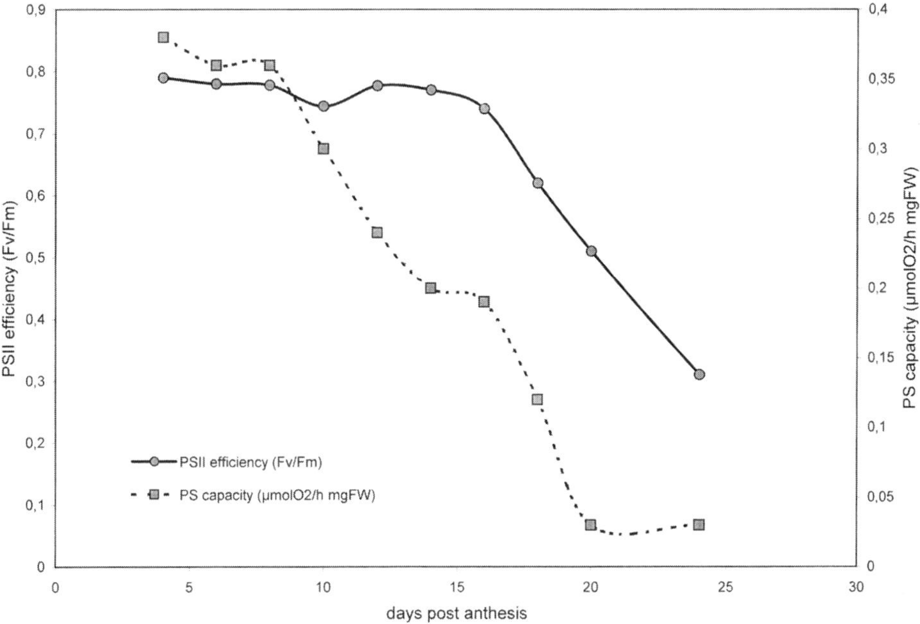

Figure 11-1. Decrease in photosynthetic capacity and in PSII efficiency during senescence of flag leaves from field-grown barley plants.

of the photosynthetic apparatus, especially in the inner light-harvesting complexes (see Section III), which affects the energy distribution within photosystem II (Humbeck and Krupinska, 1999, 2002).

C. Risk of Photoinhibition

Fast fluorescence kinetics showed a distinct change in the heterogeneity of photosystem II centers during senescence (Lu and Zhang, 1998). Analyses of fluorescence quenching parameters at steady-state photosynthesis showed that during senescence non-photochemical quenching (q_N) increased and photochemical quenching (q_P) decreased. It has been concluded that the decrease in q_P is a result of an increase in the proportion of Q_B non-reducing photosystem II reaction centers (Lu and Zhang, 1998). This would increase the risk of photoinhibition since at decreased q_P the portion of excitation energy used in the photochemical process is decreased. The surplus of absorbed light energy might lead to photoinhibition. To avoid photochemically induced damage, an increase in q_N (Lu and Zhang, 1998), reflecting the conversion of excess light energy into less harmful heat, would be required. Recently, Lu *et al.* (2001) showed that under field conditions the xanthophyll cycle-related thermal dissipation in the photosystem II antennae is significantly enhanced in senescent wheat leaves, especially at midday when light intensity is high.

In excised rye seedling leaves, an increase in photoinhibition during senescence, measured as a decline in the F_v/F_m ratio is accompanied by a decrease in protection mechanisms, e.g. catalase activity (Kar *et al.*, 1993). To what extent photoinhibition during senescence plays a major role at agricultural conditions in the field is not yet clear. Simultaneous measurements of net photosynthesis rates and fluorescence parameters of flag leaves from wheat grown in the field indicate that under these conditions photoinhibition probably is not a relevant phenomenon (DiMarco *et al.*, 1994). These authors explain an increasing discrepancy between the net photosynthetic rate and electron transport rate during senescence of the flag leaves in terms of an increasing photorespiratory activity. A role for photorespiration in sustaining electron transport and protection from photoinhibition during senescence is also discussed by Kingston-Smith *et al.* (1997) comparing a stay-green mutant of *Festuca* with the wild type, and by Lovelock and Winter (1996) comparing different tropical tree species. Another sink for electrons might be the Mehler reaction, by which electrons from photosystem I are transferred to oxygen. The Mehler reaction might be coupled to a sequence of reactions involving superoxide dismutase, ascorbate peroxidase and monoascorbate reductase (Kingston-Smith *et al.*, 1997).

III. Chloroplast Breakdown

The degradation of the chloroplast and its components is one of the most prominent phenomena of senescence and, as discussed before, is the basis of the decrease in photosynthetic activity. Chloroplasts contain a large portion of the plant's nitrogen and carbon resources which may be transported to and used in other parts of the plant after degradation of macromolecules. In order to guarantee an efficient export of the remobilized resources, the various constituents of the plastids, e.g. proteins, pigments, membrane lipids and nucleic acids, have to be degraded in a well-coordinated fashion (see also Gepstein, 1988; Noodén *et al.*, 1997).

A. Sequence of Senescence-associated Events

In order to unravel the sequence of senescence-associated events during chloroplast breakdown, Inada *et al.* (1998a,b) investigated tissues and cells of senescing rice coleoptiles by combining electron microscopy with fluorescence microscopy employing specific antibodies and dyes. Their data confirm earlier findings by Butler and Simon (1971) about the sequence of typical events: (i) degradation of chloroplast DNA; (ii) condensation of the nucleus in conjunction with a decrease in the size of the dense-chromatin region, size shrinkage of the chloroplasts, degradation of Rubisco, dilatation of thylakoid membranes, appearance of osmiophilic globules, condensation of cytoplasm; (iii) disorganization of the nucleus, degeneration of tonoplast; (iv) complete loss of cytoplasmic components, distortion of the cell wall (Inada *et al.*, 1998a,b). The same order of events obviously occurs simultaneously in all mesophyll cells of rice coleoptiles reflecting that the same program of senescence is operative in these cells, too.

B. Degradation of Plastid Constituents

1. Nucleic acids

According to Inada *et al.* (1998a,b), one of the first events during senescence-related chloroplast breakdown is the degradation of chloroplast DNA. A preferential degradation of chloroplast DNA was also observed in second leaves of rice before yellowing by Sodmergen *et al.* (1991). These authors discuss a possible involvement of a Zn^{2+}-dependent nuclease in chloroplast DNA degradation. However, such a DNAse has not yet been isolated.

The immunocytochemical studies of Inada *et al.* (1998b) suggest that chloroplast DNA is first digested to specific DNA fragments which are dispersed throughout the chloroplast and cytoplasm. Eventually, these DNA fragments are also broken down. Though it is well known that during leaf senescence also total RNA content decreases (Brady, 1988; Krause *et al.*, 1998) and total RNAse activity increases (Blank and McKeon, 1991; Taylor *et al.*, 1993; Lers *et al.*, 1998), no clearcut picture exists about senescence-related RNA degradation within the plastid. Earlier studies reported a preferential degradation of ribosomal plastid RNA, but usually both cytoplasmic and plastid ribosomal RNA decrease in parallel during senescence (Brady, 1988). At least a few ribosomes seem to persist in senescing plastids.

2. Stromal proteins

Chloroplasts contain 70–80% of the total nitrogen of mature leaves (Mae *et al.*, 1993). This nitrogen is mainly bound in protein components of the chloroplasts which are either carbon reduction cycle enzymes of the stroma or proteins of the photosynthetic apparatus located in the thylakoid membrane. The degradation of these chloroplast proteins yields about 90% of the nitrogen remobilized by a senescent leaf (Morita, 1980) and thereby it is one of the most important features of senescence (see also Chapter 10, T. Mae). Accordingly, the degradation of the most abundant protein compound, the stroma localized ribulose-1,5-bisphosphate carboxylase/oxygenase (Rubisco), has been studied intensively (see also Gepstein, 1988). A close correlation between the decline in Rubisco activity and content on the one hand and the decline in photosynthetic function on the other hand (e.g. Friedrich and Huffaker, 1980; Mae *et al.*, 1993; Jiang *et al.*, 1993; Ono *et al.*, 1995) indicates that Rubisco degradation is the key factor responsible for the reduction in photosynthetic capacity during senescence. Other stromal enzymes, which are much less abundant than Rubisco, are degraded with kinetics similar to Rubisco. This was shown for Rubisco activase and phosphoribulokinase (Ru5P) (Crafts-Brandner *et al.*, 1996). The knowledge about proteolytic activities underlying the degradation of protein compounds during senescence is summarized in Chapter 7 by U. Feller.

Environmental factors may have a severe influence on the decrease in protein components during senescence. Under natural conditions, when plants are growing and upper leaves are shading the older leaves at lower positions, a decrease in the photon fluence rate could be the primary trigger for Rubisco degradation (Hikosaka, 1996). However, experiments performed by Mae *et al.* (1993) with shaded and unshaded leaves of *Lolium tremulentum* showed that the senescence-related decrease of Rubisco was hardly affected by the light intensity, whereas the decrease in the levels of other proteins of the photosynthetic apparatus (e.g. LHCII, cytochrome f) during senescence was clearly retarded under low light condition (see Section V). Light control of senescence is treated in detail in Chapter 26 by L. D. Noodén and M. J. Schneider.

3. Thylakoid membrane components

i. Pigment–protein complexes

Pigment–protein complexes in thylakoid membranes account for more than 30% of the total protein of chloroplasts (Matile, 1992). The loss of their apoproteins is well coordinated with the degradation of the chlorophylls which are non-covalently bound to these apoproteins (Okada and Katoh, 1998). Chlorophyll breakdown is actually a prerequisite for the remobilization of these protein compounds (Matile, 1992, 1997; see also Chapter 12). Changes in light environment seem to affect chlorophyll degradation and hence degradation of pigment–protein complexes via phytochrome.

While during dark-induced senescence the chlorophyll a/b ratio stays constant, under natural senescence a decline of this ratio was observed. This indicates that the light-harvesting system, harboring all the chlorophyll b, has an enhanced stability in comparison to the reaction center complexes which contain chlorophyll a exclusively (Kura-Hotta *et al.*, 1987; Humbeck *et al.*, 1996; Miersch *et al.*, 2000). In a comparison of pine needles of the current, second and third year a significant reduction in the levels of PSI and cytochrome b6/f complexes has been observed in second and third year needles (Shinohara and Murakami, 1996). Other complexes of the photosynthetic apparatus like LHCII decayed later in senescence. Recently, Humbeck and Krupinska (1999, 2002) showed that the decline in the levels of the inner light-harvesting complexes of photosystem II (CP29) and photosystem I (LHCI), which are supposed to play an important role in energy transfer within the photosynthetic units (Bassi *et al.*, 1993), correlates with the start of the decline in photosystem II efficiency. A more comprehensive study on the disappearance kinetics of six light-harvesting proteins during dark-induced senescence clearly confirmed their differential stability (Rosiak-Figielek and Jackowski, 2000).

ii. Pigments

Chlorophyll catabolism will be reviewed in Chapter 12 by P. Matile and S. Hörtensteiner. The relative stability of carotenoids as compared to chlorophylls might be explained in terms of their association with plastoglobuli formed during senescence. It has been suggested that fatty acids released from thylakoids interact with carotenoids to form stable esters which accumulate in plastoglobuli (Tevini and Steinmüller, 1985). Another explanation for the relative stability of carotenoids over chlorophylls would be a continued biosynthesis throughout the phase of senescence comparable to the situation during fruit ripening (Fraser *et al.*, 1994). During senescence of wheat leaves under field conditions changes in the stoichiometry of carotenoids were observed. While neoxanthin and β-carotin decrease in parallel with chlorophylls the xanthophyll cycle pigments are less affected (Lu *et al.*, 2001). Compositional changes and specific degradation of carotenoids are still obscure, although some experimental data for the contribution of enzymes have been reported. So far, neither specific enzymes nor degradation products have been identified (Biswal, 1995). At least during the final stage of senescence, free radicals seem to be involved in carotenoid breakdown (Biswal, 1995).

iii. Lipids

The dismantling of thylakoids is accompanied by the formation of plastoglobuli which contain lipid metabolites originating from thylakoids (Tevini and Steinmüller, 1985). An increased formation of activated oxygen species during leaf senescence could initiate lipid peroxidation and hence membrane ridification (Thompson *et al.*, 1998). However, in contrast to the cytoplasmic membranes, thylakoids do not undergo phase changes or alterations in lipid composition during senescence (Thompson *et al.*, 1998). In the case of thylakoids the levels of unsaturated fatty acids, which are prone to peroxidation, remain essentially unchanged over the period when photosynthetic activity declines by approximately 80% and chlorophyll levels are reduced by 75% (summarized by Thompson *et al.*, 1998). During thylakoid breakdown, galactolipids do not accumulate in plastoglobuli as other lipid components do (Tevini and Steinmüller, 1985), suggesting that these lipids may be utilized as a source of carbon and energy (Matile, 1992). Galactolipids are enzymatically degraded by α-galactosidases, β-galactosidases and galactolipases (Matile, 1992). Senescence-related up-regulation of a β-galactosidase gene has been shown in asparagus (King *et al.*, 1995) and up-regulation of α-galactosidase specific genes has been shown in senescing barley leaves by Chrost and Krupinska (2000).

IV. Expression of Nuclear Genes Encoding Photosynthesis-related Proteins

Levels of chloroplast proteins decline during senescence with different kinetics. Some of the chloroplast proteins, e.g. the light-harvesting complexes of PSI and PSII, are quite stable and only decrease at later stages of senescence, while other components of the photosynthetic apparatus, e.g. components of PSI and the cytochrome b_6/f complex, decrease much faster (Gepstein, 1988). The changes in protein levels are caused by both, an increase in proteolytic processes (see also Chapter 7 by U. Feller) and a decrease in expression of the corresponding genes. This section will discuss the expression of nuclear genes related to photosynthesis.

A. Calvin Cycle-related Genes

Studies dealing with senescence-specific changes in nuclear gene expression focus mainly on the *rbcS* gene, coding for the small subunit of Rubisco. The actual level of *rbcS* mRNA per fresh weight as well as the relative portion of *rbcS* transcripts on a total RNA basis decrease when senescence proceeds (e.g. Bate *et al.*, 1991; Jiang *et al.*, 1993; Crafts-Brandner *et al.*, 1996; Humbeck *et al.*, 1996). Transcript levels for other enzymes involved in carbon reduction, e.g. plastidic glyceraldehyde-3-phosphate dehydrogenase and aldolase, were found to be coordinately down regulated with the level of *rbcS* mRNA during senescence of tobacco leaves (Ludewig and Sonnewald, 2000). The mRNA level for Rubisco activase (*rca* gene) was found to decline during senescence of bean leaves when calculated on a fresh weight basis, but is rather stable when calculated on a total RNA basis (Crafts-Brandner *et al.*, 1996). Normalizing on the basis of equal amounts of total RNA or protein will show changes in the relative portion of a specific mRNA or protein, whereas normalizing on the basis of fresh weight or leaf area will reflect changes in the actual amount of a specific mRNA. These results show that changes in RNA- or protein-levels during senescence have to be interpreted differently depending on the basis of normalization (see also discussion in Noodén *et al.*, 1997).

B. Genes Encoding Thylakoid Membrane Proteins

The levels of several nuclear transcripts specific for thylakoid membrane proteins decline during senescence. A decrease in the expression of *Lhcb* genes encoding apoproteins of the major light-harvesting complex of photosystem II was reported (Bate *et al.*, 1991; Glick *et al.*, 1995; Shinohara and Murakami, 1996). By additional pulse-labeling studies, Bate *et al.* (1991) found that the rate of synthesis of LHC proteins declines during senescence, too. The relative portion of LHC proteins is rather stable during senescence when based on equal protein amounts (see Section III). Compared to other proteins, LHC proteins seem to be less sensitive to proteolytic breakdown during senescence. In comparison to the mRNA levels for the bulk LHC (Humbeck *et al.*, 1996) the mRNA levels for the minor light-harvesting complexes, e.g. CP29 and LHCI, are decreasing at higher rates during senescence (Humbeck and Krupinska, 1999, 2002).

C. Regulatory Mechanisms Underlying the Down-regulation of Photosynthesis-related Nuclear Genes

The investigations described above reveal a preferential decrease in the abundance of several nuclear transcripts specific for proteins of the chloroplast, while other nuclear genes show either an unchanged or even higher expression level during senescence (see also Chapter 4 by M. Jones, Chapter 5 by H.R. Woo *et al.* and Chapter 6 by Y.-S. Noh *et al.*). The molecular mechanism underlying this preferential down-regulation of photosynthesis-related nuclear gene expression during senescence is not yet fully understood. An increase in sugar levels has been proposed to be an important factor contributing to the down-regulation of photosynthetic activity and the transcript levels of photosynthesis-related genes during senescence (Wingler *et al.*, 1998; Paul and Foyer, 2001). In order to clarify whether a decrease in the rate of synthesis, i.e. transcription of the corresponding genes, or an increase in the rate of degradation of these mRNAs by specific RNAses is causing the decrease in transcript levels, nuclear run-on transcription experiments have been performed. These studies showed that both transcriptional and posttranscriptional regulation contribute to the decline in mRNA abundance (Glick *et al.*, 1995). Moreover, not only expression of nuclear photosynthesis-related genes, but also expression of plastid genes is down regulated (see the following section). This indicates a well-coordinated program, which at the onset of senescence ensures a simultaneous decrease in the expression of photosynthesis-related genes and an enhanced proteolytic degradation of the corresponding proteins. Such a program would guarantee an efficient recycling of the N resources of the chloroplast and concomitantly diminish the energy-consuming synthesis of proteins which are not required at this developmental stage.

V. Plastid Gene Expression during Senescence

Plastid gene expression has been studied extensively during biogenesis of chloroplasts. These studies showed that gene expression in plastids is controlled at the level of transcriptional activity and at different posttranscriptional levels including RNA stability, translational activity and posttranslational modifications [reviewed by Gruissem and Tonkyn (1993) and by Stern *et al.* (1997)]. Very few studies, however, deal with plastid gene expression during senescence.

A. Transcriptional Activity

The decrease in plastid RNA level is due at least in part to a decrease in RNA polymerase activity (Brady, 1988; Noodén and Guiamét, 1996). To unravel whether the quantitative decline in plastid DNA transcription is accompanied by qualitative changes in transcription of different genes, run-on transcripts from barley plastids of different developmental stages were hybridized, either to DNA fragments representing the complete plastome on one Southern blot (Fig. 11-2) or to a selection of individual genes belonging to different

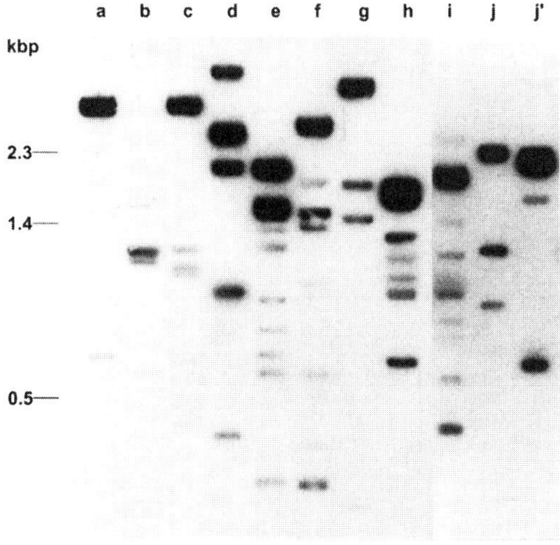

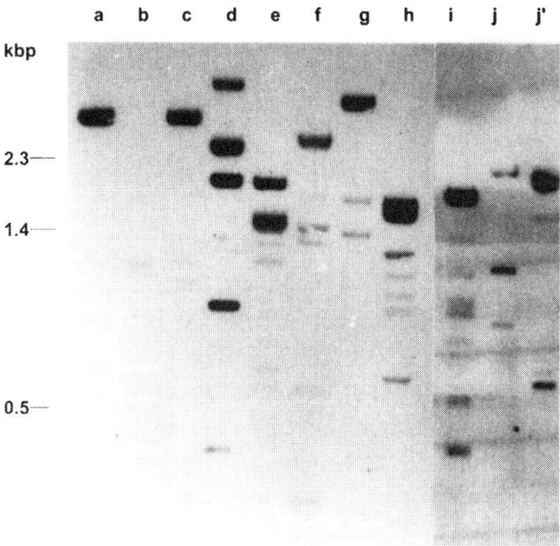

Figure 11-2. Analysis of plastid DNA transcription by run-on transcription assays with lysed barley plastids. Autoradiograms were obtained after subsequent hybridization of the same Southern blot filter with transcripts derived from fully expanded primary foliage leaves before (non-senescent) and after two days of dark incubation (senescent). Plastid DNA fragments on the filter in sum represent the entire barley plastid DNA (see Krupinska and Falk, 1994).

groups of plastid genes (Krupinska and Falk, 1994; Krause *et al.*, 1998). These analyses clearly showed that senescing chloroplasts are capable of synthesizing the complete set of transcripts which are also found in mature chloroplasts. However, the relative transcription rates of the individual plastid genes differ considerably in mature and senescing plastids (Table 11-1).

Table 11-1. Relative Transcriptional Activities of Photosynthesis-related Plastid Genes (Krause *et al.*, 1998)

	Non-senescent	Senescent
atpB	3.0	0.5
psaB	10.6	1.1
psaA	5.7	1.7
psbD	36.7	40.5
psbA	32.8	49.1
rbcL	3.8	2.0
16SrDNA	1.9	1.7

A preferential high rate of transcription of *psbA* and *psbD* in comparison to that of *psaA/B* might reflect the necessity of maintaining high synthesis rates of photosystem II core proteins which undergo a high light-dependent turnover during senescence.

B. Transcript Levels

Similar to the relative rates of transcription, the levels of mRNAs for photosynthesis-related proteins do not change synchronously during leaf senescence. While the relative amounts of *psaB* and *rbcL* transcripts decline rapidly during dark-induced senescence of barley leaves, the level of *psbA* mRNA declines much more slowly (Krause *et al.*, 1998). A preferential rapid decline in the level of the *psaA/B*-related transcript has also been observed in pine needles during aging in subsequent years under natural conditions. In this case the levels of both the *psbA* and *rbcL* transcripts stay constant during senescence (Shinohara and Murakami, 1996). In pine needles, the decrease in the level of the 5.0 kb *psaA/B* transcript was shown to be accompanied by an increase in the level of a 1.5 kb transcript which might be a degradation product of the *psaA/B* transcript (Shinohara and Murakami, 1996). Though the levels of both the photosystem II-related transcripts and the *psa* transcript decrease during natural senescence of poplar, this decrease again seems to be more prominent in the case of the photosystem I transcripts (Reddy *et al.*, 1997). In contrast to the results obtained with barley, in bean, the level of the *rbcL* transcript declines at about the same rate as the level of *psbA* and other photosystem II-related transcripts (Bate *et al.*, 1990, 1991). Variations in the time course of expression of *psbA* and *rbcL* might be due to differences in growth conditions which are known to particularly affect these genes (Klein and Mullet, 1987). Despite some discrepancies among the different studies on transcript levels, they all show a preferential decrease in the level of *psaA/B* transcripts during senescence. In addition to changes in the rate of transcription (see above), a selective development dependent stabilization of mRNAs could be responsible for the differences in mRNA levels of certain photosynthesis-related genes (Kim *et al.*, 1993).

C. Protein Synthesis

Although during continued development of barley seedlings in the light the levels of the photosystem I-related transcripts decline much faster than the level of the *psbA* mRNA, the rates of synthesis for the corresponding proteins decline in parallel (Klein and Mullet, 1987). Roberts *et al.* (1987) and Bate *et al.* (1991) reported that in bean leaves the D1 protein is synthesized continuously throughout senescence although its amount is decreasing during senescence. Since in these leaves synthesis of the D1 protein accounts for about 70% of total thylakoid protein it has been suggested that the *psbA* transcript relative to others is preferentially translated during senescence (Bate *et al.*, 1991; Droillard *et al.*, 1992). Correspondingly, these investigations showed that the portion of the D1 protein stays almost stable throughout senescence (Roberts *et al.*, 1987). However, based on the same amount of fresh weight or on leaf area, its level would decrease.

In contrast, in barley leaves as well as in poplar, it has been shown that during early leaf senescence the level of the D1 protein decreases much faster than the level of total protein (Humbeck *et al.*, 1996; Reddy *et al.*, 1997). This decline in the steady-state level of D1 clearly occurred before the contents of chlorophylls, cytochrome f and LSU start to decrease. One explanation for these contradictory results might be that the expression of D1 is influenced by environmental conditions. In contrast to the earlier studies with bean leaves from plants grown under controlled environmental conditions employing light intensities not exceeding 300 μmol s^{-1} m^{-2}, the investigations on barley and poplar leaves have been performed under field conditions which are characterized by changing photon fluence rates reaching maximal values of about 2000 μmol s^{-1} m^{-2}.

D. Expression of Plastid Genes not Related to Photosynthesis

In contrast to the photosynthesis-related plastid genes, the levels of plastid transcripts required for synthesis of the translation machinery are relatively stable during senescence when calculated on the basis of equal amounts of RNA (Bate *et al.*, 1991; Krause *et al.*, 1998). Among the other plastid genes not belonging to those encoding photosynthesis-related proteins and parts of the translation machinery, only a few have attracted attention in senescence-related studies. One example is the *clpP* gene which has been studied with regard to chloroplast protein degradation during senescence. This gene encodes the catalytic subunit of an archetypal two-component ATP-dependent protease. The protein and mRNA levels of *clpP* have been examined in senescing leaves of several plant species and were found to decrease during senescence in all cases reported so far (Crafts-Brandner *et al.*, 1996; Humbeck and Krupinska, 1996; Weaver *et al.*, 1999). These results make it rather unlikely that the Clp protease is playing an important role in senescence-related protein degradation in chloroplasts.

Other examples of plastid genes induced during senescence but not directly related to photosynthesis are the *ndh* genes which are supposed to encode a plastid NAD(P)H-dehydrogenase. Translation of *ndh* gene specific RNA in a protein synthesis system derived from senescent barley chloroplasts revealed an enhanced expression of these genes early in leaf senescence (Vera *et al.*, 1990). Correspondingly, it has been shown that plastid NAD(P)H-dehydrogenase activity increases during leaf senescence (Cuello *et al.*, 1995). The enzyme might have a function either in photosynthetic electron transport, in chlororespiration or in protection against oxidative stress (Catalá *et al.*, 1997). Immunological analyses revealed that the relative amount of the *ndhF* gene product increases during leaf senescence and in response to oxidative stress (Catalá *et al.*, 1997). A corresponding senescence-related increase in the portion of transcripts has been clearly shown in the case of *ndhB* (Martínez *et al.*, 1997). The relative increase in *ndh* gene expression occurs under conditions of dark-induced senescence (Martín *et al.*, 1996) as well as under conditions of developmentally induced senescence (Martínez *et al.*, 1997).

Cuello *et al.* (1995) measured NADH- and NADPH-dehydrogenase activities in the light and in the dark with ferricyanide as the electron acceptor, employing barley chloroplast suspensions free from contaminating mitochondria. While NADPH-dehydrogenase activity declines in parallel with photosynthetic electron transport during senescence in the dark, the NADH-dehydrogenase activity is stable during senescence both in the dark and in the light. The authors suggested that in senescent chloroplasts the NADH-dehydrogenase activity is involved in chlororespiratory oxidation of NADH which is generated by amino acid catabolism as a consequence of carbohydrate degradation.

VI. Regulation of Chloroplast Senescence

A. The Dilemma of the Chloroplast: Ongoing Photosynthesis during Senescence-associated Breakdown

Despite entering the developmental stage of senescence, chloroplasts still have to perform photosynthesis in order to produce assimilates. During senescence these assimilates represent a needed carbon source and are exported to other parts of the plant, e.g. to the reproductive organs. To produce assimilates, at least a few of the photosynthetic units have to be preserved in a well-organized structure which guarantees efficient absorption and utilization of light energy. In addition, efficient protection mechanisms against photodestruction have to be maintained in the plastids during senescence (Chapter 18 by E. Tyystjärvi).

It is evident that during senescence a tight coordination of the two processes, photosynthesis on the one hand and chloroplast breakdown on the other hand, is required. Hence, regulation of photosynthesis and regulation of senescence have to be closely connected. A reduction in photosynthetic activity as a result of an increase in carbohydrate levels (Lazan *et al.*, 1983; Noodén *et al.*, 1997) has even been suggested to be the primary trigger of senescence (Yamasaki *et al.*, 1996; Okada and Katoh, 1998).

B. Coordination of Senescence-related Degradation of Proteins

In order to come to a better understanding of the coordination of senescence-associated changes in the activity and composition of the photosynthetic apparatus, Jiang *et al.* (1999) compared two cultivars of rice, cv. Nipponbare and cv. Akenohoshi, which differed in the decline of light-saturated photosynthesis (P_{max}). Retarded senescence in the case of cv. Akenohoshi correlates with a considerable higher grain yield. Comparative analyses showed that in both cultivars Rubisco and all other proteins are degraded in a well-coordinated manner. The authors suggest that a tight coordination of protein breakdown in plastids (see Section III) is a strategy allowing an efficient use and recycling of amino acids during leaf senescence. In one rice cultivar leaf senescence has been compared between intact plants and plants with the panicles removed (Nakano *et al.*, 1995). In this case, panicle removal resulted in a retardation of senescence due to a reduced demand for nitrogen. Even under this condition different proteins are degraded with typically different kinetics as outlined in Section III. This also indicates the need for a coordinated regulation of the degradation process. Environmental conditions might modulate the degradation kinetics as shown for the most abundant chloroplast protein, Rubisco, whose degradation is regarded to be the main reason underlying the loss of the photosynthetic capacity of leaves (Jiang *et al.*, 1999). A decline in Rubisco content during senescence is caused on the one hand by a coordinated decline in the synthesis of both subunits (Brady, 1981) and on the other hand by a proteolytic degradation of both subunits (Miyadai *et al.*, 1990). As shown in flag leaves of barley collected in the field at different developmental stages, the concentration of the small subunit of Rubisco seems to be more stable during senescence than that of the large subunit (Humbeck *et al.*, 1996). Acccordingly, Thomas and Huffaker (1981) observed a much faster degradation of the large subunit than of the small subunit when Rubisco was incubated *in vitro* with an extract from senescing leaves. These results indicate a preferential degradation of the large subunit of Rubisco during senescence.

C. Photosynthesis and Senescence in Stay-Green Mutants and Transgenic Plants

A valuable tool for the investigation of regulatory mechanisms underlying senescence-specific changes in chloroplast function is the so-called stay-green mutants which are characterized by a retarded chlorophyll breakdown (reviewed by Thomas and Haworth, 2000). Mutation of the nuclear *sid* gene of *Festuca* disables chlorophyll degradation during leaf senescence. In this mutant, in spite of unchanged chlorophyll content, photosynthesis measured as the CO_2 fixation rate decreases similarily to that in the wild type (Thomas, 1987; Hauck *et al.*, 1997). While some stay-green mutants of soybean retain substantial amounts of Rubisco and other soluble proteins during senescence, others exhibit a significant retardation in the loss of soluble proteins (Guiamét and Giannibelli, 1996). Compared to the *sid* gene of *Festuca* the *cytG* gene of soybean seems to stabilize selectively peripheral light-harvesting complexes rather than all chlorophyll binding proteins in the thylakoid membrane (reviewed by Thomas and Howarth, 2000). Recently, three genetic loci have been identified in stay-green mutants of *Arabidopsis* that show a delay in a broad range of senescence-associated events including chlorophyll breakdown, increase in RNase activity and decline in Rubisco (Oh *et al.*, 1997). These mutants might be very useful for studies aiming at identifying the trigger of senescence. However, to unravel the interrelations among different processes occurring in senescing plastids, comparative studies with mutants defective in only a part of the senescence syndrome may be more useful. In the case of the different types of soybean mutants, differences in plastid protein loss seem not to be due to differences in protease content of the plastids, but rather in the targeting of the proteins for degradation or in the maintenance of reducing conditions within the plastid (Guiamét and Giannibelli, 1996). Combining these data from the studies with stay-green mutants it seems that different processes such as chlorophyll degradation, decrease in soluble proteins and the decrease in photosynthetic function are not strictly coupled during senescence and may be regulated independently. It will be important to identify the genes affected in the various types of stay-green mutants and to study their function with regard to the regulation of photosynthesis and chloroplast breakdown.

A divergence of different senescence-related processes has also been demonstrated in *ipt*-transformed plants: when the content of endogenous cytokinins is increased during senescence the chlorophyll level remains rather stable while the content of protein components of the photosynthetic apparatus decreases as in control plants (Synkova *et al.*, 1997). This result clearly shows that chlorophyll catabolism and the breakdown of the photosynthetic apparatus are regulated independently. Moreover, this study nicely demonstrates that chlorophyll should not be used as the sole parameter (e.g. Smart *et al.*, 1991; Ori *et al.*, 1999) to characterize the impact of a genetic manipulation on senescence.

Meanwhile, several transgenic plants with alterations in the transcript levels of photosynthesis-related nuclear genes have been described. With regard to senescence, plants with altered level of *rbcS* mRNA are of special interest. Fully expanded leaves of *rbcS* antisense plants have nearly normal levels of chlorophylls and chlorophyll-binding proteins, indicating that the expression of photosynthesis-related proteins in the thylakoid membrane is not negatively regulated by the Rubisco abundance, but they are impaired in their ability to fix carbon and to produce carbohydrates suitable for export (Jiang and Rodermal, 1995). Miller *et al.* (2000) showed that the senescence phase is prolonged in these plants, indicating that source strength is modulating the duration of leaf development.

D. Photosynthesis and Senescence at Elevated CO_2 Concentrations

During the past 15 years many reports on research about different aspects of photosynthesis at elevated CO_2 concentrations have been published. In many studies, it has been observed that following an initial increase photosynthesis often declines at high CO_2 levels (reviewed in Bowes, 1991). This acclimation at elevated CO_2 is, however, not a simple down-regulation of photosynthesis but rather results in a temporal shift of leaf development towards premature senescence (Miller *et al.*, 1997). In this regard, comparative studies on senescence at different CO_2 levels could give new insight into the interrelationship of photosynthesis and senescence.

The molecular reasons underlying premature senescence at elevated CO_2 are not clear. In principle a down-regulation of photosynthesis at elevated CO_2 levels may be achieved by either an end product inhibition or by a redeployment of nitrogen (Bowes, 1991). Most scientists favor the idea that at elevated CO_2 levels carbohydrate accumulation exerts a negative feedback inhibition on photosynthesis, which most likely occurs at the level of Rubisco (Bowes, 1991; Webber *et al.*, 1994; Pearson and Brooks, 1995; Miglietti *et al.*, 1996; Sicher and Bunce, 1997). Increases in sugar and starch contents indeed correlate with a clearly reduced level of the *rbcS* transcripts under conditions of high CO_2 (Van Oosten and Besford, 1995; Ono and Watanabe, 1997). An acceleration of Rubisco degradation at elevated CO_2 is not necessarily an indication of senescence but might just reflect the lower need for Rubisco at elevated CO_2 (Webber *et al.*, 1994). On the basis of determinations of yield and nitrogen distribution in barley Fangmeier *et al.* (2000) propose that premature senescence of flag leaves under elevated CO_2 is induced by a greater nitrogen demand of the reproductive tissues. Recently, studies with transgenic tobacco plants having the *ipt* gene for cytokinin biosynthesis under control of the senescence-induced SAG-12 promoter showed that under high CO_2 levels senescence and the down-regulation of photosynthesis-related genes is retarded in these plants, while levels of soluble sugars were indistinguishable from wild type (Ludewig and Sonnewald, 2000). This result strongly argues against a role for soluble sugars in the regulation of photosynthesis-related transcripts under elevated CO_2. The authors suggest that high CO_2 levels accelerate senescence by increasing the flow of sugars through hexokinase. Indeed, hexokinase overexpression leads to accelerated senescence in transgenic tomato plants (Dai *et al.*, 1999).

The effect of elevated CO_2 on photosynthesis varies with the developmental stage of the leaves (Nie *et al.*, 1995; Pearson and Brooks, 1995; Sicher and Bunce, 1997; van Oosten and Besford, 1995). While in developing wheat leaves, a substantial increase in photosynthesis was exerted by elevated CO_2, in mature leaves before the onset of grain filling the photosynthesis declined dramatically in comparison to control plants kept under ambient CO_2 concentration (Garcia *et al.*, 1998).

Several studies show that leaf development is accelerated by CO_2 only with regard to the timing of increasing and declining photosynthesis and not to the timing of leaf expansion (Miller *et al.*, 1997; Fangmeier *et al.*, 2000). Differences between developmentally induced senescence and premature senescence caused by elevated CO_2 are also obvious at the level of plastid gene expression. Under high CO_2, levels of plastid gene transcripts related to photosynthesis were only reduced when calculated on a total RNA basis, but not on a chloroplast 16S rRNA basis (Van Oosten and Besford, 1995). In comparison, during developmentally induced senescence and dark-induced senescence the concentrations of different photosynthesis-related transcripts declined while the level of 16S rRNA stayed almost constant (see Section V). With regard to the differences between natural senescence and premature senescence as a response to elevated CO_2, it has been suggested that premature senescence at elevated CO_2 concentration more resembles a photobleaching process than developmentally induced senescence (Sicher, 1998).

VII. Conclusions

During senescence chloroplasts are in a dilemma: on the one hand they are subject to degradation and on the other hand they still have to perform photosynthesis. Minimization of mutual disturbances of these two processes requires a well-tuned coordination at different levels, ranging from gene expression to degradation of proteins and pigments. When senescence is induced untimely, e.g. by growth of the plants at elevated CO_2 or by various stress situations including dark treatment, the balanced progression of senescence-associated events may be disturbed. It may be anticipated that under these conditions photosynthesis and nitrogen reallocation are less efficient than during developmentally induced senescence.

Chloroplast senescence is not a linear sequence of changes but a complex syndrome of parallel processes which are more or less tightly coupled and to some extent regulated independently. The interrelations among the different senescence-associated events occurring in plastids may be severely influenced by environmental conditions. On account of this complex situation, it is not acceptable to monitor chloroplast senescence exclusively by changes in chlorophyll levels or any other single parameter. To avoid erroneous judgements about the developmental stage of plastids, e.g. in transgenic plants or otherwise experimentally manipulated plants, a broad set of parameters including at least chlorophyll fluorescence and content of soluble proteins in addition to chlorophyll content, have to be determined. This approach will enable examination of whether a given treatment causes a shift in the entire senescence syndrome or accelerates/delays only a part of the program.

Acknowledgments

We thank Tom van der Kooij (Institute of Botany, University of Kiel) for critical reading of the manuscript.

References

Bassi, R., Pineau, B., and Marquardt, J. (1993). Carotenoid-binding proteins of photosystem II. *European J. Biochem.* **212**, 297–303.

Bate, N.J., Straus, N.A., and Thompson, J.E. (1990). Expression of chloroplast photosynthesis genes during leaf senescence. *Physiol. Plant.* **80**, 217–225.

Bate, N.J., Rothstein, S.J., and Thompson, J.E. (1991). Expression of nuclear and chloroplast photosynthesis-specific genes during leaf senescence. *J. Exp. Bot.* **42**, 801–811.

Biswal, B. (1995). Carotenoid catabolism during leaf senescence and its control by light. *J. Photochem. Photobiol. B: Biology* **30**, 3–13.

Biswal, U.C., and Biswal, B. (1988). Ultrastructural modifications and biochemical changes during senescence of chloroplasts. *Int. Rev. Cytol.* **113**, 271–320.

Blank, A., and McKeon, T.A. (1991). Three RNAses in senescent and nonsenescent wheat leaves: Characterization by activity staining in sodium dodecyl sulfate-polyacrylamide gels. *Plant Physiol.* **97**, 1402–1408.

Bowes, G. (1991). Growth at elevated CO_2: photosynthetic responses mediated through Rubisco. *Plant Cell Environ.* **14**, 795–806.

Brady, C.J. (1981). A coordinated decline in the synthesis of subunits of ribulosebisphosphate carboxylase in aging wheat leaves. I. Analyses of isolated protein, subunits and ribosomes. *Aust. J. Plant Physiol.* **8**, 591–602.

Brady, C.J. (1988). Nucleic acid and protein synthesis. In *Senescence and Aging in Plants* (L.D. Noodén and A.C. Leopold, Eds.), pp. 147–179. Academic Press, San Diego.

Butler, W., and Simon, E.W. (1971). Ultrastructural aspects of senescence in plants. *Adv. Gerontol. Res.* **3**, 73–129.

Catalá, R., Sabater, B., and Guera, A. (1997). Expression of the plastid *ndhF* gene product in photosynthetic and non-photosynthetic tissues of developing barley seedlings. *Plant Cell Physiol.* **38**, 1382–1388.

Chrost, B., and Krupinska, K. (2000). Expression of genes with homologies to known α-galactosidases in barley leaves indicates their involvement in senescence associated galactolipid metabolism. *Physiol. Plant.* **110**, 111–119.

Crafts-Brandner, S.J., Klein, R.R., Klein, P., Holzer, R., and Feller, U. (1996). Coordination of protein and mRNA abundances of stromal enzymes and mRNA abundances of the Clp protease subunits during senescence of *Phaseolus vulgaris* (L.) leaves. *Planta* **200**, 312–318.

Cuello, J., Quiles, M.J., Rosauro, J., and Sabater, B. (1995). Effects of growth regulators and light on chloroplasts NAD(P)H dehydrogenase activities of senescent barley leaves. *Plant Growth Regul.* **17**, 225–232.

Dai, N., Schaffer, A., Petreikov, M., Shahak, Y., Gillo, Y., Ratrur, K., Levine, A., and Granot, D. (1999). Overexpression of Arabidopsis hexokinase in tomato plants inhibits growth, reduces photosynthesis, and induces rapid senescence. *Plant Cell* **11**, 1253–1266.

Di Marco, G., Ianelli, M.A., and Loreto, F. (1994). Relationship between photosynthesis and photorespiration in field-grown wheat leaves. *Photosynthetica* **30**, 45–51.

Droillard, M.J., Bate, N.J., Rothstein, S.J., and Thompson, J.E. (1992). Active translation of the D-1 protein of photosystem II in senescing leaves. *Plant Physiol.* **99**, 589–594.

Fangmeier, A., Chrost, B., Högy, P., and Krupinska, K. (2000). CO_2 enrichment enhances flag leaf senescence in barley due to greater grain nitrogen sink capacity. *Environ. Exp. Bot.* **44**, 151–164.

Fraser, P.D., Truestala, M.R., Bird, C.R., Schuck, W., and Bramby, M. (1994). Carotenoid biosynthesis during tomato fruit development. *Plant Physiol.* **105**, 405–413.

Friedrich, J.W., and Huffaker, R.C. (1980). Photosynthesis, leaf resistances, and ribulose-1,5-bisphosphate carboxylase degradation in senescing barley leaves. *Plant Physiol.* **65**, 1103–1107.

Garcia, R.L., Long, S.P., Wall, G.W., Osborne, C.P., Kimball, B.A., Nie, G.Y., Pinter, P.J., Lamorte, R.L., and Wechsung, F. (1998). Photosynthesis and conductance of spring-wheat leaves: field response to continuous free-air atmospheric CO_2 enrichment. *Plant Cell Environ.* **21**, 659–669.

Gay, A.P., and Thomas, H. (1995). Leaf development in *Lolium temulentum* L.: photosynthesis in relation to growth and senescence. *New Phytol.* **130**, 159–168.

Gepstein, S. (1988). Photosynthesis. In *Senescence and Aging in Plants* (L.D. Noodén and A.C. Leopold, Eds.), pp. 85–109. Academic Press, San Diego.

Glick, R.E., Schlagnhaufer, C.D., Arteca, R.N., and Pell, E.J. (1995). Ozone-induced ethylene emission accelerates the loss of ribulose-1,5-bisphosphate carboxylase/oxygenase and nuclear-encoded mRNAs in senescing potato leaves. *Plant Physiol.* **109**, 891–898.

Grover, A. (1993). How do senescing leaves lose photosynthetic activity? *Curr. Sci.* **64**, 226–233.

Gruissem, W., and Tonkyn, J.C. (1993). Control mechanisms of plastid gene expression. *Crit. Rev. Plant Sciences* **12**, 19–55.

Guiamét, J.J., and Giannibelli, M.C. (1996). Nuclear and cytoplasmic "stay-green" mutations of soybean alter the loss of leaf soluble proteins during senescence. *Physiol. Plant.* **96**, 655–661.

Hauck, B., Gay, A.P., Macduff, J., Griffiths, C.M., and Thomas, H. (1997). Leaf senescence in a non-yellowing mutant of *Festuca pratensis*: implications of the stay-green mutation for photosynthesis, growth and nitrogen nutrition. *Plant Cell Environ.* **20**, 1007–1018.

Hikosaka, K. (1996). Effects of leaf age, nitrogen nutrition and photon flux density on the organization of the photosynthetic apparatus in leaves of a vine (*Ipomoea tricolor Cav.*) grown horizontally to avoid mutual shading of leaves. *Planta* **198**, 144–150.

Humbeck, K., and Krupinska, K. (1996). Does the Clp protease play a role during senescence associated protein degration in barley leaves? *J. Photochem. Photobiol. B: Biology* **36**, 321–326.

Humbeck, K., and Krupinska, K. (1999). Successive degradation of the light-harvesting system of the photosynthetic apparatus during senescence of barley flag leaves. In *The Chloroplast: From Biology to Biotechnology* (J. Argyroudi-Akoyunoglou, Ed.), pp. 178–183. Kluwer, Amsterdam.

Humbeck, K., and Krupinska, K. (2003). The abundance of minor chlorophyll a/b-binding proteins CP29 and LHCI of barley (*Hordeum vulgare* L.) during leaf senescence is controlled by light. *J. Exp. Bot.* **54**, 375–383.

Humbeck, K., Quast, S., and Krupinska, K. (1996). Functional and molecular changes in the photosynthetic apparatus during senescence of flag leaves from field-grown barley plants. *Plant Cell Environ.* **19**, 337–344.

Inada, N., Sakai, A., Kuroiwa, H., and Kuroiwa, T. (1998a). Three-dimensional analysis of the senescence program in rice (*Oryza sativa* L.) coleoptiles—Investigations of tissues and cells by fluorescence microscopy. *Planta* **205**, 153–164.

Inada, N., Sakai, A., Kuroiwa, H., and Kuroiwa, T. (1998b). Three-dimensional analysis of the senescence program in rice (*Oryza sativa* L.) coleoptiles—Investigations by fluorescence microscopy and electron microscopy. *Planta* **206**, 585–597.

Jiang, C.Z., and Rodermel, S.R. (1995). Regulation of photosynthesis during leaf development in *Rbcs* antisense DNA mutants of tobacco. *Plant Physiol.* **107**, 215–224.

Jiang, C.Z., Rodermel, S.R., and Shibles, R.M. (1993). Photosynthesis, Rubisco activity and amount, and their regulation by transcription in senescing soybean leaves. *Plant Physiol.* **101**, 105–112.

Jiang, C.Z., Ishihara, K., Satoh, K., and Katoh, S. (1999). Loss of the photosynthetic capacity and proteins in senescing leaves at top positions of two cultivars of rice in relation to the source capacities of the leaves for carbon and nitrogen. *Plant Cell Physiol.* **40**, 496–503.

Kar, M., Streb, P., Hertwig, B., and Feierabend, J. (1993). Sensitivity to photodamage increases during senescence in excised leaves. *J. Plant Physiol.* **141**, 538–544.

Kim, M., Christopher, D.A., and Mullet, J.E. (1993). Direct evidence for selective modulation of *psb*A, *rpo*A, *rbc*L and 16S RNA stability during barley chloroplast development. *Plant. Mol. Biol.* **22**, 447–463.

King, G.A., Davies, K.M., Stewart, R.J., and Borst, W.M. (1995). Similarities in gene expression during the post-harvest induced senescence of spears and natural foliar senescence. *Plant Physiol.* **108**, 125–128.

Kingston-Smith, A.H., Thomas, H., and Foyer, C.H. (1997). Chlorophyll *a* fluorescence, enzyme and antioxidant analyses provide evidence for the operation of alternative electron sinks during leaf senescence in a *stay-green* mutant of *Festuca pratensis*. *Plant Cell Environ.* **20**, 1323–1337.

Klein, R.R., and Mullet, J.E. (1987). Control of gene expression during higher plant chloroplast biogenesis. *J. Biol. Chem.* **262**, 4341–4348.

Krause, K., Falk, J., Humbeck, K., and Krupinska, K. (1998). Responses of the transcriptional apparatus of barley chloroplasts to a prolonged dark period and to subsequent reillumination. *Physiol. Plant.* **104**, 143–152.

Krupinska, K., and Falk, J. (1994). Changes in RNA-polymerase activity during biogenesis, maturation and senescence of barley chloroplasts. Comparative analysis of transcripts synthesized either in run-on assays or by transcriptionally active chromosomes. *J. Plant Physiol.* **143**, 298–305.

Kura-Hotta, M., Satoh, K., and Katoh, S. (1987). Relationship between photosynthesis and chlorophyll content during leaf senescence of rice seedlings. *Plant Cell Physiol.* **28**, 1321–1329.

Kura-Hotta, M., Hashimoto, H., Satoh, K., and Katoh, S. (1990). Quantitative determination of changes in the number and size of chloroplasts in naturally senescing leaves of rice seedlings. *Plant Cell Physiol.* **31**, 33–38.

Lazan, H.B., Barlow, E.W.R., and Brady, C.J. (1983). The significance of vascular connection in regulating senescence of the detached flag leaf of wheat. *J. Exp. Bot.* **34**, 726–736.

Lers, A., Khalchitsky, A., Lomaniec, E., Burd, S., and Green, P. (1998). Senescence-induced RNases in tomato. *Plant Mol. Biol.* **36**, 439–449.

Lovelock, C.E., and Winter, K. (1996). Oxygen-dependent electron transport and protection from photoinhibition in leaves of tropical tree species. *Planta* **198**, 580–587.

Lu, C., and Zhang, J.H. (1998). Modifications in photosystem II photochemistry in senescent leaves of maize plants. *J. Exp. Bot.* **49**, 1671–1679.

Lu, C., Lu, Q., Zhang, J., and Kuang, T. (2001). Characterization of photosynthetic pigment composition, photosystem II photochemistry and thermal energy dissipation during leaf senescence of wheat plants grown in the field. *J. Exp. Bot.* **52**, 1805–1810.

Ludewig, F., and Sonnewald, U. (2000). High CO_2-mediated down-regulation of photosynthetic gene transcripts is caused by accelerated leaf senescence rather than sugar accumulation. *FEBS Lett.* **479**, 19–24.

Mae, T., Kai, N., Makino, A., and Ohira, K. (1984). Relation between ribulose bisphosphate carboxylase content and chloroplast number in naturally senescing primary leaves of wheat. *Plant Cell Physiol.* **25**, 333–336.

Mae, T., Thomas, H., Gay, A.P., Makino, A., and Hidema, J. (1993). Leaf development in *Lolium temulentum*: photosynthesis and photosynthetic proteins in leaves senescing under different irradiances. *Plant Cell Physiol.* **34**, 391–399.

Martín, M., Casano, L.M., and Sabater, B. (1996). Identification of the product of *ndhA* gene as a thylakoid protein synthesized in response to photooxidative treatment. *Plant Cell Physiol.* **37**, 293–298.

Martínez, P., López, C., Roldán, M., Sabater, B., and Martín, M. (1997). Plastid DNA of five ecotypes of Arabidopsis thaliana: sequence of *ndhG* and maternal inheritance. *Plant Sci.* **37**, 113–122.

Martinoia, E., Heck, U., Dalling, M.J., and Matile, P. (1983). Changes in chloroplast number and chloroplast constituents in senescing barley leaves. *Biochem. Physiol. Pflanzen* **178**, 147–155.

Matile, P. (1992). Chloroplast senescence. In *Crop Photosynthesis: Spatial and Temporal Determinants* (N.R. Baker, and H. Thomas, Eds.), pp. 413–440. Elsevier, Amsterdam.

Matile, P. (1997). The vacuole and cell senescence. *Adv. Bot. Res.* **25**, 87–112.

Miersch, I., Heise, J., Zelmer, I., and Humbeck, K. (2000). Differential degradation of the photosynthetic apparatus during leaf senescence in barley (*Hordeum vulgare* L.). *Plant Biol.* **2**, 618–623.

Miglietti, F., Giuntoli, A., and Bindi, M. (1996). The effect of free air carbon dioxide enrichment (FACE) and soil nitrogen availability on the photosynthetic capacity of wheat. *Photosynth. Res.* **47**, 281–290.

Miller, A., Tsai, C.H., Hemphill, D., Endres, M., Rodermel, S., and Spalding, M. (1997). Elevated CO_2 effects during leaf ontogeny—A new perspective on acclimation. *Plant Physiol.* **115**, 1195–1200.

Miller, A., Schlagnhaufer, C., Spalding, M., and Rodermel, S. (2000). Carbohydrate regulation of leaf development: prolongation of leaf senescence in Rubisco antisense mutants of tobacco. *Photosynth. Res.* **63**, 1–8.

Miyadai, K., Mae, T., Makino, A., and Ojima, K. (1990). Characteristics of ribulose-1,5-bisphosphate carboxylase/oxygenase degradation by lysates of mechanically isolated chloroplasts from wheat leaves. *Plant Physiol.* **92**, 1215–1219.

Morita, K. (1980). Release of nitrogen from chloroplasts during senescence in rice (*Oryza sativa* L.). *Ann. Bot.* **46**, 297–302.

Nakano, H., Makino, A., and Mae, T. (1995). Effects of panicle removal on the photosynthetic characteristics of the flag leaf of rice plants during the ripening stage. *Plant Cell Physiol.* **36**, 653–659.

Nie, G.-Y., Long, S.P., Garcia, R.L., Kimball, B.A., Pinter, P.J., LaMorte, R.L., Wall, G.W., and Webber, A.N. (1995). Effects of free-air CO_2 enrichment on the development of the photosynthetic apparatus in wheat, as indicated by changes in leaf proteins. *Plant, Cell Environ.* **18**, 855–864.

Noodén, L.D., and Guiamét, J.J. (1996). Genetic control of senescence and aging in plants. In *Handbook of the Biology of Aging*, pp. 94–117. Academic Press, San Diego.

Noodén, L.D., Guiamét, J.J., and John, I. (1997). Senescence mechanisms. *Physiol. Plant.* **101**, 746–753.

Oh, S.A., Park, J.-H., Lee, G.I., Paek, K.H., Park, S.K., and Nam, H.G. (1997). Identification of three genetic loci controlling leaf senescence in *Arabidopsis thaliana*. *Plant J.* **12**, 527–535.

Okada, K., and Katoh, S. (1998). Two long-term effects of light that control the stability of proteins related to photosynthesis during senescence of rice leaves. *Plant Cell Physiol.* **39**, 394–404.

Ono, K., and Watanabe, A. (1997). Levels of endogenous sugars, transcripts of *rbc*S and *rbc*L, and of RuBisCO protein in senescing sunflower leaves. *Plant Cell Physiol.* **38**, 1032–1038.

Ono, K., Hashimoto, H., and Katoh, S. (1995). Changes in the number and size of chloroplasts during senescence of primary leaves of wheat grown under conditions. *Plant Cell Physiol.* **36**, 9–17.

Ori, N., Juarez, M.T., Jackson, D., Yamaguchi, J., Banowetz, G.M., and Hake, S. (1999). Leaf senescence is delayed in tobacco plants expressing the maize homeobox gene *knotted1* under the control of a senescence-activated promoter. *Plant Cell* **11**, 1073–1080.

Paul, M.J., and Foyer, C.H. (2001). Sink regulation of photosynthesis. *J. Exp. Bot.* **52**, 1383–1400.

Pearson, M., and Brooks, G.L. (1995). The influence of elevated CO_2 on growth and age-related changes in leaf gas exchange. *J. Exp. Bot.* **46**, 1651–1659.

Pleijel, H., Ojanpera, K., Danielsson, H., Sild, E., Gelang, J., Wallin, G., Skarby, L., and Sellden, G. (1997). Effects of leaf senescence in spring wheat— possible consequences for grain yield. *Phyton Annales Rei Botanicae* **37**, 227–232.

Reddy, M.S.S., Trivedi, P.K., Tuli, R., and Sane, P.V. (1997). Expression of chloroplastic genes during autumnal senescence in a deciduous tree *Populus deltiodes*. *Biochem. Mol. Biol. Int.* **43**, 677–684.

Roberts, D.R., Thompson, J.E., Dumbroff, E.B., Gepstein, S., and Mattoo, A.K. (1987). Differential changes in the synthesis and steady state level of thylakoid proteins during bean leaf senescence. *Plant. Mol. Biol.* **9**, 343–353.

Rosiak-Figielek, B., and Jackowski, G. (2000). The disappearance kinetics of Lhcb polypeptides during dark-induced senescence of leaves. *Aust. J. Plant Phys.* **27**, 245–251.

Schreiber, U., Schliwa, U., and Bilger, W. (1986). Continuous recording of photochemical and non-photochemical chlorophyll fluorescence quenching with a new type of modulation fluorometer. *Photosynth. Res.* **10**, 51–58.

Shinohara, K., and Murakami, A. (1996). Changes in levels of thylakoid components in chloroplasts of pine needles of different ages. *Plant Cell Physiol.* **37**, 1102–1107.

Sicher, R.C. (1998). Yellowing and photosynthetic decline of barley primary leaves in response to atmospheric CO_2 enrichment. *Physiol. Plant.* **103**, 193–200.

Sicher, R.C., and Bunce, J.A. (1997). Relationship of photosynthetic acclimation to changes of Rubisco activity in field-grown winter wheat and barley during growth in elevated carbon dioxide. *Photosynth. Res.* **52**, 27–38.

Smart, C.M., Scofield, S.R., Bevan, M.W., and Dyer, T.A. (1991). Delayed leaf senescence in tobacco plants transformed with *tmr*, a gene for cytokinin production in agrobacterium. *Plant Cell* **3**, 647–656.

Sodmergen, K.S., Kawano, S., Tano, S., and Kuroiwa, T. (1991). Degradation of chloroplast DNA in second leaves of rice (*Oryza sativa*) before leaf yellowing. *Protoplasma* **160**, 89–98.

Stern, D.B., Higgs, D.C., and Yang, J. (1997). Transcription and translation in chloroplasts. *Trends Plant Sci.* **2**, 308–315.

Synkova, H., van Loven, K., and Valcke, R. (1997). Increased content of endogenous cytokinins does not delay degradation of photosynthetic apparatus in tobacco. *Photosynthetica* **33**, 595–608.

Taylor, C.B., Bariola, P.A., Delcardayre, S.B., Raines, R.T., and Green, P.J. (1993). RNS2: a senescence-associated RNase of *Arabidopsis* that diverged from the S-RNase before speciation. *Proc. Natl. Acad. Sci. USA* **90**, 5118–5122.

Tevini, M., and Steinmüller, D. (1985). Composition and function of plastoglobuli. II. Lipid composition of leaves and plastoglobuli during beech leaf senescence. *Planta* **163**, 91–96.

Thomas, H. (1987). *Sid*: a Mendelian locus controlling thylakoid membrane disassembly in senescing leaves of *Festuca pratensis*. *Theor. Appl. Genetics* **73**, 551–555.

Thomas, H., and Howarth, C.J. (2000). Five ways to stay green. *J. Exp. Bot.* **51**, 329–337.

Thomas, H., and Huffaker, R.C. (1981). Hydrolysis of radioactively labelled ribulose-1,5-bisphosphate carboxylase by an endo-peptidase from the primary leaf of barley seedlings. *Plant Sci. Lett.* **20**, 251–262.

Thompson, J.E., Froese, C.D., Madey, E., Smith, M.D., and Hong, Y.D. (1998). Lipid metabolism during plant senescence. *Progr. Lipid Res.* **37**, 119–141.

Van Oosten, J.J., and Besford, R.T. (1995). Some relationship between the gas exchange, biochemistry and molecular biology of photosynthesis during leaf development of tomato plants after transfer to different carbon dioxide concentrations. *Plant, Cell Environ.* **18**, 1253–1266.

Vera, A., Tomás, R., Martín, M., and Sabater, B. (1990). Apparent expression of small single copy cpDNA region in senescent chloroplasts. *Plant Sci.* **72**, 63–67.

Weaver, L.M., Froehlich, J.E., and Amasino, R.M. (1999). Chloroplast-targeted ERD1 protein declines but its mRNA increases during senescence in Arabidopsis. *Plant Physiol.* **119**, 1209–1216.

Webber, A.N., Nie, G.-Y., and Long, S.P. (1994). Acclimation of photosynthetic proteins to rising atmospheric CO_2. *Photosynth. Res.* **39**, 413–425.

Wingler, A., von Schaewen, A., Leegood, R.C., Lea, P.J., and Quick, W.P. (1998). Regulation of leaf senescence by cytokinins, sugars and light. *Plant Physiol.* **116**, 329–335.

Yamasaki, T., Kudoh, T., Kamimura, Y., and Katoh, S. (1996). A vertical gradient of the chloroplast abundance among leaves of *Chenopodium album*. *Plant Cell Physiol.* **37**, 43–48.

12

How Leaves Turn Yellow: Catabolism of Chlorophyll

S. Hörtensteiner and P. Matile

I. Introduction

There is little doubt that the loss of chlorophyll (Chl) is the most frequently considered parameter symptomatic of leaf senescence. Color changes are eye-catching and Chl contents can easily be assessed. But beyond its scientific utility as a marker of senescence, the degreening of leaves is a widely appreciated natural phenomenon, especially in autumn, when the foliage of deciduous trees turns into polychromatic beauty. Not always does degreening take place at the very end of leaf development. In many species, petals or bracts run through a period of greening during growth and Chl is replaced by other pigments shortly before anthesis. Likewise, cotyledons of developing seeds are first green and then lose Chl during seed maturation. In species such as canola with cotyledons expanding after germination and developing into short-lived green leaves, the organ runs through two periods of synthesis and degradation, respectively. One of Gregor Mendel's pairs of characters, green and yellow pea seeds, refers to chlorophyll breakdown during seed development, whereby the variety with green seeds was a mutant defective in breakdown. Indeed, the catabolism of Chl has a prominent position in the history of classical genetics. For longer than a century, Chl breakdown has been known to be genetically controlled and yet, until quite recently, it has remained a mystery. Difficulties in tracking down the catabolic reactions underlying the disappearance of the green color were primarily due to lack of knowledge about the products which, concomitant with degreening, should occur or even accumulate. In senescing leaves, Chl seemed to disappear without leaving a trace and as products of breakdown were unknown, the search for the catabolic metabolism was hopeless. In the end, the breakthrough was achieved with the help of a stay-green mutant like Gregor Mendel's green seed variety of pea.

II. Porphyrin Macrocycle Cleavage

The green color of Chl is due to the conjugated system of π electrons in the tetrapyrrolic macrocycle and, hence, the loss of green color must be due to its abolition. Non-green catabolites were first observed in senescing leaves of *Festuca pratensis* by chromatographical comparison of extracts from leaves of a yellowing wild-type and a stay-green mutant, respectively (Düggelin *et al.*, 1988). A group of compounds was exclusively detectable in wild-type leaves and only transiently so during degreening. In their native form these water-soluble putative intermediary or primary catabolites were colorless but easily detected by their blue fluorescence. Radiolabeling of the pyrrole units during Chl biosynthesis in greening barley leaves and expanding cotyledons of canola, and subsequent tracing of the label during senescence confirmed that these fluorescing Chl catabolites (FCCs) were breakdown products of the porphyrin moiety of Chl (Matile *et al.*, 1992; Ginsburg and Matile, 1993). Small pools of FCCs were localized in senescing chloroplasts (subsequently refered to as gerontoplasts, Gpls) and production *in organello* was demonstrated to require the supply of either ATP or glucose-6-phosphate in the suspending medium (Schellenberg *et al.*, 1990; Ginsburg *et al.*, 1994). When isolated Gpls were lysed, one of the FCCs was produced if the medium was supplemented with ferredoxin (Fd) and cofactors required for keeping Fd in the reduced state (Schellenberg *et al.*, 1993; Ginsburg *et al.*, 1994). On reverse phase HPLC, this FCC was the most apolar catabolite and was considered to represent the primary (pFCC) product of cleavage of the porphyrinic macrocycle.

Meanwhile, the chemical constitution of pFCC has been elucidated (Mühlecker *et al.*, 1997). It appears from Fig. 12-1 that this linear tetrapyrrole is derived from Chl a and pheophorbide a (Pheide), by the oxygenolytic opening of the macrocycle at the α-methine bridge and reductions of double bonds in the β- and δ-methine bridges. Formally, the conversion of Pheide a into pFCC comprises the insertion of two atoms of oxygen and four atoms of hydrogen. The fluorescence of pFCC is due to the unsaturated methine bridge linking pyrroles C and D and the Schiff's-base structure associated with it.

Sensitive analysis of fluorescing catabolites was very helpful in the development of an *in vitro* system for the detailed study of macrocycle opening. It turned out that the cleavage of Pheide requires the presence of two enzymes, one being associated with Gpl membranes, the other originating from the stroma (Hörtensteiner *et al.*, 1995). As shown in Fig. 12-1, cleavage occurs in two steps: the membrane-bound oxygenase produces a red-colored intermediary catabolite (RCC) and, in a channeled reaction, the stroma enzyme RCC reductase (RCCR) reduces the double bond in the δ-methine bridge of RCC (Rodoni *et al.*, 1997a).

The oxygenase has an absolute specificity for Pheide a as substrate with Pheide b being a competitive inhibitor; accordingly, the enzyme has been named Pheide a oxygenase (PaO; Hörtensteiner *et al.*, 1995). This Fe-containing oxygenase requires dioxygen and reduced Fd for driving the redox cycle. PaO has been identified as a monooxygenase: the oxygen atom of

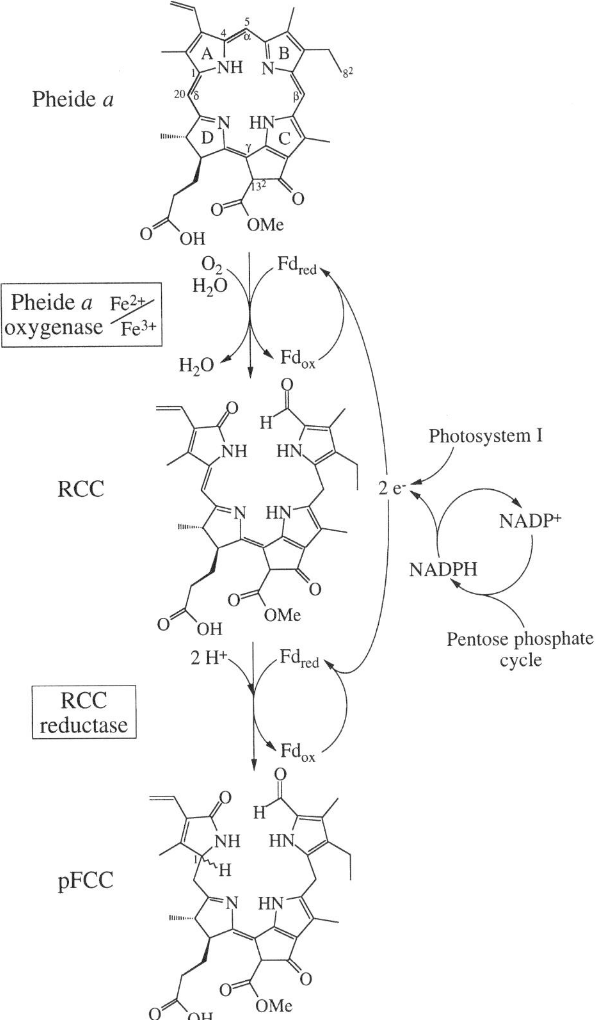

Figure 12-1. The key reaction of Chl breakdown in senescent leaves: oxygenolytic opening of the porphyrin macrocycle by the joint action of pheophorbide a oxygenase and RCC reductase. The fluorescing catabolite pFCC represents the first detectable non-green product of breakdown.

the C5-formyl group originates from dioxygen and the lactam-O probably comes from water (Hörtensteiner *et al.*, 1998a). The reaction product of PaO, RCC, is released only in the presence of RCCR which also requires reduced Fd as an electron donor (Rodoni *et al.*, 1997a).

The conversion of Pheide a into pFCC has been studied mainly in Gpls from senescing barley primary leaves and from canola cotyledons. Positive results obtained with preparations from yellowing leaves of a number of other species allow the conclusion that the PaO/RCCR system represents the common principle of macrocycle cleavage in vascular plants.

It is obvious that the action of PaO/RCCR requires perfectly intact Gpls, because Fd must be continuously reduced either in the light at photosystem I or, in the dark, through NADPH and its regeneration via the oxidative pentose-phosphate cycle (Hörtensteiner *et al.*, 1995; Rodoni *et al.*, 1997a). Activity of PaO occurs only in senescing leaves (Hörtensteiner *et al.*,

1995) and is positively correlated with rates of Chl breakdown (Rodoni *et al.*, 1998). It is also present in ripening fruits of sweet pepper (Moser and Matile, 1997) and tomato (Akhtar *et al.*, 1999). In stay-green genotypes of meadow fescue (Vicentini *et al.*, 1995a) and pea (Thomas *et al.*, 1996), deficiency of PaO activity has been identified as the genetic lesion responsible for Chl retention during leaf senescence.

RCCR has been purified from barley leaves to near homogeneity. When served with RCC as substrate, the production of pFCC *in vitro* takes place only in an anoxic milieu, suggesting that in the channeled reaction, oxygen consumption by PaO creates an oxygen-free microenvironment (Rodoni *et al.*, 1997b). It appears from the structure depicted in Fig. 12-1 that one or the other of two stereoisomers of pFCC may result from reduction of the C20/C1 double bond. These pFCCs have slightly different polarities on HPLC. Both of them are produced when RCC is enzymically reduced *in vitro* by the action of barley RCCR. However, *in vivo* the RCCR-catalyzed reaction is stereospecific, whereby the isomer is determined by the source of RCCR. Thus, the enzymes from barley and canola produce the more polar pFCC-1, whereas with RCCR from sweet pepper and other Solanaceae pFCC-2 occurs specifically (Rodoni *et al.*, 1997b). Hence, evolution of Chl breakdown has led to at least two types of RCCR in the plant kingdom. Meanwhile, some 60 species from various families have been tested (Matile group, unpublished) and the nature of RCCR turned out to be uniform within families and is possibly of some taxonomic significance.

III. Chlorophyll Catabolic Pathway

The PaO/RCCR system of macrocycle cleavage represents the core element of a catabolic machinery which, in recent years turned out to be unexpectedly complicated. First of all, catabolism via the PaO pathway does not proceed beyond the stage of linear tetrapyrroles, but the structure of final products indicates that pFCC is subjected to several modifications. The end products are still colorless but no longer fluorescing. It appears from the structures of non-fluorescing Chl catabolites (NCCs) identified so far from several species (Fig. 12-2) that the hydroxylation in $C8^2$ of the ethyl side-chain of pyrrole B is a common modification as is the tautomerization in pyrrole D which is responsible for the loss of fluorescence. Whereas the final catabolite of *Cercidiphyllum japonicum* (Curty and Engel, 1996) and *Liquidambar* species (Iturraspe and Moyano, 1995) represents a kind of basic and only NCC, in other species Chl is catabolized into several and further modified NCCs. In all three principal NCCs of canola the $C13^2$ carboxymethyl group is demethylated. One of them is esterified with malonic acid, the third one is glucosylated at the $C8^2$ hydroxyl group (Mühlecker and Kräutler, 1996). Radiolabeling of Chl in barley has yielded evidence for the occurrence of some ten NCCs in addition to the one depicted in Fig. 12-2 which has a hydroxylated vinyl side-chain at pyrrole A (Peisker *et al.*, 1990).

As to the enzymes responsible for the modifications and conjugations of pFCC, so far only a malonyltransferase of canola has been studied in some detail (Hörtensteiner, 1998). A methylesterase identified by its activity with Pheide as substrate (Shioi *et al.*, 1996a) may represent the enzyme which, in some species, is responsible for demethylation of FCCs. There is ample evidence that modifications and conjugations occur at the level of FCCs. NCCs have been localized in the vacuoles of senescent mesophyll cells (e.g. Hinder *et al.*, 1996) and the tautomerization of FCCs into NCCs appears to occur non-enzymically under the mild acidic conditions of the vacuolar sap (Hörtensteiner *et al.*, 1998b).

The occurrence in degreening tissues of trace amounts of green catabolites such as chlorophyllide and Pheide (e.g. Rise and Goldschmidt, 1990) points to the involvement in Chl

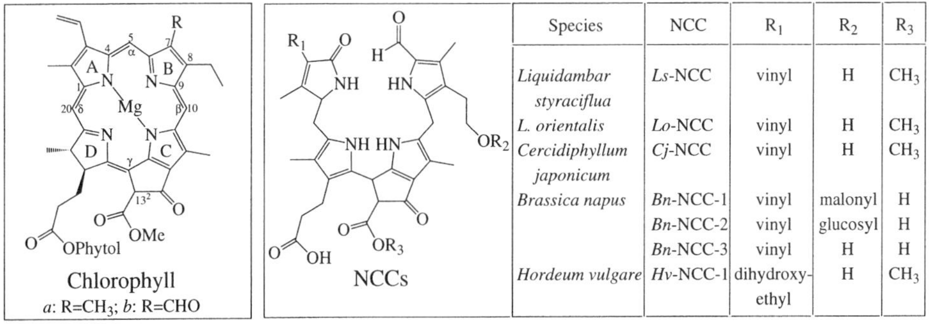

Species	NCC	R_1	R_2	R_3
Liquidambar styraciflua	*Ls*-NCC	vinyl	H	CH_3
L. orientalis	*Lo*-NCC	vinyl	H	CH_3
Cercidiphyllum japonicum	*Cj*-NCC	vinyl	H	CH_3
Brassica napus	*Bn*-NCC-1	vinyl	malonyl	H
	Bn-NCC-2	vinyl	glucosyl	H
	Bn-NCC-3	vinyl	H	H
Hordeum vulgare	*Hv*-NCC-1	dihydroxy-ethyl	H	CH_3

Figure 12-2. Chemical constitutions of final products of Chl breakdown in senescent leaves of various species. Note that these non-fluorescing catabolites (NCCs) are derivatives of Chl a. For references see text.

breakdown of enzymes acting upstream of porphyrin cleavage (Fig. 12-3). Most probably breakdown is initiated by dephytylation catalyzed by chlorophyllase, a hydrophobic protein of plastid membranes which is constitutive but functionally latent. Only upon incubation of chloroplast (Cpl) membranes in the presence of detergent or high concentrations of acetone is the endogenous Chl dephytylated (e.g. Amir-Shapira *et al.*, 1986; Matile *et al.*, 1997; Fang *et al.*, 1998). Chlorophyllase has been purified from several sources (Shioi and Sasa, 1986; Trebitsch *et al.*, 1993; Tsuchiya *et al.*, 1997) and it will probably be the first catabolic enzyme that has been tracked down to the gene (E.E. Goldschmidt, personal communication).

The existence of a dechelatase catalyzing the exchange of the central Mg^{2+} ion for two protons has been demonstrated in various ways (Bazzaz and Rebeiz, 1978; Owens and Falkowski, 1982; Ziegler *et al.*, 1988; Langmeier *et al.*, 1993). The activity is also constitutive and is associated with Cpl membranes (Vicentini *et al.*, 1995b). Very unexpectedly, it turned out upon protein fractionation that the activity is associated with a low molecular weight compound rather than with a protein (Shioi *et al.*, 1996b). The nature of this "Mg-dechelating substance" has not yet been elucidated.

All NCCs isolated so far are derivatives of Chl a. In two cases it has been demonstrated that these final catabolites represent the totality of Chl that had been present in presenescent leaves (Ginsburg and Matile, 1993; Curty and Engel, 1996). In other words, Chl b appears to be catabolized in a pathway which deals with a-forms exclusively. This is in accordance with the absolute specificity of PaO for Pheide a but implies that the breakdown of Chl b requires conversion of Chl b to Chl a, i.e. the reversion of Chl b biosynthesis from Chl a. The corresponding reduction of 7^1 formyl via 7^1 hydroxyl to 7^1 methyl has initially been discovered in etioplasts (Ito *et al.*, 1993) and is regarded as one part of a cycle by which plants adjust the Chl a/b ratio in the antenna complexes in response to light conditions (Ohtsuka *et al.*, 1997). The reduction occurs in two steps with NADPH and reduced Fd, respectively, as electron donors (Scheumann *et al.*, 1998). In the meantime, the role of Chl b to Chl a conversion has been demonstrated convincingly in senescing barley leaves: not only has the intermediary 7^1 OH Chl a been detected but also a marked transient increase of Chl b reductase activity been observed concomitant with the induction of Chl breakdown (Scheumann *et al.*, 1999). A cytoplasmic mutation in soybean, cytG, is distinguished by a high retention of Chl b during senescence (Guiamét *et al.*, 1991) and, hence, may be deficient with regard to Chl b reduction and its regulation during senescence.

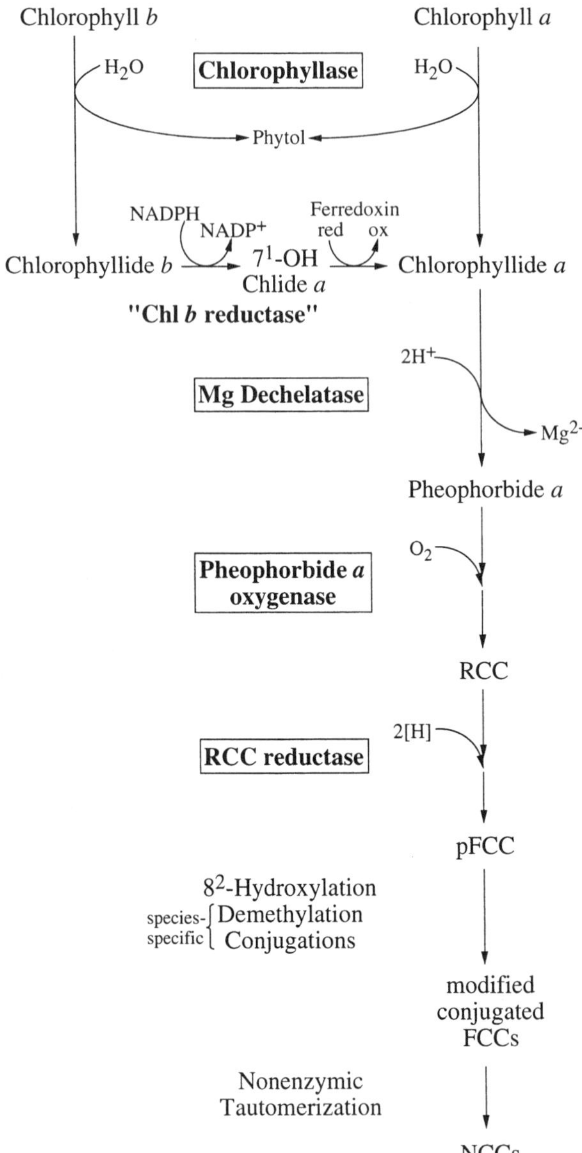

Figure 12-3. The pathway of chlorophyll degradation in senescing leaves. It is referred to as the "PaO pathway" because pheophorbide a oxygenase represents the key enzyme. Chl b to Chl a reduction is important with regard to the specificity of PaO for Pheide a as substrate.

IV. Intracellular Organization

The PaO pathway of Chl breakdown in senescing leaves extends over several subcellular compartments, starting with lipoprotein complexes in the thylakoids and ending with NCCs in the vacuoles (Fig. 12-4).

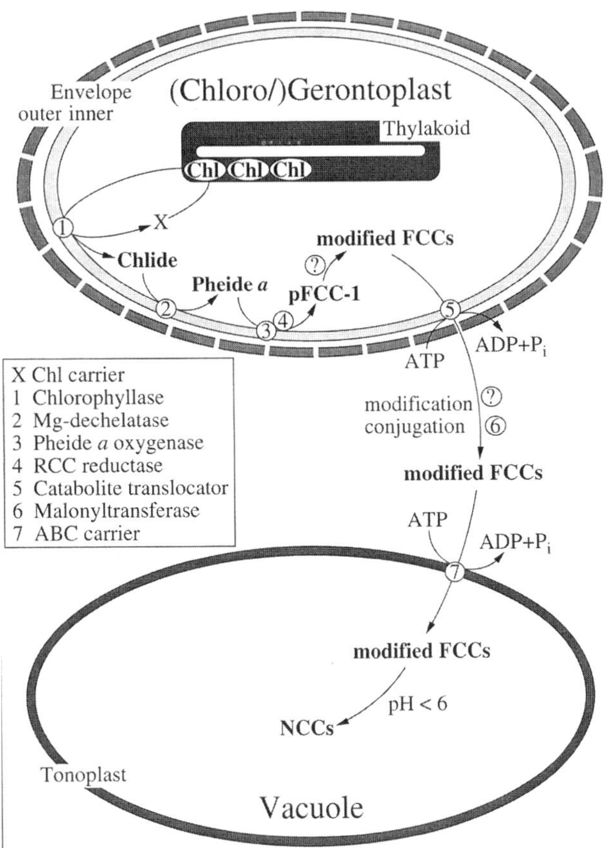

Figure 12-4. Subcellular organization of Chl breakdown in senescing mesophyll cells. For explanations and references see text.

In the course of senescence, Cpls differentiate into Gpls. Development of Gpls is controlled by nuclear genes (Yoshida, 1961). In rice leaves, chloroplast DNA has been observed to disappear even before declining Chl contents indicate the onset of senescence (Sodmergen *et al.*, 1989, 1991). Whereas thylakoids and stroma components are progressively lost during Cpl senescence, the envelope remains intact. For the understanding of Chl breakdown the intactness of the envelope is important because it has been identified as the site of chlorophyllase (Matile *et al.*, 1997), PaO (Matile and Schellenberg, 1996), and probably also Mg dechelatase. The spatial separation of Chl in the thylakoids from chlorophyllase in the inner envelope membrane seems to explain the functional latency mentioned above, but the question of how a physical contact between enzyme and substrate is achieved during breakdown has not yet been answered. In the model of Fig. 12-4, "X" marks a putative carrier which is thought to be responsible for the dismantling of Chl–apoprotein complexes in the thylakoids and transport of Chl molecules to the site of the catabolic system. This factor is likely to be a protein because, albeit active chlorophyllase is present, cytoplasmic protein synthesis is required for the initiation of dephytylation (Thomas *et al.*, 1989). A recently discovered apoprotein of a water-soluble Chl complex which is present in drought-stressed leaves of Brassicaceae (Kamimura *et al.*, 1997; Nishio and Satoh, 1997) and is able to extract Chl from thylakoidal complexes *in vitro* (Satoh *et al*., 1998) represents a promising candidate for "X". Still another possibility of intraplastidial transport of Chl is suggested by the presence of Chl in plastoglobules (Young *et al*., 1991; Picher *et al*., 1993). These lipid droplets increase during gerontoplast development in both number and size, and have been interpreted as metabolically inert accumulations of undigested lipophilic material. However, the recent discovery of a distinct set of proteins at the surface of plastoglobules (Kessler *et al*., 1999) suggests that they may play a role beyond the accumulation of phytol, carotenoids and tocopherol left behind upon the remobilization of Cpl components.

Catabolism within Gpls most probably concerns the hydroxylation of pFCC in the $C8^2$ position, possibly by the action of one of the senescence-associated cytochromes P_{450} (Buchanan-Wollaston, 1997). Modified FCCs are exported into the cytosol. In barley Gpls, ATP-dependent export, presumably catalyzed by a carrier in the inner envelope membrane has been demonstrated (Matile *et al*., 1992). Further modifications and conjugations taking place species-specifically are likely to be located in the cytosol. The final deposition of catabolites in the vacuoles has been demonstrated to be mediated by a directly energized transporter in the tonoplast (Hinder *et al*., 1996). It belongs to a group of ATP-binding cassette (ABC) transporters that are known to be responsible for the sequestration in vacuoles of xenobiotics conjugated to glutathione (Tommasini *et al*., 1998; Lu *et al*., 1998). This final step of Chl breakdown is equally reminiscent of detoxification of xenobiotics and secondary metabolites as are the modifications of FCCs which confer increased water solubility for deposition in the cell sap.

Collectively, the multifarious aspects of the PaO pathway demonstrate the necessity of intactness of subcellular organization in senescing mesophyll cells. For example, the continuous regeneration of reduced Fd implies the orderly breakdown in the thylakoids so that photosystem I remains functionally fit until Chl is completely broken down. In the stroma, all the components required for catabolism, particularly Fd, RCCR and enzymes of the pentose-phosphate cycle must either been retained or, if degraded like other stroma proteins, be replaced by newly imported proteins. Indeed, cytoplasmic protein synthesis is surprisingly active during senescence (Klerk *et al*., 1992) and the senescence-specific expression of a Fd gene in maize leaves (Smart *et al*., 1995) supports the notion that components broken down whilst required for breakdown of Chl are continuously replaced. Requirement for intactness of mitochondria appears from ATP-dependent catabolic events. In other words, the maintenance of homoeostasis in the cytoplasm is indispensable for orderly breakdown of Chl and for metabolism in senescing cells generally. The maintenance of cell viability is also demonstrated by the ability of regreening, in some species even of completely yellowed leaves (e.g. Mothes, 1960). In any case, progressive membrane deterioration, lipid peroxidation and other apparent symptoms of senescence-associated decline are very unlikely to play a role during the period of degreening and may be interpreted as post-senescence events induced upon the collapse of subcellular compartmentation.

V. Regulation of Chlorophyll Breakdown

Regulation of Chl breakdown most probably is associated with the expression of PaO. This oxygenase is the only catabolic enzyme whose activity is present exclusively in Chl-degrading tissues and is positively correlated with rates of Chl breakdown (Rodoni *et al*., 1998). In contrast, RCCR is a constitutive enzyme of plastids which is present even in Chl-free etioplasts and in roots (Rodoni *et al*., 1997a). Unfortunately, the molecular cloning of the PaO gene has so far been unsuccessful. Other catabolic enzymes such as chlorophyllase are also constitutive although activities may be modulated as in the case of *Citrus*, where chlorophyllase is synthesized in response to ethylene-induced fruit ripening (Trebitsch *et al.*, 1993). Like other senescence parameters, the loss of Chl is influenced by light and also by phytohormones. Light retards degreening in most systems but there are exceptions. One of them concerns barley leaves in which the stress hormone methyl-jasmonate causes an extremely fast breakdown of Chl in the light (Weidhase *et al.*, 1987). Light-dependent retardation of degreening is mediated by phytochrome (e.g. Rousseaux *et al.*, 1997); effects of both, photoperiod (e.g. Kar, 1986) and light dosage (Noodén *et al.*, 1996) have been demonstrated.

Perhaps the most convincing evidence for hormonal control is the inhibition of breakdown by cytokinin (e.g. Smart *et al.*, 1991; Gan and Amasino, 1995). In barley primary leaves, treatment with cytokinin causes delayed degreening and this effect is correlated with delayed appearance of PaO activity (Rodoni *et al.*, 1998).

Effects of temperature on the catabolic machinery have been studied mainly in species of crop plants in which degreening is important with regard to the value of products. In canola, for example, green seeds cause a major production problem because green oil is valued negatively. In this case, incomplete Chl breakdown during seed maturation is caused by mild freezing stress (Green *et al.*, 1998), whereas in ripening bananas, desired degreening of the peel is inhibited at tropically high temperatures (Seymour *et al.*, 1987; Blackbourn *et al.*, 1990). The parts of the catabolic system that are affected by low or high temperatures have not yet been identified.

VI. Significance of Chlorophyll Breakdown

An important lesson about the significance of Chl breakdown is taught by the various stay-green mutants. In all of them a deficiency in the catabolic system causes a marked retention not only of Chl but also of apoproteins of pigment–protein complexes in the thylakoids (Thomas, 1982; Guiamét *et al.*, 1991; Cheung *et al.*, 1993; Guiamét and Giannibelli, 1994; Bachmann *et al.*, 1994). Hence, Chl appears to protect its apoproteins from proteolysis. The apoproteins account for a substantial proportion of the total Cpl protein. In other words, stay-green genotypes pay a penalty (Hauck *et al.*, 1997) as some 30% of plastidic N is excluded from remobilization during senescence and recycling to other parts of the plant. It is not surprising, therefore, that many of the known stay-green cultivars are diazotrophic legumes which are less dependent on N recycling than are plants which depend on nitrate. The famous stay-green mutant of *Festuca pratensis*, Bf 993, has not been discovered in a natural meadow but coincidentally in a breeding program (Thomas and Stoddart, 1975). Natural stay-greens are perennials in which nutrient recycling includes microbial decomposition of leaves in litter and soil and reabsorption of nutrients by the roots. In *Alnus* (a diazotrophic genus again), Chl is not degraded in autumn and the entire senescence program is reduced to the differentiation of abscission layers (Bortlik *et al.*, 1987).

The protection of complexed Chl-binding proteins from proteolytic degradation does not explain why Chl is degraded at all. Theoretically, dismantling of the complexes with only the protein part being broken down is imaginable all the more as the four N atoms of the Chl molecule are not available for recycling. And yet, breakdown is indispensable because Chl is a photodynamically active pigment and, hence, a hazard if not protected as is the case in the thylakoids. In the complexed state, the photodynamism of Chl is compensated for by carotenoids. They represent efficient quenchers of triplet state activated Chl as well as of singlet oxygen which occur when the harvesting of quanta is not balanced by deactivation of Chl through photosynthetic electron transport. Hence, free Chl produced upon dismantling of complexes must immediately be inactivated in order to prevent photodynamic damage in the senescing cells. This vitally important process must be attributed primarily to macrocycle cleavage catalyzed by the PaO/RCCR system. In land plants, the abolition of photodynamism is particularly important as catabolites are stored intracellularly until the very end of the senescence period. It is justified, therefore, to apostrophize Chl breakdown as a kind of detoxification process.

It is understandable that in an alga such as *Chlorella*, Chl breakdown is less elaborated; enforced Chl breakdown in darkness and under heterotrophic conditions proceeds only to the level of macrocycle cleavage and the first product, RCC, is excreted into the surrounding medium (Oshio and Hase, 1969; Engel *et al.*, 1991). In the evolution of Chl catabolism, further photodynamic inactivation of RCC by RCCR probably coincides with the transition to terrestrial life and to intracellular disposal of catabolites. Indeed, the activity of RCCR has been detected in several taxa of pteridophytes including primitive ferns (Matile group, unpublished results).

VII. Chlorophyll Bleaching

Although there are good reasons to believe that in senescing leaves Chl is catabolized exclusively via the PaO pathway, alternative mechanisms of breakdown have been, and still are, considered.

Under appropriate conditions Chl is readily bleached, for example by peroxidases or, in the presence of free unsaturated fatty acids, by "Chl oxidase" associated with the photosystems (for references see Matile *et al.*, 1999). The evidence favoring peroxidative and oxidative Chl bleaching is largely based on correlations between activities and rates of degreening or on effects of inhibitors. Perhaps the strongest argument against a role of these bleaching reactions in Chl catabolism during leaf senescence is provided by results obtained with stay-green mutants: so far none of them has turned out to be deficient with regard to peroxidase or Chl oxidase.

Nevertheless, alternative mechanisms of Chl breakdown almost certainly play a role at stages of development other than senescence. Thus, turnover of Chl in developing and mature leaves (Stobart and Hendry, 1984) is not associated with the accumulation of NCCs (Matile group, unpublished results) suggesting that the PaO pathway is not involved. The stabilization of newly synthesized Chl requires the presence of apoproteins; uncomplexed Chl molecules are rapidly degraded and thereby photodynamically inactivated. Such continuous detoxification is also important with regard to photodynamically active porphyrinic intermediary products of Chl biosynthesis. Indeed, oxygen requiring enzymes catalyzing pigment bleaching *in vitro* have been demonstrated for example in developing leaves of cucumber (Whyte and Castelfranco, 1993). The properties of these enzymes as described so far are clearly different from those of the PaO/RCCR system. Unfortunately, catabolites are unknown and the work has not been pursued recently.

Convincing evidence for the existence of alternative mechanisms underlying the loss of green color is also provided by observations in the stay-green mutant of meadow fescue. During leaf senescence, the D_1 protein of the reaction center of photosystem II is retained in the mutant as are all other proteins complexed with Chl. Turnover of D_1 in the light, however, is normal suggesting that two different processes are responsible for breakdown during light-dependent turnover and senescence, respectively (Hilditch *et al.*, 1986). Whereas high Chl retention in senescing leaves of the mutant is caused by deficient PaO activity (Vicentini *et al.*, 1995a) the mechanism of bleaching in the light is unknown.

VIII. Prospect

It is easy to predict that future research on Chl breakdown will only lead to a better insight if genes encoding catabolic enzymes become available. As yet, attempts at the molecular cloning have not been overly successful. Nothing has been published so far and only personal communications announce that the genes of *Citrus* chlorophyllase (E.E. Goldschmidt) and barley RCCR (K. Wüthrich and S. Hörtensteiner) have recently been isolated. Unfortunately, the key enzyme of the catabolic pathway, PaO, has not yet been purified and tracked down to the molecular level. The gene of PaO will probably turn out to be among the few genes that are exclusively expressed in green tissues induced to catabolize Chl. In any case, the availability of the relevant genes will open marvelous opportunities to study gene expression as associated with leaf yellowing and color changes in ripening fruits. It may even provide opportunities for producing genetically engineered stay-greens which can be used for turfs and other instances where degreening during development is undesirable.

Note Added in Proof

While typesetting this manuscript the cloning of chlorophyllase (Jacob-Wilk *et al.* (1999), *Plant J.* **20**, 653–661; Tsuchiya *et al.* (1999), *Proc. Natl. Acad. Sci. USA* **96**, 15362–15367) and of RCCR (Wüthrich *et al.* (2000) *Plant J.* **21**, 189–198) has been reported. *Arabidopsis* mutants defective in RCCR develop a lesion mimic phenotype (Mach *et al.* (2001) *Proc. Natl. Acad. Sci. USA* **98**, 771–776). Several reviews on the topic have been published recently (Takamiya *et al.* (2000) *Trends Plant Sci.* **5**, 426–431; Thomas *et al.* (2001) *Adv. Bot. Res.* **35**, 1–52; Hörtensteiner and Feller (2002) *J. Exp. Bot.* **53**, 927–937).

References

Akhtar, M.S., Goldschmidt, E.E., John, I., Rodoni, S., Matile, P., and Grierson, D. (1999). Altered patterns of senescence and ripening in *gf*, a stay-green mutant of tomato. *J. Exp. Bot.* **50**, 1115–1122.

Amir-Shapira, D., Goldschmidt, E.E., and Altmann, A. (1986). Autolysis of chlorophyll in aqueous and detergent suspensions of chloroplast fragments. *Plant Sci.* **43**, 201–206.

Bachmann, A., Fernandez-Lopez, J., Ginsburg, S., Thomas, H., Bouwkamp, J.C., Solomos, T., and Matile, P. (1994). Stay-green genotypes of *Phaseolus vulgaris* L.: chloroplast proteins and chlorophyll catabolites during foliar senescence. *New Phytol.* **126**, 593–600.

Bazzaz, M.B., and Rebeiz, C.A. (1978). Chloroplast culture: the chlorophyll repair potential of mature chloroplasts incubated in a simple medium. *Biochem. Biophys. Acta* **504**, 310–323.

Blackbourn, H.D., Jeger, M.J., and John, P. (1990). Inhibition of degreening in the peel of bananas ripened at tropical temperatures. V. Chlorophyll bleaching activity measured *in vitro*. *Ann. Appl. Biol.* **117**, 175–186.

Bortlik, K., Gut, H., and Matile P. (1987). Yellowing and non-yellowing trees: a comparison of protein- and chlorophyll loss in senescent leaves. *Bot. Helv.* **97**, 323–328.

Buchanan-Wollaston, V. (1997). The molecular biology of leaf senescence. *J. Exp. Bot.* **48**, 181–199.

Cheung, A.Y., McNellis, T., and Piekos, B. (1993). Maintenance of chloroplast components during chromoplast differentiation in the tomato mutant green flesh. *Plant Physiol.* **101**, 1223–1229.

Curty, C., and Engel, N. (1996). Detection, isolation and structure elucidation of a chlorophyll a catabolite from autumnal senescent leaves of *Cercidiphyllum japonicum*. *Phytochemistry* **42**, 1531–1536.

Düggelin, T., Bortlik, K., Gut, H., Matile, P., and Thomas, H. (1988). Leaf senescence in *Festuca pratensis*: Accumulation of lipofuscin-like compounds. *Physiol. Plant.* **74**, 131–136.

Engel, N., Jenny, T.A., Mooser, V., and Gossauer, A. (1991). Chlorophyll catabolism in *Chlorella protothecoides*. Isolation and structure elucidation of a red bilin derivative. *FEBS Lett.* **293**, 131–133.

Fang, Z., Bouwkamp, J.C., and Solomos, T. (1998). Chlorophyllase activities and chlorophyll degradation during leaf senescence in non-yellowing mutant and wild type of *Phaseolus vulgaris* L. *J. Exp. Bot.* **49**, 503–510.

Gan, S., and Amasino, R.M. (1995). Inhibition of leaf senescence by autoregulated production of cytokinin. *Science* **270**, 1986–1988.

Ginsburg, S., and Matile, P. (1993). Identification of catabolites of chlorophyll-porphyrin in senescent rape cotyledons. *Plant Physiol.* **102**, 521–527.

Ginsburg, S., Schellenberg, M., and Matile, P. (1994). Cleavage of chlorophyll porphyrin. Requirement for reduced ferredoxin and oxygen. *Plant Physiol.* **105**, 545–554.

Green, B.R., Singh, S., Babic, I., Bladen, C., and Johnson-Flanagan, A.M. (1998). Relationship of chlorophyll, seed moisture and ABA levels in the maturing *Brassica napus* seed and effect of a mild freezing stress. *Physiol. Plant.* **104**, 125–133.

Guiamét, J.J., and Giannibelli, C. (1994). Inhibition of the degradation of chloroplast membranes during senescence in nuclear "stay green" mutants of soybean. *Physiol. Plant.* **91**, 395–402.

Guiamét, J.J., Schwartz, E., Pichersky, E., and Noodén, L.D. (1991). Characterization of cytoplasmic and nuclear mutations affecting chlorophyll and chlorophyll-binding proteins during senescence in soybean. *Plant Physiol.* **96**, 227–231.

Hauck, B., Gay, A.P., Macduff, J., Griffiths, C.M., and Thomas, H. (1997). Leaf senescence in a non-yellowing mutant of *Festuca pratensis*: implications of the stay-green mutation for photosynthesis, growth and nitrogen nutrition. *Plant Cell Environ.* **20**, 1007–1018.

Hilditch, P., Thomas, H., and Rogers, L.J. (1986). Two processes for the breakdown of the Q_B protein of chloroplasts. *FEBS Lett.* **208**, 313–316.

Hinder, B., Schellenberg, M., Rodoni, S., Ginsburg, S., Vogt, E., Martinoia, E., Matile, P., and Hörtensteiner, S. (1996). How plants dispose of chlorophyll catabolites. Directly energized uptake of tetrapyrrolic breakdown products into isolated vacuoles. *J. Biol. Chem.* **271**, 27233–27236.

Hörtensteiner, S. (1998). NCC malonyltransferase catalyzes the final step of chlorophyll breakdown in rape (*Brassica napus*). *Phytochemistry* **49**, 953–956.

Hörtensteiner, S., Vicentini, F., and Matile, P. (1995). Chlorophyll breakdown in senescent leaves: enzymic cleavage of phaeophorbide *a in vitro*. *New Phytol.* **129**, 237–246.

Hörtensteiner, S., Wüthrich, K.L., Matile, P., Ongania, K.-H., and Kräutler, B. (1998a). The key step in chlorophyll breakdown in higher plants. Cleavage of pheophorbide a macrocycle by a monooxygenase. *J. Biol. Chem.* **273**, 15335–15339.

Hörtensteiner, S., Wüthrich, K.L., and Matile, P. (1998b). New aspects on the catabolic pathway of chlorophyll in senescent rape cotyledons. *J. Exp. Bot.* **49**, suppl. 65.

Ito, H., Tanaka, Y., Tsuji, H., and Tanaka, A. (1993). Conversion of chlorophyll *b* to chlorophyll *a* by isolated cucumber etioplasts. *Arch. Biochem. Biophys.* **306**, 148–151.

Iturraspe, J., and Moyano, N. (1995). A new 5-formylbilinone as the major chlorophyll *a* catabolite in tree senescent leaves. *J. Org. Chem.* **60**, 6664–6665.

Kamimura, Y., Mori, T., Yamasaki, T., and Katoh, S. (1997). Isolation, properties and a possible function of a water-soluble chlorophyll a/b-protein from Brussels sprouts. *Plant Cell Physiol.* **38**, 133–138.

Kar, M. (1986). The effect of photoperiod on chlorophyll loss and lipid peroxidation in excised senescing rice leaves. *J. Plant Physiol.* **123**, 389–393.

Kessler, F., Schnell, D., and Blobel, G. (1999). Isolation of plastoglobules from pea (*Pisum sativum* L.) chloroplasts and identification of associated proteins. *Planta* **208**, 107–113.

Klerk, H., Tophof, S., and Van Loon, L.C. (1992). Synthesis of proteins during the development of the first leaf of oat (*Avena sativa*). *Physiol. Plant.* **85**, 595–605.

Langmeier, M., Ginsburg, S., and Matile, P. (1993). Chlorophyll breakdown in senescent leaves: demonstration of Mg-dechelatase activity. *Physiol. Plant.* **89**, 347–353.

Lu, Y.-P., Li, Z.-S., Drozdowicz, Y.M., Hörtensteiner, S., Martinoia, E., and Rea, P.A. (1998). AtMRP2, an Arabidopsis ATP binding cassette transporter able to transport glutathione-S-conjugates and chlorophyll catabolites: functional comparison with AtMRP1. *Plant Cell* **10**, 267–282.

Matile, P., and Schellenberg, M. (1996). The cleavage of phaeophorbide a is located in the envelope of barley gerontoplasts. *Plant Physiol. Biochem.* **34**, 55–59.

Matile, P., Schellenberg, M., and Peisker, C. (1992). Production and release of a chlorophyll catabolite in isolated senescent chloroplasts. *Planta* **187**, 230–235.

Matile, P., Schellenberg, M., and Vicentini, F. (1997). Localization of chlorophyllase in the chloroplast envelope. *Planta* **201**, 96–99.

Matile, P., Hörtensteiner, S., and Thomas, H. (1999). Chlorophyll degradation. *Annu. Rev. Plant Physiol. Plant Mol. Biol.* **50**, 67–95.

Moser, D., and Matile, P. (1997). Chlorophyll breakdown in ripening fruits of *Capsicum annuum*. *J. Plant Physiol.* **150**, 759–761.

Mothes, K. (1960). Über das Altern der Blätter und die Möglichkeit ihrer Wiederverjüngung. *Naturwissenschaften* **47**, 337–351.

Mühlecker, W., and Kräutler, B. (1996). Breakdown of chlorophyll: constitution of nonfluorescing chlorophyll-catabolites from senescent cotyledons of the dicot rape. *Plant Physiol. Biochem.* **34**, 61–75.

Mühlecker, W., Ongania, K.-H., Kräutler, B., Matile, P., and Hörtensteiner, S. (1997). Tracking down chlorophyll breakdown in plants: elucidation of the constitution of a fluorescent chlorophyll catabolite. *Angew. Chem. Int. Ed. Engl.* **36**, 401–404.

Nishio, N., and Satoh, H. (1997). Water-soluble chlorophyll protein in cauliflower may be identical to BnD22, a drought-induced, 22 kilodalton protein in rapeseed. *Plant Physiol.* **115**, 841–846.

Noodén, L.D., Hillsberg, J.W., and Schneider, M.J. (1996). Induction of leaf senescence in *Arabidopsis thaliana* by long days through a light-dosage effect. *Physiol. Plant.* **96**, 491–495.

Ohtsuka, T., Ito, H., and Tanaka, A. (1997). Conversion of chlorophyll *b* to chlorophyll *a* and the assembly of chlorophyll with apoproteins by isolated chloroplasts. *Plant Physiol.* **113**, 137–147.

Oshio, Y., and Hase, E. (1969). Studies on red pigments excreted by cells of *Chlorella protothecoides* during the process of bleaching induced by glucose or acetate. I. Chemical properties of the red pigments. *Plant Cell Physiol.* **10**, 41–49.

Owens, T.G., and Falkowski, P.G. (1982). Enzymatic degradation of chlorophyll *a* by marine phytoplankton *in vitro*. *Phytochemistry* **21**, 979–984.

Peisker, C., Thomas, H., Keller, F., and Matile, P. (1990). Radiolabelling of chlorophyll for studies on catabolism. *J. Plant Physiol.* **136**, 544–549.

Picher, M., Grenier, G., Purcell, M., Proteau, L., and Beaumont, G. (1993). Isolation and purification of intralamellar vesicles from *Lemna minor* L. chloroplasts. *New Phytol.* **123**, 657–663.

Rise, M., and Goldschmidt, E.E. (1990). Occurrence of chlorophyllides in developing, light grown leaves of several plant species. *Plant Sci.* **71**, 147–151.

Rodoni, S., Mühlecker, W., Anderl, M., Kräutler, B., Moser, D., Thomas, H., Matile, P., and Hörtensteiner, S. (1997a). Chlorophyll breakdown in senescent chloroplasts. Cleavage of pheophorbide *a* in two enzymic steps. *Plant Physiol.* **115**, 669–676.

Rodoni, S., Vicentini, F., Schellenberg, M., Matile, P., and Hörtensteiner, S. (1997b). Partial purification and characterization of RCC reductase, a stroma protein involved in chlorophyll breakdown. *Plant Physiol.* **115**, 677–682.

Rodoni, S., Schellenberg, M., and Matile, P. (1998). Chlorophyll breakdown in senescing barley leaves as correlated with phaeophorbide *a* oxygenase activity. *J. Plant Physiol.* **152**, 139–144.

Rousseaux, M.C., Ballaré, C.L., Jordan, E.T., and Vierstra, R.D. (1997). Directed overexpression of PHY A locally suppresses stem elongation and leaf senescence responses to far-red radiation. *Plant Cell Environ.* **20**, 1551–1558.

Satoh, H., Nakyama, K., and Okada, M. (1998). Molecular cloning and functional expression of a water-soluble chlorophyll-protein, a putative carrier of chlorophyll molecules in cauliflower. *J. Biol. Chem.* **273**, 30568–30575.

Schellenberg, M., Matile, P., and Thomas, H. (1990). Breakdown of chlorophyll in chloroplasts of senescent barley leaves depends on ATP. *J. Plant Physiol.* **136**, 564–568.

Schellenberg, M., Matile, P., and Thomas, H. (1993). Production of a presumptive chlorophyll catabolite *in vitro*: requirement for reduced ferredoxin. *Planta* 191, 417–420.

Scheumann, V., Schoch, S., and Rüdiger, W. (1998). Chlorophyll *a* formation in the chlorophyll *b* reductase reaction requires reduced ferredoxin. *J. Biol. Chem.* **273**, 35102–35108.

Scheumann, V., Schoch, S., and Rüdiger, W. (1999). Chlorophyll *b* reduction during senescence of barley seedlings. *Planta* **209**, 364–370.

Seymour, G.B., Thompson, A.K., and John, P. (1987). Inhibition of degreening in the peel of bananas ripened at tropical temperatures. I. Effect of high temperature on changes in the pulp and peel during ripening. *Ann. Appl. Biol.* **110**, 145–151.

Shioi, Y., and Sasa, T. (1986). Purification of solublized chlorophyllase from *Chlorella protothecoides*. *Meth. Enzymol.* **123**, 421–427.

Shioi, Y., Watanabe, K., and Takamiya, K. (1996a). Enzymatic conversion of pheophorbide *a* to the precursor of pyropheophorbide *a* in leaves of *Chenopodium album*. *Plant Cell Physiol.* **37**, 1143–1149.

Shioi, Y., Tomita, N., Tsuchiya, T., and Takamiya, K. (1996b). Conversion of chlorophyllide to pheophorbide by Mg-dechelating substance in extracts of *Chenopodium album*. *Plant Physiol. Biochem.* **34**, 41–47.

Smart, C.M., Scofield, S.R., Bevan, M.W., and Dyer, T.A. (1991). Delayed leaf senescence in tobacco plants transformed with tmr, a gene for cytokinin production in *Agrobacterium*. *Plant Cell* **3**, 647–656.

Smart, C.M., Hosken, S.E., Thomas, H., Greaves, J.A., Blair, B.G., and Schuch, W. (1995). The timing of maize leaf senescence and characterization of senescence-related cDNAs. *Physiol. Plant.* **93**, 673–682.

Sodmergen, S., Kawano, S., Tano, S., and Kuroiwa, T. (1989). Preferential digestion of chloroplast nuclei (nucleoids) during senescence of the coleoptile of *Oryza sativa*. *Protoplasma* **152**, 65–68.

Sodmergen, S., Kawano, S., Tano, S., and Kuroiwa, T. (1991). Degradation of chloroplast DNA in second leaves of rice (*Oryza sativa*) before leaf yellowing. *Protoplasma* **160**, 89–98.

Stobart, A.K., and Hendry, G.A.F. (1984). Chlorophyll turnover in greening wheat leaves. *Phytochemistry* **23**, 27–30.

Thomas, H. (1982). Leaf senescence in a non-yellowing mutant of *Festuca pratensis*. I. Chloroplast membrane polypeptides. *Planta* **154**, 212–218.

Thomas, H., and Stoddart, J.L. (1975). Separation of chlorophyll degradation from other senescence processes in leaves of a mutant genotype of meadow fescue (*Festuca pratensis*). *Plant Physiol.* **56**, 438–441.

Thomas, H., Schellenberg, M., Vicentini, F., and Matile, P. (1996). Gregor Mendel's green and yellow pea seeds. *Bot. Acta* **109**, 3–4.

Thomas, H., Bortlik, K., Rentsch, D., Schellenberg, M., and Matile, P. (1989). Catabolism of chlorophyll *in vivo*: significance of polar chlorophyll catabolites in a non-yellowing senescence mutant of *Festuca pratensis* Huds. *New Phytol.* **111**, 3–8.

Tommasini, R., Vogt, E., Fromenteau, M., Hörtensteiner, S., Matile, P., Amrhein, N., and Martinoia, E. (1998). An ABC-transporter of *Arabidopsis thaliana* has both glutathione-conjugate and chlorophyll catabolite transport activity. *Plant J.* **13**, 773–780.

Trebitsch, T., Goldschmidt, E.E., and Riov, J. (1993). Ethylene induces *de novo* synthesis of a chlorophyll degrading enzyme in *Citrus* fruit peel. *Proc. Natl. Acad. Sci. USA* **90**, 9441–9445.

Tsuchiya, T., Ohta, H., Masuda, T., Mikami, B., Kita, N., Shioi, Y., and Takamiya, K. (1997). Purification and characterization of two isozymes of chlorophyllase from mature leaves of *Chenopodium album*. *Plant Cell Physiol.* **38**, 1026–1031.

Vicentini, F., Hörtensteiner, S., Schellenberg, M., Thomas, H., and Matile, P. (1995a). Chlorophyll breakdown in senescent leaves: Identification of the lesion in a stay-green genotype of *Festuca pratensis*. *New Phytol.* **129**, 247–252.

Vicentini, F., Iten, F., and Matile, P. (1995b). Development of an assay for Mg-dechelatase of oilseed rape cotyledons, using chlorophyllin as the substrate. *Physiol. Plant.* **94**, 57–63.

Weidhase, R.A., Lehmann, J., Kramell, H., Semdner, G., and Parthier, B. (1987). Degradation of ribulose-1,5-bisphosphate carboxylase and chlorophyll in senescing barley leaf segments triggered by jasmonic acid methylester and counteraction by cytokinin. *Physiol. Plant.* **69**, 161–166.

Whyte, B.J., and Castelfranco, P.A. (1993). Breakdown of thylakoid pigments by soluble proteins of developing chloroplasts. *Biochem. J.* **290**, 361–367.

Yoshida, Y. (1961). Nuclear control of chloroplast activity in *Elodea* leaf cells. *Protoplasma* **54**, 476–492.

Young, A.J., Wellings, R., and Britton, G. (1991). The fate of chloroplast pigments during senescence of primary leaves of *Hordeum vulgare* and *Avena sativum*. *J. Plant Physiol.* **137**, 701–705.

Ziegler, R., Blaheta, A., Guha, N., and Schönegge, B. (1988) Enzymatic formation of pheophorbide and pyropheophorbide during chlorophyll degradation in a mutant of *Chlorella fusca* Shihira et Kraus. *J. Plant Physiol.* **132**, 327–332.

13

Free Radicals and Oxidative Stress

Jennifer M. Mach and Jean T. Greenberg

I. Introduction

Oxidative stress has challenged organisms since the beginning of aerobic metabolism caused an increase in the partial pressure of oxygen in the atmosphere, approximately 2.5×10^9 years ago (Reinbothe *et al.*, 1996). Since then, organisms have adapted to detoxify the reactive oxygen species produced as byproducts of metabolism or as products of environmental stress, sensing the presence of reactive oxygen species, and activating specific signal transduction pathways in response. Many organisms have also harnessed reactive oxygen species for use as signaling and defensive molecules. Indeed, both plants and animals produce reactive oxygen species as part of their defenses against pathogen attack. Thus, reactive oxygen species are a toxic metabolic byproduct that must be detoxified, an important group of signaling molecules, and a weapon in the arsenal against pathogen attack. In this review, we will examine the mechanisms of generating and detoxifying reactive oxygen species, and the roles of reactive oxygen in signal transduction and programmed cell death in plants.

II. The Molecules that Cause Oxidative Stress

Oxidative stress is caused by a group of extremely reactive molecules comprising free radicals of oxygen, singlet oxygen, and hydrogen peroxide; these are collectively known as active oxygen species (AOS). Singlet oxygen ($^1O_2$) is formed when the two unpaired

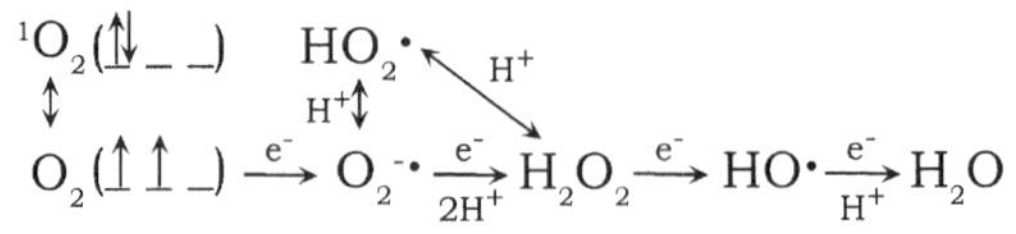

$$
\begin{array}{lllll}
{}^{1}O_2(\uparrow\downarrow\,\_\,\_) & HO_2^{\bullet} \searrow^{H^+} & & & \\
\updownarrow & H^+\updownarrow & & & \\
O_2(\uparrow\,\uparrow\,\_) \xrightarrow{e^-} & O_2^{-\bullet} \xrightarrow[2H^+]{e^-} & H_2O_2 \xrightarrow{e^-} & HO^{\bullet} \xrightarrow[H^+]{e^-} & H_2O
\end{array}
$$

Figure 13-1. Formation of active oxygen species.

electrons of molecular oxygen become paired, thus increasing the reactivity by making O_2 amenable to donation of electrons in different spin configurations. Addition of electrons to molecular oxygen forms free radicals [see Fig. 13-1; also reviewed in Halliwell and Gutteridge (1985), Cadenas (1989) and Wojtaszek (1997)]. Transfer of a single electron, often catalyzed by transition metals such as Fe and Cu, to O_2 forms the superoxide radical, $O_2^{-\bullet}$. In aqueous solutions, $O_2^{-\bullet}$ disproportionates to hydrogen peroxide, H_2O_2. The decomposition of H_2O_2 can form the extremely reactive hydroxyl radical, $^{\bullet}OH$; indeed, this is one of the main sources of H_2O_2 toxicity (Cadenas, 1989).

The reactivity of free radicals causes them to either abstract H^+ from or transfer electrons to other molecules, thus forming more free radicals. AOS have different reactivities and different abilities to move throughout the cell. For example, the hydroxyl radical is extremely reactive, but singlet oxygen is less reactive (Halliwell and Gutteridge, 1985). Of all the AOS, H_2O_2 is the least reactive, but is uncharged and thus can move through membranes. The reactivity and mobility of superoxide varies depending on the pH; the protonated species reacts faster, can move through membranes, and dismutes to H_2O_2 faster than the unprotonated form. $O_2^{-\bullet}$ is more likely to be protonated in the apoplast, which has a pH around 5 to 6, than in the neutral cytoplasm (Felle, 1998; Green and Hill, 1984; Mühling *et al.*, 1995). Thus, H_2O_2 generated at any location in the cell can move through membranes to other locations, but the subcellular location of $O_2^{-\bullet}$ generation influences its reactivity and mobility.

III. Generating Active Oxygen Species

Free radicals and reactive oxygen compounds form during normal aerobic metabolism, environmental stress, senescence, and the plant defense response to pathogen attack. For example, an oxidative burst in response to pathogen attack is produced in the apoplast (Wojtaszek, 1997). Metabolic sources of AOS include the chloroplast, the mitochondrion, and the peroxisome (Del Río *et al.*, 1992); these organelles contain antioxidant enzymes to remove the AOS produced there (Foyer and Halliwell, 1976). The defenses against AOS target each of the different reactive species and localize to the subcellular regions of AOS formation.

A. Metabolic Byproducts

1. Photosynthesis and respiration

The light reactions of photosynthesis involve a long chain of electron transfers, beginning with the removal of an electron from a water molecule and eventually depositing the electron with $NADP^+$. Molecular oxygen can compete with $NADP^+$ to receive the electron from ferredoxin. Formation of $O_2^{-\bullet}$ in the chloroplast stroma by this reaction is termed the Mehler reaction (Mehler, 1951) and is a major source of superoxide in the plant cell (Polle, 1996).

Similarly, mitochondrial electron transfers can produce superoxide when molecular oxygen competes with other electron acceptors. Also, chlorophyll and other porphyrins can interact with oxygen by triplet–triplet interchange, in which an electron donated from oxygen returns with its spin reversed, forming singlet oxygen. Thus, in an aerobic environment, reactions essential for the life of the cell also produce AOS as a byproduct.

2. Peroxisomes

Peroxisomes are small, membrane-bound organelles that conduct multiple oxidative functions, including photorespiration and the glyoxylate cycle, depending on tissue location (Del Rio *et al.*, 1992). Leaf peroxisomes contain the enzymes to conduct the oxidative photosynthetic carbon cycle of photorespiration, as well as many antioxidant enzymes to detoxify the active oxygen produced. Multiple AOS are produced in peroxisomes, including H_2O_2 from flavin oxidases and photorespiration, and $O_2^{-\bullet}$ from both xanthine oxidases in the matrix and NADH-dependent electron transport components in the membrane (Del Río *et al.*, 1992). One of the major routes of reactive oxygen formation in the peroxisome is transfer of electrons from NADH to a flavoprotein, cytochrome b5. Oxygen can compete with the normal electron acceptors, thus forming superoxide radicals.

B. Abiotic Stresses

Oxidative stress plays a major role in environmental stresses such as chilling, drought, wounding, and exposure to sulfur dioxide, heavy metals, ozone, or excess light (reviewed in (Noctor and Foyer, 1998). AOS may be generated by different mechanisms during the stress, but cause the induction of common adaptive responses, such as increased production of antioxidant enzymes. The survival of abiotic stress can depend, not on the severity of the oxidative stress, but on the alacrity of the antioxidant response. For example, in chilling stress, acclimation at 14°C induces increases in antioxidant enzyme levels and allows maize seedlings to survive subsequent exposure to the usually lethal temperature of 4°C (Prasad, 1996). Both the low and the moderate temperature cause a pulse of H_2O_2, but at the lower temperature, antioxidant defenses such as catalase and ascorbate peroxidase do not increase (Prasad *et al.*, 1994). Thus, the difference between a survivable and a lethal temperature may be the induction of protective enzymes.

Excess light causes oxidative stress in leaves when the ability of photosystem II (PSII) to use energy from the light-harvesting complex (LHC) is overloaded and electrons transfer to oxygen. AOS generated in this process can damage PSII and the LHC, causing photoinhibition of photosynthesis. The plant's response to this stress is very quick: plants grown in low light and exposed to excess light show expression of AOS scavengers such as ascorbate peroxidases (APX1 and APX2) after less than 15 minutes of stress (Karpinski *et al.*, 1997). Experiments with an APX2–luciferase fusion show that H_2O_2 is important for the response to excess excitation energy: vacuum infiltration of catalase into leaves to remove H_2O_2 blocks the induction of APX2 by excess light (Karpinski *et al.*, 1999). Treatment with exogenous H_2O_2 also induced APX2, but not to the same levels as excess light, which may reflect differences in flux between H_2O_2 applied to the plant and H_2O_2 produced by the plant, or an additional signal that acts synergistically within the plant. Thus, excess light activates the expression of antioxidant genes through H_2O_2.

The pollutant ozone (O_3) causes ozonolysis of lipids and the formation of radicals such as superoxide and the hydroxyl radical (Mudd, 1996). Ozone reacts primarily with components of the extracellular space (Van Camp *et al.*, 1994). Comparison of sensitive and insensitive cultivars of tobacco may have uncovered a different mechanism for response to abiotic stress: amplification of AOS production. Exposure of the sensitive cultivar to 150 nL L^{-1} O_3 caused the induction of cell death (Schrauder *et al.*, 1998). In both the sensitive and insensitive cultivars, ozone exposure triggered an increase in AOS in the apoplast and similar induction of antioxidant defences; only in the sensitive cultivar did the initial exposure lead to prolonged production of AOS in the apoplast and cell death (Schrauder *et al.*, 1998).

C. Oxidative Burst

As part of the plant disease resistance response, plant cells release $O_2^{-\bullet}$ and H_2O_2 in response to recognition of a specific pathogen elicitor. The reactive oxygen species produced may act directly to kill pathogens; they also induce several responses in the plant, including cell wall protein crosslinking, programmed cell death (hypersensitive response, or HR), and expression of defense genes in the surrounding area. These are discussed in the next section.

Evidence exists for two mechanisms to generate the oxidative burst. First, Bolwell *et al.* (1995) propose H_2O_2 generation by a pH-dependent, cell wall linked peroxidase. In french bean cells treated with purified elicitor, they find no detectable generation of superoxide, but do find KCN-inhibitable generation of H_2O_2. One caution in the interpretation of these results is that KCN can itself act as a H_2O_2 scavenger (Barcelo, 1998). Even disregarding inhibitor studies, the generation of H_2O_2 by a cell wall linked peroxidase whose activity is dependent on an increase in pH is supported by a body of evidence (Bestwick *et al.*, 1999). For example, in lettuce cells undergoing HR, H_2O_2 accumulates within the cell wall proximal to attached *Pseudomonas syringae* (Bestwick *et al.*, 1997).

The second mechanism for generation of AOS is generation of superoxide from a plasma-membrane NADPH oxidase, similar to the phagocyte oxidative burst. NADPH-dependent superoxide synthesis can be observed in purified plasma membrane from cultured rose cells (Murphy and Auh, 1996). Moreover, diphenylene iodonium (DPI), a suicide substrate of mammalian NADPH oxidase (Doussiere *et al.*, 1999), inhibits the production of H_2O_2 in soybean suspension cells treated with purified elicitor (Levine *et al.*, 1994). As with many inhibitor studies, others (Bestwick *et al.*, 1999) caution in the interpretation of these results, as DPI is not specific for NADPH oxidase and may be acting on a different target in plants. More evidence comes from the isolation, in both rice and *Arabidopsis*, of a protein similar to $gp91^{phox}$, a subunit of the mammalian NADPH oxidase (Keller *et al.*, 1998).

H_2O_2 is likely produced by both mechanisms, but to different degrees in different systems (Bestwick *et al.*, 1999). Moreover, both systems may operate synergistically in the oxidative burst, with the H^+ consumption required for dismutation of superoxide causing an increase in extracellular pH (Wojtaszek, 1997). This pH increase, along with that produced by open ion channels, activates the pH-dependent peroxidase on the cell wall. A further interesting suggestion is that an ATP-dependent proton pump, activated more slowly than the oxidative burst, may acidify the extracellular space and terminate production of reactive oxygen.

D. Photosensitizing Compounds

Many compounds can absorb light and emit active electrons, producing AOS; however, intriguing evidence indicates that some of these molecules may act as specific signals, rather than simply as toxic molecules. For a review of phototoxicity, refer to Tyystjärvi (this volume, Chapter 18). One striking example of a photoactive molecule is chlorophyll. Chlorophyll precursors and breakdown products also absorb light, but have no productive outlet for high-energy electrons. Mutants in chlorophyll synthesis enzymes cause the accumulation of phototoxic porphyrin compounds and a light-induced cell death phenotype (Hu *et al.*, 1998). A mutation red chlorophyll catabolite (RCC) reductase, an enzyme in the chlorophyll breakdown pathway, causes the light-dependent accelerated cell death phenotype of *acd2* mutants (Mach *et al.*, 2001). Interestingly, overexpression of the ACD2 protein, which is predicted to reduce the amount of chlorophyll breakdown products, makes *Arabidopsis* tolerant to bacterial infection: the bacteria grow, but the plant shows reduced cell death and other disease symptoms. Thus, chlorophyll breakdown products may enhance or activate cell death in disease. However, not all cases of porphyrin accumulation cause cell death; plants that accumulate the porphyrin pheophorbide a, the precursor to RCC, show delayed senescence, remaining green longer than usual (Thomas, 1987; Thomas and Matile, 1988; Thomas *et al.*, 1996).

IV. Defending Against Oxidative Stress

A. Cellular Damage by Active Oxygen Species

AOS cause protein oxidation, DNA damage, and lipid peroxidation. Oxidization of proteins occurs through multiple reactions, including side chain alterations and backbone cleavage, and disrupts protein structure, causing denaturation, aggregation, and susceptibility to degradation (reviewed in Dean *et al.*, 1997). Oxidative stress causes increased frequencies of mutation and illegitimate recombination in purple photosynthetic bacteria (Ouchane *et al.*, 1997). Lipid peroxidation can result in membrane damage. Thus, the extreme reactivity of AOS causes damage to all the structures of the cell; to protect against this damage, plants use both enzymes and small molecule antioxidants in all subcellular compartments.

B. Enzymatic Detoxification

Because of the variety of AOS types and subcellular locations, many enzymes are required to remove active oxygen species (Noctor and Foyer, 1998) (diagrammed in Fig. 13-2). Superoxide dismutase (SOD) converts superoxide into H_2O_2, which subsequently must be detoxified. Two of the major enzymatic scavengers of H_2O_2 are catalase (CAT) and ascorbate peroxidase (APX). Both enzymes remove H_2O_2, but catalyze the reaction differently and are localized to different subcellular compartments. Catalase is primarily found in peroxisomes; APX is primarily found in chloroplasts and the cytosol, indicating that they may have non-overlapping functions.

Antioxidant enzymes are essential for survival of oxidative stress. For example, under high light conditions, catalase-deficient plants develop bleached areas of cell death on their leaves and accumulate high levels of oxidized glutathione (Willekens *et al.*, 1997). Catalase-deficient leaves are also very sensitive to exogenous H_2O_2, showing photobleaching and membrane damage.

C. Small Molecule Antioxidants

Much of the control of oxidative stress in the plant cell comes from large quantities of small molecule oxidants, namely ascorbic acid (vitamin C) and glutathione (reviewed in

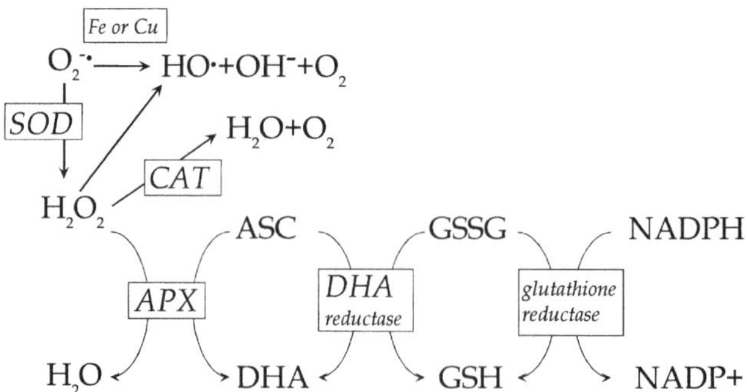

Figure 13-2. Detoxification of active oxygen species.

(Noctor and Foyer, 1998). Glutathione and ascorbate can react directly with free radicals, or can be used by antioxidant enzymes as a source of reducing power (see Fig. 13-2). Glutathione (γGlu-Cys-Gly) exists as either a reduced form (GSH) or as an oxidized form in which two molecules are joined through a disulfide bond between the cysteines (GSSG). GSH reacts readily with free radicals, forming a glutathione radical ($GS^{\bullet}$) that can react in pairs to form GSSG. The oxidized form can then be reduced to GSH by the enzyme glutathione reductase, expending NADPH and H^+ (Foyer and Halliwell, 1976).

Ascorbate also acts as a reducing agent and is regenerated by GSH oxidation. H^+ donation by ascorbate forms the monodehydroxyascorbate radical, which then disproportionates to form dehydroxyascorbate (DHA). This can then be regenerated to ascorbic acid by the enzyme dehydroascorbate reductase, which uses GSH. Thus, ascorbate and glutathione can reduce each other, further buffering the antioxidant capacity of the cell in what is called the ascorbate–glutathione cycle (Foyer and Halliwell, 1976). The small molecule antioxidants can react with active oxygen species, detoxify free radicals, and be regenerated into their reducing forms.

Although glutathione and ascorbic acid are major antioxidants, other molecules also act to protect the plant from oxidative stress, especially in specialized cellular environments. Alpha tocopherol, also known as vitamin E, can act in non-aqueous environments to detoxify lipid radicals. The vitamin E radical formed subsequently reacts with ascorbic acid, thus removing radicals from the membrane and reducing lipid peroxidation. Carotenoids act to prevent the activated triplet state of chlorophyll from reacting with oxygen to form singlet oxygen. Carotenoid-deficient mutants of a purple photosynthetic mutant showed drastic reduction of survival and increased frequency of mutations and illegitimate recombination when grown in aerobic photosynthetic conditions, as compared to anaerobic conditions (Ouchane *et al.*, 1997).

The antioxidant capacity of the cell is critical for the survival of environmental stress. For example, the ascorbate-deficient *vtc* (Vitamin C) mutant of *Arabidopsis* (Conklin *et al.*, 1996, 1997) accumulates only 30% of the normal ascorbate concentration and shows increased sensitivity to ozone stress, sulfur dioxide and UVB irradiation. Thus, small molecule antioxidants are just as critical as enzymatic antioxidants.

The localization of antioxidant defenses follows the localization of sources of AOS. For example, the chloroplast contains many antioxidants, including glutathione and glutathione reductase (Foyer and Halliwell, 1976). Also the apoplast contains many antioxidants that are modulated during the oxidative burst; these antioxidants may prevent killing of the cell by the extreme amounts of AOS produced. For example, in barley inoculated with powdery mildew, whole-leaf SOD did not change, but apoplastic SOD activity increased by 150–300% (Vanacker *et al.*, 1998).

V. Active Oxygen in Resistance and Senescence

Plants have harnessed AOS in their defense against pathogens. The oxidative burst in plants may have several functions, including the action of reactive oxygen species in killing pathogens, in crosslinking cell wall components, and in signaling plant cells to induce endogenous defenses.

A. Crosslinking of Cell Wall Proteins

One of the defenses mediated by H_2O_2 is crosslinking of proteins of the cell wall. In cultured soybean cells treated with fungal elicitors, a glycine-rich cell wall protein with simple repeats (Val-Tyr-Lys-Pro-Pro), disappears from the SDS-extractable fraction on polyacrylamide gels, but is still detectable by *in situ* immunofluorescence (Bradley *et al.*, 1992). This activity can also be stimulated by H_2O_2, and is blocked by the addition of catalase or ascorbate. Crosslinking toughens cell walls and may limit pathogen ingress; indeed, elicitor-treated soybean culture cells are highly resistant to protoplasting by microbial cellulysin and pectolyase (Brisson *et al.*, 1994).

B. Triggering of Programmed Cell Death

AOS from the oxidative burst trigger programmed cell death in the immediate vicinity of the infection and trigger gene expression in a wider area and at lower concentrations. In soybean suspension cultured cells, cell death was induced at 6–8 mM exogenous H_2O_2 and glutathione S-transferase (GST) expression was induced at 1–2 mM (Levine *et al.*, 1994), indicating that reactive oxygen can act as a long-range signal. Moreover, H_2O_2 acts synergistically with NO (Delledone *et al.*, 1998) in the induction of cell death. It is unclear whether this PCD is triggered by overwhelming of the cell's antioxidative defenses or by activation of a specific genetic program, as H_2O_2 and NO are both oxidative stressors. Suggestively, a brief pulse of H_2O_2 is all that is required to trigger cell death, and additional pulses do not potentiate the cell death response (Levine *et al.*, 1994).

C. Lesion Mimic Mutants

Mutations with a phenotype that resembles a hypersensitive response may help elucidate the genetic control of cell death and the relationship between AOS and cell death. These mutants, called lesion mimics, have been isolated in many plants [*Les* and other mutants of maize, and *accelerated cell death*, or *acd* mutants, and *lesions simulating disease resistance*, or *lsd* mutants of *Arabidopsis* (Dietrich *et al.*, 1994; Greenberg and Ausubel, 1993; Greenberg *et al.*, 1994; Walbot *et al.*, 1983)]. Lesion mimic mutants can be roughly classified by whether the patches of dead cells spread or not; this may reflect an effect on two different processes, lesion initiation and lesion propagation (Walbot *et al.*, 1983).

The connection between AOS and the lesion mimic phenotype has been explored in the case of *lsd1*. The *lsd1* phenotype includes spreading patches of cell death that appear spontaneously in mutant plants under restrictive conditions; these lesions can be induced under permissive conditions by infiltration of a superoxide-generating system (Jabs *et al.*, 1996). Also, superoxide accumulates at the border of spreading lesions. Interestingly, infiltration of a H_2O_2-generating system does not induce the lesions. lsd1 mutants do not induce copper–zinc superoxide dismutase properly in response to the signal molecule salicylic acid (Kliebenstein *et al.*, 1999). The inability to detoxify superoxide likely triggers cell death.

Another intriguing gene with a lesion mimic phenotype is the *lethal leaf spot* (*lls1*) gene of maize, which encodes a protein similar to aromatic ring-hydroxylating dioxygenases (Gray *et al.*, 1997). LLS1 may enzymatically remove an aromatic compound; another possibility is that the iron centers act to sense the redox state. Indeed, the bacterial regulator SoxR contains iron–sulfur clusters whose redox state regulates the activity of the protein (Ding and Demple, 1997). A more complete understanding of the relationship between AOS and the lesion mimic phenotype awaits the development of better tools for visualizing and manipulating AOS, so that both the localization and the timing of AOS production can be examined and perturbed in these mutants.

D. Systemic Acquired Resistance and Salicylic Acid

The formation of H_2O_2 in the epicenter of the infection induces gene expression in the surrounding area and in distal tissues, such as uninfected leaves on the same plant. The distal leaves will subsequently be resistant to pathogen challenge, in a process known as systemic acquired resistance, or SAR. The induction of SAR may act in a reiteration of the induction of gene expression in the tissue surrounding an infection. Indeed, the presence of an HR induces "microbursts" of reactive oxygen production and cell death in distal tissues, which then induce systemic gene expression (Alvarez *et al.*, 1998). This phenomenon can also be triggered by the infiltration of an H_2O_2-generating system.

SAR also requires the signaling molecule salicylic acid (SA) (Gaffney *et al.*, 1993). SA acts to potentiate the hypersensitive burst in response to pathogen: application of physiological concentrations of SA to cultured soybean cells does not induce H_2O_2 production, but when pathogen is also added, the oxidative burst reaches its maximum much faster (Shirasu *et al.*, 1997). SA binds one form of tobacco catalase and reduces its activity (Chen *et al.*, 1993); this may account for the potentiation of H_2O_2 production by SA. Thus, SA acts synergistically with other signals in pathogen recognition to increase the oxidative burst.

E. Active Oxygen and Senescence

Leaf senescence can be induced by multiple environmental, developmental, and hormonal stimuli, the most important of which may be a decrease in photosynthetic capacity. Whether this is caused by lack of light, or lack of photosystem capacity, it induces a series of events, starting with breakdown of chloroplasts, then loss of proteins, and finally, nuclear breakdown and programmed cell death (reviewed in Gan and Amasino, 1997; Noodén, 1988). The orderly disassembly of cellular components allows the cellular nutrients to be recycled into growing leaves, seeds, or storage tissue.

During senescence, the production of AOS increases and some antioxidant defenses are increased while others are decreased. The peroxisome may play a key role in this process: peroxisomes increase in senescence and their enzyme composition changes. For example, during senescence in pea leaves, purified peroxisomes contained increased activity of superoxide-producing enzymes such as xanthine oxidase, as well as increased superoxide dismutase activity, but decreased catalase activity (Pastori and del Rio, 1997). This study also found increased production of superoxide from the peroxisomal membrane, as well as increased production of H_2O_2. This increase of AOS, specifically H_2O_2, suggests that AOS may act as a signal to induce gene expression and trigger programmed cell death in senescence similarly to disease resistance. One striking difference is the speed of execution of cell death in these two processes, with the rapid cell death in disease resistance allowing for leaf dessication and halting of pathogen spread, and the delayed cell death in senescence allowing for the dismantling and recycling of cellular components.

VI. Mechanisms of Signal Transduction

Although reactive oxygen species are often seen as toxic byproducts of stress or metabolism, they also induce many cellular responses. Indeed, the role of antioxidants in the plant may be both to detoxify these compounds and to modulate their levels for signaling. In a complex signaling network such as this, it is often difficult to separate cause from effect; is the plant responding to reactive oxygen, or responding through reactive oxygen? Many possible mechanisms for sensing the levels of active oxygen species can be imagined, such as receptors for oxidized cell components, or the degradation of the oxidized form of a particularly sensitive protein. One mechanism that has been seen in bacteria, mammals, and plants, is transcriptional control through sensing the redox state of a specific protein. Although the specific mechanisms of sensing reactive oxygen are unknown, the following examples may prove instructive.

A. Control of Transcription by Plastoquinone Redox State

Plastoquinone is one of a chain of electron carriers that transfer electrons from photosystem II (PSII), which removes them from water, to photosystem I (PSI), which eventually transfers them to $NADPH^+$. If an excess of electrons is being pumped in from PSII, then the reduced form of plastoquinone, PQH_2, accumulates; if an excess of electrons is being removed by PSI, then the oxidized form, PQ, accumulates. This balance can be changed by changing the light wavelength to favor either PS I or II. Transfer of electrons to plastoquinone can also be blocked by the inhibitor DCMU [3-(3′,4′-dichlorophenyl)-1,1′-dimethyl urea], leading to an accumulation of PQH_2. Transfer of electrons from plastoquinone can be blocked by the inhibitor DBMIB (2,5-dibromo-3-methyl-6-isopropyl-p-benzoquinone), leading to an accumulation of PQ. Using inhibitors and light changes, Pfannschmidt *et al.* (1999) found that the redox state of plastoquinones affects the transcription of the chloroplast-encoded genes for proteins in the reaction center of PSI, psaAB. The transcription of psaAB decreased with DCMU treatment, which inhibits electron flow into the plastoquinone pool, and increased with DBMIB treatment, which inhibits electron flow out of the plastoquinone pool. This transcriptional regulation serves to maintain the balance between PS I and II; in conditions favoring the passage of electrons through PSII, reduced plastoquinones will accumulate, causing an increase in transcription of PSI genes to compensate.

B. MAP Kinase Pathway Signaling

Although the exact mechanisms of sensing oxidative stress remain unknown, that signal is transduced into gene transcription and other cellular responses by mitogen-activated protein kinase (MAPK) signaling. This mechanism involves a cascade of protein phosphorylation, starting with the activation of upstream protein kinases and is involved in the plant response to abiotic stresses, pathogens, and plant hormones (Jonak *et al.*, 1999). The upstream protein kinases NPK (Nicotiana protein kinase) from tobacco and ANP (Arabidopsis NPK-like protein kinase) activate stress-induced MAPK expression and gene expression (Kovtun *et al.*, 2000). ANP activity is specifically induced by H_2O_2 treatment. Interestingly, transformation of constitutively active NPK into tobacco produced enhanced tolerance to environmental stress without the activation of some known drought and cold response pathways (Kovtun *et al.*, 2000). This approach can tease apart cause from effect in the complex response to oxidative stress.

VII. Conclusions

Active oxygen species form during normal metabolism, during the response to abiotic stresses, during senescence and during the response to pathogen attack. Maintaining oxidative balance is essential for the survival of the cell, so in addition to having antioxidant defenses localized to the area of AOS production, the cells also have mechanisms for sensing and responding to oxidative imbalances. These mechanisms control many aspects of metabolism, such as the balance of photosystem I and II components. Beyond metabolism, AOS act in the plant defense response, to induce defence responses such as crosslinking of cell wall proteins, programmed cell death, and gene transcription. Response to AOS may involve such diverse mechanisms as direct chemical modifications of cell wall proteins, indirect triggering of general responses, and possibly PCD, by disruption of the oxidative balance of the cell, and signal transduction in response to changes in the redox state of unknown signaling molecules. Determination of the mechanisms of signal transduction in response to AOS provides many opportunities for exciting future research.

References

Alvarez, M.E., Pennell, R.I., Meijer, P.-J., Ishikawa, A., Dixon, R.A., and Lamb, C. (1998). Reactive oxygen intermediates mediate a systemic signal network in the establishment of plant immunity. *Cell* **92**, 773–784.

Barcelo, A.R. (1998). Use and misuse of peroxidase inhibitors. *Trends in Plant Science* **3**, 418.

Bestwick, C.S., Brown, I.R., Bennett, M.H.R., and Mansfield, J.W. (1997). Localization of hydrogen peroxide accumulation during the hypersensitive reaction of lettuce cells to *Pseudomonas syringae* pv *phaseolicola*. *The Plant Cell* **9**, 209–221.

Bestwick, C., Bolwell, P., Mansfield, J., Nicole, M., and Wojtaszek, P. (1999). Generation of the oxidative burst—scavenging for the truth. *Trends in Plant Science* **4**, 88–89.

Bolwell, G.P., Butt, V.S., Davies, D.R., and Zimmerlin, A. (1995). Origin of the oxidative burst in plants. *Free Radical Research* **23**, 517–532.

Bradley, D.J., Kjellbom, P., and Lamb, C.J. (1992). Elicitor- and wound-induced oxidative cross-linking of a proline-rich plant cell wall protein: a novel, rapid defense response. *Cell* **70**, 21–39.

Brisson, L.F., Tenhaken, R., and Lamb, C. (1994). Function of oxidative cross-linking of cell wall structural proteins in plant disease resistance. *The Plant Cell* **6**, 1703–1712.

Cadenas, E. (1989). Biochemistry of oxygen toxicity. *Annual Review of Biochemistry* **58**, 79–110.

Chen, Z., Silva, H., and Klessig, D.F. (1993). Active oxygen species in the induction of plant systemic acquired resistance by salicylic acid. *Science* **262**, 1883–1886.

Conklin, P.L., Williams, E.H., and Last, R.L. (1996). Environmental stress sensitivity of an ascorbic acid-deficient *Arabidopsis* mutant. *Proceedings of the National Academy of Sciences of the USA* **93**, 9970–9974.

Conklin, P.L., Pallanca, J.E., Last, R.L., and Smirnoff, N. (1997). L-ascorbic acid metabolism in the ascorbate-deficient *Arabidopsis* mutant vtc1. *Plant Physiology* **115**, 1277–1285.

Dean, R.T., Fu, S., Stocker, R., and Davies, M.J. (1997). Biochemistry and pathology of radical-mediated protein oxidation. *Biochemical Journal* **324**, 1–18.

Del Rio, L.A., Sandalio, L.M., Palma, J.M., Bueno, P., and Corpas, F.J. (1992). Metabolism of oxygen radicals in peroxisomes and cellular implications. *Free Radical Biology and Medicine* **13**, 557–580.

Delledone, M., Xia, Y., Dixon, R.A., and Lamb, C. (1998). Nitric oxide functions as a signal in plant disease resistance. *Nature* **394**, 585–588.

Dietrich, R.A., Delaney, T.P., Uknes, S.J., Ward, E.R., Ryals, J.A., and Dangl, J.L. (1994). Arabidopsis mutants simulating disease resistance response. *Cell* **77**, 565–577.

Ding, H., and Demple, B. (1997). In vivo kinetics of a redox-regulated transcriptional switch. *Proceedings of the National Academy of Sciences of the USA* **94**, 8445–8449.

Doussiere, J., Gaillard, J., and Vignais, P.V. (1999). The heme component of the neutrophil NADPH oxidase complex is a target for aryliodonium compounds. *Biochemistry* **38**, 3694–3703.

Felle, H. (1998). The apoplastic pH of the *Zea mays* root cortex as measured with pH-sensitive microelectrodes: aspects of regulation. *Journal of Experimental Botany* **49**, 987–995.

Foyer, C.H., and Halliwell, B. (1976). The presence of glutathione and glutathione reductase in chloroplasts: a proposed role in ascorbic acid metabolism. *Planta* **133**, 21–25.

Gaffney, T., Friedrich, L., Vernooij, B., Negrotto, D., Nye, G., Uknes, S., Ward, E., Kessmann, H., and Ryals, J. (1993). Requirement of salicylic acid for the induction of systemic acquired resistance. *Science* **261**, 754–756.

Gan, S., and Amasino, R.M. (1997). Making sense of senescence. *Plant Physiology* **113**, 313–319.

Gray, J., Close, P.S., Briggs, S.P., and Johal, G.S. (1997). A novel suppressor of cell death in plants encoded by the *Lls1* gene of maize. *Cell* **89**, 25–31.

Green, M.J., and Hill, H.A.O. (1984). Chemistry of dioxygen. *Methods in Enzymology* **105**, 3–22.

Greenberg, J.T., and Ausubel, F.M. (1993). *Arabidopsis* mutants compromised for the control of cellular damage during pathogenesis and aging. *The Plant Journal* **4**, 327–341.

Greenberg, J.T., Guo, A., Klessig, D.F., and Ausubel, F.M. (1994). Programmed cell death in plants: a pathogen-triggered response activated coordinately with multiple defense functions. *Cell* **77**, 551–563.

Halliwell, B., and Gutteridge, J.M.C. (1985). *Free Radicals in Biology and Medicine*. Clarendon Press, Oxford.

Hu, G., Yalpani, N., Briggs, S.P., and Johal, G.S. (1998). A porphyrin pathway impairment is responsible for the phenotype of a dominant disease lesion mimic mutant of maize. *The Plant Cell* **10**, 1095–1105.

Jabs, T., Dietrich, R.A., and Dangl, J.L. (1996). Initiation of runaway cell death in an *Arabidopsis* mutant by extracellular superoxide. *Science* **273**, 1853–1856.

Jonak, C., Lingerink, W., and Hirt, H. (1999). MAP kinases in plant signal transduction. *Cell and Molecular Life Sciences* **55**, 204–213.

Karpinski, S., Escobar, C., Karpinska, B., Creissen, G., and Mullineaux, P. (1997). Photosynthetic electron transport regulates the expression of cytosolic ascorbate peroxidase genes in *Arabidopsis* during excess light stress. *The Plant Cell* **9**, 627–640.

Karpinski, S., Reynolds, H., Karpinska, B., Wingsle, G., Creissen, G., and Mullineaux, P. (1999). Systemic signaling and acclimation in response to excess excitation energy in *Arabidopsis*. *Science* **284**, 654–657.

Keller, T., Damude, H., Werner, D., Doerner, P., Dixon, R., and Lamb, C. (1998). A plant homologue of the neutrophil NADPH oxidase gp91^{phox} subunit gene encodes a plasma membrane protein with Ca^{2+} binding motifs. *The Plant Cell* **10**, 255–266.

Kliebenstein, D.J., Dietrich, R.A., Martin, A.C., Last, R.L., and Dangl, J.L. (1999). LSD1 regulates salicylic acid induction of copper zinc superoxide dismutase in *Arabidopsis thaliana*. *Molecular-Plant Microbe Interactions* **12**, 1022–1026.

Kovtun, Y., Chiu, W.L., Tena, G., and Sheen, J. (2000). Functional analysis of oxidative stress-activated mitogen-activated protein kinase cascade in plants. *Proceedings of the National Academy of Sciences of the USA* **97**, 2940–2945.

Levine, A., Tenhaken, R., Dixon, R., and Lamb, C. (1994). H_2O_2 from the oxidative burst orchestrates the plant hypersensitive disease resistance response. *Cell* **79**, 583–593.

Mach, J.M., Castillo, A.R., Hoogstraten, R., and Greenberg, J.T. (2001). The Arabidopsis accelerated cell death gene ACD2 encodes red chlorophyll catabolite reductase and suppresses the spread of disease symptoms. *Proceedings of the National Academy of Sciences of the USA* **98**, 771–776.

Mehler, A.H. (1951). Studies on reactions of illuminated chloroplasts. I. Mechanism of the reduction of oxygen and other Hill reagents. *Archives of Biochemistry and Biophysics* **33**, 65–77.

Mudd, J.B. (1996). Biochemical basis for the toxicity of ozone. In *Plant Response to Air Pollution* (M. Yunus and M. Iqbal, Ed.), pp. 267–283. John Wiley and Sons, Chichester, UK.

Mühling, K.H., Plieth, C., Hansen, U.-P., and Sattelmacher, B. (1995). Apoplastic pH of intact leaves of *Vinca faba* as influenced by light. *Journal of Experimental Botany* **46**, 377–382.

Murphy, T.M., and Auh, C.-K. (1996). The superoxide synthases of plasma membrane preparations from cultured rose cells. *Plant Physiology* **110**, 621–629.

Noctor, G., and Foyer, C.H. (1998). Ascorbate and glutathione: keeping active oxygen under control. *Annual Review of Plant Physiology and Plant Molecular Biology* **49**, 249–279.

Noodén, L.C. (1988). The phenomenon of senescence and aging. Whole plant senescence. In *Senescence and Aging in Plants* (L.C. Noodén and A. Leopold, Eds.), pp. 1–50, 391–439. Academic Press, San Diego, CA.

Ouchane, S., Picaud, M., Vernotte, C., and Astier, C. (1997). Photooxidative stress stimulates illegitimate recombination and mutability in carotenoid-less mutants of *Rubrivivax gelatinosus. The EMBO Journal* **16**, 4777–4787.

Pastori, G.M., and del Rio, L.A. (1997). Natural senescence of pea leaves: an activated oxygen-mediated function for peroxisomes. *Plant Physiology* **113**, 411–418.

Pfannschmidt, T., Nilsson, A., and Allen, J.F. (1999). Photosynthetic control of chloroplast gene expression. *Nature* **397**, 625–628.

Polle, A. (1996). Mehler reaction: friend or foe in photosynthesis? *Botanica Acta* **109**, 84–89.

Prasad, T.K. (1996). Mechanisms of chilling-induced oxidative stress injury and tolerance in developing maize seedlings: changes in antioxidant system, oxidation of proteins and lipids, and protease activities. *The Plant Journal* **10**, 1017–1026.

Prasad, T.K., Anderson, M.D., Martin, B.A., and Stewart, C.R. (1994). Evidence for a chilling-induced oxidative stress in maize seedlings and a regulatory role for hydrogen peroxide. *The Plant Cell* **6**, 65–74.

Reinbothe, S., Reinbothe, C., Apel, K., and Lebedev, N. (1996). Evolution of chlorophyll biosynthesis—the challenge to survive photooxidation. *Cell* **86**, 703–705.

Schrauder, M., Moeder, W., Wiese, C., Van Camp, W., Inzé, D., Langebartels, C., and Sandermann, H.J. (1998). Ozone-induced oxidative burst in the ozone biomonitor plant, tobacco Bel W3. *The Plant Journal* **16**, 235–245.

Shirasu, K., Nakajima, H., Rajasekhar, V.K., Dixon, R.A., and Lamb, C. (1997). Salicylic acid potentiates an agonist-dependent gain control that amplifies pathogen signals in the activation of defense mechanisms. *The Plant Cell* **9**, 261–270.

Thomas, H. (1987). Sid: a Mendelian locus controlling thylakoid membrane disassembly in senescing leaves of Festuca pratensis. *Theoretical and Applied Genetics* **73**, 551–555.

Thomas, H., and Matile, P. (1988). Photobleaching of chloroplast pigments in leaves of a non-yellowing mutant genotype of Festuca pratensis. *Phytochemistry* **27**, 345–348.

Thomas, H., Schellenberg, M., Vicentini, F., and Matile, P. (1996). Gregor Mendel's green and yellow pea seeds. *Botanica Acta* **109**, 3–4.

Van Camp, W., Willekens, H., Bowler, C., Van Montagu, M., Inzé, D., Reupold-Popp, P., Sandermann, H.J., and Langebartels, C. (1994). Elevated levels of superoxide dismutase protect transgenic plants against ozone damage. *Bio/Technology* **12**, 165–168.

Vanacker, H., Carver, T.W., and Foyer, C.H. (1998). Pathogen-induced changes in the antioxidant status of the apoplast in barley leaves. *Plant Physiology* **117**, 1103–1114.

Walbot, V., Hoisington, D.A., and Neuffer, M.G. (1983). Disease lesion mimic mutations. In *Genetic Engineering of Plants, An Agricultural Perspective* (T. Kosuge and C.P. Meredith, Eds.), pp. 431–442. Plenum, New York.

Willekens, H., Chamnongpol, S., Davey, M., Schraudner, M., Langebartels, C., Van Montagu, M., Inzé, D., and Van Camp, W. (1997). Catalase is a sink for H_2O_2 and is indispensable for stress defence in C_3 plants. *The EMBO Journal* **16**, 4806–4816.

Wojtaszek, P. (1997). Oxidative burst: an early plant response to pathogen infection. *Biochemical Journal* **322**, 681–692.

14

Nutrient Resorption

Keith T. Killingbeck

I. Introduction

Nutrient resorption, the withdrawal of nutrients from senescing plant tissues, is a complex process whose physiology, ecology, and evolution have been scrutinized widely (Chapin and Kedrowski, 1983; Noodén and Leopold, 1988; Aerts, 1996; Killingbeck, 1996). Nonetheless, resorption remains somewhat of an enigma. Even rudimentary questions such as whether site fertility and moisture availability have significant, predictable impacts on resorption remain unresolved. For a keystone process that critically influences nutrient allocation patterns in virtually all woody plants, and that provides a significant percentage of the nitrogen and phosphorus used annually by a variety of plant communities (Switzer and Nelson, 1972; Turner, 1977; Ryan and Bormann, 1982), such ignorance is not bliss.

Despite the existence of gaps in our knowledge of resorption, recent theoretical and empirical advances have led to a suite of valuable insights into the mechanisms that account for interspecies and intersite variability in resorption. What has not emerged in the years following the classic paper by Chapin and Kedrowski (1983) that sparked renewed interest in resorption, are models that portray the relationships between resorption and the factors by which it is influenced. Two such qualitative models (Figs. 14-1 and 14-2) are introduced here to serve as a framework for an analysis of the mechanisms that control resorption. These predictive models are comprised primarily of a series of testable hypotheses that consider the most important factors known to, or suspected to influence resorption.

Short of complete plant mortality, cell death in perennial plants is most closely tied to senescence and abscission of tissues such as leaves. To be clear from the outset, the models presented here refer specifically to resorption from leaves. Resorption of substances from other tissues has been studied sporadically (e.g., nutrients from roots, Gordon and Jackson, 2000; nectar from flowers, Búrquez and Corbet, 1991; nutrients from heartwood, Andrews and Siccama, 1995), but the preponderance of research on resorption has concentrated on leaves. Considering the fact that leaves contain larger pools of limiting nutrients than any other seasonally-abscised plant tissues (e.g., Duvigneaud and Denaeyer-De Smet, 1970; Whittaker *et al.*, 1979), such bias has been well-placed.

II. Potential and Realized Resorption

The observation that resorption in an individual plant or plant population can vary significantly from year to year led to the realization that plants often resorb amounts of nutrients lower than amounts that could be resorbed under optimum conditions. To differentiate between resorption that is not physiologically maximal, and that which is, Killingbeck *et al.* (1990) introduced the terms potential and realized resorption. Potential resorption is the maximum amount of resorption biochemically attainable, and realized resorption is the actual amount of resorption in a given year.

The importance of this dichotomy surfaces when one attempts to ascertain the specific cause–effect relationships between resorption and several parameters to which resorption is linked. For example, several fertilization experiments have resulted in the conclusion that short-term changes in soil fertility have no significant impact on nutrient resorption (e.g., Birk and Vitousek, 1986; Lajtha, 1987). This has clear implications for our understanding of realized resorption, but it does not shed light on whether the long-term fertility of a site plays a role in the evolution of potential resorption. The plasticity needed to rapidly adjust realized resorption to short-term changes in soil fertility may not have sufficient selective advantage to warrant its evolution. Conversely, results from intersite comparisons (e.g., Ralhan and Singh, 1987; Pugnaire and Chapin, 1993) suggest that potential resorption may track long-term fertility. Therefore, separate models for potential and realized resorption were developed so as not to miss critical complexities in the relationships between resorption and its determinants.

III. Resorption Efficiency and Proficiency

Resorption can be expressed in two additional ways: as resorption efficiency or resorption proficiency. Resorption efficiency, formerly termed "reabsorption coefficient" (Miller, 1938), is the percentage of green-leaf nutrient content that is withdrawn from senescing leaves before abscission. It is most accurately calculated for any nutrient as area-specific mass in green leaves (mass of nutrient per cm^2 of leaf surface) minus area-specific mass in senesced leaves divided by area-specific mass in green leaves, the quantity multiplied by 100. An area-specific nitrogen resorption efficiency of 60% indicates that 60% of the nitrogen in green leaves was withdrawn before abscission.

Resorption efficiency has also been calculated and reported as the change in nutrient concentration between green and senesced leaves. It is calculated as nutrient concentration in green leaves (mass of nutrient per mass of leaf) minus nutrient concentration in senesced leaves divided by nutrient concentration in green leaves, the quantity multiplied by 100.

This method of reporting resorption efficiency is inferior to the "area-specific mass" method because changes in total leaf mass during senescence could artificially amplify or depress changes in nutrient concentrations, even in the complete absence of net movement into or out of leaves by the nutrient in question (Stenlid, 1958; Guha and Mitchell, 1966). However, because surface area of dicotyledonous plant leaves remains relatively constant throughout the growing season after leaf expansion has been completed (Woodwell, 1974), reporting resorption efficiency on an "area-specific mass" basis accurately represents net nutrient movements in senescing leaves.

Historically, resorption was measured, reported, and discussed as resorption efficiency until 1996 when a new measure of resorption was introduced: resorption proficiency (Killingbeck, 1996). Proficiency is simply the amount of a nutrient that remains in fully senesced leaves. From a biological perspective, an important advantage of measuring resorption as proficiency rather than efficiency is that proficiency is a more unequivocal measure of the degree to which selection has acted to minimize nutrient loss in ephemeral leaves. From a methodological perspective, the advantages of measuring resorption as proficiency rather than efficiency include: (1) leaves need to be collected just once in a growing season, (2) vagaries in green-leaf nutrient content do not influence the calculation of proficiency, (3) timing of the collection of green leaves, which is critical to the calculation of efficiency, is eliminated as a source of variability or error, and (4) measurement of leaf surface area becomes optional.

The last point appears to be at odds with the statement above cautioning against the use of non-area-specific concentration data to calculate efficiency. However, it is not. Because the calculation of proficiency does not rely on the comparison of green-leaf nutrients to senesced-leaf nutrients, intervening changes in leaf mass are not an issue. Further, proficiency is most accurately expressed on a concentration basis (mass of nutrient per mass of leaf) because interspecies or interplant comparisons of proficiency expressed on an area-specific basis would be invalid if leaf thickness varied among species or individuals.

To ensure that the following models are as generally applicable as is feasible, no distinction was made between resorption efficiency and proficiency. Because proficiency has only recently been considered, relationships between resorption and its determinants offered in the model are primarily supported by studies that focused exclusively on resorption efficiency. In most cases, however, the nature and direction of the relationships described apply equally to resorption efficiency and proficiency (but see the hypothesis relating green-leaf nitrogen to realized resorption). Therefore, except where noted in the text, the phrase "resorption decreases" in the model implies that resorption becomes both less efficient and less proficient, and the phrase "resorption increases" implies that resorption becomes both more efficient and more proficient.

IV. Macronutrients and Trace Metals

Many macronutrients and trace metals are capable of being resorbed from senescing leaves. Among the alkali metals and alkaline earth elements, only potassium has been repeatedly reported to be resorbed in significant quantities (e.g., Woodwell, 1974; Hill *et al.*, 1979; Ostman and Weaver, 1982; Staaf, 1982; Ralhan and Singh, 1987; Scott *et al.*, 1992). However, the ease with which precipitation leaches potassium from leaves (Gosz *et al.*, 1975), especially senescing leaves (Tukey, 1970), places in doubt the reported quantities of potassium resorbed from plants growing outdoors. Magnesium and sodium have not often been considered in studies of resorption, but when they have, they have most often been found to be accreted into senescing leaves (accretion = net movement of a substance into, rather than out of, senescing leaves: magnesium, Killingbeck, 1992, 1993; Scott *et al.*, 1992; sodium, Woodwell, 1974; Ralhan and Singh, 1987; Killingbeck, 1993). Calcium has repeatedly been found to be accreted into senescing leaves (e.g., Woodwell, 1974; Hill *et al.*, 1979; Ostman and Weaver, 1982; Ralhan and Singh, 1987; Killingbeck *et al.*, 1990; Killingbeck, 1992, 1993; Scott *et al.*, 1992), yet several studies have reported calcium resorption (Ostman and Weaver, 1982; Scott *et al.*, 1992).

Among trace metals, copper (Mukherjee, 1969; Hill *et al.*, 1979; Killingbeck and Costigan, 1988; Killingbeck *et al.*, 1990, 2002; May and Killingbeck, 1992; Killingbeck, 1993), manganese (Mukherjee, 1969; May and Killingbeck, 1992), and zinc (Mukherjee, 1969; Killingbeck, 1985, 1993; Killingbeck and Costigan, 1988; May and Killingbeck, 1992) have been resorbed, but not consistently so. Each of these trace metals has also been accreted into senescing leaves (copper, Killingbeck, 1985; manganese, Hill *et al.*, 1979; Killingbeck, 1985, 1992, 1993; Killingbeck *et al.*, 1990, 2002; zinc, Guha and Mitchell, 1966; Killingbeck *et al.*, 1990, 2002; Killingbeck, 1992). Likewise, the macronutrient iron has also been resorbed (Killingbeck, 1985; Killingbeck and Costigan, 1988) and accreted (Chapin *et al.*, 1975; Killingbeck, 1985, 1992, 1993; Killingbeck *et al.*, 1990).

Net influx and outflow of trace metals in senescing leaves appears to be so variable that both resorption and accretion of the same metal have occurred among different species growing at the same site (Whittaker *et al.*, 1979; Killingbeck and Costigan, 1988). Individual plants have also resorbed and accreted the same trace metal in different years (May and Killingbeck, 1992; Killingbeck, 1993).

By contrast, nitrogen and phosphorus have consistently been resorbed in relatively high quantities by the vast majority of deciduous and evergreen plants in which resorption has been measured (Chapin and Kedrowski, 1983, Table 1; Aerts, 1996, Appendix 1; Killingbeck, 1996, Tables 1–3; however, see later discussion concerning species that harbor nitrogen-fixing symbionts). The preeminently important roles these two macronutrients play in a wide variety of essential plant functions (Marschner, 1986), along with the fact that they are the nutrients most often responsible for limiting plant growth (Chapin, 1980), have undoubtedly influenced evolutionary selection for their efficient and proficient resorption. Therefore, the two resorption models that follow were developed primarily from the literature on nitrogen and phosphorus resorption, and apply most directly to resorption of these two essential macronutrients.

V. Model I: Determinants of Potential Resorption

Unlike realized resorption that can vary widely from year to year, potential resorption remains more fixed during the lifetime of a plant. It is essentially an evolved, nutrient-specific set-point, the magnitude of which is ultimately determined for each nutrient by biochemical limitations. These biochemical limitations prevent complete evacuation of a nutrient from a senescing leaf. Nutrient recovery from structural, recalcitrant compounds may be impossible, or energetically unfeasible, and nutrients that are essential components of the enzymatic machinery responsible for disassembling foliar biochemicals may be unavailable for export.

Biochemical limitations are in turn influenced by an array of factors, most notably phylogenetic relationships, species-specific plant characteristics related to both source–sink

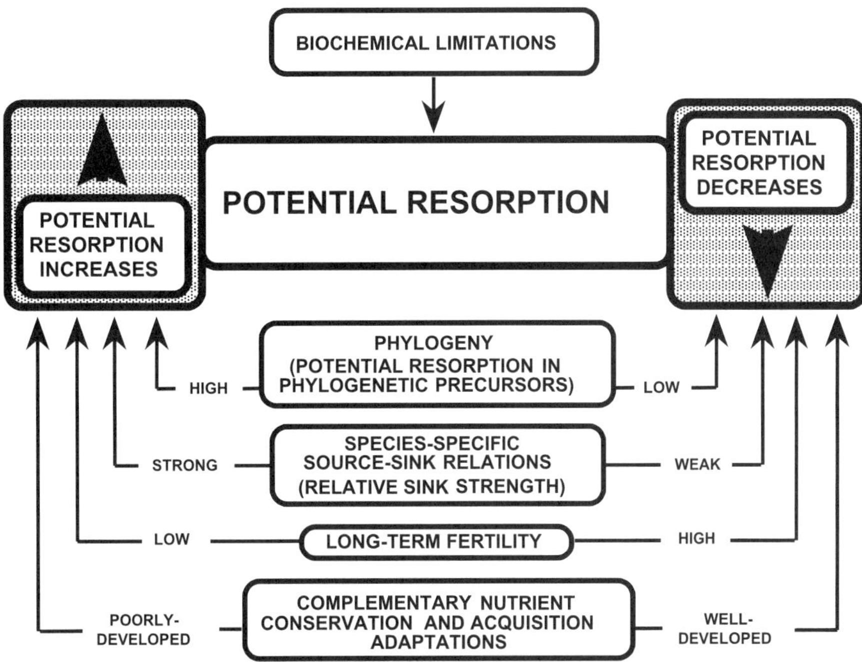

Figure 14-1. Model I: determinants of potential resorption.

relations and nutrient use strategies, and long-term site fertility (Fig. 14-1). Hypotheses describing these determinants of potential resorption are outlined below. Other factors also influence potential and realized resorption efficiency and proficiency (e.g., CO_2 enrichment, Norby *et al.*, 2000; leaf size and specific leaf mass, Killingbeck and Whitford, 2001; life-form class, Killingbeck *et al.*, 2002; temperature, Nordell and Karlsson, 1995), but because it is beyond the scope and intent of these models to consider every factor that has ever been related to resorption, only the best documented, or theoretically most compelling relationships will be addressed.

Potential resorption that is low (inefficient and unproficient) in phylogenetic precursors decreases potential resorption, and potential resorption that is high in phylogenetic precursors increases potential resorption. Similarities in resorption among related taxa, and dissimilarities among less related taxa, suggest that phylogeny (taxonomic relatedness) influences potential resorption (Killingbeck, 1996). Unusually high or low potential resorption in a species could conceivably be the result of a close taxonomic affiliation with precursors in which evolutionary selection was particularly strong, or weak, for efficient and proficient resorption.

Weak nutrient sinks present during senescence decrease potential resorption, and strong sinks present during senescence increase potential resorption. Species have inherent differences in the strength of nutrient sinks present during leaf senescence. Strong sinks, such as large masses of maturing fruits or developing roots, should act to elevate potential resorption, while weak sinks should decrease potential resorption. Phenology is a critical component of this relationship because, for example, flowers and fruits produced in early summer add nothing to autumnal sink-strength.

High long-term site fertility decreases potential resorption, and low long-term site fertility increases potential resorption. In seven of 11 studies that examined the effects of fertility on resorption in natural sites that differed in long-term fertility, resorption was higher on low fertility sites than on high fertility sites (Small, 1972; Stachurski and Zimka, 1975; Boerner, 1984a, 1984b; Kost and Boerner, 1985; Ralhan and Singh, 1987; Pugnaire and Chapin, 1993: but see Ostman and Weaver, 1982; Staaf, 1982; del Arco *et al.*, 1991; Minoletti and Boerner, 1994). This suggests the existence of evolutionary tradeoffs between resorption and long-term site fertility.

Well-developed nutrient conservation and acquisition adaptations decrease potential resorption, and poorly-developed nutrient conservation and acquisition adaptations increase potential resorption. Species differ considerably in their abilities to acquire and conserve nutrients. Well-developed nutrient acquisition and/or conservation adaptations such as exceptionally efficient nutrient uptake or symbiotic nitrogen-fixation should decrease potential resorption, while the lack of such adaptations should increase potential resorption. Plants harboring nitrogen-fixing symbionts have consistently low nitrogen resorption efficiency (Dawson and Funk, 1981; Rodriguez-Barrueco *et al.*, 1984, Côté and Dawson, 1986; Côté *et al.*, 1989; Killingbeck, 1993) suggesting that the evolution of nutrient acquisition and nutrient resorption may be guided by a degree of reciprocity. Additionally, embedded within this determinant is a cadre of other plant adaptations that may influence potential resorption. Leaf habit is perhaps the most studied and debated of these, yet the questions of if and how potential resorption differs between evergreen and deciduous species remain unresolved (Killingbeck, 1996, 1998; Craine and Mack, 1998; Aerts and Chapin, 2000).

VI. Model II: Determinants of Realized Resorption

Of all the parameters that have an impact on nutrient resorption, the nine that appear to be most strongly linked to realized resorption in empirical studies, and/or in theory, constitute the determinants of realized resorption presented in Model II (Fig. 14-2). These determinants are of particular significance to the understanding of changes in realized resorption because they themselves are susceptible to significant year-to-year variation.

Early abscission results in abnormally low resorption, and late abscission results in normal or elevated resorption. In a clone of aspen (*Populus tremuloides*), resorption was lowest in ramets that dropped leaves early, and highest in ramets that dropped leaves late (Killingbeck *et al.*, 1990). Even on individual ramets, resorption was lower in leaves that were dropped early from a ramet than those dropped later. Experiments that prolonged leaf attachment in paper birch (*Betula papyrifera*) had no effect on resorption (Chapin and Moilanen, 1991).

Low amounts of energy available for metabolic processes during leaf senescence result in abnormally low resorption, and high amounts of energy result in normal or elevated resorption. The complete energetic cost of resorption has yet to be quantified, but the energy needed to produce catabolic enzymes and to transport molecules out of leaves must be substantial. Therefore, resorption must be subnormal when available energy declines below some unknown threshold. Resorption was lower in leaves of paper birch (*Betula papyrifera*) that were shaded to reduce photosynthetic capture of energy than in unshaded leaves (Chapin and Moilanen, 1991).

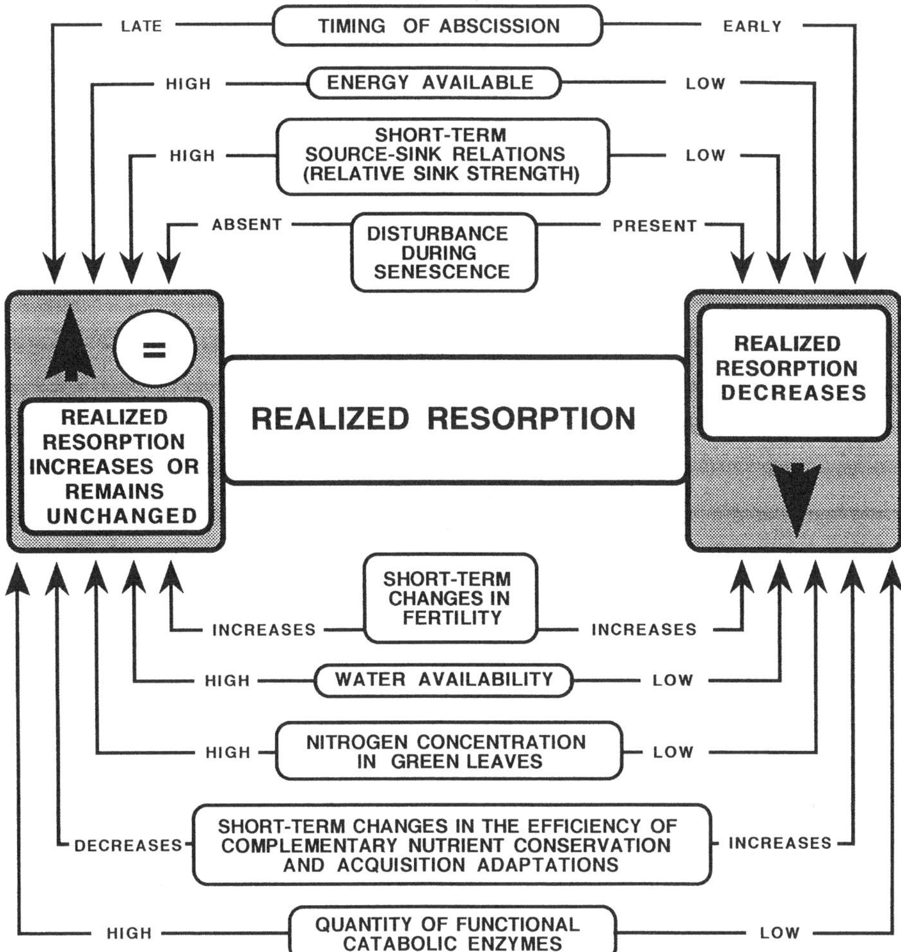

Figure 14-2. Model II: determinants of realized resorption.

Low nutrient sink strength during leaf senescence results in abnormally low resorption, and high nutrient sink strength results in normal or elevated resorption. During senescence, leaves act as strong nutrient sources. The concomitant presence of strong sinks should promote resorption. For example, in plants that produce maturing fruits at the time of leaf senescence, unusually small fruit crops might lower resorption unless other sinks compensate for this loss in sink strength. The presence of "active sinks" increased nutrient resorption in barley (*Hordeum valgare*, Pugnaire and Chapin, 1992), and source–sink relations strongly influenced resorption in potato (*Solanum tuberosum*, Berchtold *et al.*, 1993).

Severe disturbance during senescence results in abnormally low resorption, and the absence of disturbance results in normal resorption. Disruption or termination of the normal sequence of physiological events that take place during leaf senescence can decrease resorption. Non-defoliating wind and salt damage inflicted by Hurricane Gloria in 1985 on the senescing leaves of a population of aspen (*Populus tremuloides*) completely eliminated foliar resorption of nitrogen and phosphorus (Killingbeck, 1988).

Short-term increases in fertility induced by fertilization decrease resorption, increase resorption, or have no effect on resorption. The entire range of responses in resorption has been elicited by fertilization experiments that have changed the short-term fertility of sites. Fertilization had no significant impact on resorption in creosote bush (*Larrea tridentata*, Lajtha, 1987; Lajtha and Klein, 1988; Lajtha and Whitford, 1989) and loblolly pine (*Pinus taeda*, Birk and Vitousek, 1986), yet reduced resorption in three species of marsh plants (*Calamagrostis canadensis*, *Carex lacustris*, and *Typha latifolia*, Shaver and Melillo, 1984), Douglas-fir (*Pseudotsuga menziesii*, Turner, 1977), sawgrass (*Cladium jamaicense*, Richardson *et al.*, 1999), spring beauty (*Claytonia virginica*, Anderson and Eickmeier, 2000), and Scots pine (*Pinus sylvestris*, Näsholm, 1994). By contrast, resorption increased in response to fertilization of radiata pine (*Pinus radiata*, Nambiar and Fife, 1987). The lack of a convincing, mechanistic explanation for this diversity of responses leaves us without a general, predictive hypothesis relating resorption to short-term changes in site fertility.

The more consistent linkage between potential resorption and long-term fertility at first seems to be incongruous with the inability of some plants to quickly, predictably, and consistently adjust realized resorption in response to short-term changes in site fertility. However, the evolution of one is not inextricably linked to the other. At present, it appears that adjusting realized resorption in response to short-term changes in fertility is not universally adaptive and has not been selected for consistently. Conversely, set-points of potential resorption have apparently evolved to maximize fitness through compensation for unusually high or low long-term site fertility.

Low water availability results in abnormally low resorption, and high water availability results in normal or elevated resorption. Water stress reduces resorption. In studies comparing resorption to natural or artificial gradients in available moisture, water stress has been directly linked to decreased resorption (Hocking, 1982; del Arco *et al.*, 1991, Pugnaire and Chapin, 1992; Minoletti and Boerner, 1994; but see Lajtha and Whitford, 1989). Even though drought can apparently trigger the onset of resorption (Heckathorn and DeLucia, 1994), other effects of drought, such as premature leaf abscission (Escudero and del Arco, 1987; Hocking, 1982) and reduced phloem-loading and transport (Pugnaire and Chapin, 1992) also act to decrease resorption.

Low concentrations of nitrogen in green leaves result in abnormally low resorption efficiency, and high concentrations of nitrogen result in normal or elevated resorption efficiency. By necessity, generality was sacrificed in this hypothesis because of the existence of major differences between nitrogen and phosphorus, and between efficiency and proficiency, in relationship to green-leaf nutrient concentrations. Significant, positive relationships between green-leaf nitrogen and nitrogen resorption efficiency have been found in a wide variety of species and ecosystems (Chapin and Kedrowski, 1983; Lajtha, 1987; del Arco *et al.*, 1991; Nordell and Karlsson, 1995; but see Aerts, 1996). In the two studies above in which phosphorus was also measured, green-leaf phosphorus was not significantly related to phosphorus resorption efficiency (Chapin and Kedrowski, 1983; Lajtha, 1987). By contrast, nitrogen resorption proficiency was inversely related to green-leaf nitrogen in four species of desert shrubs (Killingbeck and Whitford, 2001). That is, increases in green-leaf nitrogen resulted in decreased nitrogen resorption proficiency (i.e., elevated nitrogen in senesced leaves). The most plausible explanation for this clear difference between efficiency and proficiency is that increases in leaf nitrogen come from both resorbable and recalcitrant nitrogen. Resorption efficiency increases because the ratio between resorbed nitrogen and leaf mass increases, yet proficiency falls because higher amounts of recalcitrant nitrogen-containing compounds remain in senesced leaves.

Short-term increases in the efficiency of complementary nutrient conservation and acquisition adaptations decrease resorption, and short-term decreases in the efficiency of complementary nutrient conservation and acquisition adaptations increase resorption. Consistently low resorption in plants harboring nitrogen-fixing symbionts (Dawson and Funk, 1981; Rodriguez-Barrueco *et al.*, 1984, Côté and Dawson, 1986; Côté *et al.*, 1989; Killingbeck, 1993; Killingbeck and Whitford, 2001) has provided evidence for a reciprocal linkage between nutrient acquisition and resorption. If resorption and other mechanisms of nutrient conservation, such as high nutrient-use efficiency, form a "shifting, complementary complex of nutrient conservation adaptations" in which "the complex, rather than the components" are strongly linked to nutrient availability (Killingbeck and Costigan, 1988), then changes in the efficiency of either nutrient acquisition or complementary conservation adaptations may force reciprocal changes in resorption.

Low quantities of functional enzymes available to disassemble foliar macromolecules result in abnormally low resorption, and high quantities of functional catabolic enzymes result in normal or elevated resorption. Trace metals are essential cofactors for some of the enzymes needed to disassemble foliar macromolecules during senescence. Disassembly of macromolecules such as proteins is a prerequisite for resorption because nitrogen is moved out of leaves as amino acids (Chapin and Kedrowski, 1983; Chapin *et al.*, 1986). Zinc, an essential component of some proteinases and peptidases (Price *et al.*, 1972; Kabata-Pendias and Pendias, 1984), was positively linked to resorption in the desert shrub ocotillo (*Fouquieria splendens*, Killingbeck, 1992). Although trace metals are sometimes resorbed (as discussed earlier), their active transport into senescing leaves may be in response to the need for functional catabolic enzymes (Killingbeck, 1985).

VII. Synthesis

Resorption is a keystone plant process that conserves nutrients by retrieving them from senescing leaves. The quantity of a nutrient that physiologically can be resorbed from a senescing leaf (potential resorption) is determined by constraints imposed on nutrient retrieval by biochemical limitations, and by genetic controls shaped by phylogeny and a suite of plant and site characteristics (Fig. 14-1). The quantity of a nutrient that is actually resorbed (realized resorption) is influenced by potential resorption and a wide and varied array of determinants that are primarily environmental, not genetic (Fig. 14-2). Although realized resorption appears to reach potential resorption in some cases, it is highly likely that realized resorption is frequently reduced below potential resorption by conditions that terminate, or slow down the retrieval of nutrients from senescing leaves. In any given year, realized resorption will be less than, or equal to potential resorption, but never higher. Because there appears to be no obvious advantage to inefficient or unproficient realized resorption, it seems likely that selection continually favors adaptations that allow realized resorption to reach, or closely approach potential resorption.

References

Aerts, R. (1996). Nutrient resorption from senescing leaves of perennials: are there general patterns? *J. Ecol.* **84**, 597–608.

Aerts, R., and Chapin, F.S., III (2000). The mineral nutrition of wild plants revisited: a re-evaluation of processes and patterns. *Advan. Ecol. Res.* **30**, 1–67.

Anderson, W.B., and Eickmeier, W.G. (2000). Nutrient resorption in *Claytonia virginica* L.: implications for deciduous forest nutrient cycling. *Can. J. Bot.* **78**, 832–839.

Andrews, J.A., and Siccama, T.G. (1995). Retranslocation of calcium and magnesium at the heartwood-sapwood boundary of Atlantic white cedar. *Ecology* **76**, 659–663.

Berchtold, A., Besson, J.-M., and Feller, U. (1993). Effects of fertilization levels in two farming systems on senescence and nutrient contents in potato leaves. *Plant Soil* **154**, 81–88.

Birk, E.M., and Vitousek, P.M. (1986). Nitrogen availability and nitrogen use efficiency in loblolly pine stands. *Ecology* **67**, 69–79.

Boerner, R.E.J. (1984a). Foliar nutrient dynamics and nutrient use efficiency of four deciduous tree species in relation to site fertility. *J. Appl. Ecol.* **21**, 1029–1040.

Boerner, R.E.J. (1984b). Nutrient fluxes in litterfall and decomposition in four forests along a gradient of soil fertility in southern Ohio. *Can. J. For. Res.* **14**, 794–802.

Búrquez, A., and Corbet, S.A. (1991). Do flowers reabsorb nectar? *Funct. Ecol.* **5**, 369–379.

Chapin, F.S., III (1980). The mineral nutrition of wild plants. *Ann. Rev. Ecol. Syst.* **11**, 233–260.

Chapin, F.S., III, and Kedrowski, R.A. (1983). Seasonal changes in nitrogen and phosphorus fractions and autumnal retranslocation in evergreen and deciduous taiga trees. *Ecology* **64**, 376–391.

Chapin, F.S., III, and Moilanen, L. (1991). Nutritional controls over nitrogen and phosphorus resorption from Alaskan birch leaves. *Ecology* **72**, 709–715.

Chapin, F.S., III, Van Cleve, K., and Tieszen, L.L. (1975). Seasonal nutrient dynamics of tundra vegetation at Barrow, Alaska. *Arct. Alp. Res.* **7**, 209–226.

Chapin, F.S., III, Shaver, G.R., and Kedrowski, R.A. (1986). Environmental controls over carbon, nitrogen and phosphorus fractions in *Eriophorum vaginatum* in Alaskan tussock tundra. *J. Ecol.* **74**, 167–195.

Côté, B., and Dawson, J.O. (1986). Autumnal changes in total nitrogen, salt-extractable proteins and amino acids in leaves and adjacent bark of black alder, eastern cottonwood and white basswood. *Physiol. Plant.* **67**, 102–108.

Côté, B., Vogel, C., and Dawson, J.O. (1989). Autumnal changes in tissue nitrogen of autumn olive, black alder and eastern cottonwood. *Plant Soil* **118**, 23–32.

Craine, J.M., and Mack, M.C. (1998). Nutrients in senesced leaves: keys to the search for potential resorption and resorption proficiency: comment. *Ecology* **79**, 1818–1820.

Dawson, J.O., and Funk, D.T. (1981). Seasonal change in foliar nitrogen concentration of *Alnus glutinosa*. *Forest Sci.* **27**, 239–243.

del Arco, J.M., Escudero, A., and Vega Garrido, M. (1991). Effects of site characteristics on nitrogen retranslocation from senescing leaves. *Ecology* **72**, 701–708.

Duvigneaud, P., and Denaeyer-De Smet, S. (1970). Biological cycling of minerals in temperate deciduous forests. In *Analysis of Temperate Forest Ecosystems* (D.E. Reichle, Ed.), pp. 199–225. Springer-Verlag, New York.

Escudero, A., and del Arco, J.M. (1987). Ecological significance of the phenology of leaf abscission. *Oikos* **49**, 11–14.

Gordon, W.S., and Jackson, R.B. (2000). Nutrient concentrations in fine roots. *Ecology* **81**, 275–280.

Gosz, J.R., Likens, G.E., and Eaton, J.S. (1975). Leaching of nutrients from leaves of selected tree species in New Hampshire. In *Mineral Cycling in Southeastern Ecosystems* (F.G. Howell, J.B. Gentry and M.H. Smith, Eds.), pp. 630–641. US Department of Commerce, Springfield, VA.

Guha, M.M., and Mitchell, R.L. (1966). The trace and major element composition of the leaves of some deciduous trees II. Seasonal changes. *Plant Soil* **24**, 90–112.

Heckathorn, S.A., and DeLucia, E.H. (1994). Drought-induced nitrogen retranslocation in perennial C_4 grasses of tallgrass prairie. *Ecology* **75**, 1877–1886.

Hill, J., Robson, A.D., and Loneragan, J.F. (1979). The effect of copper supply on the senescence and the retranslocation of nutrients of the oldest leaf of wheat. *Ann. Bot.* **44**, 279–287.

Hocking, P.J. (1982). Effect of water stress on redistribution of nutrients from leaflets of narrow-leaved lupin (*Lupinus angustifolius* L.). *Ann. Bot.* **49**, 541–543.

Kabata-Pendias, A., and Pendias, H. (1984). *Trace Elements in Soils and Plants*. CRC Press, Boca Raton, FL.

Killingbeck, K.T. (1985). Autumnal resorption and accretion of trace metals in gallery forest trees. *Ecology* **66**, 283–286.

Killingbeck, K.T. (1988). Hurricane-induced modification of nitrogen and phosphorus resorption in an aspen clone: an example of diffuse disturbance. *Oecologia* **75**, 213–215.

Killingbeck, K.T. (1992). Inefficient nitrogen resorption in a population of ocotillo (*Fouquieria splendens*), a drought-deciduous desert shrub. *Southwest. Nat.* **37**, 35–42.

Killingbeck, K.T. (1993). Inefficient nitrogen resorption in genets of the actinorhizal nitrogen fixing shrub *Comptonia peregrina*: physiological ineptitude or evolutionary tradeoff? *Oecologia* **94**, 542–549.

Killingbeck, K.T. (1996). Nutrients in senesced leaves: keys to the search for potential resorption and resorption proficiency. *Ecology* **77**, 1716–1727.

Killingbeck, K.T. (1998). Nutrients in senesced leaves: reply. *Ecology* **79**, 1820–1821.

Killingbeck, K.T., and Costigan, S.A. (1988). Element resorption in a guild of understory shrub species: niche differentiation and resorption thresholds. *Oikos* **53**, 366–374.

Killingbeck, K.T., and Whitford, W.G. (2001). Nutrient resorption in shrubs growing by design, and by default in Chihuahuan Desert arroyos. *Oecologia* **128**, 351–359.

Killingbeck, K.T., May, J.D., and Nyman, S. (1990). Foliar senescence in an aspen (*Populus tremuloides*) clone: the response of element resorption to interramet variation and timing of abscission. *Can. J. For. Res.* **20**, 1156–1164.

Killingbeck, K.T., Hammen-Winn, S.L., Vecchio, P.G., and Goguen, M.E. (2002). Nutrient resorption efficiency and proficiency in fronds and trophopods of a winter-deciduous fern, *Dennstaedtia punctilobula*. *Int. J. Plant Sci.* **163**, 99–105.

Kost, J.A., and Boerner, R.E.J. (1985). Foliar nutrient dynamics and nutrient use efficiency in *Cornus florida*. *Oecologia* **66**, 602–606.

Lajtha, K. (1987). Nutrient reabsorption efficiency and the response to phosphorus fertilization in the desert shrub *Larrea tridentata* (DC.) Cov. *Biogeochemistry* **4**, 265–276.

Lajtha, K., and Klein, M. (1988). The effect of varying nitrogen and phosphorus availability on nutrient use by *Larrea tridentata*, a desert evergreen shrub. *Oecologia* **75**, 348–353.

Lajtha, K., and Whitford, W.G. (1989). The effect of water and nitrogen amendments on photosynthesis, leaf demography, and resource-use efficiency in *Larrea tridentata*, a desert evergreen shrub. *Oecologia* **80**, 341–348.

Marschner, H. (1986). *Mineral Nutrition of Higher Plants*. Academic Press, San Diego, CA.

May, J.D., and Killingbeck, K.T. (1992). Effects of preventing nutrient resorption on plant fitness and foliar nutrient dynamics. *Ecology* **73**, 1868–1878.

Miller, E.C. (1938). *Plant Physiology* (2nd ed.), McGraw-Hill, New York.

Minoletti, M.L., and Boerner, R.E.J. (1994). Drought and site fertility effects on foliar nitrogen and phosphorus dynamics and nutrient resorption by the forest understory shrub *Viburnum acerifolium* L. *Am. Midl. Nat.* **131**, 109–119.

Mukherjee, K.L. (1969). Microelement composition of sugarcane leaves during their growth and senescence. *J. Indian Bot. Soc.* **48**, 180–184.

Nambiar, E.K.S., and Fife, D.N. (1987). Growth and nutrient retranslocation in needles of radiata pine in relation to nitrogen supply. *Ann. Bot.* **60**, 147–156.

Näsholm, T. (1994). Removal of nitrogen during needle senescence in Scots pine (*Pinus sylvestris* L.). *Oecologia* **99**, 290–296.

Noodén, L.D., and Leopold, A.C. (1988). *Senescence and Aging in Plants*. Academic Press, New York.

Norby, R.J., Long, T.M., Hartz-Rubin, J.S., and O'Neill, E.G. (2000). Nitrogen resorption in senescing tree leaves in a warmer, CO_2-enriched atmosphere. *Plant Soil* **224**, 15–29.

Nordell, K.O., and Karlsson, P.S. (1995). Resorption of nitrogen and dry matter prior to leaf abscission: variation among individuals, sites and years in the mountain birch. *Funct. Ecol.* **9**, 326–333.

Ostman, N.L., and Weaver, G.T. (1982). Autumnal nutrient transfers by retranslocation, leaching, and litter fall in a chestnut oak forest in southern Illinois. *Can. J. For. Res.* **12**, 40–51.

Price, C.A., Clark, H.E., and Funkhouser, E.A. (1972). Functions of micronutrients in plants. In *Micronutrients in Agriculture* (J. Mortvedt, P. Giordano and W. Lindsay, Eds.), pp. 231–242. Soil Science Society of America, Madison, WI.

Pugnaire, F.I., and Chapin, F.S., III (1992). Environmental and physiological factors governing nutrient resorption efficiency in barley. *Oecologia* **90**, 120–126.

Pugnaire, F.I., and Chapin, F.S., III (1993). Controls over nutrient resorption from leaves of evergreen Mediterranean species. *Ecology* **74**, 124–129.

Ralhan, P.K., and Singh, S.P. (1987). Dynamics of nutrients and leaf mass in central Himalayan forest trees and shrubs. *Ecology* **68**, 1974–1983.

Richardson, C.J., Ferrell, G.M., and Vaithiyanathan, P. (1999). Nutrient effects on stand structure, resorption efficiency, and secondary compounds in Everglades sawgrass. *Ecology* **80**, 2182–2192.

Rodriguez-Barrueco, C., Miguel, C., and Subramaniam, P. (1984). Seasonal fluctuations of the mineral concentration of alder (*Alnus glutinosa* (L.) Gaertn.) from the field. *Plant Soil* **78**, 201–208.

Ryan, D.F., and Bormann, F.H. (1982). Nutrient resorption in northern hardwood forests. *BioScience* **32**, 29–32.

Scott, D.A., Proctor, J., and Thompson, J. (1992). Ecological studies on a lowland evergreen rain forest on Maraca Island, Brazil II. Litter and nutrient cycling. *J. Ecol.* **80**, 705–717.

Shaver, G.R., and Melillo, J.M. (1984). Nutrient budgets of marsh plants: efficiency concepts and relation to availability. *Ecology* **65**, 1491–1510.

Small, E. (1972). Photosynthetic rates in relation to nitrogen recycling as an adaptation to nutrient deficiency in peat bog plants. *Can. J. Bot.* **50**, 2227–2233.

Staaf, H. (1982). Plant nutrient changes in beech leaves during senescence as influenced by site characteristics. *Acta Ecol.* **3**, 161–170.

Stachurski, A., and Zimka, J. (1975). Methods of studying forest ecosystems: leaf area, leaf production, and withdrawal of nutrients from leaves of trees. *Ekol. Polska* **23**, 637–648.

Stenlid, G. (1958). Salt losses and redistribution of salts in higher plants. In *Encyclopedia of Plant Physiology. Vol. 4: Mineral Nutrition of Plants* (W. Ruhland, Ed.), pp. 615–637. Springer-Verlag, New York.

Switzer, G.L., and Nelson, L.E. (1972). Nutrient accumulation and cycling in loblolly pine (*Pinus taeda* L.) plantation ecosystems: the first twenty years. *Soil Sci. Soc. Amer. Proc.* **36**, 143–147.

Tukey, H.B., Jr. (1970). The leaching of substances from plants. *Ann. Rev. Plant Phys.* **21**, 305–324.

Turner, J. (1977). Effect of nitrogen availability on nitrogen cycling in a Douglas-fir stand. *Forest Sci.* **23**, 307–316.

Whittaker, R.H., Likens, G.E., Bormann, F.H., Eaton, J.S., and Siccama, T.G. (1979). The Hubbard Brook ecosystem study: forest nutrient cycling and element behavior. *Ecology* **60**, 203–220.

Woodwell, G.M. (1974). Variations in the nutrient content of leaves of *Quercus alba*, *Quercus coccinea*, and *Pinus rigida* in the Brookhaven Forest from bud-break to abscission. *Amer. J. Bot.* **61**, 749–753.

15

Whole Plant Senescence

Larry D. Noodén, Juan J. Guiamét
and Isaac John

I. It was six men of Hindustan To learning much inclined, Who went to see the Elephant (Though all of them were blind). That each by observation Might satisfy his mind.	II. The FIRST approached the Elephant And happening to fall Against his broad and sturdy side At once began to bawl: 'God bless me, but the Elephant Is very like a wall!'
III. The SECOND, feeling of the tusk, Cried 'Ho! What have we here So very round and smooth and sharp? To me 'tis mightly clear This wonder of an Elephant Is very like a spear.'	IV. The THIRD approached the animal, And happening to take The squirming trunk within his hands, Thus boldly up and spake: 'I see, 'quoth he, 'the Elephant Is very like a snake!'
V. The FOURTH reached out his eager hand, And felt about the knee, 'What most this wondrous beast is like Is mightly plain,' quoth he: 'Tis clear enough the Elephant Is very like a tree!'	VI. The FIFTH, who chanced to touch the ear, Said: 'E'en the blindest man Can tell what this resembles most, Deny the fact who can, This marvel of an Elephant Is very like a fan!'
VII. The SIXTH no sooner had begun About the beast to grope, Than, seizing on the swinging tail That fell within his scope, 'I see,' quoth he, 'the Elephant Is very like a rope!'	VIII. And so these men of Indostan Disputed loud and long, Each in his own opinion Exceeding stiff and strong. Though each was partly in the right And all were in the wrong.

IX. So, oft in theologic wars The disputants, I ween, Rail on in utter ignorance Of what each other mean And prate about an Elephant Not one of them has seen!	*The Blind Men and the Elephant: A Hindoo Fable* by John Godfrey Saxe, a 19th century poet

I. Introduction

A. Objectives

All who analyze integrative, whole organism processes (and maybe all biologists) share some of the difficulties facing the six blind men trying to determine the shape of an elephant by examining at its separate parts. It is particularly important to recognize the message about the need for an integrated view in this age-old parable when dealing with the complexities of whole organism integrative processes such as whole plant senescence. The component processes of whole plant senescence need to be viewed together, and interconnection is an important theme here. Whole plant senescence is an important and interesting phenomenon whose analysis is tractable; however, progress can be hindered by unnecessary controversy and monolithic ideas. Some of these problems will be discussed specifically below (whole plant senescence is covered by Section II) under rules of evidence.

This is a huge and very complex literature, much of it out of reach of (older than) searchable digital databases. Compounding these difficulties, important information relevant to whole plant senescence is often buried within papers where senescence is not an obvious theme or even a keyword. This field cannot be covered in depth with a short summary. Nonetheless, an overview with some updating on key issues seems useful, especially for non-specialists.

B. Patterns

In terms of longevity, plants fall mostly into three categories: annual, biennial, and perennial. When considering whole plant senescence, however, it seems better to classify them as monocarpic (semelparous) or polycarpic (iteroparous) which emphasizes their senescence pattern and its relationship to reproduction. Monocarpy(ic) refers to a life cycle with one reproductive episode followed by death, and polycarpy(ic) to more than one reproductive episode (Hildebrand, 1882; Molisch, 1938). The terms monocarpy and polycarpy predate semelpary and iteropary by about a century and would normally take precedence (Noodén, 1980). Generally, monocarpic plants show a rapid, distinctive degeneration (monocarpic senescence) toward the end of their reproductive phase, while polycarpic plants undergo a more gradual decline. Usually, monocarpic plants are annual or biennial; however, a few are monocarpic perennials, and there are some variable (facultative) types (Noodén, 1988). Far more is known about monocarpic senescence than about the decline of polycarpic plants (Noodén, 1988), yet even this is quite incomplete. This summary will deal primarily with monocarpic senescence which is senescence in the physiologists' sense (Chapters 1 and 23). Senescence of polycarpic plants is less clear and also less tractable to physiological analysis; this is discussed below in Section IV.

C. What is Whole Plant Senescence?

The scope of whole plant senescence needs to be considered. For example, is it a cellular phenomenon or a system phenomenon? Considerable evidence indicates that most individual cells or even most parts such as shoot cuttings can continue to live long beyond the source plant (system) if they are excised and cultured under favorable conditions (Noodén, 1980, 1988). Therefore, whole plant senescence is primarily a system phenomenon.

It is not known whether monocarpic plants die as a result of organ failure or global failure, but one could argue that leaf senescence, and therefore organ failure, is particularly important (Noodén, 1988). Nonetheless, the different organs influence each other and are therefore tied together in the whole organism. Thus, it is not entirely clear which organ should be measured, but given the importance of the leaves, it is reasonable to continue using them and photosynthesis-related parameters as measures of senescence, at least until better criteria are established.

D. Senescence Parameters

Traditionally, whole plant senescence has been viewed primarily in terms of leaf senescence and that is measured mainly through chloroplastic parameters, e.g., chlorophyll loss, decreased photosynthesis, total leaf protein (mostly in the chloroplast protein) or nitrogen (Chapter 1). This emphasis on chloroplastic parameters at the whole plant level is understandable, because (a) these changes are the most conspicuous manifestations of the senescence syndrome and (b) many of the whole plant studies have been directed at developmental changes of these parameters. While chloroplast degeneration is certainly a prominent feature of the broad senescence syndrome, it may not be part of the primary (causal) senescence pathway, i.e., leaves do not die just because their chloroplasts have degenerated (Chapter 1). If we knew precisely the primary steps of the senescence process, these steps would be good measures of senescence, but that will have to wait (Chapter 1).

Since death is the end result of senescence, that could serve as a measure. However, so far, little effort has gone into defining criteria for organism death, and that may be due to the difficulty of defining the end point (Noodén and Penney, 2001). Ideally, that end point would be the death of the last living cell, but this seems impossible to measure. Often, however, the dying organs, e.g., leaves, collapse quickly, so their death can be measured easily.

E. Importance of Chronological Age

The issue of the extent to which monocarpic senescence is linked to age has already been discussed in Chapter 1; however, it has special implications in whole plant senescence. While it is true that monocarpic senescence progresses with age, it has been known for a long time that senescence is not simply a result of growing older (Leopold, 1961; Noodén and Leopold, 1978). Perhaps, the simplest demonstration of this is the prolongation of the life of leaves that have been excised and rooted (Molisch, 1938). Also, individual leaves on a plant may senesce in an order different ("nonsequential") from their chronological age (Resende, 1964; Pate *et al.*, 1983; Mondal and Choudhuri, 1984). In addition, the correlative controls are able to override age and accelerate or retard senescence of the whole plant or its parts. For example, using soybean explants (a leaf with one or more pods and a subtending stem segment), it can be shown that monocarpic senescence is related to the developmental stage of the pod, not the age of the leaf (Noodén, 1985). In other words, leaf senescence is a product of internal controls, not simply passive aging. These degenerative changes should not be called aging (Chapter 1; Noodén and Leopold, 1978).

Evidence indicates that some important processes including regenerative ability (Noodén, 1980; Grbic and Bleecker, 1995; Weaver *et al.*, 1998) do diminish with age. Furthermore, Arabidopsis leaves, and presumably other short-lived plants, seem to be built to last only a limited time (Bleecker and Patterson, 1997; Noodén and Penney, 2001), and therefore, age may be an important factor in their decline. Nonetheless, senescing Arabidopsis leaves do show changes in gene expression characteristic of leaf senescence (Quirino *et al.*, 1999), so a typical senescence program appears to be taking place anyhow.

F. Correlative Controls

Anyone who has watched plants develop over their life cycle will readily accept that there must be a lot of internal coordination, i.e., many correlative controls (the influence of one part of an organism on another). Often, hormones mediate these controls (Leopold and Noodén, 1984). The reproductive and senescence phases of monocarpic plants are also subject to a network of correlative controls, and an effort to summarize what is known about these controls has been made for soybean (Noodén, 1984).

In monocarpic plants where senescence and death closely follow reproductive development, it should not be surprising that this senescence (monocarpic senescence) is often controlled by the developing reproductive structures (see Molisch, 1938; Sax, 1962; Noodén, 1980). In these cases, removal of the reproductive structures or prevention of their development (e.g., sterility mutations) usually prolongs the life of the plant. Soybean, cucumber and mignonette are prime examples. On the other hand, there are some prominent exceptions (Noodén, 1988). For example, grain head removal may not always delay senescence in some of the grasses such as corn and wheat; indeed, this may actually accelerate senescence. In Arabidopsis, the longevity of the rosette leaves is not controlled by the reproductive structures even though plant longevity is (Hensel *et al.*, 1993; Noodén and Penney, 2001). While control by the reproductive structures will be emphasized here, it is important to note that other patterns may exist in particular species, particularly in the grass family. In addition, it follows that treatments altering reproductive development, particularly fruit development, likewise affect monocarpic senescence. Failure to recognize this important correlative control can lead to unnecessary controversy.

G. The Causes of Whole Plant Senescence

In monocarpic senescence, a central question is how do the developing reproductive structures cause the death of the plant. While the reproductive structures may be the most important correlative controllers in monocarpic senescence, other parts also participate. For example, the roots play an important role in maintaining leaves, apparently mainly through their production of cytokinins (Molisch, 1938; Van Staden *et al.*, 1988). The flux of cytokinins from the roots to the leaves has been shown to decrease during monocarpic senescence in a variety of plants ranging from rice to beans (Van Staden *et al.*, 1988). In soybean, it can be shown that the decrease in cytokinin production by the roots is necessary, but not itself sufficient, for monocarpic senescence, and the developing pods induce the decline in cytokinin production (Noodén and Letham, 1993). The ways by which the developing fruit can induce monocarpic senescence will be discussed further below in Section III.

H. Evolution and Differences among Monocarpic Species

Another special issue surrounding whole plant senescence has to do with differences among species. Clearly, monocarpy has evolved independently several times from polycarpy, and there are some significant differences, particularly in the correlative and hormonal controls (Stebbins, 1974; Noodén, 1980; Young and Augspurger, 1991). These differences pose a dilemma with regard to making generalizations about monocarpic senescence in different species. It seems best to try to formulate a general picture covering monocarpic senescence in all species, but to recognize that differences may exist.

II. Complexity and the Rules of Evidence

Of all the fields that we are acquainted with, this one has suffered most from unnecessary controversy. No doubt, this is due to the complexity of whole plant senescence, and it seems necessary to discuss some of these cases in very specific terms in order to clarify the issues looming in this field and to prevent reoccurrence. Three specific cases will be described and the evidence will be laid out; the reader can judge for himself/herself. Without a doubt, the most significant problems surround the mechanism or cause of monocarpic senescence.

First, some rules of evidence for causality. In 1876, Koch laid an important foundation for these rules that have become known as Koch's postulates (see e.g., Thimann, 1955) which grew out of his studies on the causes of disease. Just the correlation of the presence or absence of a bacterium with the occurrence/absence of a disease is not sufficient to establish that the bacterium is the cause of the disease. It must also be shown that introducing the bacterium into a healthy organism produces the disease. A similar set of problems exist in establishing which hormones control particular plant processes. Even if the application of a hormone causes an effect, this does not prove that the hormone normally regulates that process. Jacobs (1979) has developed an extension of Koch's postulates for analyzing hormonal controls, and this requires additional manipulations to establish a causal connection between a hormone and a putative effect in a whole plant. For example, the putative source of the hormone should be removed to block the effect on the target and then exogenous hormone is added back in place of the hormone source to reinstate the effect on the target. Some analogous manipulations have been used in studying causality in monocarpic senescence, but the advent of molecular biology and better genetic tools will offer additional criteria.

Monocarpic senescence has been viewed as exhaustion death or nutrient deficiency death based on observations that nutrients were redistributed from the leaves to the developing fruits in monocarpic plants (Hildebrand, 1882; Molisch, 1938). In many monocarpic plants, there is a good correlation between nutrient accumulation in the fruits and whole plant senescence. Thus, the exhaustion death idea does have some intuitive appeal. Some are thoroughly convinced by simple correlation evidence; others are not (Noodén, 1988; Grabau, 1995). Significantly, this correlation does not hold for many situations (Noodén, 1980, 1988). As in the cases cited above, more than simple correlation is required to establish a causal connection.

Second, probably as a result of complexity, one may encounter conflicting data within the same paper. For example, in one of the most widely cited papers on whole plant senescence, depodding of soybean plants appeared to decrease photosynthesis, yet did not decrease overall dry weight accumulation (Wittenbach, 1983). Since the decrease in photosynthesis in depodded plants has been interpreted to show that monocarpic senescence (if measured simply by photosynthetic rate) is not controlled by the reproductive structures, it seems important to resolve a contradiction like this (Noodén and Guiamét, 1989).

Third, non-use of available data—the literature in this field is widely scattered, and often highly relevant data are buried within papers not focused on senescence. While these factors add to the challenges posed by complexity, this is the kind of drawing together that reviewers are supposed to do (or at least try). One prominent case of particular importance here is the proposal that the developing fruits create a deficiency in carbohydrates, i.e., sugars, that leads to the death of monocarpic plants at the end of the reproductive phase (Kelly and Davies, 1988; Sklensky and Davies, 1993). Much of this discussion focuses on the garden pea with emphasis on senescence of the shoot apex as the cause of the death of the plant. This hypothesis is, like most of the earlier nutrient withdrawal and diversion hypotheses, based primarily on a correlation between the distribution/redistribution of dry matter and photosynthate. The following lines of contrary evidence, mostly in the mainstream senescence literature, are not reconciled with this hypothesis: (1) Application of sucrose to the pea shoot apex does not prevent its death (Lockhart and Gottschall, 1961). (2) Sucrose applications may actually promote senescence rather than retard it (see Section IIIB below). (3) Carbohydrate levels often increase rather than decrease during senescence (see Section IIIB below). These observations seem incompatible with the carbohydrate deficiency mechanism proposed.

The problems outlined here reflect the status of the field of whole plant senescence and to some extent its past culture, and it seems necessary to consider them in order to move forward.

III. What is the Cause of Monocarpic Senescence?

Over the past 70 years, there have been many ideas about what causes monocarpic plants to die. They can be grouped as follows, but these are not mutually exclusive and may work in combination:

A. Loss of vegetative regenerative capacity
B. Nutrient (and hormone) deficiencies
    1. Nutrient withdrawal
    2. Nutrient (and hormone) diversion
C. Death hormone (senescence signal).

A. Loss of Vegetative Regenerative Capacity

Often, monocarpic plants cease their vegetative growth fairly abruptly early in their reproductive phase (Noodén, 1980; Kelly and Davies, 1988; Reekie, 1999). Conversely, the perennial polycarpic pattern requires continued vegetative growth (Thomas *et al.*, 2000). This prominent shift (diversion) in growth-related allocation of resources in monocarpy seems to be part of a reproductive strategy that optimizes reproductive output for these plants (Section V below). Not only does this shift (divert) the allocation of assimilates from the vegetative parts to the reproductive structures, but there is often redistribution (withdrawal) of nutrients already invested in the vegetative parts. This diversion/withdrawal is often quite prominent, and it has given rise to the nutrient exhaustion explanation for the cause of monocarpic senescence which will be discussed below.

Cessation of growth may be caused by a simple arrest of the apex growth or by senescence and death of the shoot apex (Noodén, 1980). Pea is a prime example of apex senescence (Lockhart and Gottschall, 1961; Kelly and Davies, 1988).

On the surface, it seems that nutrient diversion or withdrawal might cause or contribute directly to the cessation of growth in the reproductive phase of monocarpic plants, but numerous inconsistencies exist (Noodén, 1980). For example, parthenocarpic cucumber fruits do not cause growth cessation as seed-bearing fruits do even though they seem to represent about the same amount of diversion/withdrawal as seed-bearing fruits (McCollum, 1934). Furthermore, fruit removal does not reinstate shoot growth in many species, e.g., soybean (Noodén, 1980). Thus, the shutdown of vegetative growth seems to be under a more direct control, separate from nutrient diversion.

Implicit in the review by Kelly and Davies (1988) is the idea that monocarpic plants, particularly peas, die because the shoot apex dies. The evidence against this idea is discussed in greater detail elsewhere (Noodén, 1980), but even in peas, the locus *Veg* causes death of the apex without inducing the corresponding monocarpic senescence (Reid and Murfet, 1984). Moreover, decapitation does not induce senescence in plants; in fact, it usually inhibits senescence of the lower leaves (Noodén, 1980). So, what is the role of apex death in monocarpic senescence? These look like separate processes, and careful distinction between them will prevent controversy.

Nonetheless, cessation of apex growth and of continued growth or regeneration of the plant, e.g., leaf production, may be necessary for monocarpic senescence; that is the plant will not die if it keeps growing and producing new vegetative structures. For example, in *Arabidopsis*, continued regeneration of photosynthetic parts extends the life of the plant (Noodén and Penny, 2001).

B. Nutrient (and Hormone) Deficiencies

Variations of the deficiency (exhaustion death) idea have been the dominant explanations for causality in monocarpic senescence. Because nutrients were observed to accumulate in the seeds at the same time that the vegetative parts, particularly the leaves, were losing nutrients and dying, it was believed that the developing fruits caused a nutrient deficiency in the vegetative parts (Molisch, 1938). The nutrient deficiency ideas in turn fall into two subgroups: (1) withdrawal and (2) diversion. These are not entirely different, and they may work together; however, it seems simpler to discuss them separately.

1. Nutrient withdrawal (redistribution)

The basic idea here is that the developing fruits withdraw minerals and/or carbohydrates from the vegetative parts thereby killing them. This is the original form of the exhaustion death hypothesis (Hildebrand, 1882; Molisch, 1938), and it has emphasized N and P, which are often limiting from the plant's environment and are quite mobile within the plant. There certainly is an enormous old literature showing correlations to support this idea in many monocarpic plants. As the fruits develop, carbohydrates, N, P, and other mobile minerals exit from the vegetative parts, particularly the leaves, and reciprocally accumulate in the fruits, particularly the seeds (Molisch, 1938; Noodén, 1980; Marschner, 1995). However, this interplay between source (e.g., leaves) and sink (e.g., seeds) is more complex than a simple withdrawal (Noodén, 1988). For example, leaves may continue to move nutrients toward their base after detachment, which physically disconnects them from the sinks.

The exhaustion death idea and this mainly correlational evidence were questioned long ago (Mothes and Engelbrecht, 1952; Leopold *et al.*, 1959), mainly because male plants show the same monocarpic senescence as female plants in some species such as hemp and spinach even though their reproductive sinks are quite small. Although foliar applications of mineral nutrients sometimes does delay monocarpic senescence, numerous non-correlations such as the failure of foliar fertilizing to prevent monocarpic death have usually been ignored (Noodén, 1980, 1988). Nonetheless, the nutrient exhaustion hypothesis is widely accepted.

Rather little direct testing of the nutrient withdrawal ideas has been carried out as contrasted to simple gathering of correlational data and that testing is mostly with soybean. Soybean seems to be the classic example of massive nutrient redistribution causing death during monocarpic senescence, yet it offers some of the clearest evidence against the nutrient deficiency ideas. Some of these experiments are more complex and are discussed elsewhere (Noodén, 1988), but three simple tests with clear results are from:

a. Soybean explants. When these cuttings, which include a leaf, one or more pods and a subtending stem (a nutritional unit), are fed a solution of minerals resembling xylem sap through the transpiration stream, the minerals have only a slight delaying effect, not prevention, of leaf senescence (Noodén, 1988; Mauk *et al.*, 1990). This is not what the nutrient withdrawal hypothesis would predict.
b. Steam-girdling (phloem destruction) the leaf petiole in pod-bearing soybeans does not block pod induction of leaf senescence even though it blocks nutrient withdrawal (Fig. 15-1; Wood *et al.*, 1986). Indeed, the blocked leaves build up high concentrations of starch (Fig. 15-2). Again, these data seem contradictory to the nutrient withdrawal idea. The control, a steam-girdled leaf on a depodded plant, also accumulates starch but does not turn yellow, so the phloem blockage does not cause accumulation of toxins.
c. Depodding (or removal of the seeds) after all or most of the nutrient withdrawal has occurred still prevents leaf senescence and plant death in soybean (Noodén, 1980, 1988). The lethal effect of the seeds occurs very late, when the seeds are yellowing. Similar observations have been reported for cucumber and bean (Leopold *et al.*, 1959).

Interestingly, even in soybean where the filling pods seem to place a particularly heavy demand on the parent plant, the growing seeds can be supplied from current assimilation rather than redistribution if senescence is prevented (Noodén *et al.*, 1979; Noodén, 1985, 1988). Some interesting observations made on wheat warrant mention here. Steam-girdling or severing the connection between the sink (filling grains) and the source (flag leaves) causes carbohydrate accumulation in the leaves and promotes senescence (Lazan *et al.*, 1983; Frölich and Feller, 1991). The sugar-deficiency idea is generally entertained as a facet of nutrient diversion, so it will be discussed below in that context.

2. Nutrient (and hormone) diversion

There are many variations of this idea, but in general, the developing fruits (or other large sinks) are believed to divert the supply of essential nutrients (or hormones) creating a deficiency(ies) which ultimately proves lethal to the vegetative parts.

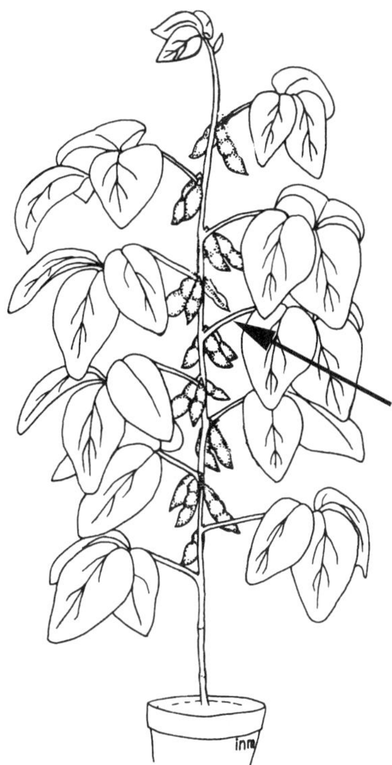

Figure 15-1. Steam-girdling in the petiole of a pod-bearing soybean plant (Noodén and Murray, 1982). This treatment blocks nutrient withdrawal from the leaf but not the induction of leaf senescence (yellowing) by the pods.

Among the hormones that could be involved here, the best candidate is cytokinin which plays a well-established antisenescence role (Van Staden *et al.*, 1988), but there is direct evidence against this kind of diversion of cytokinins in soybean (Noodén and Letham, 1993).

One avenue for diversion is the transpiration stream, but the transpiration rate of the fruits is small relative to the leaves (Noodén, 1988). In addition, defruiting causes the stomata in the leaves to close and thereby diminishes their transpiration. Therefore, the fruits do not compete with the leaves for hormones and nutrients coming up from the roots by diverting the transpiration stream.

Understandably, the major candidate for nutrient diversion is photosynthate, specifically sucrose, and that is the major theme of the reviews by Kelly and Davies (1988) and Sklensky and Davies (1993). Likewise, in soybean, it has been concluded that the "high rate of photosynthate utilization by developing fruits decreases the photosynthate available for leaf maintenance processes" (Egli and Crafts-Brandner, 1996). If sucrose deprivation caused monocarpic senescence, then exogenous applications of sugar should be effective in preventing leaf senescence, but sugar applications actually promote senescence in a variety of tissues (Table 15-1) except for some leaves placed in darkness (Goldthwaite and Laetsch, 1967; Thimann, Tetley and Krivak, 1977). Similarly, sucrose applications do not prevent pea apex senescence (Lockhart and Gottschall, 1961), which is central to the mechanism of monocarpic senescence envisioned by Davies *et al.* (Kelly and Davies, 1988;

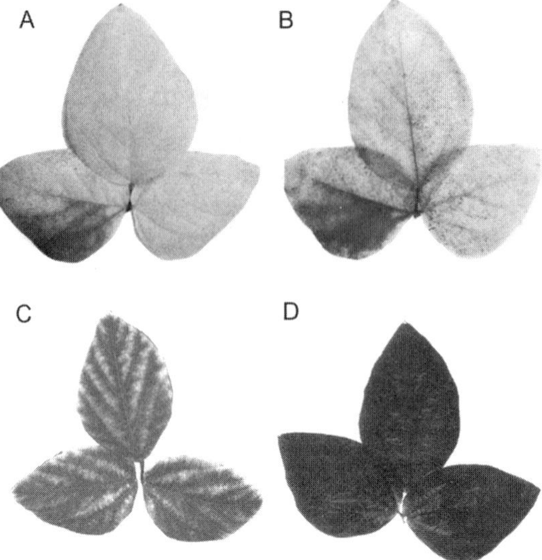

Figure 15-2. Effects of phloem destruction (steam-girdling) treatments on senescence (chlorophyll loss) and starch accumulation in the leaves of pod-bearing soybean plants (Wood *et al.*, 1986). A and C are fresh leaves taken from control plants at late podfill (A) or from plants treated as shown in Fig. 15-1 (C). B and D are the same leaves (A and C respectively) after clearing and staining with I_2 for starch. These figures show that steam-girdling the petiole blocks the withdrawal of starch from that leaf but does not block the yellowing (chlorophyll loss) induced by the pods.

Sklensky and Davies, 1993). Furthermore, numerous studies on a wide range of species including soybean (Egli *et al.*, 1980) show that the sugar concentrations in the leaves of plants undergoing monocarpic senescence may actually increase rather than decrease as required for the sugar deficiency mechanism [see Noodén *et al.* (1997) for some of these references]. Indeed, several reports attribute leaf senescence in a wide range of species to elevated carbohydrate levels (Allison and Weinmann, 1970; Mandahar and Garg, 1975; Hall and Milthorpe, 1977; Lazan *et al.*, 1983; Araus and Tapia, 1987; Ceppi *et al.*, 1987; Schaffer *et al.*, 1991; Prioul and Schwebel-Dugue, 1992; Fischer *et al.*, 1998). Some of these reports involve experimental manipulation of senescence, so they go beyond simply correlating with the normal progression of senescence. For some time, carbohydrate accumulation has been believed to exert a feed-back influence on photosynthetic capacity, and this can now be attributed to repression of photosynthesis-related genes (Azcon-Bieto, 1986; Sheen, 1994; Koch, 1996; Paul and Foyer, 2001). Proceeding a step further, Wood *et al.* (1986) found that steam-girdling a petiole in a pod-bearing soybean plant does not block the pod-induced senescence of that leaf (Fig. 15-1, discussed above) even as it blocks nutrient flux out of the leaf and causes a massive accumulation of starch (Fig. 15-2). This array of data seems incompatible with the sugar deficiency mechanism proposed by Davies *et al.* Further discussion of the pros and cons of the nutrient deficiency ideas can be found elsewhere (Noodén, 1980, 1988; Grabau, 1995).

Table 15-1. Senescence-promoting Effects of Sugars

Part used	Species (common name)	Light conditions	Observations	References
Leaf disks	*Phaseolus vulgaris* (common bean)	Darkness	0.01–0.1 M sucrose, glucose or fructose accelerates chlorophyll but apparently not protein loss	Goldthwaite and Laetsch, 1967
Leaf disks	*Xanthium pennsylvanicum* (cocklebur)	3000 lux continuous	0.01 M sucrose or glucose promotes chlorophyll loss, but NaCl does not	Khudairi, 1970
Detached leaves	*Spinacia oleracea* (spinach)	9 h per day, 16 μm light m^{-1} s^{-2}	0.05 M glucose feed through the transpiration stream promotes loss of chlorophyll, protein, Rubisco and several Calvin cycle enzymes	Krapp *et al.*, 1991
Leaf disks from young and mature leaves	*Nicotinia tobacum* (tobacco)	16 h per day, 20 μm light m^{-1} s^{-2}	50 mM glucose causes decreases in chlorophyll and hydroxypyruvate reductase; an osmotic agent (sorbitol) does not	Wingler *et al.*, 1998

C. Death Hormone (Senescence Signal)

As outlined above (Section IF), monocarpic senescence is under correlative controls, and usually, the developing fruits, particularly the seeds, exert some senescence-promoting influence. Thus, it makes sense to look for a simple solution like production of a senescence hormone (a positive signal) from the fruits (Wilson, 1997).

Except for soybean, there has not been much effort to characterize the *in vivo* behavior of the senescence signal (Noodén, 1988). In soybean, it seems to be exerted by the developing seeds on the leaves fairly late in seed development, i.e., at a time, when the seeds are able to live independently from the parent plant. The movement of the senescence signal is somewhat restricted by the vascular architecture (orthostichies), and it can act on the leaf nearest the pod via the xylem, i.e., it is not blocked by steam girdling of the petiole on that leaf (Figs. 15-1 and 15-2).

There is some direct evidence for an actual chemical factor that behaves like the senescence signal in soybean (Noodén *et al.*, 1990). Basically, axenic leachate from seeds excised at their senescence-inducing stage induces senescence resembling the *in vivo* pattern when it is fed into an excised leaf through the base of the petiole. A wide range of known senescence-inducing compounds including abscisic acid have been tested, but none matches the senescence signal.

Another line of evidence warrants some discussion here, because it has been offered as a test of the senescence hormone hypothesis and implied evidence against the senescence hormone idea (Hamilton and Davies, 1988). ^{14}C-Labeled CO_2 was supplied to the pods of pea plants in an effort to determine if labeled metabolites are exported to the shoot apex where the pods cause senescence. There are a number of technical problems with these experiments, but the simplest and most relevant here is isotopic dilution. Given the extent of isotopic dilution (i.e., lowering the specific activity of the ^{14}C-CO_2) by the ^{12}C-CO_2 produced by the pods or from the atmosphere and then further dilution by metabolites within the source and target tissues, one cannot expect to get detectable levels of ^{14}C into hormone-like molecules which function at very low concentrations (e.g., $<10^{-6}$ M). It should also be noted that much of the ^{14}C from ^{14}C-CO_2 appeared in sucrose, which as noted above, can promote senescence; however, pods are unlikely to export the large amounts required even if they actually do export some small quantities of organic materials.

So, where does the senescence hormone idea stand? It remains a feasible option, and it would explain the behavior of the senescence signal in soybean at least. Nonetheless, the senescence signal hypothesis cannot be proven by eliminating the alternatives, and direct evidence remains very limited.

IV. Senescence in Polycarpic Plants and Clones

Although polycarpic plants and clones may decline in their vigor, e.g., growth rates, etc., and they may even have characteristic longevities, they may not experience whole plant senescence in the sense of endogenously programmed degeneration (physiologists' sense), and usually, this decline is gradual (Noodén, 1988). It seems necessary to differentiate here between unitary organisms and clones. A unitary organism would be a single individual (integrated physiological unit) with a single root–shoot complex even if the shoot is branched. A clone, on the other hand, would be a genet, all the products of a single zygote independent of their size or degree of subdivision/separation. Where clonal organisms form multiple shoots (ramets), these can become separated into individual organisms, each with its own roots, i.e., unitary organisms. For example, all the Cabernet Sauvignon grape vines in the world belong to one genet, even though they have been carried far from each other (Penning-Roswell, 1971). It will only cause confusion if these clones consisting of disparate parts are not distinguished from individual organisms. Thus, senescence in clones may not be the same as that of individuals (e.g., a unitary individual). Likewise, longevity of clones is a different matter from longevity of unitary organisms.

Do clones or populations senesce? Some clonal populations do lose their vigor with age, even though they continue to produce new ramets, and sometimes, but probably not always, these declines may be attributed to accrual of pathogens, e.g., viruses (see Sax, 1962; Noodén, 1980, 1988) or genetic load (Kleckowski, 1997). Even whole communities may decline in their biomass growth and productivity as they age (Leopold, 1980; Huettl and Mueller-Dombois, 1993; Ryan *et al.*, 1997), and they may be reinvigorated after destruction by burning, etc. (Hilbert and Larigauderie, 1990; Ryan *et al.*, 1997). Sometimes, pollution is a good suspect for causing the degeneration of plant clones or communities, but other times, as in the case of ohi'a trees on Hawaii, the cause remains obscure and it could even be endogenous (Mueller-Dombois, 1992; Huettl and Mueller-Dombois, 1993). Some large populations such as bamboo species are clonal and also monocarpic perennials whose distantly disconnected individuals may die off synchronously (Janzen, 1976). While some of these declines may be endogenous (Noodén, 1988), the cause is unclear for most, and therefore whether or not they senesce in the physiologists' sense is very uncertain.

The question of senescence in polycarpic plants, especially trees, is also a complex matter with uncertain conclusions, and these have been addressed more completely elsewhere (Noodén, 1988; Watkinson, 1992; Noodén and Guiamét, 1996; Silvertown *et al.*, 2001; Roach, Chapter 23). While plants, even unitary plants, owing to their special growth patterns and regenerative abilities, can be viewed as a sort of clonal system or a metapopulation of cells, they still form a distinctive organism and that poses some limits for the community of cells. In other words, they are defined organisms, and they do have a finite, species-characteristic longevity (Noodén, 1988; Noodén and Guiamét, 1996).

In polycarpic perennials, whole organism senescence is not as conspicuous as that of monocarpic plants, yet a degeneration phenomenon can be demonstrated in longevity patterns and the different species have characteristic longevities which does suggest that there may be some sort of genetic (endogenous) control even if it affects mainly regenerative capacity and environmental resistance (Roach, Chapter 23).

Numerous and diverse ideas have been set forth to explain what limits the life of trees and causes their decline (see Noodén, 1988; Watkinson, 1992; Huettl and Mueller-Dombois, 1993; Larson, 2001; Lanner and Connor, 2001). Most are concerned with declining growth rate and diminished regenerative capacity in older organisms. These declines may be due to various factors or combinations of factors, none of which are clearly implicated, but some of these may be endogenous. Being part of an old organism with its accumulated bulk (necromass and unproductive living tissues) could become an increasing liability with age. In general, excising a part with growth potential (i.e., a shoot cutting) often can produce a vigorous new plant flourishing after its "parent" has died, and this process can be repeated indefinitely as witnessed by the many perennial cultivars (Sax, 1962; Noodén, 1988) that have been propagated vegetatively, some for hundreds of years, e.g., Cabernet Sauvignon grape (Penning-Roswell, 1971). On the other hand, accumulated genetic load could also be a factor in the decline of some polycarpic plants (Kleckowski, 1997), and that may be why vegetative regenerative capacity diminishes with age in some species (see Noodén,

1980, 1988). Here, a periodic sexual cycle seems to be required to restore vigor of the plant.

One particularly remarkable aspect of longevity in polycarpic plants is the ability of environmental stress to prolong it (Noodén, 1988). For example, a more stressful high, dry alpine environment greatly increases the longevity of bristlecone pines (Schulman, 1954), mild mineral nutrient deficiency extends the life of duckweed plants (Wangerman, 1965), and living in the stark environment on a cliff face similarly prolongs white cedar longevity (Larson, 2001). The underlying causes of these life extensions are unknown; however, acute stress can, of course, overwhelm and kill an organism.

Polycarpic (usually perennial) plants have evolved into monocarpic (mostly annual and biennial) plants, and this process has apparently occurred independently in several different taxonomic groups (Section IH above). It is also important to note the environmental factors or forces that drive the selection of an annual life cycle over a perennial life cycle. Ultimately, the pattern that yields the best reproductive output for a given environment is the one that evolves (Fox, 1990; Begon *et al.*, 1996; Noodén and Guiamét, 1996; Barbour *et al.*, 1999). Annuals (monocarpy) tend to evolve in environments where the survival of mature or established plants is low, that is where small, short-lived plants reproduce more successfully than large, long-lived plants. Thus, environments that are highly variable (e.g., deserts or disturbed areas) or habitats subject to heavy predation tend to favor annuals, but many crop plants have been selected for a clear, synchronous annual life cycle. By contrast, annuals may be disfavored (perennials favored) in stable environments such as mature forests. Perennials are also favored (annuals disfavored) in environments with very short growing seasons (McKenna and Houle, 2000).

V. Reproductive Yield

The question of how senescence affects reproductive yield (or even vegetative yield for certain crops) is an important underlying theme in much of whole plant physiology. Senescence is also an important issue in reproductive development. Here again, it is important to recognize physiological differences among monocarpic species but nonetheless to look for general principles.

In monocarpic species, there is a tradeoff or shift in allocation of resources from vegetative to reproductive growth, and this shift can be fairly abrupt. This allocation shift appears to be part of a strategy that optimizes the reproductive output (Begon *et al.*, 1996; Barbour *et al.*, 1999; Reekie, 1999). This shift in partitioning of resources is reflected in the cessation of shoot growth described above in Section IIIA, and the loss of assimilatory power (senescence) usually occurs later, apparently as a separate process. Field crops appear to have been selected not only to sharpen this allocation shift but also to synchronize seed maturation and facilitate harvest by dying (Hancock, 1992).

A key question is what limits seed yield, and there are two general perspectives on this; (1) sink limitation (e.g., seeds) and (2) source limitation (e.g., photosynthesis in leaves). Neither seed growth nor assimilation are simple one-component processes, and pinning down precisely which limits seed yields is difficult. There certainly is evidence to support both ideas. Indeed, some field crops, e.g., wheat (Kruk *et al.*, 1997) may have selected for sink strength, so they are now source limited. Nonetheless, there has been a strong belief that sink limitation is the key, particularly in soybean. However, using morphactin (an auxin transport inhibitor) treatments, it is possible to increase the number of seeds set in soybean by about 40%, yet there is no increase in yield (Noodén and Noodén, 1985). The total dry weight of seeds produced is the same; there are more seeds, but they are smaller. Another viewpoint (Bloom *et al.*, 1985) sees the multiple processes that go into reproductive yield as fairly well balanced, a sort of "just-in-time" system for a developing plant. This is also reflected in the complex, integrated correlative controls that regulate reproductive development in soybean (Noodén, 1984). From an economic or resource-use perspective, this balance is very efficient, but from a practical perspective, it means that increasing crop yields beyond exploitation of factors like disease resistance and lodging may be very challenging, requiring a better understanding of the controls at a whole plant level.

Nonetheless, delaying leaf senescence during monocarpic senescence does sometimes result in an increase in seed yield in rice (Dingkuhn and Kropff, 1996), corn (Prioul, 1996) and soybean (Noodén, 1985; Guiamét *et al.*, 1990; Egli and Crafts-Brandner, 1996); however, these increases may not materialize under all circumstances, particularly in field conditions (Luquez and Guiamét, 2001).

References

Allison, J.C., and Weinmann, H. (1970). Effects of the absence of developing grain on carbohydrate content and senescence of maize leaves. *Plant Physiology* **46**, 435–436.

Araus, J.L., and Tapia, L. (1987). Photosynthetic gas exchange characteristics of wheat flag leaf blades and sheaths during grain filling: The case of a spring crop grown under Mediterranean climate conditions. *Plant Physiology* **85**, 667–673.

Azcon-Bieto, J. (1986). The control of photosynthetic gas exchange by assimilate accumulation in wheat. In *Biological Control of Photosynthesis* (R. Marcelle, H. Clijsters, and M. Van Poucke, Eds.), pp. 231–240. Dordrecht: Martinus Nijhoff.

Barbour, M.G., Burk, J.H., Pitts, W.D., Gilliam, F.S., and Schwartz, M.W. (1999). *Terrestrial Plant Ecology* (3rd ed.). Benjamin/Cummings, Menlo Park, CA.

Begon, M., Harper, J.L., and Townsend, C.R. (1996). *Ecology: Individuals, Populations, and Communities* (3rd ed.). Blackwell Science, Oxford, UK.

Bleecker, A.B. (1998). The evolutionary basis of leaf senescence: Method to the madness? *Current Opinion in Plant Biology* **1**, 73–78.

Bleecker, A.B., and Patterson, S.E. (1997). Last exit: senescence, abscission, and meristem arrest in Arabidopsis. *Plant Cell* **9**, 1169–1179.

Bloom, A.J., Chapin, III, F.S., and Mooney, H.A. (1985). Resource limitation in plants—An economic analogy. *Annual Review of Ecology and Systematics* **16**, 363–392.

Ceppi, D., Sala, M., Gentinetta, E., Verderio, A., and Motto, M. (1987). Genotype-dependent leaf senescence in maize. Inheritance and effects of pollination-prevention. *Plant Physiology* **85**, 720–725.

Dingkuhn, M., and Kropff, M. (1996). Rice. In *Photoassimilate Distribution in Plants and Crops* (E. Zamski, and A.A. Schaffer, Eds.), pp. 519–547. Marcel Dekker, New York.

Egli, D.B., and Crafts-Brandner, S.J. (1996). Soybean. In *Photoassimilate Distribution in Plants and Crops* (E. Zamski and A.A. Schaffer, Eds.), pp. 595–623. Marcel Dekker, New York.

Egli, D.B., Leggett, J.E., and Cheniae, A. (1980). Carbohydrate levels in soybean leaves during reproductive growth. *Crop Science* **20**, 468–473.

Fischer, A., Brouquisse, R., and Raymond, P. (1998). Influence of senescence and of carbohydrate levels on the pattern of leaf proteases in purple nutsedge (*Cyperus rotundus*). *Physiologia Plantarum* **102**, 385–395.

Fox, G.A. (1990). Perennation and the persistence of annual life histories. *American Naturalist* **135**, 829–840.

Fröhlich, V., and Feller, U. (1991). Effect of phloem interruption on senescence and protein remobilization in the flag leaf of field-growth wheat. *Biochemie Und Physiologie Der Pflanzen* **187**(2), 139–147.

Goldthwaite, J.J., and Laetsch, W.M. (1967). Regulation of senescence in bean leaf disks by light and chemical growth regulators. *Plant Physiology* **42**, 1757–1762.

Grabau, L.J. (1995). Physiological mechanisms of plant senescence. In *Handbook of Plant and Crop Physiology.* (M. Pessarakli, Ed.), pp. 483–496. Marcel Dekker, New York.

Grbic, V., and Bleecker, A.B. (1995). Ethylene regulates the timing of leaf senescence in Arabidopsis. *Plant Journal* **8**, 595–602.

Guiamét, J.J., Teeri, J.A., and Noodén, L.D. (1990). Effects of nuclear and cytoplasmic genes altering chlorophyll loss on gas exchange during monocarpic senescence in soybean. *Plant and Cell Physiology* **31**, 1123–1130.

Hall, A.J., and Milthorpe, F.L. (1977). Assimilate source-sink relationships in *Capsicum annuum* L. III. The effects of fruit excision on photosynthesis and leaf and stem carbohydrates. *Australian Journal of Plant Physiology* **5**, 1–13.

Hamilton, D.A., and Davis, P.J. (1988). Sucrose and malic acid as the compounds exported to the apical bud of pea following carbon-14 carbon dioxide labelling of the fruit: No evidence for a senescence factor. *Plant Physiology* **88**, 466–472.

Hancock, J.F. (1992). *Plant Evolution and the Origin of Crop Species*. Prentice Hall, Englewood Cliffs, NJ.

Hensel, L.L., Grbic, V., Baumgarten, D.A., and Bleecker, A.B. (1993). Developmental and age-related processes that influence the longevity and senescence of photosynthetic tissues in Arabidopsis. *Plant Cell* **5**, 553–564.

Hilbert, D.W., and Larigauderie, A. (1990). The concept of stand senescence in chaparral and other Mediterranean type ecosystems. *Acta Oecologica-International Journal of Ecology* **11**, 181–190.

Hildebrand, F. (1882). Die Lebensdauer und Vegetationsweise der Pflanzen, ihre Ursache und ihre Entwicklung. *Bot. Jahrb. Syst. Pflanzengesch. Pflanzengeogr.* **2**, 51–135.

Huettl, R.F., and Mueller-Dombois, D. (Eds.) (1993). *Forest Decline in the Atlantic and Pacific Region*. Springer-Verlag, Berlin.

Jacobs, W.P. (1979). *Plant Hormones and Plant Development*, 339 pp. Cambridge University Press, Cambridge, UK.

Janzen, D.H. (1976). Why bamboos wait so long to flower. *Annual Review of Ecology and Systematics* **7**, 347–391.

Kelly, M.O., and Davies, P.J. (reviewer H.W. Woolhouse). (1988). The control of whole plant senescence. *CRC Critical Reviews in Plant Sciences* **7**, 139–173.

Khudairi, A.K. (1970). Chlorophyll degradation by light in leaf discs in the presence of sugar. *Plant Physiology* **23**, 613–622.

Kleckowski, E.J., Jr. (1997). Somatic theory of clonality. In *The Ecology and Evolution of Clonal Plants* (H. van Groenendael and J. de Kroon, Eds.), pp. 227–241. Backhuys, Leiden.

Koch, K.E. (1996). Carbohydrate-modulated gene expression in plants. *Annual Review of Plant Physiology and Plant Molecular Biology* **47**, 509–540.

Krapp, A., Quick, W.P., and Stitt, M. (1991). Ribulose-1,5-bisphosphate carboxylase-oxygenase, other Calvin-cycle enzymes, and chlorophyll decrease when glucose is supplied to mature spinach leaves via the transpiration stream. *Planta* **186**, 58–69.

Kruk, B.C., Calderini, D.F., and Slafer, G.A. (1997). Grain weight in wheat cultivars released from 1920 to 1990 as affected by post-anthesis defoliation. *Journal of Agricultural Science* **128**, 273–281.

Lanner, R.M., and Connor, K.F. (2001). Does bristlecone pine senesce? *Experimental Gerontology* **36**, 675–685.

Larson, D.W. (2001). The paradox of great longevity in a short-lived tree species. *Experimental Gerontology* **36**, 651–673.

Lazan, H.B., Barlow, E.W.R., and Brady, C.J. (1983). The significance of vascular connection in regulating senescence of the detached flag leaf of wheat. *Journal of Experimental Botany* **34**, 726–736.

Leopold, A.C. (1980). Aging and senescence in plant development. In *Senescence in Plants* (K.V. Thimann, Ed.), pp. 1–12. CRC Press, Boca Baton, FL.

Leopold, A.C. (1961). Senescence in plant development. *Science* **134**, 1727–1732.

Leopold, A.C., and Noodén, L.D. (1984). Hormonal regulatory systems in plants. In *Encyclopedia of Plant Physiology, New Series*. Vol. 10. *Hormonal Regulation of Development II* (T. Scott, Ed.), pp. 4–22. Springer Verlag, Berlin.

Leopold, A.C., Niedergang-Kamien, E., and Janick, J. (1959). Experimental modification of plant senescence. *Plant Physiology* **34**, 570–573.

Lockhart, J.A., and Gottschall, V. (1961). Fruit-induced and apical senescence in *Pisum sativum*. *Plant Physiology* **36**, 389–398.

Luquez, V.M., and Guiamét, J.J. (2001). Effects of the 'stay green' genotype Ggd1d1d2d2 on leaf gas exchange, dry matter accumulation and seed yield in soybean (*Glycine max* L. Merr.). *Annals of Botany* **87**, 313–318.

Mandahar, C.L., and Garg, I.D. (1975). Effect of ear removal on sugars and chlorophylls of barley leaves. *Photosynthetica* **9**, 407–409.

Marschner, H. (1995). *Mineral Nutrition of Higher Plants* (2nd ed.). Academic Press, London.

Mauk, C.S., Brinker, A.M., and Noodén, L.D. (1990). Probing monocarpic senescence and pod development through manipulation of cytokinin and mineral supplies in soybean explants. *Annals of Botany* **66**, 191–201.

McCollum, J.P. (1934). Vegetative and reproductive responses associated with fruit development in cucumber. *Memoirs Cornell Agricultural Experiment Station* 163.

McKenna, M.F., and Houle, G. (2000). Why are annual plants rarely spring ephemerals? *New Phytologist* **148**, 295–302.

Mishra, S.D., and Gaur, B.K. (1972). Control of senescence in betel leaves by depetiolation. *Experimental Gerontology* **7**, 31–35.

Molisch, H. (1938). The Longevity of Plants (*Die Lebensdauer der Pflanze*), 226 pp. Science Press, Lancaster, PA.

Mondal, W.A., and Choudhuri, M.A. (1984). Sequential and non-sequential pattern of monocarpic senescence in two rice cultivars. *Physiologia Plantarum* **61**, 287–292.

Mothes, K., and Engelbrecht, L. (1952). Über geschlectsverschiedenen Stoffwechsel zweihäusiger einjähriger Pflanzen I. Untersuchungen über den Stickstoffumsatz beim Hanf (*Cannabis sativa* L.). *Flora* **139**, 1–27.

Mueller-Dombois, D. (1992). A global perspective on forest decline. *Environmental Toxicology and Chemistry* **11**, 1069–1076.

Neumann, P.M., and Noodén, L.D. (1984). Pathway and regulation of phosphate translocation to the pods of soybean explants. *Physiologia Plantarum* **60**, 166–170.

Noodén, L.D. (1980). Senescence in the whole plant. In *Senescence in Plants* (K.V. Thimann, Ed.), pp. 219–258. CRC Press, Boca Raton, FL.

Noodén, L.D. (1984). Integration of soybean pod development and monocarpic senescence. *Physiologia Plantarum* **62**, 273–284.

Noodén, L.D. (1985). Regulation of soybean senescence. In *World Soybean Research Conference III: Proceedings.* (R. Shibles, Ed.), pp. 891–900. Westview Press, Boulder, CO.

Noodén, L.D. (1988). Whole plant senescence. In *Senescence and Aging in Plants* (L.D. Noodén and A.C. Leopold, Eds.), pp. 391–439. Academic Press, San Diego, CA.

Noodén, L.D., and Guiamét, J.J. (1996). Genetic control of senescence and aging in plants. In *Handbook of the Biology of Aging.* (4th ed.) (E.L. Schneider and J.W. Rowe, Eds.), pp. 94–118. Academic Press, San Diego, CA.

Noodén, L.D., and Guiamét, J.J. (1989). Regulation of assimilation and senescence by the fruit in monocarpic plants. *Physiologia Plantarum* **77**, 267–274.

Noodén, L.D., and Leopold, A.C. (1978). Phytohormones and the endogenous regulation of senescence and abscission. In *Phytohormones and Related Compounds: A Comprehensive Treatise* (D.S. Letham, P.B. Goodwin, and T.J.V. Higgins, Eds.), Vol, II, pp. 329–369. Elsevier/North-Holland Biomedical Press, Amsterdam.

Noodén, L.D., and Letham, D.S. (1993). Cytokinin metabolism and signaling in the soybean plant. *Australian Journal of Plant Physiology* **20**, 639–653.

Noodén, L.D., and Murray, B.J. (1982). Transmission of the monocarpic senescence signal via the xylem in soybean. *Plant Physiology* **69**, 754–756.

Noodén, L.D., and Noodén, S.M. (1985). Effects of morphactin and other auxin transport inhibitors on soybean (*Glycine max* cultivar Anoka) soybean senescence and pod development. *Plant Physiology* **78**, 263–266.

Noodén, L.D., and Penney, J.P. (2001). Correlative controls of senescence and plant death in *Arabidopsis thaliana* (*Brassicaceae*). *Journal of Experimental Botany* **52**, 2151–2159.

Noodén, L.D., Guiamét, J., Singh, S., Letham, D.S., Tsuji, J., and Schneider, M.J. (1990). Hormonal control of senescence. In *Plant Growth Substances 1988* (R. Pharis and S. Rood, Eds.), pp. 537–546. Springer-Verlag, Berlin.

Noodén, L.D., Guiamét, J.J., and John, I. (1997). Senescence mechanisms. *Physiologia Plantarum* **101**, 746–753.

Pate, J.S., Peoples, M.B., and Atkins, C.A. (1983). Post-anthesis economy of carbon in cultivar of cowpea. *Journal of Experimental Botany* **34**, 544–562.

Paul, M.J., and Foyer, C.H. (2001). Sink regulation of photosynthesis. *Journal of Experimental Botany* **52**, 1383–1400.

Penning-Roswell, E. (1971). *The Wines of Bordeaux*. International Wine and Food Publishing Co., London.

Prioul, J.L., and Schwebel-Dugue, N. (1992). Source-sink manipulations and carbohydrate metabolism in maize. *Crop Science* **32**, 751–756.

Prioul, J.-P. (1996). Corn. In *Photoassimilate Distribution in Plants and Crops* (E. Zamski and A.A. Schaffer, Eds.), pp. 549–594. Marcel Dekker, New York.

Quirino, B.F., Normanly, J., and Amasino, R.M. (1999). Diverse range of gene activity during Arabidopsis thaliana leaf senescence includes pathogen-independent induction of defense-related genes. *Plant Molecular Biology* **40**, 267–278.

Reekie, E.G. (1999). Resource allocation, trade-offs, and reproductive effort in plants. In *Life History Evolution in Plants* (T.O. Vuorisalo and K.P. Mutikainen, Eds.), pp. 173–193. Kluwer, Dordrecht.

Reid, J.B., and Murfet, I.C. (1984). Flowering in *Pisum*—a 5th Locus, *Veg. Annals of Botany* **53**, 369–382.

Resende, F. (1964). Senescence induced by flowering. *Port. Acta Biol., Ser. A* **8**, 248–266.

Ryan, M.G., Binkley D., and Powers, J.H. (1997). Age-related decline in forest productivity: Pattern and process. *Advances in Ecological Research* **27**, 213–262.

Sax, K. (1962). Aspects of aging in plants. *Annual Review of Plant Physiology* **13**, 489–506.
Schaffer, A.A., Nerson, H., and Zamski, E. (1991). Premature leaf chlorosis in cucumber associated with high starch accumulation. *Journal of Plant Physiology* **138**, 186–190.
Schulman, E. (1954). Longevity under adversity in conifers. *Science* **119**, 396–399.
Sheen, J. (1994). Feedback control of gene expression. *Photosynthesis Research* **39**, 427–438.
Silvertown, J., Franco, M., and Perez-Ishiwara, R. (2001). Evolution of senescence in iteroparous perennial plants. *Evolutionary Ecology Research* **3**, 393–412.
Sklensky, D.E., and Davies, P.J. (1993). Whole plant senesence: Reproduction and nutrient partitioning. *Horticultural Reviews* **15**, 335–366.
Stebbins, G.L. (1974). *Flowering Plants: Evolution above the Species Level.* Harvard University Press, Cambridge, MA.
Thimann, K.V. (1955). *The Life of Bacteria*. Macmillan, New York.
Thimann, K.V., Tetley, R.M., and Krivak, B.M. (1977). Metabolism of oat leaves during senescence. V. Senecence in light. *Plant Physiology* **59**, 448–454.
Thomas, H., Thomas, H.M., and Ougham, H. (2000). Annuality, perenniality and cell death. *Journal of Experimental Botany* **51**, 1781–1788.
Van Staden, J., Cook, E.L., and Noodén, L.D. (1988). Cytokinins and senescence. In *Senescence and Aging in Plants* (L.D. Noodén and A.C. Leopold, Eds.), pp. 281–328. Academic Press, San Diego, CA.
Wangerman, E. (1965). Longevity and aging in plants and plant organs. In *Handbuch der Pflanzenphysiologie* (A. Lang, Ed.), Vol. XV/2, pp. 1026–1057. Springer-Verlag, Berlin.
Watkinson, A. (1992). Plant senescence. *Trends in Ecology and Evolution* **7**, 417–420.
Weaver, M., Gan, S., Quirino, B., and Amasino, R.M. (1998). A comparison of the expression patterns of several senescence-associated genes in response to stress and hormone treatment. *Plant Molecular Biology* **37**, 455–469.
Wilson, J.B. (1997). An evolutionary perspective on the 'death hormone' hypothesis in plants. *Physiologia Plantarum* **99**, 511–516.
Wingler, A., Von Schaewen, A., Leegood, R.C., Lea, P.J., and Quick, W.P. (1998). Regulation of leaf senescence by cytokinin, sugars, and light. Effects of NADH-dependent hydroxypyruvate reductase. *Plant Physiology* **116**, 329–335.
Wittenbach, V.A. (1983). Effect of pod removal on leaf photosynthesis and soluble protein composition of field-grown soybeans. *Plant Physiology* **73**, 121–124.
Wood, L.J., Murray, B.J.,Okatan, Y., and Noodén, L.D. (1986). Effect of petiole phloem disruption on starch and mineral distribution in senescing soybean leaves. *American Journal of Botany* **73**, 1377–1383.
Young, T.P., and Augspurger, C.K. (1991). Ecology and evolution of long-lived semelparous plants. *TREE* **6**, 285–289.

16

Autumn Coloration, Carbon Acquisition and Leaf Senescence

Takayoshi Koike

I. Introduction

Autumn coloration of deciduous trees signals the final stage of the growing season before winter and preparation for the next growing season (Koike *et al.*, 1992; Feild *et al.*, 2001). Cool temperate forests are usually represented by deciduous broad-leaved trees with various kinds of autumn coloration (e.g., Chabot and Hicks, 1982; Koike, 1990; Feild *et al.*, 2001). Because of its beauty, autumn leaf coloration has attracted a lot of interest. In fact, the Chinese character for maple means "autumn coloration". Maple trees produce their best autumn color under full sunlight. The most beautiful autumn coloring is closely related to the timing of low temperature and the difference in daily temperature change of around 5 to 7°C during the night (Harada, 1954; Sato, 1989). Photoperiods, mainly short days or long nights, also induce autumnal leaf senescence (Chapter 26; Kozlowski and Pallardy, 1997). In general, short days and low temperatures trigger leaf senescence and chlorophyll degradation (Escudero and del Arco, 1987; del Arco *et al.*, 1991; Smart, 1994). Clear, bright days and dry weather produce maximal color development. In addition, the nutritional status of the growing site of the maple trees affects directly the color of the crown by way of producing additional shoot formation, such as lammas shoots, i.e., shoots developing late in the growing season. The new shoots have young leaves that bring a mixture of leaf color to the crown, because the younger leaves remain green longer.

Deciduous trees in the temperate region of the northern hemisphere usually change leaf color in autumn and shed their leaves before the winter season. The phenological progression of the shoot development in temperate woody plants is strongly affected by local weather. At first, we are fascinated by the various autumn colors of trees and think about the mechanisms producing different autumn colors; however, the patterns of autumn coloration may reflect the adaptive developmental patterns of trees in some interesting ways. The contrasting autumn coloration of trees may imply the adaptive growth characteristics of each species, e.g., deciduousness v.s. evergreeness. Moreover, initiation and progress of autumn coloration of a canopy is related to the growth characteristics of tree species (Koike, 1991).

Leaf senescence has been studied in order to increase plant yield (e.g., Thomas, 1992). Prolonging the green or photosynthetic period of monocarpic crop plants is one of the goals of plant breeding (Thomas and Smart, 1993); however, most tree species and other polycarpic plants live in natural habitats. Not only are forests important ecosystems, but also many forests are in effect important crops. It is important to understand the ecological aspects of leaf senescence or autumn coloration in the basic growth pattern of wild plants including woody species possibly to increase their yields.

In this chapter, I will first explain the progress pattern of autumn coloration of a crown of deciduous broad-leaved trees in cool temperate forests. Second, the progression of autumn coloring is related to the shoot development of tree species in successional forests. Finally, the relationship between leaf structure and shoot growth pattern will be analyzed with special references to pigment degradation patterns from leaf level to crown level. Based on this information, the ecological significance of leaf life-span is reviewed as a growth characteristic of carbon acquisition of tree species.

II. Progress of Autumn Coloration of a Tree Crown

Autumn coloring of deciduous broad-leaved trees begins in the outer part of a crown for many early-successional species such as larch, birch, poplar and willow in cool temperate climates. By contrast, it starts in the inner part of a crown for most late-successional species, e.g., maple, beech and basswood (Gower and Richards, 1990; Koike, 1990). These patterns of autumn coloration of crowns are closely related to the pattern of leaf emergence (Koike, 1990, 1991; Fig. 16-1). In North American hardwood forests, after commercial harvesting, some exploitive tree species, such as pin cherry (*Prunus pensylvanica*) keep green leaves until late in the growing season. This retention of green leaves by pin cherry may contribute to its high growth rate as a gap species in a secondary forest (Marks, 1974). A similar role of green leaves during the late growing season in hybrid poplar and crops was found to produce a high growth capacity in an open habitat (Nelson and Isebrands, 1983; Thomas and Smart, 1993).

Not all deciduous broad-leaved tree species change leaf color in autumn. For example, alder species (*Alnus hirsuta* and *A. japonica*) remain green until frost comes (Bortlik *et al.*, 1987; Neave *et al.*, 1989; Matile, 2000). Of course, older leaves of alders are shed from the inner part of a crown in the early summer (Kikuzawa, 1978, 1983; Koike, 1990). In some temperate forests, alders and woody legumes (e.g., *Robinia* sp., *Maakia amuruensis* var. *buergeri*) are planted to stabilize the soil and as a "biological fertilizer". Because they usually have symbioses with the nitrogen-fixing microorganisms in their roots, they may have less need to reclaim the nitrogen invested in their leaves. In fact, the remobilization of nitrogen from alder leaves was estimated to be only 30% maximum, in contrast to many other

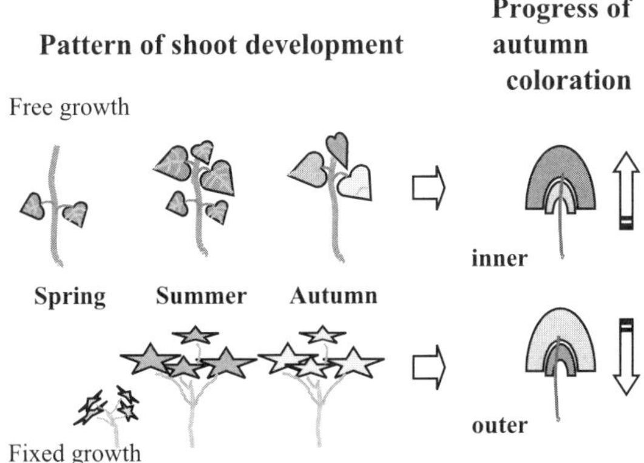

Figure 16-1. Pattern of autumn coloration of deciduous broad-leaved trees is affected by the pattern of leaf emergence. The darkest stippling represents full green color, the intermediate partially green (as in growing leaves) and the lightly stippled a change in color.

species, e.g., birch, maple, beech, larch, which reclaim 60–70% (Killingbeck, 1996; Craine and Mark, 1998). A similar pattern also occurs in *Duschekia fruticosa* (a similar species of alder) in Siberia (Matsuura and Abaimov, 1999). This ample nitrogen supply may be the reason why alders or woody legumes shed green leaves starting in early summer, and this early shedding may be induced by mutual shading of newly developing leaves. This shedding of leaves with a high N content may increase soil fertility, especially in areas with nutrient-poor soils. If maple species (e.g., *Acer palmatum*) grow on a highly fertile soil, the autumn color is not as beautiful due to many green leaves in the canopy top. This phenomenon is directly linked with the development of lammas shoots in several species growing on fertile soil.

It is considered that tree shoot development is closely correlated with the progress of its functions (e.g., growth, photosynthesis and respiration), which is strongly related to nutrient conditions (Amthor, 1989). In the next section, we will examine the event of shoot development and its functional aspects from the viewpoint of leaf senescence.

III. Shoot Development and Leaf Senescence

A. Leaf Phenology

Phenological plant research on both herbaceous and woody plants has focused on the description of seasonal events using pheno-diagrams which are useful (Lieth, 1971); however, this method may provide limited inference about the functional and dynamic aspects of plant life affected by environmental and endogenous conditions. More quantitative description is needed to show the dynamics of tree growth. Description of functional events of tree shoots has provided us with an understanding of their mode of growth.

The pattern of shoot development is usually related to the successional traits of deciduous broad-leaved trees (Marks, 1974; Maruyama, 1978). For example, based on morphological observation of the structure of a shoot, the early-successional tree species were found to elongate their shoots through a year, while late-successional trees species may keep leaves for a long time in Malaysian tropical rain forests (Koriba, 1958; Ninomiya *et al.*, 2000).

Similar leaf dynamics of successional tree species are found in the Indian mountain region with seasonal climatic conditions (Boojh and Ramakrishnan, 1982).

A similar relationship between the shoot elongation and diameter growth is also found for deciduous broad-leaved trees in the Appalachian mountain forests (Bicknell, 1982). Among these observations, the pattern of shoot elongation seems to directly relate to the forest dynamics. Early-successional species produce shoots over a long period, while late-successional species flush shoots all at the same time in order to occupy growth space. Demographic analysis of the individual leaves of a plant is a useful tool for addressing functional aspects of leaf phenology (Harper, 1987; Kikuzawa, 1983, 1984).

Deciduous broad-leaved trees with pioneer traits usually have the longer period of shoot elongation in order to secure growth space. The duration of shoot elongation of established late-successional species is usually short (Maruyama, 1978). However, some conifers, such as pine species (*Pinus densiflora*) with growth traits of the early-successional species flush their shoots at the beginning of the growing season (Nagata, 1983). By contrast, late-successional type species, such as Sugi cedar (*Cryptomeria japonica*) and Hinoki cypress (*Chamecyparis obtusa*) elongate shoots for a long time (Koike, 1982). Therefore, the pattern of shoot elongation is not always reflected from successional growth characteristics in tree species.

Leaf longevity and successional traits of deciduous broad-leaved tree species are closely correlated with the decreasing rate of net photosynthesis after it reaches the maximum value (Koike, 1988). Most early-successional trees or light-demanding species show a rapidly decreasing rate, while late-successional trees or shade-tolerant species have a slower decreasing rate (Koike, 1988; Reich *et al.*, 1995).

B. Leaf Shedding

Shedding of leaves is considered to occur to avoid environmental stresses, such as shading, cold and desiccation (Axelrod, 1966; Kozlowski and Pallardy, 1997; Addicott, 1982). In some tree species (e.g., cinnamon tree, *Cinnamomum camphora*), the old leaves turn orange in spring just before new leaf unfolding; thus, the life-span of cinnamon tree leaves is almost one year (Sato, 1989). Before leaf shedding in the beginning of early summer, one-year-old leaves change their color and nutrients are recycled from aged leaves to new leaves or branches. Similar phenomena are found for the evergreen shrub *Daphniphyllum macropodum* (the Japanese common name is "replacing leaves"). In heavy snow areas, it sheds older colored leaves after new leaves have unfolded (Hayashi, 1985). Snow cover usually acts as protection for plants against low temperature and winter desiccation, which may prolong leaf longevity. However, the physiological activities of photosynthetic production of those aged leaves decrease because of shading by shoots with new leaves. This shedding of aged leaves is believed to minimize the cost of photosynthesis due to respiration loss by new leaves under shady conditions and minimizing the leaf area to reducing evapotranspiration.

In a temperate forest, leaf senescence and shedding in small trees sometimes shows unique patterns relative to the canopy cover. For example, seedlings of *Prunus ssiori* unfold all leaves before canopy closure in deciduous broad-leaved forests of northern Japan. This cherry species efficiently uses incident light flux during the leafless period of the overstory in spring. In contrast to this species, *Carpinus cordata* seedlings flush all leaves at almost the same time as the canopy closure of tall trees but usually keep green leaves after leaf shedding of the overstory and have dead leaves in the next spring (Kitaoka *et al.*, 2000).

An interesting example of shade avoidance is found for *Daphne kamtschatica*, a summer deciduous shrub in the forest of northern Japan (Kikuzawa, 1984; Lei and Koike, 1998). In this shrub, leaf senescence starts with leaf yellowing after canopy closure of overstory trees in late spring, and all leaves are shed by mid-summer. This summer deciduousness seems to be complete shade avoidance. The photosynthetic rate of *D. kamtschatica* tended to be saturated at less than 400 μmol m^{-2} s^{-1} (Lei and Koike, 1998). However, photosynthetic photon flux density (PPFD) at the forest floor after canopy closure is 80 and 200 μmol m^{-2} s^{-1} depending on the density of canopy layers (Lei *et al.*, 1998). Although the shading cloth only reduced the PPFD and did not alter the spectrum, the leaves cannot carry out net photosynthesis under these conditions. This treatment with an artificial shading cloth accelerated senescence of the leaves. This shrub starts to unfold leaves at the beginning of autumn when most of the trees of the overstory begin to shed leaves.

Deciduousness may serve to avoid water deficits in the tropics (Axelrod, 1966). In the dry season of tropical areas, most trees shed leaves to avoid the desiccation period (Addicott, 1982). This may also be true for the shedding of needles of pine species in various seasons. For example, pine species originating from the high latitudes change needle color to yellow in autumn and shed needles before the winter season to minimize exposure to winter cold and desiccation. By contrast, pines in tropical regions sometimes shed their needles throughout the year. The intermediate case is found in middle latitude pines that change color and shed needles twice per year, namely during spring and autumn (Oohata, 1986).

Larch is well known as a deciduous conifer (Gower and Richards, 1990). However, lower branches of alpine larch (*Larix lyallii*) keep needles over winter when the moisture content in the soil is adequate (Richards, 1984). Similar overwintering capacity is also found in needles of larch seedlings (*L. gmelinii* and *L. decidua*) growing in heavy snow regions (Miyoshi, 1934; Koike, 2002). Before needle shedding, aged needles (usually one-year-old needles) change color from green to yellow and mobilize their nitrogen to new needles. This summer yellowing of overwintering needles of larch seedlings seems to be induced by desiccation in the soil rhizosphere. Leaf abscission starts in most tree species of different leaf habit (e.g., deciduous, evergreen, and marcescent) at a time of insufficient soil moisture (Escudero and del Arco, 1987). Marcescent tree species, such as oak (*Quercus dentata*) and carpinus (*Carpinus cordata*), usually hold their dead leaves until new leaves are unfolding and are apparent exceptions; however, it is believed that the transpiration stream is sealed off in these dead leaves and they protect leaf primordia from desiccation by the strong dry winds during winter and salt spray near seashores. In fact, *Q. dentata* is a major component of forest stands close to seashores.

Under mild stress conditions, leaf longevity usually increases. For instance, leaf longevity is greater in plants growing on infertile soils (Escudero and Arco, 1987; Koike and Sanada, 1989; Reich *et al.*, 1992, 1995; Kayama *et al.*, 2002). In fact, Sobrado (1991) found a ratio of leaf construction cost to maximum assimilation (cost/A_{max} where A_{max} is the maximum photosynthetic rate at light saturation) as a function of leaf life-span. In general, the decline in photosynthetic rate once it reaches its maximum value is rapid for most light-demanding species, while that in shade-tolerant species is slow (Koike, 1988, 1991; Reich *et al.*, 1995). Moreover, based on the rate of decline in photosynthesis with increase in time, the longevity of individual leaves was estimated as a function of the cost and benefit of construction of a leaf (Kikuzawa, 1991; Chapter 25). Therefore, leaf shedding should be found when the surplus production of older leaves would be less than the maintenance cost of older leaves and new leaf production.

However, leaf longevity is sometimes diminished by various kinds of stresses. Leaf habit (i.e., evergreeness or deciduousness) is dependent on the specific survival strategy in a forest stand or tundra where the availability of resources is different (Chapin *et al.*, 1990; Chapin and Kedrowski, 1983).

C. Leaf Nitrogen Content and Light Environment

Leaf nitrogen reflects photosynthetic activities (e.g., Evans, 1989; Ono *et al.*, 2001). The foliar nitrogen content in leaves of various deciduous broad-leaved trees native to North America decreases with age (Cote and Dawson, 1986). As the salt-extractable protein of leaves in eastern cottonwood (*Populus deltoides*) and white basswood (*Tilia heterophylla*) decreases, the free amino acids increase. Usually, plants keep more of their nutrients in their body when they grow under infertile soil conditions. In fact, the remobilization rate of leaf nitrogen was lower for black alder and cottonwood grown in fertile conditions compared with those in infertile conditions (Cote *et al.*, 1989).

Apple (*Malus domestica*) also shows no special autumn coloration, but its green period is elongated by nitrogen supplied at the beginning of autumn (Millard and Thomson, 1989). In the active phase, allocation of leaf nitrogen to ribulose-1,5-bisphosphate carboxylase/oxygenase (Rubisco) is usually around 20% for wild plants (Evans, 1989; Evans and Seeman, 1989) or 15–30% for several crop plants (Makino *et al.*, 1992). In autumn, degradation of Rubisco in leaves accounted for between 32 and 48% of the nitrogen subsequently remobilized and stored for leaf growth in the following spring. Most leaf protein is degraded and amino acids are exported from the leaf for storage over the winter as protein in bark (e.g., Chapin *et al.*, 1990; Ono *et al.*, 2001). This mobilization process of nitrogen is recognized as internal cycling of nitrogen (Chapter 14) and is more prominent in infertile conditions (Killingbeck, 1996).

Leaf nitrogen concentration is positively correlated with the maximum photosynthetic rate in many plant species (e.g., Evans, 1989). Generally, plants arrange their active leaves with high nitrogen content at the surface of the canopy, exposed to maximal sunlight. Field (1983) simulated that distribution of leaf nitrogen content that maximizes photosynthetic carbon gain over the canopy of an entire plant. The optimal nitrogen content increases with increasing daily photosynthetically active photon irradiance. Studies on *Solidago* sp. (Hirose and Werger, 1987) supported these results. Furthermore, the distribution of leaf nitrogen in a plant depends on the light environment for the leaf (Hikosaka *et al.*, 1994). If an aged leaf of the herbaceous vine *Ipomoea tricolor* receives adequate light flux under fertile conditions, the nitrogen content of the leaf is almost the same as that of active young leaves.

Similar nitrogen distribution patterns have been reported for a canopy of Scotch pine (*Pinus sylvestris*) grown in eastern Siberia where the tree density was very low (Koike *et al.*, 1999). The sparse pine stand on permafrost provides enough light flux in the deeper part of a pine crown, which means that most of the needles of a crown can receive enough light flux for photosynthesis. In fact, the nitrogen content of 8-year-old needles was almost the same as that of 2-year-old needles. Content of calcium in needles is found to increase with increasing needle age.

However, in autumn, the nitrogen concentration of individual leaves showed a specific pattern in the canopy of deciduous, broad-leaved trees (Koike *et al.*, 1992). The leaf nitrogen content of early-successional trees with inderminant type (e.g., birch and alder) decreased from leaves located at the base of a shoot or stem, while that of late-successional trees with determinant type (e.g., maple, basswood and beech) decreased from the surface of a crown or the canopy top (Fig. 16-2). That is, the nitrogen content in the basal leaves of determinant type species of leaf flush is less than that in the apical leaves. The former pattern is commonly found for most crop plants and many conifers. By contrast, the latter pattern is only observed for late-successional tree species with a deciduous leaf habit, especially for leaf flush type species.

In the crown of an alder (*Alnus hirsuta*) sapling, aged leaves positioned near the stem start to shed from mid-July without any yellowing (Kikuzawa, 1978; Koike *et al.*, 2001). The rest of the leaves usually remain green until frost comes, at which time they are shed green. This earlier leaf shedding is recognized as a structural specialization in shoots of alder species (Kikuzawa, 1983), but this unique trait is partly stimulated by self-shading from younger leaves. Moreover, the above-mentioned pattern of nitrogen distribution in a canopy of early

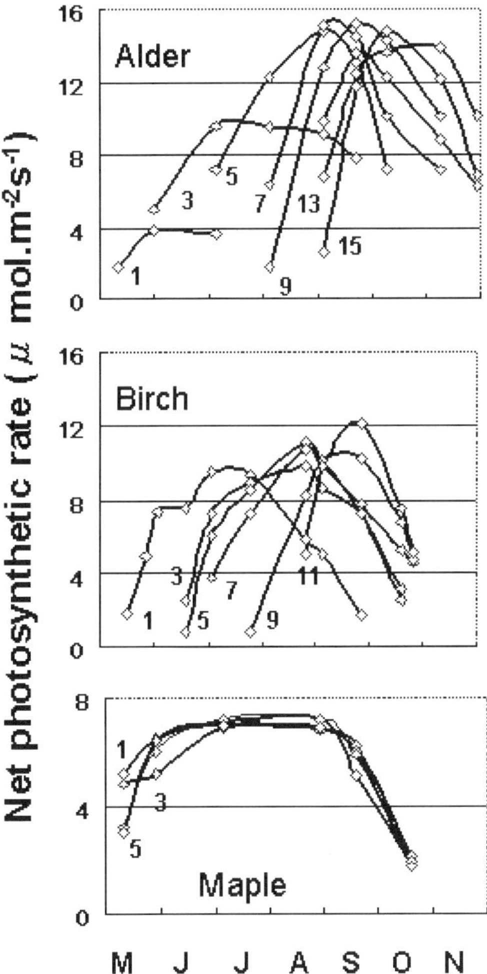

Figure 16-2. Seasonal changes in net photosynthetic rates of individual leaves with different leaf emergence patterns. Numbers in figures mean the leaf order from shoot base.

and late successional tree species and progress of autumn coloration are usually found for a whole canopy of trees exposed to full sunlight. However, trees in a forest sometimes do not clearly show this pattern of autumn coloration in their crowns, because their lower canopy receives little strong sunlight. In fact, full autumn coloration requires strong sunlight to promote anthocyanin synthesis (Feild *et al.*, 2001). The development of autumn coloration in tree crowns seems to be adapted to maximize photosynthetic production of the canopy of the tree.

In the next section, we will consider the relationship between the progress pattern of autumn coloration and leaf structure in deciduous broad-leaved trees.

IV. Leaf Structure and Function

Leaf structural differences may directly or indirectly affect photosynthetic capacity (Nobel *et al.*, 1975; Carpenter and Smith, 1981; Koike, 1988; Terashima, 1989). Alder (*Alnus hirsuta*) leaves that emerge at the beginning of autumn have two layers of epidermis, while leaves emerging earlier have only one layer. The two layers help to reduce the strong PPFD at low temperature in autumn (Koike, unpublished data). By contrast, late-successional species usually flush leaves only once, in the beginning of the growing season, and they have thinner leaves without any well-developed epidermis at the adaxial side of a leaf. These tree species with thinner leaves usually have high susceptibility to photoinhibition as compared with early-successional species (Kitao *et al.*, 2000).

The light utilization capacity of deciduous broad-leaved trees usually correlates well with leaf anatomical structure (Nobel *et al.*, 1975; Koike, 1988). Specifically, the maximum photosynthetic rate correlates positively with leaf thickness for deciduous broad-leaved tree species (Carpenter and Smith, 1981; Koike, 1988; Koike *et al.*, 1997; Castro-Diez *et al.*, 2000). The late-successional tree species have thinner leaves with a low light saturation point and lower photosynthetic rate as compared with early-successional species (Koike, 1988; Bazzaz, 1996). Under strong light flux in the top of canopy, the leaves are thinner in most late-successional tree species. In fact, maples (*Acer mono* or *A. plmatsumu*) usually suffer from photoinhibition when they reach the top of the canopy (Kitao *et al.*, 2000; Kitaoka *et al.*, 2001; Koike *et al.*, 2001). The thinner leaves seem unable to manage such strong light flux.

Most early-successional species in cool temperate forests continuously produce many leaves in order to cover a large growing space (Figs. 16-1, 16-3). They usually have thick leaves with well-developed palisade parenchyma are adapted for using high light flux (Dale, 1982, 1988; Koike, 1988; Koike *et al.*, 1997; Kimura *et al.*, 1998). Pearcy and Sims (1994) suggested that a high value of leaf mass per area (LMA) might be an adaptation of leaves acclimating to strong light flux, because a large amount of photosynthetic apparatus, especially in the palisade parenchyma, per unit area is present for absorbing photons (Terashima and Hikosaka, 1995). A similar idea has been proposed for several types of trees (Reich *et al.*, 1995).

Early-successional species usually have a high photosynthetic rate with high light saturation, while late-successional species have a low photosynthetic rate with low saturation point (Bazzaz, 1979; Koike, 1988). In addition, the former species showed a high turnover rate of leaves, i.e., the leaves are short-lived. By contrast, the latter species have leaves with a long leaf life-span (Koike, 1987; 1988; Kikuzawa, 1991). Autumn coloration of late-successional species usually starts from the surface of a crown where leaves have suffered from high light flux and a large temperature fluctuation. High light doses can induce

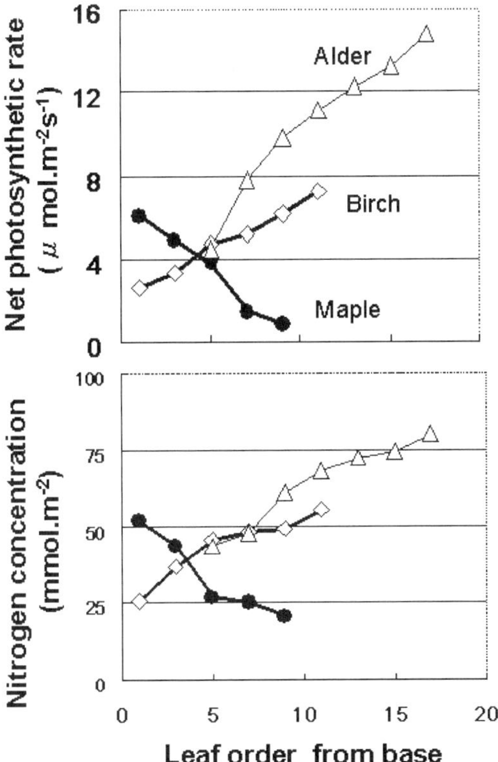

Figure 16-3. Net photosynthetic rate and leaf nitrogen concentration.

leaf senescence (Noodén *et al.*, 1996; Noodén and Leopold, 1988). These environmental stresses reduce the longevity of surface leaves because the flush leaves usually last through the season with high photosynthetic rate (Fig. 16-1).

Leaves positioned in the inner crown can photosynthesize later in the growing season (Koike *et al.*, 2001). Chlorophyll, particularly Chl b, acts as an antenna for photon absorption and for carrying photons to the reaction center. During leaf senescence, the Chl a/b decreases (Koike *et al.*, 1992; Thomas, 1997), and this enables senescing leaves to efficiently harvest the weak light flux when they are shaded (Fig. 16-4). Chl b stays at a relatively high level even in the final stages of leaf senescence, because this pigment is tightly bound to the light-harvesting complex proteins in thylakoid membranes and these complexes usually persist longer than the reaction centers in senescing leaves (Terashima and Inoue, 1985; Evans, 1989; Makino *et al.*, 1994, 1997). The rate of degradation of Chl b in most late-successional trees was slower than that in early-successional trees (Koike, 1990), and this may reflect the growth habitat of late-successional species in forests.

V. Leaf Pigments and their Ecological Role

During autumn, deciduous broad-leaf trees display various colors, such as yellow, orange, brown and red. With degradation of chlorophyll in autumn, the yellow carotenoids are

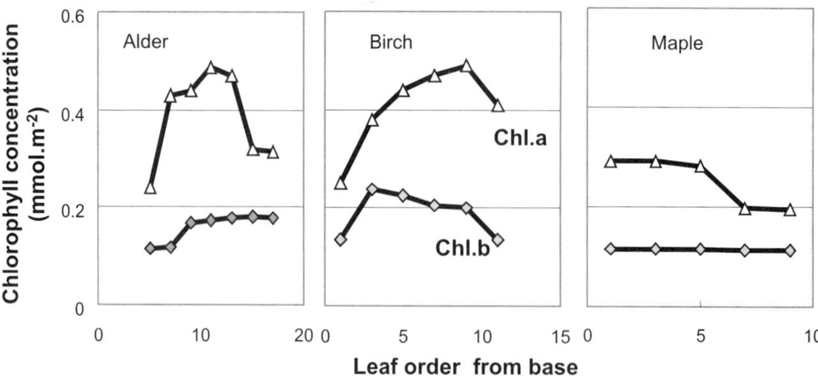

Figure 16-4. Chlorophyll a and b concentration of different leaf positions of a crown. The content of chlorophyll b of aged leaves is rather constant because it acts in accumulating weak light. Leaf order is numbered from the shoot base.

exposed and other colors such as red are due to anthocyanin (e.g., Gould *et al.*, 1995; Feild *et al.*, 2001). It has been proposed that red autumn color has coevolved with the anthocyanin in response to aphids as a warning color to discourage herbivores (Archetti, 2000). However, in general, anthocyanin is considered to act as protection against high sunlight and is synthesized from carbohydrate also resulting from the sunlight (Long *et al.*, 1994; Matile *et al.*, 1999; Smillie and Hetherington, 1999; Gould *et al.*, 2000).

Carotenoids also absorb photons and provide a protective effect against excessive blue-light irradiance (Merzlyak and Gitelson, 1995). They may help to delocalize the electrons kicked out of chlorophyll in excess light and dissipate the energy. At high photon flux density under full sunlight, the carotenoid content of leaves is higher than that of leaves in the shade (Demming-Adams, 1990). Tree canopies acclimated to high light dissipate excess energy through reactions mediated by a particular group of carotenoids. The xanthophyll cycle expends the excess energy of photon flux that would otherwise be passed on to oxygen via chlorophyll causing photo-oxidative damage (Demming-Adams, 1990; Han and Mukai, 1999).

"Autumn coloration" is also found on some conifers. In the beginning of winter, needles of Sugi cedar (*Cryptomeria japonica*) and Hinoi cypress (*Chamaecyparis obtusa*) usually change color from green to purple (Koike, 1982), which apparently protects against strong irradiance under low temperature (Ida, 1981). The role of rhodoxanthin accumulation in Sugi cedar was examined with a mutant lacking synthesis of the pigment (Han and Mukai, 1999). Net photosynthetic rate of the "purple-needle" type was not as strongly suppressed during winter as compared with the "green" needles. Similar phenomena are found in young leaves of most evergreen broad-leaved trees in temperate and tropical forests. At the leaf flushing, several evergreen broad-leaved trees including cinnamon trees flush new leaves with red color and little chlorophyll. The "spring coloration" or colored young leaves are recognized as "delayed greening" which is also believed to protect photosynthetic activities in leaves from high photon flux density (PPFD) and avoid grazing damage by herbivores (Coley and Kursar, 1996). These evergreen species realize their maximum photosynthetic rate after full expansion of leaves (Miyazawa *et al.*, 1998).

In conclusion, autumn coloration of deciduous broad-leaved trees has the very important role of maintaining a high photosynthetic capacity during the autumn when PPFD is high and temperatures are low. Among deciduous broad-leaved trees, except for alder, early-successional tree species have thick leaves and change canopy color from the inner part of a crown. By contrast, late-successional tree species have thinner leaves and change canopy color from the outer part of a canopy. The latter phenomenon seems to extend the green period and photosynthesis. Autumn coloration with anthocyanin seems to protect the photosynthetic apparatus against strong irradiance under cool temperatures. Autumn coloration in needles of conifers also acts to protect against strong PPFD with low temperature. Autumn coloration has positive value for trees growing in field conditions.

Acknowledgments

Thanks are due to Prof. H. Thomas, Dr. M. Kitao, Dr. Y. Matsuura, Dr. Y. Mukai and Dr. Y. Maruyama for constructive discussion or collecting references, and the Japanese FFPRI (MAFF) and MEXT for financial support.

References

Addicott, F.T. (1982). *Abscission*. University of California Press, Berkeley, CA.

Amthor, J.S. (1989). *Respiration and Crop Productivity*. Springer-Verlag, Berlin.

Archetii, M. (2000). The origin of autumn colors by coevolution. *Journal of Theoretical Biology* **205**, 625–630.

Axelrod, D.I. (1966). Origin of the deciduous and evergreen habits in temperate forests. *Evolution* **20**, 1–15.

Bazzaz, F.A. (1979). The physiological ecology of plant succession. *Annual Review of Ecology and Systematics* **10**, 351–371.

Bazzaz, F.A. (1996). *Plants under Changing Environment*. Cambridge University Press, Cambridge, UK.

Bicknell, S.H. (1982). Development of canopy stratification during early succession in northern hardwoods. *Forest Ecology and Management* **4**, 41–51.

Boojh, R., and Ramakrishnan, P.S. (1982). Growth strategy of trees related to successional status II. Leaf dynamics. *Forest Ecology and Management* **4**, 375–386.

Bortlik, K., Gut, H., and Matile, P. (1987). Yellowing and non-yellowing trees: A comparison of protein loss and chlorophyll loss in senescent leaves. *Botanica Helvetica* **97**, 323–328.

Carpenter, S.B., and Smith, N.D. (1981). A comparative study of leaf thickness among southern Apparition hardwood. *Canadian Journal of Botany* **59**, 1393–1396.

Castro-Diez, P., Puyravaud, J.P., and Cornelissen, J.H.C. (2000). Leaf structure and anatomy as related to leaf mass per area variation in seedlings of a range of woody plant species and types. *Oecologia* **124**, 476–486.

Chabot, B.F., and Hicks, D.J. (1982). The ecology of leaf life-spans. *Annual Review of Ecology and Systematics* **13**, 229–259.

Chapin, III, F.S., and Kedrowski, R.A. (1983). Seasonal changes in nitrogen and phosphorous fractions and autumn retranslocation in evergreen and deciduous taiga trees. *Ecology* **64**, 376–391.

Chapin, III, F.S., Schulze, E.-D., and Mooney, H.A. (1990). The ecology and economics of storage in plants. *Annual Review of Ecology and Systematics* **21**, 423–447.

Coley, P.D., and Kursar, T.A. (1996). Anti-herbivore defenses of young tropical leaves: physiological constraints and ecological trade-offs. In *Tropical Forest Plant Ecophysiology* (R.I. Mulkey, R.L. Chazdon, and A.P. Smith, Eds.), pp. 305–336. Chapman & Hall, New York.

Cote, B., and Dawson, J.O. (1986). Autumnal changes in total nitrogen, salt-extractable proteins and amino acids in leaves and adjacent bark of black alder, eastern cottonwood and white basswood. *Physiologia Plantarum* **67**, 102–108.

Cote, B., Vogel, C.S., and Dawson, J.O. (1989). Autumnal changes in tissue nitrogen of autumn olive, black alder and eastern cottonwood. *Plant and Soil* **118**, 23–32.

Craine, J.H., and Mark, M.C. (1998). Nutrients in senesced leaves: comment. *Ecology* **79**, 1818–1820.

Dale, J.E. (1982). *The Growth of Leaves*. Arnold Biol. Ser. 137. Edward Arnold, London.

Dale, J.E. (1988). The control of leaf expansion. *Annual Review of Plant Physiology and Plant Molecular Biology* **39**, 267–295.

del Arco, J.M., Escudero, A., and Garrido, M.V. (1991). Effects of site characteristics on nitrogen retranslocation from senescing leaves. *Ecology* **72**, 701–708.

Demming-Adams, B. (1990). Carotenoids and photoprotection in plants. A role for the xanthophyll zeaxanthin. *Biochemistry and Biophysical Acta* **1020**, 1–24.

Escudero, A., and del Arco, J.M. (1987). Ecological significance of the phenology of leaf abscission. *Oikos* **49**, 11–14.

Evans, J. R. (1989). Photosynthesis and nitrogen relationships in leaves of C3 plants. *Oecologia* **78**, 9–19.

Evans, J.R., and Seeman, J.R. (1989). The allocation of protein nitrogen in the photosynthetic apparatus: costs, consequences, and control. In *Photosynthesis* (W.R. Briggs, Ed.), pp. 183–205. Alan R. Liss, New York.

Feild, T.S., Lee, D.W., and Holbrook, N.M. (2001). Why leaves turn red in autumn. The role of anthocyanins in senescing leaves of Red-Osier dogwood. *Plant Physiology* **127**, 566–574.

Field, C. (1983). Allocating leaf nitrogen for the maximization of carbon gain: Leaf age as a control on the allocation program. *Oecologia* **56**, 341–347.

Gower, S.T., and Richards, J.H. (1990). Larches; deciduous conifers in evergreen world. *BioScience* **40**, 818–826.

Gould, K.S., Kuhn, D.N., Lee, D.W., and Oberbauer, S.F. (1995). Why leaves are sometimes red. *Nature* 378, 241–242.

Gould, K.S., Markham, K.R., Smith, R.H., and Goris, J.J. (2000). Functional role of anthocyanins in leaves of *Quintinia serrata* A. Cunn. *Journal of Experimental Botany* **51**, 1107–1115.

Han, Q., and Mukai, Y. (1999). Cold acclimation and photoinhibition of photosynthesis accompanied by needle color changes in *Cryptomeria japonica* during the winter. *Journal of Forest Research* **4**, 229–243.

Harada, Y. (1954). [*Forest and environment*], 159 pp. Hokkaido Association of Silviculture Promotion, Sapporo (in Japanese).

Harper, J. (1987). The value of a leaf. *Oecologia* **80**, 53–58.

Hayashi, Y. (1985). [*Trees in Japan*] Yama-Keikoku Co., Ltd, Tokyo (in Japanese).

Hikosaka, K., Terashima, I., and Katoh, S. (1994). Effects of leaf age, nitrogen nutrition and photon flux density on the distribution of nitrogen among leaves of a vine (*Ipomea tricolor* Cav.) grown horizontally to avoid mutual shading of leaves. *Oecologia* **97**, 451–457.

Hirose, T., and Werger, M.J.A. (1987). Maximizing daily canopy photosynthesis with respect to the leaf nitrogen allocation pattern in the canopy. *Physiologia Plantarum* **72**, 520–526.

Ida, K. (1981). Ecophysiological studies of the responses of taxodiaceous conifers to shading, with special references to the behavior of leaf pigments. I. Distribution of carotenoids in green and autumnal reddish-brown leaves of gymnosperms. *Botanical Magazine Tokyo* **94**, 41–54.

Kayama, M., Sasa, K., and Koike, T. (2002). Needle life span, photosynthetic rate, and nutrient concentration of *Picea glehnii*, *P. jezoensis*, and *P. abies* planted on serpentine soil in northern Japan. *Tree Physiology* **22**, 707–716.

Kikuzawa, K. (1978). Emergence, defoliation and longevity of alder (*Alnus hirsuta* Turcz.). leaves in a deciduous hardwood forest stand. *Japanese Journal of Ecology* **28**, 299–306.

Kikuzawa, K. (1983). Leaf survival of woody plants in deciduous broad-leaved forests. 1. Tall trees. *Canadian Journal of Botany* **61**, 2133–2139.

Kikuzawa, K. (1984). Leaf survival of woody plants in deciduous broad-leaved forests. 2. Small trees and shrubs. *Canadian Journal of Botany* **62**, 2551–2556.

Kikuzawa, K. (1991). A cost-benefit analysis of leaf habitat and leaf longevity of trees and their geographical pattern. *American Naturalist* **138**, 1250–1263.

Killingbeck, K.T. (1996). Nutrients in senesced leaves: Keys to the search for potential resorption and resorption proficiency. *Ecology* **77**, 1716–1727.

Kimura, K., Ishida, A., Uemura, A., Matsumoto, Y., and Terashima, I. (1998). Effect of current-year and previous-year PPFDs on shoot gross morphology and leaf properties in *Fagus japonica*. *Tree Physiology* **18**, 459–466.

Kitao, M., Lei, T.T., Koike, T., Tobita, H., and Maruyama, Y. (2000). Susceptibility to photoinhibition of three deciduous broadleaf tree species with different successional traits raised under various light regimes. *Plant, Cell and Environment* **23**, 81–89.

Kitaoka, S., Naniwa, A., Okuyama, S., and Koike, T. (2000). [Photosynthetic characteristics of seedlings of deciduous broad-leaved trees invading to old larch plantation.] *Proceedings of the Japanese Forestry Society* **111**, 211–212 (in Japanese).

Kitaoka, S., Kitahashi, Y., Shimizu, K., Hiura, T., and Koike, T. (2001). Canopy photosynthesis and transpiration in deciduous trees with special references to stomatal and non-stomatal regulation. In *Proceedings of Canopy Biology in Tropical Rain Forests. Sarawak, Malaysia*, pp. 108–114.

Koike, T. (1982). The formation of new leaves on seedlings of *Chamaecyparis obtusa* S. et Z. treated photoperiodically from summer to winter. *Journal of the Japanese Forestry Society* **64**, 275–279.

Koike, T. (1987). Photosynthesis and expansion in leaves of the early, mid, and late successional tree species, birch, ash and maple. *Photosynthetica* **21**, 503–508.

Koike, T. (1988). Leaf structure and photosynthetic performance as related to the forest succession of deciduous broad-leaved trees. *Plant Species Biology* **3**, 77–87.

Koike, T. (1990). Autumn coloring, photosynthetic performance and leaf development of deciduous broad-leaved trees in relation to forest succession. *Tree Physiology* **7**, 21–32.

Koike, T. (1991). [Pattern of autumn coloration and leaf emergence.] *Hoppo Ringyo (Northern Forestry)* **43**, 213–216 (in Japanese).

Koike, T. (2002). [Vertical distribution of tree species in relation to their biodiversity.] In *Environmental Physiology of Woody Plants* (H. Nagata and S. Sasaki, Eds.), pp. 89–112. Bun-Ei-do, Tokyo (in Japanese).

Koike, T., and Sanada, M. (1989). Photosynthesis and leaf longevity in alder, birch, and ash seedlings grown under different nitrogen levels. *Annales des Sciences Forestieres* **46S**, 295–297.

Koike, T., Sanada, M., Lei, T.T., Kitao, M., and Lechowicz, M.J. (1992). Senescence and the photosynthetic performance of individual leaves of deciduous broad-leaved trees as related to forest dynamics. *Research in Photosynthesis* **4**, 703–706.

Koike, T., Miyashita, N., and Toda, H. (1997). Effects of shading on leaf structural characteristics in successional deciduous broad-leaved tree seedlings and their silvicultural meaning. *Forest Resources Environment* **35**, 9–25.

Koike, T., Haibara, K., Matsuura, Y., Takahashi, K., Maximov, T.C., and Ivanov, B.I. (1999). Microsite- and age-dependent shoot development and needle nutrient levels of Scotch pine growing in Yakutian Permafrost region, eastern Siberia. *Proceedings of Joint Siberian Permafrost Studies* **4**, 17–24.

Koike, T., Kitao, M., Maruyama, Y., Mori, S., and Lei, T.T. (2001). Leaf morphology and photosynthetic adjustments among deciduous broad-leaved trees within the vertical canopy profile. *Tree Physiology* **21**, 951–958.

Koriba, K. (1958). On the periodicity of tree-growth in the tropics, with reference to the mode of branching, the leaf-fall, and the formation of the resting bud. *Gardens Bulletin* **17**, 11–81.

Kozlowski, T.T., and Pallardy, S.D. (1997). *Physiology of Woody Plants.* Academic Press, San Diego.

Lei, T.T., and Koike, T. (1998). Some observations of phenology and ecophysiology of *Daphne kamtschatica* Maxim. var. *jezoensis* (Maxim.). Ohwi, a shade deciduous shrub, in the forest of northern Japan. *Journal of Plant Research* **111**, 207–212.

Lei, T.T., Tabuchi, R., Kitao, M., Takahashi, K., and Koike,T. (1998). Effects of season, weather and vertical position on the light variation in quantity and quality in a Japanese deciduous broadleaf forest. *Journal of Sustainable Forestry* **6**, 35–55.

Lieth, H. (1971). *Seasonality and Phenological Modeling.* Ecological Studies. Vol. 12, Springer-Verlag, Berlin.

Long, S.P., Humpheries, S., and Fakowski, P.C. (1994). Photoinhibition of photosynthesis in nature. *Annual Review of Plant Physiology and Plant Molecular Biology* **45**, 633–662.

Makino, A., Sakashita, H., Hidema, J., Mae, T., Ojima, K., and Osmond, B. (1992). Distinctive responses of ribulose-1,5-bisphosphate carboxylase and carbonic anhydrase in wheat leaves to nitrogen nutrition and their possible relationships to CO_2 transfer resistance. *Plant Physiology* **100**, 1737–1743.

Makino, A., Nakano, H., and Mae, T. (1994). Effects of growth temperature on the responses of ribulose-1,5-bisphosphate carboxylase, electron transport components, and sucrose synthesis enzymes to leaf nitrogen in rice, and their relationships to photosynthesis. *Plant Physiology* **105**, 1231–1238.

Makino, A., Sato, T., Nakano, H., and Mae, T. (1997). Leaf photosynthesis, plant growth and nitrogen allocation in rice under different irradiances. *Planta* **203**, 390–398.

Marks, P.L. (1974). The role of pin cherry (*Prunus pensylvanica* L.) in the maintenance of stability in northern hardwood ecosystems. *Ecological Monograph* **44**, 73–88.

Maruyama, K. (1978). Ecological studies on beech forests (32). *Bulletin of Niigata University Forest* **11**, 1–30 (in Japanese with English summary).

Matile, P. (2000). Biochemistry of indian summer: Physiology of autumnal leaf coloration. *Experimental Gerontology* **35**, 145–158.

Matile, R., Hörtensteiner, S., and Thomas, H. (1999). Chlorophyll degradation. *Annual Review of Plant Physiology and Plant Molecular Biology* **50**, 67–95.

Matsuura, Y., and Abaimov, A.P. (1999). Nitrogen mineralization in larch forest soils of continuous permafrost region, Central Siberia—An implication for nitrogen economy of a larch forest stand. *Proceedings of Joint Siberian Permafrost Studies* **8**, 129–134.

Merzlyak, M.N., and Gitelson, A. (1995). Why and what for are the leaves yellow in autumn? On the interpretation of optical spectra of senescing leaves (*Acer platanoides* L.). *Journal of Plant Physiology* **145**, 315–320.

Millard, P., and Thomson, C.M. (1989). The effect of the autumn senescence of leaves on the internal cycling of nitrogen for the spring growth of apple trees. *Journal of Experimental Botany* **40**, 1285–1289.

Miyazawa, S.-I., Satomi, S., and Terashima, I. (1998). Slow leaf development of evergreen broad-leaved tree species in Japanese warm temperate forests. *Annals of Botany* **82**, 859–869.

Miyoshi, (1934). [Needle survival of larch seedlings in Sakhalin Island.] *Bulletin of Tokyo University Forest* **12**, 34–44 (in Japanese).

Nagata, H. (1983). [Leaf shedding, its phenological characteristics.] *Plant and Nature* **17**, 10–15 (in Japanese).

Neave, I.A., Dawson, J.O., and DeLucia, E.H. (1989). Autumnal photosynthesis is expanded in nitrogen-fixing European black alder compared with white basswood: possible adaptation significance. *Canadian Journal of Forest Research* **19**, 12–17.

Nelson, N.D., and Isebrands, J.G. (1983). Late season photosynthesis and photosynthate distribution in an intensively-cultured *Populus nigra* x *laurifolia* clone. *Photosynthetica* **17**, 537–549.

Ninomiya, I., Hiromi, T., Yoneda, T., Ichie, T., Kamita, K., Kohira, M., Lee, H.S., and Ogino, K. (2000). Phenology of shoot elongation and leaf dynamics in a tropical rain forest in Sarawak. In *Proceedings of the Workshop on Forest Ecosystem Rehabilitation. Kuching, Sarawak, Malaysia* pp. 182–186.

Nobel, P.S., Zaragoza, L.J., and Smith, W.K. (1975). Relation between mesophyll surface area, photosynthetic rate and illumination level during development for leaves of *Plectranthus parviforus* Henckel. *Plant Physiology* **55**, 1067–1070.

Noodén, L.D., and Leopold, A.C. (1988). *Senescence and Aging of Plants*. Academic Press, San Diego.

Noodén, L.D., Hillsberg, J.W., and Schneider, M.J. (1996). Introduction of leaf senescence in *Arabidopsis thaliana* by long days through a light-dosage effect. *Physiologia Plantarum* **96**, 491–495.

Ono, K., Nishi, Y., Watanabe, A., and Terashima, I. (2001). Possible mechanisms of adaptive leaf senescence. *Plant Biology* **3**, 234–243.

Oohata, S. (1986). [Seasonal changes in leaf shedding in Pine species.] *Bulletin of Kyoto University Forest* **45**, 123–142 (in Japanese with English summary).

Pearcy, R.W., and Sims, D.A. (1994). Photosynthetic acclimation to changing light environments: scaling from the leaf to the whole plant. In *Exploitation of Environmental Heterogeneity by Plants* (M.M. Caldwell, and R.W. Pearcy, Eds.), pp. 145–174. Academic Press, San Diego.

Reich P.B., Walters, M.B., and Ellsworth, D.S. (1992). Leaf life-span in relation to leaf, plant and stand characteristics among diverse ecosystems. *Ecological Monograph* **62**, 365–392.

Reich, P.B., Koike, T., Gower, S.T., and Schottele, A. (1995). Causes and consequences of leaf life-span. In *Ecophysiology of Coniferous Forests* (W.K. Smith and T.M. Hinckey, Eds.), pp. 225–254. Academic Press, San Diego.

Richards, J.H. (1984). Ecophysiological characteristics of seedling and sapling subalpine larch, *Larix lyallii*, in the winter environment. *Eidg. Anst. forstl. Versuchswese, Ber.* **270**, 103–112.

Sato, Y. (1989). [*Biological Observation of Autumn Coloration*], 56 pp. Akane-shobo, Tokyo (in Japanese).

Sobrado, M.A. (1991). Cost-benefit relationships in deciduous and evergreen leaves of tropical dry forest species. *Functional Ecology* **5**, 608–616.

Smart, C.M. (1994). Gene expression during leaf senescence. *New Phytologist* **126**, 419–448.

Smillie, R.M., and Hetherington, S.E. (1999). Photoabatement by anthocyanin shields photosynthetic systems from light stress. *Photosynthetica* **36**, 451–463.

Terashima, I. (1989). Productive structure of a leaf. In *Photosynthesis* (W.R. Briggs, Ed.), pp. 207–226. Alan R. Liss, New York.

Terashima, I., and Inoue, Y. (1985). Vertical gradient in photosynthetic properties of spinach chloroplasts dependent on intra-leaf light environment. *Plant Cell Physiology* **26**, 781–785.

Terashima, I., and Hikosaka, K. (1995). Comparative ecophysiology of leaf and canopy photosynthesis. *Plant, Cell and Environment* **18**, 1111–1128.

Thomas, H. (1992). Canopy survival. In *Crop Photosynthesis: Spatial and Temporal Determinants* (N.R. Baker and H. Thomas, Eds.), pp. 11–42. Elsevier, Amsterdam.

Thomas, H. (1997). Chlorophyll: A symptom and a regulator of plastid development. *New Phytologist* **136**, 163–181.

Thomas, H., and Smart, C.M. (1993). Crops that stay green. *Annals of Applied Biology* **123**, 193–219.

17

Annual Shoot Senescence in Perennials

Maxine A. Watson and Ying Lu

I. Introduction

Many perennial plants bear annual leaves or shoots that senesce at the end of the growing season, leaving behind perennating structures from which new annual growth emerges the following year. This growth pattern is widespread. It is found in many species of the deciduous forest floor including: *Aster acuminatus*, *Clintonia borealis* (Brown *et al.*, 1985), *Erythronium americanum*, *Trillium erectum* (Lapointe, pers. commun.), *Arisaema* species (Richardson, 1999), and *Tipularia discolor* (Zimmerman and Whigham, 1992), many perennial grasses and prairie plants, the leaves of deciduous shrubs and trees (Aerts, 1996), as well as a number of well-studied agronomically and horticulturally important plant species, among them potatoes, onions and spring bulbs (Noodén, 1988).

The timing of senescence of the annual shoots of perennial plants, a phenomenon termed top senescence by Noodén (1988), can be influenced by both internal and external factors that vary in their predictability. Many external factors, such as herbivory, disease, sporadic drought, frost or the like (e.g., Casper *et al.*, 2001), are unpredictable in their timing and, thus, cannot be physiologically or developmentally anticipated. We term senescence induced by such factors imposed senescence. Imposed senescence can result in loss of nutrients by plants when leaves are rapidly shed. Senescence timing also can respond to

Plant Cell Death Processes

predictable (e.g., seasonal) changes in the external environmental or in internal resource states, like those that accompany changes in life history status (e.g., from a vegetative to a reproductive condition), a form of senescence that we term endogenous senescence. We view endogenous senescence timing as a genetically controlled evolved response. While the mechanisms of imposed and endogenous senescence, once triggered, may be the same, the timing of endogenous senescence can be anticipated in a developmental and physiological sense and therefore may proceed over a longer time period, allowing for the ordered recovery of leaf nutrients and, hence, the potential to minimize nutrient loss.

If the timing of endogenous senescence is a plant-regulated process, then endogenous senescence timing may reflect a trade-off between plants' need to acquire carbon versus their need to recycle nutrients held in the leaf (Williams, 1955; Feller, 1979; Titus and Kang, 1982; Millard and Proe, 1991; Chapter 14). When leaves senesce relatively early in the growing season, mineral nutrients, particularly nitrogen, can be reallocated to other simultaneously occurring growth processes but at the expense of potential carbon acquisition by those leaves (Small, 1972; Chapin and Kedrowski, 1983; Millard and Thomson, 1989; Marshall and Watson, 1992). Conversely, when leaves senesce later in the season, the period for assimilate acquisition is lengthened, but at the expense of nutrient reallocation to other developing structures (Mooney and Gulmon, 1982; Noodén, 1988). Plants should reabsorb mobile nutrients when the costs of maintaining the leaf are greater than the carbon gained by it (Chabot and Hicks, 1982).

Plants also should reabsorb the nutrients contained in a leaf when the nutrients in the leaf are more important for the fitness of the individual than is continued photosynthesis by the leaf (Field, 1983, 1987; Harper and Sellek, 1987; Harper, 1989; Evans, 1993). (See also Chapters 14, 17 and 23.) Genetic variability in the timing of endogenous top shoot senescence should allow for selection to optimize the acquisition and reutilization of limiting resources. This should be particularly true in long-lived perennials bearing annual structures, because selection on mean endogenous senescence time can occur on the same genotype year after year through its potential influence on growth rates and frequency of sexual organ formation and, hence, seed production.

Plasticity in the expression of endogenous senescence timing within individuals (i.e., within genotypes) also is a genetically influenced trait that can contribute to optimization of plants' responses to short term environmental variation (physiologically plastic component as discussed in Huber *et al.*, 1999). Plasticity in senescence timing can result from changes in internal carbon or nutrient status brought on either as a result of variation in the external environment or in the internal resource environment, due to changes in development or in life history expression (i.e., whether the plant is vegetative or sexual). Sexual structures, particularly fruit, are generally believed to be costly for plants (Williams, 1966; Stearns, 1992), although this need not be the case (reviewed in Watson, 1984; Watson and Casper, 1984). In those instances where sexual structures are nutritionally costly, we predict that, all else being equal, plants bearing fruit should differ from purely vegetative plants in their timing of leaf senescence. Whether early or late senescence is favored should reflect, in turn, the nature of the limiting resource. If carbon is more limiting than recyclable mineral nutrients, then we predict that leaf senescence should be delayed in fruiting plants (Watson and Lu, 1999).

In this chapter, we examine differences in endogenous leaf senescence time in the annual shoots (top shoots) of the long-lived perennial understory herb, the mayapple, *Podophyllum peltatum* (Berberidaceae). Mayapple, with its persistent rhizome, annual shoots and simple architecture, provides a good model system for such studies. In addition, because mayapple development is characterized by preformation (Jones and Watson, 2001) we also are able to evaluate the influence of the life history status of multiple shoot generations on the timing of senescence of the current year's aerial shoot. We tested the following hypotheses. (1) Among-colony differences in shoot senescence timing are genetically rather than environmentally determined. Plastic differences in the timing of shoot senescence are correlated with (2) rhizome demography, vigor and the life history status of the current year's shoot and (3) the life history status of next year's shoot bud. (4) Differences in shoot senescence time influence clone fitness through effects on branching rates and frequencies of sexual shoot production. We also tested whether increase in environmental nitrogen levels delays shoot senescence time.

II. Mayapple as a Study Organism

Mayapples are perennial clonal herbs of deciduous forests of the northeastern United States; they are found growing in discrete patches or colonies containing one or more genotypes (genets) (Policansky, 1983; Whisler and Snow, 1992). Mayapples are morphologically simple and have a limited number of developmental options that can directly influence their demography (their pattern of birth, death and ramet population increase) and their life history expression (whether they are sexually reproductive or not). These developmental options include annual decisions regarding rhizome branching, a developmental process that results in an increase in the number of ramets present, and whether to differentiate a vegetative or a sexual aerial shoot (Fig. 17-1) (Watson *et al.*, 1997). Nitrogen and carbon availability influence the pattern of mayapple development and, hence, demographic and life history expression (Benner and Watson, 1989; de Kroon *et al.*, 1991; Landa *et al.*, 1992; Geber *et al.*, 1997 a,b). Carbon appears to be the more limiting resource (Sohn and Policansky, 1977; Benner and Watson, 1989; de Kroon *et al.*, 1991); it is extensively stored throughout the rhizome system (Landa *et al.*, 1992; Jónsdóttir and Watson, 1997; Watson *et al.*, unpublished).

Mayapple rhizome systems are composed of series of physically (Sohn and Policansky, 1977) and physiologically (Landa *et al.*, 1992; Jónsdóttir and Watson, 1997; Watson *et al.*, unpublished) interconnected rhizome segments (ramets) that may persist for more than ten years. The life history status of the rhizome system in a given year reflects the morphology and function of its current aerial shoot (Fig. 17-1). Sexual shoots tend to be produced by larger rhizome segments (Sohn and Policansky, 1977; Benner and Watson, 1989; Geber *et al.*, 1997a).

The timing of mayapple seasonal phenology varies as a function of latitude and seasonal rainfall patterns. Northern populations emerge and senesce later than southern ones (Sohn and Policansky, 1977; Lu, pers. obs.; Geber, pers. obs.). Both late season drought (Watson and Lu, 1999) and high rainfall early in the growing season (Watson and Jones, pers. obs.) accelerate the onset of senescence. In Indiana, shoots emerge in early April. Senescence is first observed in early May. However, this early senescence usually is associated with the presence of either a noctuid larva (probably *Papaipema cerina*) or a rust (*Puccinia podophylii*) (Sohn and Policansky, 1977; Parker, 1988; Lu, pers. obs.), both of which we consider to be agents of imposed senescence. Uninfected plants generally do not senesce until later in the season, and it is this larger, later wave of senescence that we consider to be under endogenous control.

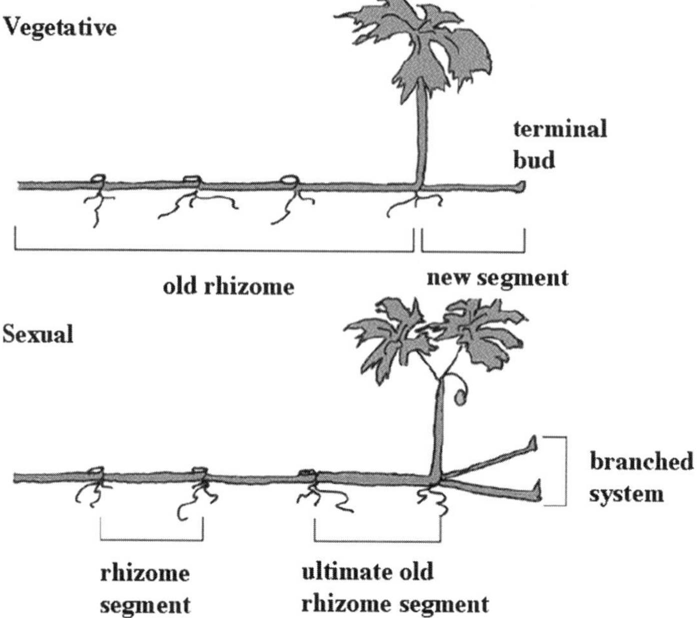

Figure 17-1. Diagrams of vegetative and sexual rhizome systems of mayapple, *Podophyllum peltatum*, as they would appear in early July in south-central Indiana. The terminal bud of the new rhizome segment contains next year's preformed shoot, which may be either vegetative or sexual. The ultimate old rhizome segment is the one that gave rise to the current year's shoot. (Reproduced with permission from Watson and Lu, 1999.)

III. Sources of Variation in Endogenous Shoot Senescence Time in Mayapple

A. Among-Colony Variation in Mean Endogenous Shoot Senescence Time

We regularly note variation among colonies in a number of phenological attributes of mayapple seasonal development, including the timing of shoot emergence and anthesis, as well as senescence (Lu, 1996). This variation may reflect microsite differences among colonies or genetically based and hence selectable variation among genotypes, or some combination of the two.

We examined the relative roles of genotype and environment in a reciprocal transplant experiment involving four colonies that differed significantly in mean shoot senescence time, two each on north- and south-facing slopes. Two years following transplant, we found significant effects of colony ($p < 0.001$) and transplant site ($p < 0.0385$), as well as an interaction between the two ($P < 0.0311$); the ANOVA model accounted for 37% of the total variation in shoot senescence time (Lu and Watson, unpubl.). The significant colony effect and interaction term indicate that genetic differences in senescence time exist, but that microsite variation also plays a role.

B. Relationship between Endogenously Controlled Shoot Senescence Time and Current Demographic and Life History Expression and Rhizome Vigor

If life history functions differ in their carbon and nutrient costs to the plant and if the timing of senescence of the annual shoots of perennial plants reflects an adaptive optimization of plants' requirements for carbon versus nutrient resources, then, in a given year, the life history status of the current shoot should influence the timing of its senescence.

We found that endogenous senescence time was plastic; it varied with plant vigor and life history status. The sexual shoots of mayapple senesce on average 21 days later than vegetative shoots. The life history status (i.e., vegetative or sexual) of next year's shoot—already present as a predetermined bud—also exerted a significant influence. Current shoots of rhizome systems that would give rise to an aerial sexual shoot in the following year senesced on average 18 days later than those that would give rise to a vegetative shoot. The combined effects of the life history status of the current year's shoot and next year's shoot explained 21% of the variation in mean senescence time of the current year's shoot (Watson and Lu, 1999). Purely vegetative systems were the first to senesce. Sexual shoots that failed to set fruit and formed a vegetative new shoot bud senesced about 10 days later. Systems that initiated a new sexual shoot bud, and whose current shoot was morphologically vegetative, or had become functionally vegetative as the result of flower abortion, senesced about 10 days later still (Table 17-1).

Fruit formation by current sexual shoots was associated with the greatest delay in senescence (30 days), regardless of the demographic status of next year's shoot (Table 17-1). However, sexual plants that failed to mature fruit, but produced a sexual new shoot bud also senesced later. These data suggest that while the presence of fruit is important in delaying onset of senescence of the current year's shoot, other combinations of current and future life history expression result in similar delays. This conclusion is reinforced by our path analyses that are discussed below.

Table 17-1. Variation in mean leaf senescence time in 1989 as a function of current reproductive status[1] and 1990 shoot bud type (reprinted with permission from Watson and Lu, 1999).

1989 reproductive status	1990 shoot bud type	Mean[2] ± SE (N)[3] (days)
Vegetative	Vegetative	187.94 ± 0.70 (955)[3]
Sexual, fl or frab	Vegetative	198.44 ± 7.29 (16)[2,3]
Sexual, flab	Vegetative	198.59 ± 3.21 (37)[2]
Vegetative	Sexual	204.66 ± 1.29 (207)[2]
Sexual, flab	Sexual	208.91 ± 1.91 (76)[1,2]
Sexual, fr	Vegetative	217.97 ± 1.80 (70)[1]
Sexual, fl or frab	Sexual	218.41 ± 2.55 (34)[1]
Sexual, fr	Sexual	220.39 ± 2.43 (31)[1]

[1] 1989 reproductive status abbreviations. There were four classes of sexual shoots: flab = shoots with aborted flowers, fl = shoots that bore healthy flowers but did not set fruit, frab = shoots whose fruit aborted, fr = shoots that successfully matured fruit.

[2] Mean senescence time measured as days since January 1.

[3] Different superscript letters indicate significant differences in mean senescence time between plants differing in reproductive status and shoot bud type (Tukey tests among the eight demographic shoot categories, $p < 0.05$).

Rhizome size or vigor also affected senescence timing. Fourteen percent of the variation in shoot senescence time was explained by the size of the new rhizome segment (Watson and Lu, 1999); shoots that gave rise to a larger new segment or to a branched rhizome also senesced later (de Kroon *et al.*, 1991; Watson and Lu, 1999).

The presence of larger organs (i.e., fruits, branches and new rhizome segments) and more persistent carbon sinks (i.e., fruits) appears to cause heightened demand for current assimilate, which delays shoot senescence (Watson and Lu, 1999; Chapter 13, this volume). Noodén (1988) reported a similar pattern of delayed senescence in fruit-bearing plants for a number of agronomically important species. The presence of these larger sinks does not appear to alter the photosynthetic efficiency of mayapple leaves (Basha, Watson and Griffith, unpubl.). Instead, mayapples maintain positive net photosynthesis following canopy closure (Watson, Basha and Griffith, unpubl; Taylor and Pearcy, 1976), which allows them to take advantage of the increased shoot life span to continue positive net carbon assimilation, at least until leaf nitrogen levels fall too low for positive net rates to be maintained (see below and Field, 1983, 1987).

C. Historical Effects of Life History Expression Demography and Growth on Endogenously Controlled Shoot Senescence Time

We found that events that occur in prior years also influence senescence timing, probably through their influence on the carbon and mineral nutrient status (vigor) of rhizome systems. If carbon is the critical limiting resource, and carbon stored in prior years supports growth in the current year then, all else being equal, we would expect larger rhizome systems, which contain more stored carbon, to senesce earlier than smaller rhizome systems. Instead we found that shoots produced by larger ultimate rhizome segments (Fig. 17-1) tended to senesce later than those produced by smaller ones; ultimate rhizome segment size explained about 9% of the variation in shoot senescence time (Watson and Lu, 1999). This departure from our prediction is likely due to the strong influence of shoot type on senescence timing. Larger ultimate rhizome segments are more likely to give rise to sexual shoots in the current year (Sohn and Policansky, 1977; Benner and Watson, 1989; Geber *et al.*, 1997a), and sexual shoots senesce significantly later than vegetative ones (Watson and Lu, 1999).

A more detailed analysis of the effects of rhizome system history on the timing of shoot senescence was performed on three years of data using path analysis (Mitchell, 1992; Wootton, 1994). Construction of the path diagram was based on the assumption that events early in developmental phenology affect later events but not vice versa (Diggle, 1994; Watson *et al.*, 1995; Geber *et al.*, 1997a,b). In our path model, leaf senescence was the terminal event in the current growing season. The path analysis demonstrated that several relationships among measured variables were consistent and strong among years (e.g., the obvious relationship between the formation of flowers and then fruit by sexual shoots), whereas others were not, suggesting the importance of annually varying environmental background on the strength and direction of interactions among variables (Lu, 1996).

The results of the path analysis were consistent with the hypothesis, and our earlier analysis (Watson and Lu, 1999), that shoot senescence time is affected by the life history status of the current shoot and of next year's shoot. Surprisingly, the path analysis indicated that flowering, rather than fruiting, is associated with the delay in shoot senescence timing; shoots that bore healthy flowers but failed to mature fruit, and which gave rise to a new sexual shoot bud, were among the last shoots to senesce. This pattern suggests that successful anthesis and the production of a sexual shoot bud by the terminal rhizome segment exert the same influence on the timing of shoot senescence as does fruit maturation.

Earlier findings suggested strong associations between shoot senescence time and the likelihood of producing a branched rhizome system (de Kroon *et al.*, 1991; Lu, 1996), a demographically important event that leads to increased ramet population size in following years. The shoots of branched systems senesced about 12 days later than those of unbranched systems [1989 senescence dates for branched (206.5 ± 2.6 days) and unbranched (194.4 ± 0.6 days) systems (mean ± SE, $p = 0.0001$) (Lu, 1996; Watson and Lu, 1999)]. A similar pattern was noted by de Kroon *et al.* (1991), who found that late senescing sexual shoots of mayapple produced branched rhizomes at higher frequency than did earlier senescing sexual shoots; the branches were heavier and longer, but lower in nitrogen and phosphorus concentrations. Taken together, these observations are consistent with our hypothesis that the timing of shoot senescence reflects the balance between plants' requirements for carbon versus nutrient resources, and support our contention that senescence timing is affected by both carbon sink demands and the mineral nutrient richness of plants' internal environments.

However, the results of our path analysis indicated that the effect of rhizome branching on shoot senescence time within the same growing season was indirect (Lu, 1996). The significant correlation between rhizome branching and late shoot senescence time was due to the strong correlation between shoot size and rhizome branching and between shoot size and late shoot senescence, with shoot size mirroring the life history shift from a vegetative to a sexual state.

We used the same path analysis to examine whether variation in senescence timing affects *subsequent* demographic performance and life history expression by examining between year effects. We found that shoot senescence time does not directly affect shoot type, shoot size, or rhizome branching in the following year. Rather, we noted a weak indirect effect of shoot senescence time on these demographic and life history traits; the indirect effect was exerted through its influence on shoot emergence time in the following year.

D. The Effect of Nitrogen and Phosphorus Status on the Timing of Top Shoot Senescence

We explored this issue by examining seasonal changes in the nitrogen (N) and phosphorus (P) content of leaves of unmanipulated mayapple rhizome systems and in rhizome systems grown on native soils in which nitrogen and phosphorus levels were supplemented.

1. Seasonal changes in plant nitrogen and phosphorus content in unmanipulated rhizome systems

We found that nutrient reallocation from leaves to other parts of the plant, in the case of mayapple the rhizome system, occurs throughout the season but is accelerated in sexual systems during the period of seed maturation and fruit fill. Similar patterns are reported by Field (1983), Noodén and Leopold (1988), and Killingbeck (1996 and Chapter 14, this volume) for a wide array of other species.

We found no correlation between outward signs of senescence (leaf color change) and leaf nutrient content, but there were significant differences in the mineral content of still green mayapple leaves as they aged. Sexual leaves were more costly to build, but they also were a larger reservoir of recyclable mineral nutrients. While sexual shoots lost nitrogen more rapidly than did vegetative ones, both nitrogen and phosphorus content and concentration were higher throughout the season in sexual shoots. Phosphorus levels were low in vegetative shoots throughout the season, but declined through time in the leaves of sexual shoots (Lu and Watson, unpubl.).

If early leaf senescence is driven by the rhizome system's need for nutrients held in the leaf as we propose, then earlier senescing leaves should be more fully depleted of minerals than are later senescing ones. However, we found that mineral content was higher in early-senesced as compared to late-senesced brown leaves (Lu and Watson, unpubl.). The fact that the expected pattern was not observed suggests that early senescence is not driven by the sink demands of the rhizome system for nitrogen stored in the leaves. Similarly, Aerts (1996), in a comparative study of the absorption efficiencies of the senescing leaves of a wide array of perennial deciduous shrubs and trees, found no evidence for nutritional controls on nutrient absorption efficiency.

2. Effects of nutrient supplementation on the timing of endogenous top shoot senescence

We tested two predictions. First, if growth is limited by either nitrogen or phosphorus, then soil nutrient supplementation should cause a delay in the timing of leaf senescence because plants' dependency on recycled nutrients should be reduced. Second, if plants' capacities to access soil nutrients vary seasonally, or if plants' requirements for recycled nutrients vary seasonally, then the effect of nutrient feeding on shoot senescence time should be affected by the timing of nutrient feeding.

Plants were fed twice either in April, June or August, with 0.2 M $(NH_4)_2HPO_4$ and then observed consecutively for two years. We found no significant effects of nutrient supplementation on senescence timing irrespective of when in the season soil nutrients were supplemented, a finding similar to that of de Kroon *et al.* (1991) also working with mayapple. The mayapple observations are consistent with data obtained from a number of agricultural species that experience monocarpic senescence (Noodén, 1988). Recently acquired nutrients appear to be stored rather than used to support growth or to affect senescence timing of the current year's shoot. Instead, support and buffering of current growth seems to rely on resources acquired and stored in prior years (Watson *et al.*, unpublished). This pattern is suggested in the behavior of a diversity of plant types, including: understory clonal herbs (e.g., Zimmerman and Whigham, 1992; Geber *et al.*, 1997a,b), aclonal herbs of other seasonal environments like the alpine (e.g., Wyka, 1999) and plants found in patchy, resource-poor habitats (Jónsdóttir *et al.*, 1996; Price *et al.*, 2001).

IV. General Discussion of Factors Regulating the Timing of Top Senescence in Other Species

Remarkably little information exists regarding the causes and consequences of variation in the timing of senescence of the annual shoots of perennial plants. Brown *et al.* (1985) appear to be the first to report variation in the timing of top senescence for a native plant species *in situ*. In an analysis of within and between species variation in vegetative phenology for two co-occurring clonal forest herbs, *Aster acuminatus* and *Clintonia borealis*, they found a strong positive correlation between the timing of what they called ramet death (and which is equivalent to our definition of endogenously controlled top shoot senescence) and ramet size, in both species. The correlation was stronger in *A. acuminatus* than in

C. boreali. However, variation in the timing of shoot senescence was less among ramets of *A. acuminatus* than among those of *C. borealis*. They suggested that the stronger correlations noted in *A. acuminatus* reflected the smaller amount of clonal integration in that species [see Hutchings and Barkham (1976) and Suzuki and Hutchings (1997) for discussions of clonal integration and storage]. Brown *et al.* (1985) also were the first to note the difficulties of distinguishing between cause and effect in the correlation between ramet size and senescence timing. That is, do large ramets senesce later because they are larger or are they larger because of their phenological characteristics? They suggested that a positive feedback system that includes reproductive success, like that reported by us for mayapple, was likely to be involved.

Additional data come from work on other perennial understory herbs of the deciduous forest floor. A study of sex change in two *Arisaema* species reports delayed senescence in fruit bearing as opposed to male or non-fruit bearing female plants (Richardson, 1999). Lapointe (pers. commun.) also reports delay in top shoot senescence of fruit bearing versus vegetative shoots of *Erythronium americanum* and *Trillium erectum*, suggesting that ephemeral clonal herbs of the northeast US deciduous forest understory behave in similar ways.

In sum, the study of causes of variation in the timing of top senescence of perennial shoots promises to provide insight into plant resource dynamics and the evolution of phenological traits. Consistent patterns of variation are identified, and their causal bases explored.

References

Aerts, R. (1996). Nutrient resorption from senescing leaves of perennials: are there general patterns? *Journal of Ecology* **84**, 597–608.

Benner, B.L., and Watson. M.A. (1989). Developmental ecology of mayapple: Seasonal patterns of resource distribution in sexual and vegetative rhizome systems. *Functional Ecology* **3**, 539–547.

Brown, R.L., Ashmum, J.W., and Pitelka, L.F. (1985). Within- and between-species variation in vegetative phenology in two forest herbs. *Ecology* **66**, 251–258.

Casper, B.B., Forseth, I.N., Kempenich, H., Seltzer, S., and Xavier, K. (2001). Drought changes leaf demography in *Cryptantha flava* (Boraginaceae). *Functional Ecology* **15**, 740–747.

Chabot, B.F., and Hicks, D.J. (1982). The ecology of leaf life spans. *Annual Review of Ecology and Systematics* **13**, 229–259.

Chapin, F.S., III, and Kedrowski, R.A. (1983). Seasonal changes in nitrogen and phosphorus fractions and autumn re-translocation in evergreen and deciduous taiga trees. *Ecology* **64**, 376–391.

de Kroon, H., Whigham, D.F., and Watson, M.A. (1991). Developmental ecology of mayapple: Effects of rhizome severing, fertilization and timing of shoot senescence. *Functional Ecology* **5**, 360–368.

Diggle, P.K. (1994). The expression of andromonoecy in *Solanum hirtum* (Solanaceae), phenotypic plasticity and ontogenetic contingency. *American Journal of Botany* **81**, 1354–1365.

Evans, J.R. (1993). Photosynthetic acclimation and nitrogen partitioning within a lucerne canopy. Stability through time and comparison with a theoretical optimum. *Australian Journal of Plant Physiology* **20**, 69–82.

Feller, U. (1979). Effect of changed source/sink relations on proteolytic activities and on nitrogen mobilization in field-grown wheat (*Triticum aestivum* L.). *Plant and Cell Physiology* **20**, 1577–1583.

Field, C.B. (1983). Allocating leaf nitrogen for the maximization of carbon gain: leaf age as a control on the allocation program. *Oecologia* **56**, 341–347.

Field, C.B. (1987). Leaf age effects on stomatal conductance. In *Stomatal Function* (E. Zeiger, G.D. Farquhar, and I.R. Cowan, Eds.), pp. 368–384. Stanford Univ. Press, Stanford, CA.

Geber, M.A., de Kroon, H., and Watson, M.A. (1997a). Organ preformation in mayapple as a mechanism for historical effects on demography. *Journal of Ecology* **85**, 211–223.

Geber, M.A., Watson, M.A., and de Kroon, H. (1997b). Development and resource allocation in perennial plants: the significance of organ preformation. In *Plant Resource Allocation* (F.A. Bazza and J. Grace, Eds.), pp. 113–141. Academic Press, New York.

Harper, J.L. (1989). The value of a leaf. *Oecologia* **80**, 53–58.

Harper, J.L., and Sellek, C. (1987). The effects of severe mineral deficiencies on the demography of leaves. *Proceedings of the Royal Society of London Series B* **232**, 137–157.

Huber, H., Lukács, S., and Watson, M.A. (1999). Spatial structure of stoloniferous herbs: an interplay between structural blue-print, ontogeny and phenotypic plasticity. *Plant Ecology* **141**, 107–115.

Hutchings, M.J., and Barkham, J.P. (1976). An investigation of shoot interactions in *Mercurialis perennis* L., a rhizomatous perennial herb. *Journal of Ecology* **64**, 723–743.

Jones, C.S., and Watson, M.A. (2001). Heteroblastic shoot development in mayapple (*Podophyllum peltatum*, Berberidaceae), a rhizomatous forest herb with limited options. *American Journal of Botany* **88**, 1340–1358.

Jónsdóttir, I.S., and Watson, M.A. (1997). Ecological significance of physiological integration in plants. In *The Ecology and Evolution of Clonal Plants* (H. de Kroon and J. van Groenendael, Eds.), pp. 109–136. Backhuys, Leiden.

Jónsdóttir, I.S., Callaghan, T.V., and Headley, A.D. (1996). Resource dynamics within arctic clonal plants. *Ecological Bulletin* **45**, 53–64.

Killingbeck, K.T. (1996). Nutrients in senesced leaves: keys to the search for potential resorption and resorption proficiency. *Ecology* **77**, 1716–1727.

Landa, K., Benner, B., Watson, M.A., and Gartner, J. (1992). Physiological integration for carbon in mayapple (*Podophyllum peltatum*), a clonal perennial herb. *Oikos* **63**, 348–356.

Lu, Y. (1996). The developmental ecology of mayapple, *Podophyllum peltatum*. Ph.D. dissertation, Indiana Univ., Bloomington, IN.

Marshall, C., and Watson, M.A. (1992). Ecological and physiological aspects of reproductive allocation. In *Fruit and Seed Production* (C. Marshall and J. Grace, Eds.), pp. 173–202. Cambridge Univ. Press, Cambridge, UK.

Millard, P., and Proe, M.F. (1991). Leaf demography and the seasonal internal cycling of nitrogen in sycamore (*Acer pseudoplatanus* L.) seedlings in relation to nitrogen. *New Phytologist* **117**, 587–596.

Millard, P., and Thomson, C.M. (1989). The effect of the autumn senescence of leaves on the internal cycling of nitrogen for the spring growth of apple trees. *Journal of Experimental Botany* **40**, 1285–1298.

Mitchell, R.J. (1992). Testing evolutionary and ecological hypotheses using path analysis and structural equation modeling. *Functional Ecology* **6**, 123–129.

Mooney, H.A., and Gulmon, S.L. (1982). Constraints on leaf structure and function in reference to herbivory. *BioScience* **32**, 198–206.

Noodén, L.D. (1988). Whole plant senescence. In *Senescence and Aging in Plants* (L.D. Noodén, and A.C. Leopold, Eds.), pp. 392–439. Academic Press, New York.

Noodén, L.D., and Leopold, A.C. (1988). *Senescence and Aging in Plants*. Academic Press, New York.

Parker, M.A. (1988). Genetic uniformity and disease resistance in a clonal plant. *American Naturalist* **132**, 538–549.

Policansky, D. (1983). Patches, clones, and self-fertility of mayapples (*Podophyllum peltatum*). *Rhodora* **85**, 253–256.

Price, E.A.C., Gamble, R., Williams, G.G., and Marshall, C. (2002). Seasonal patterns of partitioning and remobilization in the invasive rhizomatous perennial Japanese knotweed (*Fallopia joponica* (Houtt.) Ronse Decraene. *Evolutionary Ecology* **15**, 347–362.

Richardson, C. (1999). Sex-ratio evolution in gender-diphasic *Arisaema*. Ph.D. dissertation, Indiana Univ., Bloomington, IN.

Small, E. (1972). Photosynthetic rates in relation to nitrogen recycling as an adaptation to nutrient deficiency in peat bog plants. *Canadian Journal of Botany* **50**, 2227–2233.

Sohn, J.J., and Policansky, D. (1977). The cost of reproduction in the mayapple *Podophyllum peltatum* (Berberidaceae). *Ecology* **58**, 1366–1374.

Stearns, S. (1992). *The Evolution of Life Histories*. Oxford University Press, New York.

Suzuki, J., and Hutchings, M.J. (1997). Interactions between shoots in clonal plants and the effects of stored resources on the structure of shoot populations. In *The Ecology and Evolution of Clonal Plants* (H. de Kroon and J. van Groenendael, Eds.), pp. 311–329. Backhuys, Leiden.

Taylor, R.J., and Pearcy, R.W. (1976). Seasonal patterns of the CO_2 exchange characteristics of understory plants from deciduous forests. *Canadian Journal of Botany* **54**, 1094–1103.

Titus, J.S., and Kang, S. (1982). Nitrogen metabolism, translocation, and recycling in apple trees. *Horticultural Review* **4**, 204–246.

Watson, M.A. (1984). Developmental constraints. Effects on population growth and patterns of resource allocation in a clonal plant. *American Naturalist* **123**, 411–426.

Watson, M.A., and Casper, B.B. (1984). Morphogenetic constraints on patterns of carbon distribution in plants. *Annual Review of Ecology and Systematics* **15**, 233–258.

Watson, M.A., and Lu, Y. (1999). Timing of shoot senescence and demographic expression in the clonal perennial *Podophyllum peltatum* (Berberidaceae). *Oikos* **86**, 67–78.

Watson, M.A., Geber, M.A., and Jones, C.S. (1995). Ontogenetic contingency and the expression of plant plasticity. *Trends in Ecology Evolution* **10**, 474–475.

Watson, M.A., Hay, M.J.M., and Newton, P.C.D. (1997). The role of developmental phenology and the timing of meristem determination on clone fitness. In *The Ecology and Evolution of Clonal Plants* (H. de Kroon and J. van Groenendael, Eds.), pp. 109–136. Backhuys, Leiden.

Whisler, S.L., and Snow, A.A. (1992). Potential for the loss of self-incompatibility in pollen-limited populations of mayapple (*Podophyllum peltatum*). *American Journal Botany* **79**, 1273–1278.

Williams, G.C. (1966). Natural selection, the costs of reproduction, and a refinement of Lack's principle. *American Naturalist* **100**, 687–690.

Williams, R.F. (1955). Redistribution of mineral elements during development. *Annual Review of Plant Physiology* **6**, 25–42.

Wootton, J.T. (1994). Predicting direct and indirect effects, An integrated approach using experiments and path analysis. *Ecology* **75**, 151–165.

Wyka, T. (1999). Carbohydrate storage and use in an alpine population of the perennial herb, *Oxytropis sericea*. *Oecologia* **120**, 198–208.

Zimmerman, J.K., and Whigham, D.F. (1992). Ecological functions of carbohydrates stored in corms of *Tipularia discolor* (Orchidaceae). *Functional Ecology* **6**, 575–581.

18

Phototoxicity

Esa Tyystjärvi

I. Introduction

A. History of Phototoxicity

Phototoxicity can be defined as a phenomenon in which a substance becomes damaging when activated by light. Primary phototoxins are photosensitizers, i.e. light absorption by the toxin makes it poisonous. Secondary phototoxins affect the metabolism of the target organism, causing the accumulation of an endogenous photosensitizer.

Study on phototoxicity started in autumn 1897, when Professor von Tappeiner from Pharmacological Institute of Ludwig-Maximilians University, Munich, Germany gave Oscar Raab a topic for his thesis: to study whether acridine would kill paramecia (*Paramecium caudatum*). Due to the weather conditions, the illumination in the laboratory varied from day to day, and soon Raab found that the killing efficiency of acridine depends on light (Raab, 1900). Phototoxic poisoning of grazing animals has long been of practical importance. In the 16th century, only black sheep were raised in Tarentine fields infested with the phototoxic herb *Hypericum* (Blum, 1941), because dark fur and skin afford tolerance by limiting the passage of activating light. Plants containing phototoxic furanocoumarins have been used to treat vitiligo for at least 3300 years (Fitzpatrick and Pathak, 1959).

This chapter focuses on phototoxicity reactions in which a plant is either the source of the phototoxin or the subject of the phototoxic poisoning. Light-activated veterinary diseases

(Ivie, 1982) and human porphyrias (Kappas, 1995) will not be discussed here. The rapidly growing area of the medical use of phototoxic reactions (Dougherty *et al.*, 1998; Bethea *et al.*, 1999) is also out of the scope of this review. Even within the subject, phototoxicity in plants, literature coverage is selective.

B. Targets and Sources of Phototoxins

Phototoxic poisoning occurs in all types of organisms including viruses, bacteria, fungi, plants, and animals. Indiscriminate toxicity results from the ubiquity of the biomolecules, mainly membranes and DNA, readily attacked by phototoxins. The plant kingdom is the most important source for phototoxins, with 75–100 compounds isolated from flowering plants. For reviews on phototoxicity, see Towers, 1984; Laustriat, 1986; Spikes and Straight, 1987; Downum, 1992; Berenbaum, 1995; and Arfsten *et al.*, 1996.

II. Chemical and Biochemical Basis of Phototoxicity

A. Chemistry of Reactive Oxygen Species (ROS)

Phototoxicity is closely associated with the production of reactive oxygen species (ROS), mainly free radicals and singlet oxygen ($^1O_2$). Oxygen toxicity and antioxidative defenses have been reviewed extensively (Elstner, 1982; Foyer *et al.*, 1994; Ahmad, 1995; Bartosz, 1997; Halliwell and Gutteridge, 1999; see also Chapter 13). ROS include both highly reactive species like the hydroxyl radical $HO^{\bullet}$ and singlet oxygen $^1O_2$ and the partially reduced forms of oxygen (H_2O_2, $O_2{}^{2-}$, $O_2{}^{-}$ and $HO_2^{\bullet}$). $^1O_2$ is reactive towards compounds containing double bonds, oxidizing histidine, tryptophan, methionine and cysteine, cholesterol, unsaturated fatty acids, NADPH, and the bases of DNA (Halliwell and Gutteridge, 1999). $^1O_2$ can also transfer its excitation energy to a quencher molecule such as α-tocopherol, β-carotene or azide (Foote, 1987). The lifetime of $^1O_2$ is 1–4 μs in H_2O, 55–68 μs in D_2O (Ogilby and Foote, 1983; Valenzeno, 1987), and 20–25 μs in lipid micelles (Valenzeno, 1987). Superoxide is toxic mainly because its presence can lead to formation of the highly reactive hydroxyl radical.

B. Type I and Type II Reactions

Several competing reaction routes are available for most phototoxins, and the actual mechanism(s) by which a given toxin functions in a particular biological system, is never trivial to deduce (Foote, 1987). Photosensitization starts when the sensitizer S absorbs a quantum, and internal conversion to an excited triplet ^3S leads to a long-lived reactive molecule. The reactions that follow are classified as Type I reactions starting with electron transfer involving ^3S and Type II reactions starting with energy transfer from ^3S to $^3O_2$, forming $^1O_2$ (Gollnick, 1968) (Fig. 18-1).

Type I reactions. Excited states can enter both oxidative and reductive one-electron chemistry (Foote, 1987), both yielding radical products that are easily oxygenated. Type I reactions are often complex combinations of radical reactions producing an oxygenated or otherwise inactivated substrate (Fig. 18-1). Oxidation of ^3S by $^3O_2$, yielding $O_2{}^{-}$, is a special Type I reaction.

Type II reactions. Excitation energy donation from ^3S to another triplet species occurs readily if certain energetic requirements are filled and if the other triplet is close enough. Energy transfer from ^3S to molecular oxygen ($^3O_2$) produces the singlet excited state of

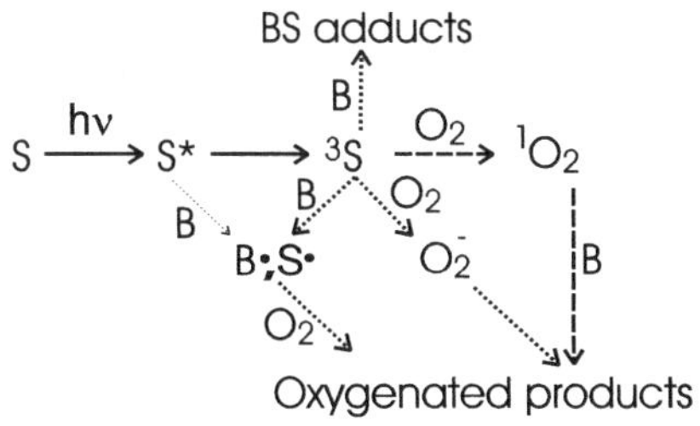

Figure 18-1. Photosensitization of a biomolecule B. The sensitizer S is excited by light to an excited singlet state S*. Chemical reactions usually start from long-lived triplet excited state ^3S which results from intersystem crossing in S*. One-electron transfer reactions between ^3S and a B or between ^3S and O_2 yield radical products, and oxygenation of B or another biomolecule follows (Type I reactions; dotted arrows). Adduct formation between S and B does not require oxygen. Energy transfer reaction between ^3S and $^3O_2$ yields the reactive singlet oxygen $^1O_2$ which may oxidize several biomolecules (Type II reaction, dashed arrow).

oxygen ($^1O_2$) and the ground state of S (Fig. 18-1). Cellular membranes are particularly sensitive to Type II oxidation.

Type I and Type II pathways are always in competition in natural systems (Ahmad, 1992). The Type I/II character of a photodynamic reaction can be probed with inhibitors specific to a certain ROS, by comparing reaction rates between aerobic and anaerobic conditions, and by using D_2O to affect the lifetime of $^1O_2$. All methods have limitations. Most $^1O_2$ quenchers and chemical traps may also react with oxygen radicals or quench the excited state of the sensitizer. The use of D_2O, likewise, is limited to cases where quenching by the H_2O or D_2O limits the rate of $^1O_2$ decay (Foote, 1987).

Photosensitizers have an extensive π-electron system (Downum, 1992). High absorptivity, high intersystem crossing rate and a long ^3S lifetime are expected to favor phototoxicity. In the polyacetylene group, at least three conjugated triple bonds are required for phototoxicity (Marchant and Cooper, 1987). Most phototoxic effects also require that both the sensitizer and light penetrate the target cell.

III. Phototoxic Compounds and their Roles in Nature

A. Acridine and Xanthene Dyes and Polyaromatic Hydrocarbons

The acridine dyes like acridine orange [1] (see Fig. 18-2 for the formulae) and the xanthene dyes like rose bengal [2] are primarily Type II sensitizers. In light, these compounds inactivate bacteriophage ØX147, kill bacteria (Houba-Herin *et al.*, 1982), cause loss of chlorophyll from plant leaves and inactivate the photosynthetic electron transfer chain (Knox and Dodge, 1985a,b; Chung and Jung, 1995). The absorption maxima are in the visible region, and the quantum yield of $^1O_2$ production in water is very high for rose bengal, methylene blue and eosin Y (Houba-Herin *et al.*, 1982). The toxicity of low molecular weight polyaromatic hydrocarbons (e.g. anthracene [3]) is also enhanced by ultraviolet-A (UVA) light (Arfsten *et al.*, 1996).

B. Plant Secondary Metabolites

Furanocoumarins (psoralens) and related compounds. Furanocoumarins, found mainly in Apiaceae and Rutaceae (Pathak *et al.*, 1962) are the most intensively studied group

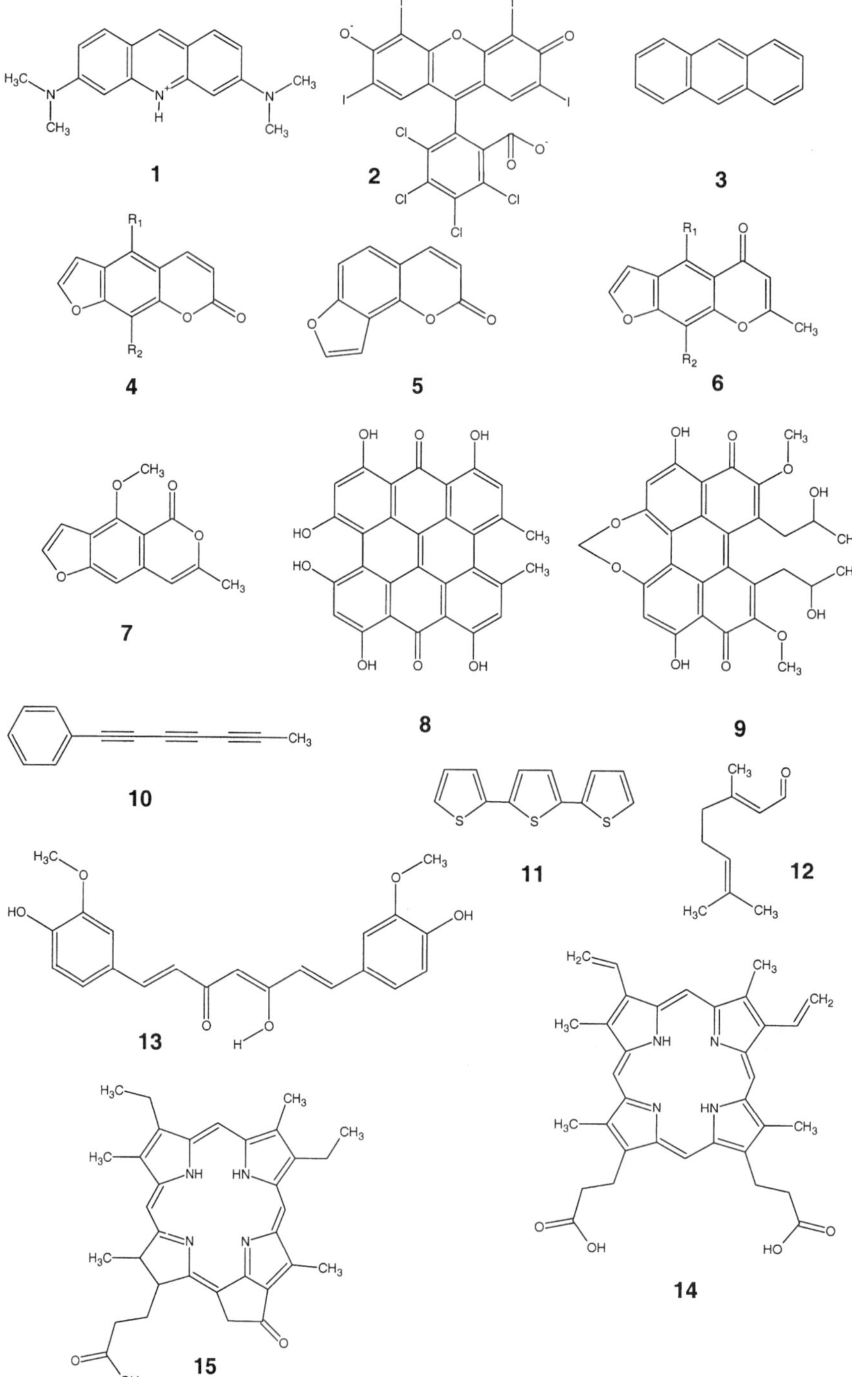

Figure 18-2. Structural formulae of some photosensitizers. 1, Acridine orange; 2, Rose bengal; 3, Anthracene; 4, Psoralen (R_1=H; R_2=H); 8-Methoxypsoralen (R_1=H; R_2=OMe); 5-Methoxypsoralen (R_1=OMe; R_2=H); 5, Angelicin; 6, Khellin (R_1=R_2=OMe); Visnagin (R_1=OMe; R_2=H); 7, Coriandrin; 8, Hypericin; 9, Cercosporin; 10, Phenylheptatriyne; 11, α-Terthienyl; 12, Citral; 13, Curcumin; 14, Protoporphyrin IX; 15, Phylloerythrin.

of phototoxins. Linear furanocoumarins (e.g. psoralen, 8-methoxypsoralen or 8-MOP and 5-methoxypsoralen [4]) or angular ones (e.g. angelicin [5]) absorb UVA light (320–400 nm), and numerous variations of the psoralen theme are known (Pathak *et al.*, 1962; Becker *et al.*, 1993). Furanocoumarin-containing plants like *Ammi majus* are known to cause photosensitization of livestock and poultry (Clare, 1955; Dollahite *et al.*, 1978; Ivie, 1982). Handling of furanocoumarin-containing plants like celery (Ashwood-Smith *et al.*, 1985) or parsnip (Pedersen and Arlés, 1997) may also sensitize human skin to UVA; the reaction is often followed by increased pigmentation of the irritated area (Pathak *et al.*, 1962). The phototoxicity of furanocoumarins has been demonstrated in bacteria (Fowlks *et al.*, 1958) and viruses (Hudson *et al.*, 1993). Interestingly, 8-MOP caused UVA-dependent wilting of a fern when fed in the transpiration stream, while *Herachleum*, an 8-MOP-containing plant, was insensitive (Camm *et al.*, 1976). Furanochromones, e.g. visnagin and khellin [6] and furoisocoumarins like coriandrin [7] resemble furanocoumarins in their biological action (Abeysekera *et al.*, 1983; Hudson *et al.*, 1993).

Phototoxicity of furanocoumarins is largely based on their reactions with DNA (Mathews, 1963; Ivie, 1987; Bethea *et al.*, 1999). Furanocoumarins intercalate, or bind reversibly between two pyrimidine bases of a double-stranded DNA. Under UVA light, a covalent bond is formed between a furanocoumarin and one or two pyrimidine bases; the diadduct may form an interstrand cross-link. Angular furanocoumarins form mainly monoadducts (Cimino *et al.*, 1985). These Type I photosensitization reactions do not require oxygen.

As expected from their reactions with DNA, 8-MOP and 5-MOP cause mutations in bacteria (Mathews, 1963). Treatment of psoriasis with furanocoumarins and UVA increases the risk of skin cancer (Stern *et al.*, 1998). Furanocoumarins also produce ROS in UVA light (Cannistraro and van de Vorst, 1977) and interact with cellular membranes and proteins (Beijersbergen van Hegenouwen *et al.*, 1989).

Furanocoumarins were the first plant-derived compounds with a demonstrated light-dependent toxicity in the larvae of a herbivorous insect (Berenbaum, 1978). The production of furanocoumarins is linked to the openness of the plant habitat in Apiaceae (Berenbaum, 1981), suggesting a role for furanocoumarins in plant defense.

Hypericin. Photosensitization of livestock grazing on St. John's wort (*Hypericum*) (Clare, 1955) is caused by the red, fluorescent pigment hypericin [8] (Horsley, 1934), located in special glands of the plant. Hypericin absorbs between 530 and 600 nm (Knox and Dodge, 1985c), and illumination in the presence of hypericin kills insect larvae (Sandberg and Berenbaum, 1989), causes pigment loss and membrane damage in leaf disks (Knox and Dodge, 1985c) and inhibits protein kinases (Agostinis *et al.*, 1996). Hypericin is predominantly a Type II photosensitizer (Knox and Dodge, 1985c), but Type I reactions may also occur (Durán and Song, 1986). Insects specialized to feeding on *Hypericum* avoid the toxic effects either by avoiding the hypericin-containing glands or by avoiding the activation of the poison by light (Sandberg and Berenbaum, 1989; Fields *et al.*, 1990). Hypericin is also found in the photosensory pigment of *Stentor* (Song, 1981), and the ciliate is sensitive to light because of the endogenous hypericin (Yang *et al.*, 1986).

Polyacetylenes (polyines) and thiophenes. Polyines and the biosynthetically related thiophenes form a family of more than 750 UVA-absorbing phytochemicals (Bohlmann *et al.*, 1973), with all phototoxic members in Asteraceae (Marchant and Cooper, 1987). The most studied compounds are phenylheptatriyne (PHT) [10] and α-terthienyl (α-T) [11].

The UVA-dependent biological effects of α-T range from inactivation of enzymes *in vitro* (Bakker *et al.*, 1979), photokilling of bacteria (Downum *et al.*, 1982) and inactivation of viruses (Hudson, 1989) to phototoxicity to the green alga *Chlorella* (Arnason *et al.*, 1981).

The UVA level in sunlight and the α-T content in plant tissues are high enough to control lepidopteran larvae (Champagne *et al.*, 1984). In higher plants, α-T and UVA cause growth retardation (Campbell *et al.*, 1982), inhibit photosynthesis (Brennan, 1994) and promote decarboxylation of IAA (Brennan, 1996).

α-T is a strictly oxygen-dependent phototoxin and the production of $^1O_2$ during illumination of α-T (Bakker *et al.*, 1979; Kagan *et al.*, 1984a) probably causes the UVA-dependent biological effects (Marchant and Cooper, 1987). The nematocidal activity of α-T in *Tagetes* roots (Uhlenbroek and Bijloo, 1958) may be caused by $^1O_2$ produced after chemical excitation of α-T (Gommers and Bakker, 1988).

The biological activities of phototoxic polyines resemble those of α-T (Downum *et al.*, 1982; Kagan *et al.*, 1984b; Gong *et al.*, 1988). Low concentrations of O_2 are sufficient for PHT-sensitized phototoxicity (Kagan and Tuveson, 1988). Phototoxic polyines and thiophenes probably restrict insect herbivory on plants containing these compounds by acting both as feeding deterrents and photosensitizers (Champagne *et al.*, 1984). Insects feeding on phototoxic Asteraceae have both behavioral (Guillet *et al.*, 1995, 1997) and metabolic adaptations (Aucoin *et al.*, 1990; Berenbaum, 1994; Guillet *et al.*, 1997).

Other plant secondary metabolites. Several other classes of plant-derived chemicals like β-carbolines have phototoxic properties (Downum, 1992). New phototoxins like citral [12] (Asthana *et al.*, 1992) and curcumin [13] (Chignell *et al.*, 1994) are also found.

C. Porphyrins and Related Compounds

Porphyrins, precursors and constituents of heme and chlorophyll, are visual light absorbing pigments mainly capable of Type II reactions. Metabolic accumulation of protoporphyrin IX [14] mediates phototoxic poisoning of bacteria (Żołądek *et al.*, 1996) and plants (Kunert and Dodge, 1989), and causes the photosensitivity of skin in the erythropoietic protoporphyria disease in human (Kappas *et al.*, 1995). Phylloerythrin [15], a chlorophyll degradation product, is the primary photosensitizer in hepatogenous photosensitivity of ruminants (Ivie, 1982), and hematoporphyrin derivatives are used in photodynamic cancer therapy (Dougherty *et al.*, 1998). Chlorophylls sensitize the formation of ROS, and chloroplasts have a remarkable arsenal of antioxidant defenses (Foyer *et al.*, 1994; Bartosz, 1997). The light-harvesting chlorophyll–protein complex of Photosystem II (PS II) may structurally avoid contact between triplet chlorophyll a and oxygen (Siefermann-Harms and Angerhofer, 1998). Derivatives of bacteriochlorophyll are promising new candidates for photodynamic cancer therapy (Scherz *et al.*, 1997).

D. Cercosporin

A plant pathogenic fungus *Cercospora* produces a photosensitizer, cercosporin [9], and uses it to attack its host plant (Kuyama and Tamura, 1957; Daub and Ehrenshaft, 1993). This red pigment absorbs at 400–600 nm and emits red fluorescence. Illumination of plant tissues in the presence of cercosporin causes leakage of electrolytes and other cell constituents (Macri and Vianello, 1979). Illuminated cercosporin produces both $^1O_2$ and O_2^- *in vitro* (Daub and Hangarter, 1983), and Type II photosensitization is probably the most important mechanism in its pathogenic action (Daub and Ehrenshaft, 1993). *Cercospora* tolerates its own phototoxin, and the resistance is at least partially based on transient reduction of cercosporin by the mycelia (Daub *et al.*, 1992).

IV. Herbicides and Heavy Metals Mediating Phototoxic Responses

Several herbicides mediate phototoxic responses, acting as secondary phototoxins in plants. *Bipyridinium* herbicides, like paraquat (methyl viologen) act as electron acceptors of Photosystem I and the subsequent rapid autooxidation of the paraquat anion radical creates O_2^-. The hydroxyl radical formed in the subsequent Fenton chemistry is the primary reactant causing the destruction of the photosynthetic apparatus. The herbicidal effect of *carotenoid biosynthesis inhibitors*, like norflurazon, is based on herbicide-induced removal of photoprotective compounds (Sandmann and Böger, 1989). *Inhibitors of the plastidic protoporphyrinogen oxidase* (diphenyl ether herbicides) cause accumulation of protoporphyrin IX on the plasmalemma where the plastidic protoporphyrinogen is translocated (Kunert and Dodge, 1989; Duke and Rebeiz, 1994), causing the accumulation of protoporphyrin IX. Herbicidal application of *δ-aminolevulinate*, in turn, by-passes a regulatory step in porphyrin synthesis causing accumulation of photodynamically active tetrapyrroles (Rebeiz *et al.*, 1987; Duke and Rebeiz, 1994). DCMU (diuron) and other *PS II* herbicides bind to the Q_B site in the D1 protein of PS II (Mets and Thiel, 1989). Their primary effect is inhibition of electron flow through PS II, but several lines of evidence show that the killing of the plant finally occurs via phototoxic reactions sensitized by the chlorophylls (Pallett and Dodge, 1980).

Excess concentrations of heavy metals in the soil cause oxidative damage to plants that take up the metal ions (reviewed by Dietz *et al.*, 1999). The damage is often light-dependent (Luna *et al.*, 1994; Sandmann and Böger, 1980; Pätsikkä *et al.*, 1998). Mechanisms of heavy metal phototoxicity include participation of the metal ions in Fenton chemistry (Dietz *et al.*, 1999), inhibition of enzymes protecting against oxidative damage, and depletion of antioxidant pools (Luna *et al.*, 1994).

V. Is Photoinhibition of Photosynthesis a Phototoxicity Mechanism?

Photoinhibition of PS II, or simply photoinhibition, is a reaction in which visible light induces the loss of the electron transfer activity of PS II (Aro *et al.*, 1993). The photodamage is a simple first-order process both *in vitro* (Jones and Kok, 1966) and *in vivo* (Tyystjärvi and Aro, 1996). Continuous repair of the photodamage explains why plants do not suffer from photoinhibitory symptoms under their normal growth light intensities.

The role of ROS in photoinhibition is poorly understood. Photoinhibition leads to an increase in the $^1O_2$ content of leaves (Hideg *et al.*, 1998), and $^1O_2$ is involved in various hypothetical mechanisms of photoinhibition (Jung and Kim, 1990; Vass *et al.*, 1992; Hideg *et al.*, 1994; Keren *et al.*, 1997; Santabarbara *et al.*, 2001). The high sensitivity of PS II to $^1O_2$ generated from the outside (Knox and Dodge, 1985b) is compatible with the suggestions (Kim *et al.*, 1993; Chung and Jung, 1995; Santabarbara *et al.*, 2001) that photoinhibition is a Type II reaction in which the sensitizer may not be a functional part of PS II. However, photoinhibition occurs rapidly under anaerobic conditions (Hundal *et al.*, 1990; Trebst and Depka, 1990), indicating that ROS play a secondary role. The partial reversibility of anaerobic photoinhibition prompted Vass *et al.* (1992) to suggest that $^1O_2$ is involved in an irreversible step of aerobic photoinhibition. The photooxidative damage caused by herbicides like DCMU (Pallett and Dodge, 1980) that block PS II electron transport but do not enhance formation of chlorophyll triplet by the "acceptor-side" mechanism of Vass *et al.* (1992) may suggest that light-induced $^1O_2$ production is a general consequence of any block in PS II electron transport.

The action spectra of photoinhibition and D1 protein degradation show a high efficiency of ultraviolet-C to violet, and a low-efficiency visible tail with shallow peaks in red and possibly in green (Jones and Kok, 1966; Jung and Kim, 1990; Renger *et al.*, 1989; Greenberg *et al.*, 1989). Interpretation of the action spectrum has been difficult, and virtually every peak has been explained with a separate photosensitizer, including chlorophyll, semiquinone, tyrosine, iron–sulfur centers, cytochromes and manganese (Jones and Kok, 1966; Jung and Kim, 1990; Greenberg *et al.*, 1989; Renger *et al.*, 1989; Santabarbara *et al.*, 2001). The quantum yield of photoinhibition is independent of light intensity (Jones and Kok, 1966; Tyystjärvi and Aro, 1996), and the rate of photoinhibition can be lowered neither by reducing the size of the light-harvesting antenna of PS II nor by quenching of PS II excitations by artificial quenchers (Tyystjärvi *et al.*, 1994, 1999). These data suggest that PS II antenna chlorophylls do not sensitize PS II to photoinhibition. The similarity of the action spectrum of photoinhibition (Jones and Kok, 1966) with the absorption spectrum of manganese gluconates (Bodini *et al.*, 1976) led us to suggest that inactivation of the oxygen-evolving complex due to light absorption by manganese ions of the oxygen-evolving complex triggers photoinhibition not only in ultraviolet but also under visible light (Tyystjärvi *et al.*, 2001).

Also PS I is light-sensitive to some extent (Sonoike, 1996). The susceptibility of PS I to photodamage depends strongly on temperature and plant species, and the inhibition saturates at moderate light intensity. Active oxygen produced by the PS I electron transfer chain is apparently involved in the reaction mechanism.

VI. Phototoxicity and Programmed Cell Death

The production of ROS by phototoxicity mechanisms links programmed cell death to photosensitization phenomena. Chlorophyll-sensitized production of $^1O_2$ occurs during senescence (Biswal, 1995), and ROS are important signals in the hypersensitive response (Pontier *et al.*, 1998), but literature on phototoxicity as a mediator of apoptosis in plants is scarce. Phototoxic poisoning of flax cotyledons by rose bengal speeds up the symptoms of senescence (Percival and Dodge, 1983). Interestingly, photodynamic therapy of human cancers is thought to function via an apoptotic route rather than by a direct photokilling of the cancer cells by the phototoxicity reaction (Dougherty *et al.*, 1998).

Acknowledgments

Marja Hakala and Saija Sirkiä are thanked for technical help. The author was supported by the Academy of Finland.

References

Abeysekera, B.F., Abramowski, Z., and Towers, G.H.N. (1983). Genotoxicity of the natural furanochromones, khellin and visnagin and the identification of a khellin-thymine photoadduct. *Photochem. Photobiol.* **38**, 311–315.

Agostinis, P., Donella-Deana, A., Cuveele, J., Vanderbogaerde, A., Sarno, S., Merlevede, W., and de Witte, P. (1996). A comparative analysis of the photosensitized inhibition of growth-factor regulated protein kinases by hypericin-derivatives. *Biochem. Biophys. Res. Commun.* **220**, 613–617.

Ahmad, S. (1992). Biochemical defence of pro-oxidant plant allelochemicals by herbivorous insects. *Biochem. Syst. Ecol.* **20**, 269–296.

Ahmad, S. (Ed.) (1995). *Oxidative Stress and Antioxidant Defenses in Biology*. Chapman & Hall, New York.

Arfsten, D.P., Schaeffer, D.J., and Mulveny, D.C. (1996). The effects of near ultraviolet radiation on the toxic effects of polycyclic aromatic hydrocarbons in animals and plants: a review. *Ecotoxicol. Environ. Safety* **33**, 1–24.

Arnason, T., Stein, J.R., Graham, E., Wat, C.-K., Towers, G.H.N., and Lam, J. (1981). Phototoxicity to selected marine and freshwater algae of polyacetylenes from species in the Asteraceae. *Can. J. Bot.* **59**, 54–58.

Aro, E.-M., Virgin, I., and Andersson, B. (1993). Photoinhibition of Photosystem II—inactivation, protein damage and turnover. *Biochim. Biophys. Acta* **1143**, 113–134.

Ashwood-Smith, M.J., Ceska, O., and Chandhary, S.K. (1985). Mechanism of photosensitivity reactions to diseased celery. *Br. Med. J.* **290**, 1249.

Asthana, A., Larson, R.A., Marley, K.A., and Tuveson, R.W. (1992). Mechanisms of citral phototoxicity. *Photochem. Photobiol.* **56**, 211–222.

Aucoin, R.R., Fields, P., Lewis, M.A., Philogène, B.J.R., and Arnason, J.T. (1990). The protective effect of antioxidants to a phototoxin-sensitive insect herbivore, *Manduca sexta*. *J. Chem. Ecol.* **16**, 2913–2924.

Bakker, J., Gommers, F.J., Nieuwenhuis, I., and Wynberg, H. (1979). Photoactivation of the nematicidal compound α-terthienyl from roots of marigolds (*Tagetes* species). A possible singlet oxygen role. *J. Biol. Chem.* **254**, 1841–1844.

Bartosz, G. (1997). Oxidative stress in plants. *Acta Physiol. Plant.* **19**, 47–64.

Becker, R.S., Chakravorti, S., Gartner, C.A., and de Gracia Miguel, M. (1993). Photosensitizers: comprehensive photophysics/photochemistry and theory of coumarins, chromones, their homologues and thione analogues. *J. Chem. Soc. Faraday Trans.* **89**, 1007–1019.

Beijersbergen van Hegenouwen, G.M.J., Wijn, E.T., Schoonderwoerd, S.A., and Dall'acqua, F. (1989). A method for the determination of PUVA-induced binding of 8-methoxypsoralen (8-MOP) to epidermal lipids, proteins and DNA/RNA. *J. Photochem. Photobiol. B: Biol.* **3**, 631–635.

Berenbaum, M. (1978). Toxicity of a furanocoumarin to armyworms: A case of biosynthetic escape from insect herbivores. *Science* **201**, 532–534.

Berenbaum, M. (1981). Patterns of furanocoumarin distribution and insect herbivory in the Umbelliferae: plant chemistry and community structure. *Ecology* **62**, 1254–1266.

Berenbaum, M.R. (1994). Metabolic detoxification of plant prooxidants. In *Oxidative Stress and Antioxidant Defenses in Biology* (S. Ahmad, Ed.), pp. 181–209. Chapman & Hall, New York.

Berenbaum, M.R. (1995). Phototoxicity of plant secondary metabolites. Insect and mammalian perspectives. *Arch. Insect Biochem. Physiol.* **29**, 119–134.

Bethea, D., Fullmer, B., Syed, S., Seltzer, G., Tiano, J., Rischko, C., Gillespie, L., Brown, D., and Gasparro, F.P. (1999). Psoralen photobiology and photochemotherapy: 50 years of science and medicine. *J. Derm. Sci.* **19**, 78–88.

Biswal, B. (1995). Carotenoid catabolism during leaf senescence and its control by light. *J. Photochem. Photobiol. B: Biol.* **30**, 3–13.

Blum, H.F. (1941). *Photodynamic Action and Diseases Caused by Light*. Reinhold, New York.

Bodini, M.E., Willis, L.A., Riechel, T.L., and Sawyer, D.T. (1976). Electrochemical and spectroscopic studies of Mn(II), -(III), and -(IV) gluconate complexes. 1. Formulas and oxidation-reduction stoichiometry. *Inorg. Chem.* **15**, 1538–1543.

Bohlmann, F., Burkhardt, T., and Zdero, C. (1973). *Naturally Occurring Acetylenes*. Academic Press, London.

Brennan, T.M. (1994). Photosensitized inhibition of photosynthetic $^{14}CO_2$ fixation by α-terthienyl and ultraviolet-A. *Photochem. Photobiol.* **59**, 631–636.

Brennan, T.M. (1996). Decarboxylation of indole-3-acetic acid and inhibition of growth in *Avena sativa* seedlings by plant-derived photosensitizers. *Photochem. Photobiol.* **64**, 1001–1006.

Camm, E.L., Wat, C.-K., and Towers, G.H.N. (1976). Assessment of the roles of furanocoumarins in *Heracleum lanatum*. *Can. J. Bot.* **54**, 2562–2566.

Campbell, G., Lambert, J.D.H., Arnason, T., and Towers, G.H.N. (1982). Allelopathic properties of α-terthienyl and phenylheptatriyne, naturally occurring compounds from species of Asteraceae. *J. Chem. Ecol.* **8**, 961–972.

Cannistraro, S., and van de Vorst, A. (1977). ESR and optical absorption evidence for free radical involvement in the photosensitizing action of furanocoumarin derivatives and for their singlet oxygen production. *Biochim. Biophys. Acta* **476**, 166–177.

Champagne, D.E., Arnason, J.T., Philogène, B.J.R., Campbell, G., and McLachlan, D.G. (1984). Photosensitization and feeding deterrence of *Euxoa messoria* (Lepidoptera: Noctuidae) by α-terthienyl, a naturally occurring thiophene from the Asteraceae. *Experientia* **40**, 577–578.

Chignell, C.F., Bilski, P., Reszka, K.J., Motten, A.G., Sik, R.H., and Dahl, T.A. (1994). Spectral and photochemical properties of curcumin. *Photochem. Photobiol.* **59**, 295–302.

Chung, S.K., and Jung, J. (1995). Inactivation of the acceptor side and degradation of the D1 protein of photosystem II by singlet oxygen photogenerated from the outside. *Photochem. Photobiol.* **61**, 383–389.

Cimino, G.D., Gamper, H.B., Isaacs, S.T., and Hearst, J.E. (1985). Psoralens as photoactive probes of nucleic acid structure and function: organic chemistry, photochemistry, and biochemistry. *Ann. Rev. Biochem.* **54**, 1151–1193.

Clare, N.T. (1955). Photosensitization in animals. *Adv. Vet. Sci.* **2**, 182–211.

Daub, M.E., and Ehrenshaft, M. (1993). The photoactivated toxin cercosporin as a tool in fungal photobiology. *Physiol. Plant.* **89**, 227–236.

Daub, M.E., and Hangarter, R.P. (1983). Light-induced production of singlet oxygen and superoxide by the fungal toxin, cercosporin. *Plant Physiol.* **73**, 855–857.

Daub, M.E., Leisman, G.B., Clark, R.A., and Bowden, E.F. (1992). Reductive detoxification as a mechanism of fungal resistance to singlet oxygen-generating photosensitizers. *Proc. Natl. Acad. Sci. USA* **89**, 9588–9592.

Dietz, K.-J., Baier, M., and Krämer, U. (1999). Free radicals and reactive oxygen species as mediators of heavy metal toxicity in plants. In *Heavy Metal Stress in Plants from Molecules to Ecosystems* (M.N.V. Prasad, and J. Hagemeyer, Eds.), pp. 73–97. Springer-Verlag, Berlin.

Dollahite, J.W., Younger, R.L., and Hoffman, G.O. (1978). Photosensitization in cattle and sheep caused by feeding *Ammi majus* (Greater Ammi; Bishop's weed). *Am. J. Vet. Res.* **39**, 193–197.

Dougherty, T.J., Gomer, C.J., Henderson, B.W., Jori, G., Kessel, D., Korbelik, M., Moan, J., and Peng, Q. (1998). Photodynamic therapy. *J. Nat. Cancer Inst.* **90**, 889–905.

Downum, K.R. (1992). Light-activated plant defense. *New Phytol.* **122**, 401–420.

Downum, K.R., Hancock, R.E.W., and Towers, G.H.N. (1982). Mode of action of α-terthienyl on *Escherichia coli*: evidence for a photodynamic effect on membranes. *Photochem. Photobiol.* **36**, 517–523.

Duke, S.O., and Rebeiz, C.A. (Eds.) (1994). *Porphyric Pesticides. Chemistry, Toxicology and Pharmaceutical Applications.* ACS Symp. Ser. 559. Am. Chem. Soc., Washington.

Durán, N., and Song, P.-I. (1986). Hypericin and its photodynamic action. *Photochem. Photobiol.* **43**, 677–680.

Elstner, E.F. (1982). Oxygen activation and oxygen toxicity. *Ann. Rev. Plant Physiol.* **33**, 73–96.

Fields, P.G., Arnason, J.T., and Philogène, B.J.R. (1990). Behavioural and physical adaptations of three insects that feed on the phototoxic plant *Hypericum perforatum. Can. J. Zool.* **68**, 339–346.

Fitzpatrick, T.B., and Pathak, M.A. (1959). Historical aspects of methoxalen and other furocoumarins. *J. Invest. Dermatol.* **32**, 229–231.

Foote, C.S. (1987). Type I and Type II mechanisms of photodynamic action. In *Light-Activated Pesticides* (J.R. Heitz and K. Downum, Eds.), pp. 22–38. ACS Symp. Ser. 339. Am. Chem. Soc., Washington.

Fowlks, W.L., Griffith, D.G., and Oginsky, E.L. (1958). Photosensitization of bacteria by furocoumarins and related compounds. *Nature* **181**, 571–572.

Foyer, C.H., Lelandais, M., and Kunert, K.J. (1994). Photooxidative stress in plants. *Physiol. Plant.* **92**, 696–717.

Gollnick, K. (1968). Type II photooxygenation reactions in solution. *Advan. Photochem.* **6**, 1–122.

Gommers, F.J., and Bakker, J. (1988). Mode of action of α-terthienyl and related compounds may explain the suppressant effect of *Tagetes* species on populations of free living endoparasitic plant nematodes. In *Chemistry and Biology of Naturally Occurring Acetylenes and Related Compounds (NOARC)* (J. Lam, H. Breteler, T. Arnason and L. Hansen, Eds.), pp. 61–69. Elsevier, Amsterdam.

Gong, H.-H., Kagan, J., Seitz, R., Stokes, A.B., Meyer, F.A., and Tuveson, R.W. (1988). The phototoxicity of phenylheptatriyne: oxygen-dependent hemolysis of human erythrocytes and inactivation of *Escherichia coli. Photochem. Photobiol.* **47**, 55–63.

Greenberg, B.M., Gaba, V., Canaani, O., Malkin, S., Mattoo, A.K., and Edelman, M. (1989). Separate photosensitizers mediate degradation of the 32-kDa photosystem II reaction center protein in the visible and UV spectral regions. *Proc. Natl. Acad. Sci. USA* **86**, 6617–6620.

Guillet, G., Lavigne, M.-È., Philogène, B.J.R., and Arnason, J.T. (1995). Behavioral adaptations of two phytophagous insects feeding on two species of phototoxic Asteraceae. *J. Insect Behav.* **8**, 533–546.

Guillet, G., Chauret, D., and Arnason, J.T. (1997). Phototoxic polyacetylenes from *Viguiera annua* and adaptations of a chrysomelid beetle, *Zygogramma continua*, feeding on this plant. *Phytochemistry* **45**, 695–699.

Halliwell, B., and Gutteridge, J.M.C. (1999). *Free Radicals in Biology and Medicine.* Oxford University Press, Oxford, UK.

Hideg, É., Spetea, C., and Vass, I. (1994). Singlet oxygen production in thylakoid membranes during photoinhibition as detected by EPR spectroscopy. *Photosynth. Res.* **39**, 191–199.

Hideg, É., Kálai, T., Hideg, K., and Vass, I. (1998). Photoinhibition of photosynthesis in vivo results in singlet oxygen production detection via nitroxide-induced fluorescence quenching in broad bean leaves. *Biochemistry* **37**, 11405–11411.

Horsley, C.H. (1934). Investigation into the action of St. John's wort. *J. Pharm. Exp. Therap.* **50**, 310–322.

Houba-Herin, N., Carlberg-Bacq, C.M., Piette, J., and Van de Vorst, A. (1982). Mechanisms for dye-mediated photodynamic action: singlet oxygen production, deoxyguanosine oxidation and phage inactivating efficiencies. *Photochem. Photobiol.* **36**, 297–306.

Hudson, J.B. (1989). Plant photosensitizers with antiviral properties. *Antiviral Res.* **12**, 55–74.

Hudson, J.B., Graham, E.A., Harris, L., and Ashwood-Smith, M.J. (1993). The unusual UVA-dependent antiviral properties of the furoisocoumarin, coriandrin. *Photochem. Photobiol.* **57**, 491–496.

Hundal, T., Aro, E.-M., Carlberg, I., and Andersson, B. (1990). Restoration of light induced photosystem II inhibition without de novo protein synthesis. *FEBS Lett.* **267**, 203–206.

Ivie, G.W. (1982). Chemical and biochemical aspects of photosensitization in livestock and poultry. *J. Nat. Cancer Inst.* **69**, 259–262.

Ivie, G.W. (1987). Biological actions and metabolic transformations of furanocoumarins. In *Light-Activated Pesticides* (J.R. Heitz and K.R. Downum, Eds.), pp. 217–230. ACS Symp. Ser. 339. Am. Chem. Soc., Washington.

Jones, L.W., and Kok, B. (1966). Photoinhibition of chloroplast reactions. I. Kinetics and action spectra. *Plant Physiol.* **41**, 1037–1043.

Jung, J., and Kim, H.-S. (1990). The chromophores as endogeous sensitizers involved in the photogeneration of singlet oxygen in spinach thylakoids. *Photochem. Photobiol.* **52**, 1003–1009.

Kagan, J., and Tuveson, R. (1988). Are there any photocytotoxic reactions of phenylheptatriyne that are not oxygen dependent? In *Chemistry and Biology of Naturally Occurring Acetylenes and Related Compounds (NOARC)* (J. Lam, H. Breteler, T. Arnason and L. Hansen, Eds.), pp. 71–84. Elsevier, Amsterdam.

Kagan, J., Prakash, I., Dhawan, S.N., and Jaworski, J.A. (1984a). The comparison of several butadiyne and thiophene derivatives to 8-methoxypsoralen and methylene blue as singlet oxygen sensitizers. *Photobiochem. Photobiophys.* **8**, 25–33.

Kagan, J., Tadema-Wielandt, K., Chan, G., Dhawan, S.N., Jaworsky, J., Prakash, I., and Arora, S.K. (1984b). Oxygen requirement for near-UV mediated cytotoxicity of phenylheptatriyne to *Escherichia coli*. *Photochem. Photobiol.* **39**, 465–467.

Kappas, A., Sassa, S., Galbraith, R.A., and Nordmann, Y. (1995). The porphyrias. In *The Metabolic and Molecular Basis of Inherited Disease* (C.R. Scriver, A.L. Beaudet, W.S. Sly and D. Valle, Eds.), pp. 2103–2159. McGraw-Hill, New York.

Keren, N., Berg, A., van Kan, P.J.M., Levanon, H., and Ohad, I. (1997). Mechanism of photosystem II photoinactivation and D1 protein degradation at low light: The role of back electron flow. *Proc. Natl. Acad. Sci. USA* **94**, 1579–1584.

Kim, C.S., Han, G.H., Kim, J.M., and Jung, J. (1993). *In situ* susceptibilities of photosystems I and II to photosensitized inactivation via singlet oxygen mechanism. *Photochem. Photobiol.* **57**, 1069–1074.

Knox, J.P., and Dodge, A.D. (1985a). The photodynamic action of eosin, a singlet-oxygen generator. Some effects on leaf tissue of *Pisum sativum* L. *Planta* **164**, 22–29.

Knox, J.P., and Dodge, A.D. (1985b). The photodynamic action of eosin, a singlet-oxygen generator. The inhibition of photosynthetic electron transport. *Planta* **164**, 30–34.

Knox, J.P., and Dodge, A.D. (1985c). Isolation and activity of the photodynamic pigment hypericin. *Plant Cell Env.* **8**, 19–25.

Kunert, K.J., and Dodge, A.D. (1989). Herbicide-induced radical damage and antioxidative systems. In *Target Sites of Herbicide Action* (P. Böger and G. Sandmann, Eds.), pp. 45–63. CRC Press, Boca Raton.

Kuyama, S., and Tamura, T. (1957). Cercosporin. A pigment from *Cercosporina kikuchii Matsumoto et Tomoyasu*. I. Cultivation of fungus, isolation and purification of pigment. *J. Am. Chem. Soc.* **79**, 5725–5726.

Laustriat, G. (1986). Molecular mechanisms of photosensitization. *Biochimie* **68**, 771–778.

Luna, C.M., González, C.A., and Trippi, V.S. (1994). Oxidative damage caused by excess copper in oat leaves. *Plant Cell Physiol.* **35**, 11–15.

Macri, F., and Vianello, A. (1979). Photodynamic activity of cercosporin on plant tissues. *Plant Cell Env.* **2**, 267–271.

Marchant, Y.Y., and Cooper, G.K. (1987). Structure and function relationships in polyacetylene photoactivity. In *Light-Activated Pesticides* (J.R. Heitz and K.R. Downum, Eds.), pp. 241–254. ACS Symp. Ser. 339. Am. Chem. Soc., Washington.

Mathews, M.M. (1963). Comparative study of lethal photosensitization of *Sarcina lutea* by 8-methoxypsoralen and by toluidine blue. *J. Bacteriol.* **85**, 322–328.

Mets, L., and Thiel, A. (1989). Biochemistry and genetic control of the Photosystem-II herbicide target site. In *Target Sites of Herbicide Action*, L.P. Böger and G. Santarius, Eds.), pp. 1–24. CRC Press, Boca Raton.

Ogilby, P.R., and Foote, C.S. (1983). Chemistry of singlet oxygen, 42. Effect of solvent isotopic substitution and temperature on the lifetime of singlet molecular oxygen ($^1\Delta g$). *J. Am. Chem. Soc.* **105**, 3423–3430.

Pallett, K.E., and Dodge, A.D. (1980). Studies into the action of some photosynthetic inhibitor herbicides. *J. Exp. Bot.* **31**, 1051–1066.

Pathak, M.A., Daniels Jr., F., and Fitzpatrick, T.B. (1962). The presently known distribution of furocoumarins (psoralens) in plants. *J. Invest. Dermatol.* **39**, 225–249.

Pätsikkä, E., Aro, E.-M., and Tyystjärvi, E. (1998). Increase in the quantum yield of photoinhibition contributes to copper toxicity in vivo. *Plant Physiol.* **117**, 619–627.

Pedersen, N.B., and Arlés, U.-B.P. (1997). Phototoxic reaction to parsnip and UV-A sunbed. *Contact Dermatitis* **39**, 97.

Percival, M.P., and Dodge, A.D. (1983). Photodynamic effects of rose bengal on senescent flax cotyledons. *J. Exp. Bot.* **34**, 47–54.

Pontier, D., Balagué, C., and Roby, D. (1998). The hypersensitive response. A programmed cell death associated with plant resistance. *C.R. Acad. Sci. Paris* **321**, 721–734.

Raab, O. (1900). Ueber die Wirkung fluorescirender Stoffe auf Infusorien. *Z. Biol.* **21**, 525–546.

Rebeiz, C.A., Montazer-Zouhoor, A., Mayasich, J.M., Tripathy, B.C., Wu, S.M., and Rebeiz, C.C. (1987). Photodynamic herbicides and chlorophyll biosynthesis modulators. In *Light-Activated Pesticides* (J.R. Heitz and K. Downum, Eds.), pp. 295–328. ACS Symp. Ser. 339. Am. Chem. Soc., Washington.

Renger, G., Völker, M., Eckert, H.J., Fromme, R., Hohm-Veit, S., and Gräber, P. (1989). On the mechanism of Photosystem II deterioration by UV-B irradiation. *Photochem. Photobiol.* **49**, 97–105.

Sandberg, S.L., and Berenbaum, M.R. (1989). Leaf-tying by tortricid larvae as an adaptation for feeding on phototoxic *Hypericum perforatum*. *J. Chem. Ecol.* **15**, 875–885.

Sandmann, G., and Böger, P. (1980). Copper-mediated lipid peroxidation processes in photosynthetic membranes. *Plant Physiol.* **66**. 797–800.

Sandmann, G., and Böger, P. (1989). Inhibition of carotenoid biosynthesis by herbicides. In *Target Sites of Herbicide Action* (P. Böger and G. Sandmann, Eds.), pp. 25–44, CRC Press, Boca Raton.

Santabarbara, S., Neverov, K.V., Garlaschi, F.M., Zucchelli, G., and Jennings, R.C. (2001). Involvement of uncoupled antenna chlorophylls in photoinhibition in thylakoids. *FEBS Lett.* **491**, 109–113.

Scherz, A., Salomon, Y., and Fiedor, L. (1997). Chlorophyll and bacteriochlorophyll derivatives, their preparation and pharmaceutical compositions comprising them. U.S. Pat. No. 5,650,292.

Siefermann-Harms, D., and Angerhofer, A. (1998). Evidence for an O_2-barrier in the light-harvesting chlorophyll-*a/b*-protein complex LHC II. *Photosynth. Res.* **55**, 83–94.

Spikes, J.D., and Straight, R.C. (1987). Biochemistry of photodynamic action. In *Light-Activated Pesticides* (J.R. Heitz and K.R. Downum, Eds.), pp. 98–108. ACS Symp. Ser. 339. Am. Chem. Soc., Washington.

Song, P.-S. (1981). Photosensory transduction in *Stentor coeruleus* and related organisms. *Biochim. Biophys. Acta* **639**, 1–29.

Sonoike, K. (1996). Photoinhibition of Photosystem I: Its physiological significance in the chilling sensitivity of plants. *Plant Cell Physiol.* **37**, 239–247.

Stern, R.S., Liebman, E.J., Väkevä, L., and the PUVA follow-up study (1998). Oral psoralen and ultraviolet-A light (PUVA) treatment of psoriasis and persistent risk of nonmelanoma skin cancer. *J. Natl. Cancer Inst.* **90**, 1278–1284.

Towers, G.H.N. (1984). Interactions of light with phytochemicals in some natural and novel systems. *Can. J. Bot.* **62**, 2900–2911.

Trebst, A., and Depka, B. (1990). Degradation of the D-1 protein subunit of Photosystem II in isolated thylakoids by UV-light. *Z. Naturforsch.* **45c**, 765–771.

Tyystjärvi, E., and Aro, E.-M. (1996). The rate constant of photoinhibition, measured in lincomycin-treated leaves, is directly proportional to light intensity. *Proc. Natl. Acad. Sci. USA* **93**, 2213–2218.

Tyystjärvi, E., Kettunen, R., and Aro, E.-M. (1994). The rate constant of photoinhibition in vitro is independent of the antenna size of Photosystem II but depends on temperature. *Biochim. Biophys. Acta* **1186**, 177–185.

Tyystjärvi, E., King, N., Hakala, M., and Aro, E.-M. (1999). Artificial quenchers of chlorophyll fluorescence do not protect against photoinhibition. *J. Photochem. Photobiol. B: Biol.* **48**, 142–147.

Tyystjärvi, E., Kairavuo, M., Pätsikkä, E., Keränen, M., Khriachtchev, L., Tuominen, I., Guiamet, J.J., and Tyystjärvi, T. (2001). The quantum yield of photoinhibition is the same in flash light and under continuous illumination—implications for the mechanism. In *PS2001 Proceedings: 12th International Congress on Photosynthesis*, CSIRO Publishing Company, Melbourne. CD-ROM.

Uhlenbroek, J.H., and Bijloo, J.D. (1958). Investigations on nematicides I. Isolation and structure of a nematicidal principle occurring in *Tagetes* roots. *Rec. Trav. Chim.* **77**, 1004–1009.

Valenzeno, D.P. (1987). Photomodification of biological membranes with emphasis on singlet oxygen mechanisms. *Photochem. Photobiol.* **46**, 147–160.

Vass, I., Styring, S., Hundal, T., Koivuniemi, A., Aro, E.-M., and Andersson, B. (1992). Reversible and irreversible intermediates during photoinhibition of photosystem II. Stable reduced Q_A species promote chlorophyll triplet formation. *Proc. Natl. Acad. Sci. USA* **89**, 1408–1412.

Yang, K.-C., Prusti, R.K., Walker, E.B., Song, P.-S., Watanabe, M., and Furuya, M. (1986). Photodynamic action in *Stentor coeruleus* sensitized by endogenous pigment stentorin. *Photochem. Photobiol.* **43**, 305–310.

Żołądek, T., Nhi, N.B., and Rytka, J. (1996). *Saccharomyces cerevisiae* mutants defective in heme biosynthesis as a tool for studying the mechanism of phototoxicity of porphyrins. *Photochem. Photobiol.* **64**, 957–962.

19

Ultraviolet Effects

Lars Olof Björn

I. Introduction

Studies of the effects of ultraviolet radiation on organisms have played an important part in the history of science. In particular I would like to point to the work by Frederick L. Gates. Before him, nobody had any idea of the role of nucleic acids. Gates (1928, 1929, 1930) found that the ability of ultraviolet radiation to kill cells of bacteria varies with wavelength in the same way as does the ability of nucleic acid to absorb radiation. This was the first indication of the fundamental importance of nucleic acid to life and a key experiment at the entrance to molecular biology.

Gates had identified one of the most important cellular targets for the deleterious action of ultraviolet radiation, but this target is not the only important one. Neither is it true that "a single reaction" is involved in the damaging effect of ultraviolet radiation on DNA; several direct photochemical reactions caused by absorption of radiation in DNA are involved, but also indirect effects caused by absorption of radiation in other molecules and the ensuing formation of radicals and active oxygen species which attack DNA. The latter indirect mechanisms are more important with the longer wavelength components which dominate in daylight.

The curve for the maximal killing or inactivation effect of ultraviolet radiation has a maximum near 260 nm. This is in the so-called UV-C spectral band (wavelength less than

280 nm). The sun emits UV-C radiation, but also UV-B (280–315 nm) and UV-A (315–400 nm) radiation, as well as electromagnetic radiation of shorter and longer wavelength, and particles such as neutrinos. The reason for the division of the ultraviolet (UV) band into three regions is a practical one. UV-C radiation does not penetrate to the Earth's surface and is not part of the natural environment in the biosphere. UV-B penetrates to an appreciable but variable extent. It is absorbed by ozone, mainly present in the stratosphere, the amount of which is highly variable. Although less biologically active than UV-C, it still has an appreciable effect, and so the amount of ozone present in the atmosphere is of great biological importance. UV-A is only weakly absorbed by ozone, and in many respects has biological effects similar to what we call visible light or photosynthetically active radiation (400–700 nm).

During the past quarter of a century attention has been focused on the UV-B band, and the reason for this is the depletion of stratospheric ozone that has taken place as a result of man's pollution of the atmosphere. This is not the place to spend much space on the problem of stratospheric ozone depletion, but the interested reader is referred to recent reviews of this important topic (UNEP, 1995, 1998, 2002).

II. Importance of Ultraviolet Radiation as an Environmental Factor for Plants

A. Evolutionary Aspects and the Role of Ozone

Life has probably existed on Earth for about 4 billion years, while the Earth itself is thought to be approximately 4.6 billion years old. It seems as if life was already here as soon as the surface had cooled to a life-compatible temperature and the worst bombardment by meteorites, comets and asteroids had subsided. The first life was aquatic, and life has remained under water for 90% of the time that has elapsed since then. As surprising as the rapid appearance of the first life on the young Earth, equally surprising at first glance is its slow journey on to land. One has to look for a specific reason for the slow conquest of terra firma. Such a reason could be the radiation climate. Before life there was no photosynthesis, before photosynthesis there was hardly any free oxygen in the atmosphere, and without free oxygen no ozone could be formed. And for a long time after the advent of oxygenic photosynthesis, oxygen did not accumulate in the atmosphere, since reductants such as divalent iron and hydrogen sulfide must first be oxidized.

With the exception of trace amounts of sulfur dioxide produced by man or by volcanic activity, ozone is the only gas present in the atmosphere which is capable of absorbing UV-B radiation. The amount of ozone protecting us from the sun's UV-B radiation is surprisingly small. If all the ozone in the atmosphere could be condensed to a liquid and smeared over the surface of the Earth, the layer would be only 4 micrometers thick. This is one reason for the vulnerability of the ozone layer, another one being the fact that our pollutants act as catalysts, so that for instance a single atom of bromine can decompose thousands of ozone molecules.

As for plants, the shift from aquatic to terrestrial life was coupled to the emergence of new biochemical pathways. One such pathway is the shikimic acid pathway. It produces lignin, a component of the wood necessary to support land plants, but also flavonoids and other substances serving as UV-B protective filters, especially in the outer parts of the plants.

B. Damage and Inhibition by UV

The first observation of ultraviolet radiation damage in plants was probably made by Siemens (1880). Arthur and Newell (1929) review the older literature and try to define the active spectral region using short-wavelength cut-off filters.

With sufficient exposure, ultraviolet radiation produces damage to organisms of all species. It was mentioned that DNA is one of the main molecular targets, but that also other targets are important. Let us first take a look at DNA as a target for the damaging effects of UV.

It is quite clear that the most common lesion to DNA caused by ultraviolet radiation is formation of cyclobutane dimers (CPDs) of the pyrimidine bases, and the second most common is the formation of so-called (6-4)-photoproducts. The pyrimidine bases involved are thymine (T) and cytosine (C). They can react in all combinations, and since the DNA strands are vectorial (with "downstream" and "upstream" directions) there are for both dimerization and (6-4) photoproduct formation four possibilities, i.e. TT, TC, CT and CC. In addition, thymine dimers can have either *cis*-syn or *trans*-syn configuration, depending on whether similar groups (CH_3 and H, respectively) are on the same or opposite sides of the rings. *Trans*-syn dimers, however, are of little importance in double stranded DNA (Smith *et al.*, 1996). All these lesions occur with different frequencies and produce different effects (Smith *et al.*, 1996; Jiang and Taylor, 1993; Horsefall and Lawrence, 1994). Some methods to assay for them, however, distinguish clearly only between dimers and (6-4) photoproducts. Among CPDs, the TT dimers are regarded as the least important, as they do not terminate replication and result in adenine insertion into the new strand.

Horsefall and Lawrence (1994) found the mutation frequencies per DNA lesion to increase in the order CPD < (6-4) TC < (6-4) TT, and the so-called Dewar isomer, which arises from the photoconversion of (6-4) photoproducts by UV-A or long-wavelength UV-B, to be even more mutagenic. It has also been claimed (Mitchell *et al.*, 1991) that (6-4) photoproducts, in contrast to CPDs, are formed mainly in actively transcribed parts of the genome. All this adds up to a view that, although CPDs arise approximately four times as often as (6-4) photoproducts (Britt, 1996), the latter may be of greater importance.

However, CPD formation may be a trigger for UV-C-induced apoptosis, as in experiments by Nishigaki *et al.* (1998) on mammal cells. These investigators did not discriminate between different kinds of CPD, but from the foregoing one might assume that the CPDs involving cytosine would be the most important.

In addition to pyrimidine dimers, (6-4) photoproducts, and Dewar isomers, a number of other lesions can, at a lower frequency, be induced in DNA by irradiation of cells with ultraviolet radiation. Pyrimidine dimers and (6-4) photoproducts differ from other lesions in that under the influence of blue light or ultraviolet-A radiation they can be very rapidly repaired by enzymes called photolyases, a process called photoreactivation. After conversion of (6-4) photoproducts to Dewar isomers, they cannot be photoreactivated, only repaired by much slower dark repair processes, involving several enzymes. Thus some of these other lesions [other than pyrimidine dimers and (6-4) photoproducts] which occur in low frequency, may be the most important ones, due to slow or inefficient repair.

The reader is referred to Cadet *et al.* (1992) for a primer of DNA photochemistry. Since that review was written, increased attention has been drawn to oxidative damage to DNA, caused by absorption of ultraviolet radiation in molecules other than DNA.

As already mentioned, it is quite clear that radiation targets other than DNA play an important role in plant life (Björn, 1996, 1999). Probably such non-DNA targets can also be involved in UV-induced apoptosis, as is the case in mammalian cells (Schwarz, 1998; Kulms *et al.*, 1999).

Both DNA and other molecules, in particular lipids, can be attacked by free radicals and reactive oxygen species produced by ultraviolet radiation. Particularly in the presence of both free iron and molecular oxygen, oxidative damage to DNA is easily initiated also with UV-A radiation (Audic and Giacomeni, 1993).

The information in this section is based to a large extent on experiments with other organisms than plants, and with isolated DNA. A review covering DNA damage and repair in plants has been written by Britt (1996).

C. Regulation by UV

Increasingly it has become clear, that ultraviolet radiation has not only damaging and inhibiting effects, but also regulative ones. A number of genes are activated by UV, others are down-regulated, and the phosphorylation pattern of proteins is changed. We shall not deal here with effects of UV-A, mediated by the same receptors (cryptochromes) as blue light effects. The regulatory effects of UV-B and UV-C radiation can be divided into two broad categories: Either they are a secondary effect when the plant is stressed, usually by damage to DNA, or they are evoked by activation of a specific UV-B receptor (Björn 1999).

There are many different signal transduction pathways for UV-induced modulation of gene activity. A.-H.-Mackerness *et al.* (1999) have shown that reactive oxygen species are involved in the down-regulation by UV-B of genes coding for photosynthetic enzymes.

The existence of a UV-B receptor is evident from many experiments, but the molecular nature of the receptor has not yet been defined. The earlier literature on the subject is discussed by Björn (1999). One group of molecules which have been advocated in this context are the pteridines, of which some have absorption spectra resembling the action spectra for UV-B regulations. I would like to propose here as another possibility provitamin D (D_2 or D_3, both of which occur in plants). The action spectrum for photochemical conversion of provitamin D_3 (dehydrocholesterol) in human skin to previtamin D_3 (the direct precursor of vitamin D_3) (MacLaughlin *et al.*, 1982) peaks at about 295 nm, just as many action spectra for regulative effects in plants do.

D. Stimulation by UV

It is not surprising that growth, photosynthesis and other activities of plants can be inhibited by UV, neither is it surprising that plants have a sensing system for UV-B and can adjust themselves to a changed UV-B climate. More surprising is the fact that some plant growth and other processes can be stimulated by increased UV-B, and currently no explanation exists for this.

Thus photosynthetic oxygen production in the South African desert plant *Dimorphotheca* is 37% higher on a chlorophyll basis in plants grown under UV-B compared to those grown without UV-B (Yu and Björn, 1999). UV-B can also have an ameliorating effect on the impact of other stress factors, e.g., ozone poisoning (Schnitzler *et al.*, 1999), and it is required for rapid recovery from midday depression of photosynthesis in the brown alga *Dictyota dichotoma* (Flores-Moya *et al.*, 1999). Takács *et al.* (1999) found an indicator of "potential thylakoid membrane activity" to be stimulated by UV-B in the moss *Dicranum scoparium*. Our group has found that the longitudinal growth of the moss *Hylocomium splendens*, under conditions of good water supply, can be stimulated by up to 31% by increased UV-B radiation (Gehrke *et al.*, 1996; Björn *et al.*, 1997). These are just a few of the positive effects of UV-B that can be found in the literature, so it would not be correct to consider UV-B radiation as a negative factor for plant life processes under all conditions.

III. UV and Plant Senescence

Senescence is a phenomenon distinct from apoptosis, but still related. Especially in plants, apoptosis is often a phenomenon of specific cells, not whole tissues. For instance, in xylem differentiation, or differentiation of sclereid idioblasts, cells undergoing apoptosis are embedded among cells not undergoing apoptosis. Apoptosis is also a relatively rapid process, taking place within 24 hours. On the contrary, senescence is a process of whole tissues, and often whole organs, and occurs gradually over many days. Still, both senescence and apoptosis imply programmed death, and therefore we have reason to say also something about senescence in the context of apoptosis.

A. UV Acceleration of Senescence

Arthur and Newell (1929) may have been the first to note that exposure to ultraviolet radiation causes plant leaves to turn yellow. Since this effect of ultraviolet radiation is counteracted by visible light, Tanada and Hendricks (1953) attributed this to dimer formation in DNA, a lesion which can be repaired by photoreactivation (Chessin and Cohen, 1962). Indeed, two different types of photolyase have, as already mentioned, been demonstrated in plants. However, the phenomenon may also have another explanation. Several investigators, and most clearly Wu (1971) and Wu *et al.* (1973) have produced evidence that a diffusable factor is produced in the epidermis as a result of irradiation, and that this factor upon entering the mesophyll speeds up breakdown of chlorophyll in excised leaves. If the epidermis is removed immediately after irradiation of the leaves, and also if the factor is leached out by allowing the leaves to float on water, irradiated epidermis down, the UV effect is canceled. If a sufficiently high exposure to UV is administered, cells die immediately (necrosis), and the yellowing process does not take place even if the epidermis is left intact.

B. UV Retardation of Senescence

Contrary to UV-B, UV-A (315–400 nm wavelength) can retard senescence (Cuello *et al.*, 1994). The opposite effects of UV-A and UV-B on senescence could possibly be related to their opposite effects on formation of hydrogen peroxide (see Section VI below).

IV. Evidence for UV-induced PCD in a Plant

The only publication dealing with a case of clear-cut apoptosis induced by ultraviolet is that by Danon and Gallois (1998). They irradiated either seedlings of *Arabidopsis thaliana*, or protoplasts prepared from them, with 1–50 kJ m^{-2} of 254 nm radiation, and observed fragmentation of DNA which increased both with UV exposure and with time after exposure up to 12 h. Southern analysis revealed the "ladder" of DNA fragments with sizes in multiples of 180 base pairs, typical of apoptosis. Using the TUNEL reaction, fragmentation of DNA in protoplasts could be shown as soon as 2 h after irradiation. Also morphological changes of the nuclei similar to those observed in animal cells took place. It should be noted that the radiation doses needed to induce apoptosis in these plant cells were several hundred times larger than those needed in experiments with animal cell cultures.

Danon and Gallois (1998) propose that the apoptosis is triggered by DNA damage. One should not read into this that the primary radiation target necessarily is DNA. This question would have to be decided by action spectroscopy. Scheuerlein *et al.* (1995) studied UV-B-induced DNA degradation in *Euglena*. If this degradation, as the authors seem to believe, depends on the same photoreaction as UV-inhibition of motility in the same organism, the action spectrum (Häder and Liu, 1990) rules out DNA as the primary target. The DNA degradation observed in*Euglena* has in common with apoptosis in animal cells that it depends on nucleases activated by calcium and other divalent metal ions, and elevation of cytoplasmic $[Ca^{2+}]$ is probably one step in signal transduction.

V. Pathways for UV-induced PCD in Animal Cells

Early literature on apoptosis in animals, in the 1980s, concentrated on the nematode *Caenorhabditis elegans*. In the 1990s attention, especially as regards UV-induced PCD, turned to human and other mammal cells, mainly skin cells (fibroblasts, melanocytes, keratinocytes). Much of the effort has gone into identifying "apoptosis on" genes, i.e. genes which are switched on when apoptosis is induced. The most famous one of these is p53, which encodes a transcription factor. When a cell is exposed to UV radiation or certain other environmental factors, the normal p53 protein either temporarily halts cell division or causes apoptosis. If p53 has mutated, this may fail, and the result can be skin cancer.

Other "apoptosis on" as well as "apoptosis off" genes have been identified both in *Caenorhabditis elegans* and in human cells.

Apparently conflicting opinions can be found in the literature, both concerning the initial events in UV-induced apoptosis, and concerning the ensuing course of events. To some extent the conflicts can be resolved by distinguishing between UV-A, UV-B, and UV-C radiations, which may all cause apoptosis. It now seems fairly clear that there are several primary radiation targets for UV-induced apoptosis in animal cells.

Nishigaki *et al.* (1998) found that UV-C-induced apoptosis could be prevented by irradiating the cells with photoreactivating light after the UV-C administration. They drew the conclusion that the apoptosis was triggered by the formation of pyrimidine dimers. If mammal cells, as plant and insects cells do, can photoreactivate (6-4) photoproducts as well, they could also be the cause of apoptosis.

Sheikh *et al.* (1998), on the other hand write: "... plasma membrane initiated events play a predominant role in mediating UV-irradiation-induced apoptosis". Kulms *et al.* (1999) point out that UV-induced apoptosis is mediated by nuclear as well as by membrane effects. As for the membrane events, it is clear that one form of apoptosis is associated with an opening of a "megapore" in the outer mitochondrial membrane, and the ensuing release of cytochrome c. However, in other cases apoptosis takes place without such an event. Megapore-associated apoptosis can be recognized by its inhibition by cyclosporin A, which prevents the megapore from opening. Megapore-associated apoptosis has been observed under circumstances when ultraviolet radiation (particularly of longer wavelength) causes the production of singlet oxygen (Godar, 1999).

In the next section we shall refer to some further literature on UV-induced apoptosis in animal cells which has particular relevance for interpretation of similar events in plants.

VI. Possible Signaling Mediators between UV and PCD in Plants

UV irradiation stimulates endogenous ceramide production in animal cells by activation of sphingomyelinase (Pushkareva *et al.*, 1995; Spiegel *et al.*, 1996; Kolesnick and Krönke, 1998). Although sphingolipid concentrations in plants are reported to be low, Yoshida and Uemura (1986) report that 17% of the tonoplast lipids in mung bean are a ceramide glycoside. A role for ceramide as a signal mediator for apoptosis in plant cells has been suggested by Gilchrist (1998); see also Abbas *et al.* (1994).

Regarding the connection between reactive oxygen species and apoptosis the reader is referred to Chapter 13 and to Overmyer *et al.* (2003). Reactive oxygen species can certainly form in plants as a result of ultraviolet irradiation; we have already mentioned that A.-H.-Mackerness *et al.* (1999) invoked such a process in explaining down-regulation of certain genes by UV-B. T.M. Murphy and collaborators (see Murphy, 1990), Nottaris *et al.* (1997) and others have demonstrated UV-C-induced hydrogen peroxide production in plant cells. UV-B and UV-A do not have this effect (Murphy, 1990), and a very interesting observation is that the production is inhibited by blue light as well as by UV-A or UV-B radiation (Murphy and Vu, 1996). This latter observation can harmonize the previously described results by Tanada and Hendricks (1953), Cuello *et al.* (1994) and Wu *et al.* (1973). Perhaps the protective effect of longer wavelength radiation that Tanada and Hendricks (1953) as well as Cuello *et al.* (1994) observed was not photoreactivation of DNA, but inhibition of H_2O_2 production, and H_2O_2 the diffusible factor that Wu *et al.* (1973) postulated the existence of.

On the other hand, it seems clear that, in mammal cells, apoptosis can take place also under conditions when the formation of reactive oxygen species is minimal (Shimuzu *et al.*, 1995; Jacobson and Raff, 1995). Most probably, there are also in plant cells pathways to apoptosis dependent on reactive oxygen species, and others that are not.

Ethylene production in plants is increased by ultraviolet irradiation (Khalilova *et al.*, 1993; Predieri *et al.*, 1995). On the other hand, as described in Chapter 8, ethylene can induce apoptosis in plant cells. Thus it seems possible that ultraviolet radiation can induce apoptosis via ethylene production.

Based on experiments with $^{86}Rb^+$ as a tracer, Negash *et al.* (1987) drew the conclusion that irradiation with UV-C radiation causes leakage of potassium ions from guard cells, while UV-A stimulates potassium accumulation. In this context it is interesting to note that Wang *et al.* (1999) ascribe one form of UV-induced apoptosis in mammal cells to the UV-activation of potassium channels. Thus it is possible that UV radiation can induce apoptosis in plant cells by a similar mechanism.

VII. Conclusion

The three wavelength bands of ultraviolet radiation, i.e. UV-A, UV-B and UV-C, all have different effects on plants. UV-C, which is not a component of the natural environment, has predominantly negative effects, UV-A positive effects from the perspective of plant development and survival. The effects of UV-B can be positive or negative. Of course this is a generalization, and the official spectral limits between the bands, 280 and 315 nm, do not mark sudden changes in biological actions.

True apoptosis induced by ultraviolet radiation in plants has so far been observed only in one case, while there are numerous observations of UV effects on the related phenomenon of senescence.

References

Abbas, H.K., Tanaka, T., Duke, S.O., Porter, J.K., Wray, E.M., *et al.* (1994). Fumonisin and AAAL-toxin induced disruption of sphingolipid metabolism with accumulation of free sphingoid bases. *Plant Physiol.* **106**, 1085–1093.

A.-H.-Mackerness, S., Jordan, B.R., and Thomas, B. (1999). Reactive oxygen species in the regulation of photosynthetic genes by ultraviolet-B radiation (UV-B: 280–320 nm) in green and etiolated buds of pea (*Pisum sativum* L.). *J. Photochem. Photobiol. B: Biology* **48**, 180–188.

Arthur, J.M., and Newell, J.M. (1929). The killing of plant tissue and the inactivation of tobacco mosaic virus by ultraviolet radiation. *Am. J. Bot.* **16**, 338–353.

Audic, A., and Giacomoni, P.U. (1993). DNA nicking by ultraviolet radiation is enhanced in the presence of iron and of oxygen. *Photochem. Photobiol.* **57**, 508–512.

Benda, T.A. (1955). Some effects of ultra-violet radiation on leaves of French bean (*Phaseolus vulgaris* L.). *Ann. Appl. Biol.* **43**, 71–85.

Björn, L.O. (1996). Effects of ozone depletion and increased UV-B on terrestrial ecosystems. *Int. J. Environ. Stud.* **51**, 217–243.

Björn, L.O. (1999). UV-B effects: Receptors and targets. In *Concepts in Photobiology: Photosynthesis and Photomorphogenesis* (G.S. Singal, *et al.*, Eds.), pp. 821–832. Narosa Publishing House, New Delhi, and Kluwer, Dordrecht.

Björn, L.O., Callaghan, T.V., Johnsen, I., Lee, J.A., Manetas, Y., Paul, N.D., Sonesson, M., Wellburn, A., Coop, D., Heide-Jørgensen, H.S., Gehrke, C., Gwynn-Jones, D., Johanson, U., Kyparissis, A., Levizou, E., Nikolopoulos, D., Petropoulou, Y., and Stephanou, M. (1997). The effects of UV-B radiation on European heathland species. *Plant Ecology* **128**, 252–264.

Britt, A.B. (1996). DNA damage and repair in plants. *Annu. Rev. Plant Physiol. Plant Mol. Biol.* **47**, 75–100.

Cadet, J., Anselmino, C., Douki, T., and Voiturez, L. (1992). Photochemistry of nucleic acids in cells. *J. Photochem. Photobiol. B: Biol.* **15**, 277–298.

Cuello, J., Sanchesz, M.D., and Sabater, B. (1994). Retardation of senescence by UV-A light in barley (*Hordeum vulgare* L.) leaf segments. *Envir. Exp. Botany* **34**, 1–8.

Danon, A., and Gallois, P. (1998). UV-C radiation induces apoptotic-like changes in *Arabidopsis thaliana*. *FEBS Lett.* **437**, 131–136.

Flores-Moya, A., Hanelt, D., Figueroa, F.-L., Altamirano, M., Viñegla, B., and Salles, S. (1999). Involvement of solar UV-B radiation in recovery of inhibited photosynthesis in the brown alga *Dictyota dichotoma* (Hudson) Lamouroux. *J. Photochem. Photobiol. B: Biology* **49**, 129–135.

Gates, F.L. (1928). On nuclear derivatives and the lethal action of ultra-violet light. *Science* **68**, 479–480.

Gates, F.L. (1929). A study of the batericidal action of ultra violet light. I. The reaction to monochromatic radiations. *J. Gen. Physiol.* **13**, 231–248.

Gates, F.L. (1930). A study of the action of ultra violet light III. The absorption of ultra violet light by bacteria. *J. Gen. Physiol.* **14**, 31–42.

Gehrke, C., Johanson, U., Gwynn-Jones, D., Björn, L.O., Callaghan, T.V., and Lee, J.A. (1996). Effects of UV-B radiation on a subarctic heathland. *Ecol. Bull.* **45**, 192–203.

Gilchrist, D.G. (1998). Programmed cell death in plant disease: The purpose and promise of cellular suicide. *Annu. Rev. Phytopathol.* **36**, 393–414.

Godar, D.E. (1999). UVA1 radiation triggers two different final apoptotic pathways. *J. Invest. Dermatol.* **12**, 3–12.

Häder, D.-P., and Liu, S.M. (1990). Motility and gravitactic orientation of the flagellate Euglena gracilis impaired by artificial and solar UV-B radiation. *Curr. Microbiol.* **21**, 161–168.

Horsefall, M., and Lawrence, C.W. (1994). Accuracy of replication past the T-C (6-4) adduct. *J. Mol. Biol.* **235**, 465–471.

Jacobson, M.D., and Raff, M.C. (1995). Programmed cell death and Bcl-2 protection in very low oxygen. *Nature* **374**, 814–816.

Jiang, N., and Taylor, J.S. (1993). In vivo evidence that UV-induced C fwdarw T mutations at dipyrimidine sites could result from the replicative bypass of cis-syn cyclobutane dimers or their deamination products. *Biochemistry* **32**, 472–481.

Khalilova, F. Kh., Rakitina, T.Ya., Vlasov, P.V., and Kefeli, V.I. (1993). Effects of ultraviolet radiation (UV-B) in the growth and ethylene production of three genetic lines of *Arabidopsis thaliana*. *Fiziologia Rastenii* (Moscow) **40**, 764–769.

Kolesnick, R.N., and Krönke, M. (1998). Regulation of ceramide production and apoptosis. *Annu. Rev. Physiol.* **60**, 643–665.

Kulms, D., Pöppelmann, B., Yarosh, D., Luger, T.A., Krutmann, J., and Schwarz, T. (1999). Nuclear and cell membrane effects contribute independently to the introduction of apoptosis in human cells exposed to UVB radiation. *Proc. Natl. Acad. Sci. USA* **96**, 7974–7979.

MacLaughlin, J.A., Anderson, R.R., and Holick, M.F. (1982). Spectral character of sun light modulates photosynthesis of pre vitamin D_3 and its photoisomers in human skin. *Science* **216**, 1001–1003.

Mitchell, D.L., Nguyen, T.D., and Cleaver, J.E. (1991). Nonrandom induction of pyrimidine-pyrimidone (6-4) photoproducts in ultraviolet-irradiated human chromatin. *J. Biol. Chem.* **265**, 5353–5356.

Murphy, T.M. (1990). Effect of broadband and visible radiation on hydrogen peroxide formation by cultured rose cells. *Physiol. Plant.* **80**, 63–68.

Murphy, T.M., and Vu, H. (1996). Photoinactivation of superoxide synthases of the plasma membrane from rose (*Rosa damascena* Mill.) cells. *Photochem. Photobiol.* **64**, 106–109.

Negash, L., Jensén, P., and Björn, L.O. (1987). Effect of ultraviolet radiation on accumulation and leakage of ^{86}Rb in guard cells of *Vicia faba*. *Physiol. Plant.* **68**, 200–204.

Nishigaki, R., Mitani, H., and Shima, A. (1998). Evasion of UVC-induced apoptosis by photorepair of cyclobutane pyrimidine dimers. *Exp. Cell Res.* **244**, 43–53.

Nottaris, D., Crespi, P., Greppin, H., and Penel, C. (1997). Effect of UV-C on two cell lines from sugarbeet. *Arch. Sci.* **50**, 223–232.

Overmyer, K., Brosché, M., and Kangasjärvi, J. (2003). Reactive oxygen species and hormonal control of cell death. *Trends Plant Sci.* **8** (in press, available online as of 24 June 2003).

Predieri, S., Norman, H.A., Krizek, D.T., Pillai, P., Mirecki, R.M., and Zimmermann, R.H. (1995). Influence of UV-B radiation on membrane lipid composition and ethylene evolution in 'Doyénne d'hiver' pear shoots grown in vitro under different photosynthetic photon fluxes. *Env. Exp. Botany* **35**, 151–160.

Pushkareva, M., Obeid, L.M., and Hannum, Y.A. (1995). Ceramide: an endogenous regulator of apoptosis and growth suppression. *Immunol. Today* **16**, 294–297.

Scheuerlein, R., Treml, S., Thar, B., Tirlapur, U.K., and Häder, D.-P. (1995). Evidence for UV-B-induced DNA degradation in *Euglena gracilis* mediated by activation of metal-dependent nucleases. *J. Photochem. Photobiol. B: Biology* **31**, 113–123.

Schnitzler, J.-P., Langebartels, C., Heller, W., Liu, J., Lippert, M., Döhring, T., Bahnweg, G., and Sandermann, H. (1999). Ameliorating effect of UV-B radiation on the response of Norway spruce and Scots pine to ambient ozone concentrations. *Global Change Biology* **5**, 83–94.

Schwarz, T. (1998). UV light affects cell membrane and cytoplasmic targets. *J. Photochem. Photobiol. B: Biology* **44**, 91–96.

Sheikh, M.S., Antinore, M.J., Huang, Y., and Fornace, A.J., Jr. (1998). Ultraviolet-irradiation-induced apoptosis is mediated via ligand independent activation of tumor necrosis factor receptor 1. *Oncogene* **17**, 2555–2563.

Shimuzu, S., Eguchi, Y., Kosaka, H., Kamiike, W., Matsuda, H., and Tsujimoto, Y. (1995). Prevention of hypoxia-induced cell death by Bcl-2 and Bcl-xL. *Nature* **374**, 811–813.

Siemens, C.W. (1880). On the influence of electric light upon vegetation, and on certain physical principles involved. *Proc. Roy. Soc. London* **30**, 210–219.

Smith, C.A., Wang, M., Jioang, N., Che, L., Zhao, W., and Taylor, J.S. (1996). Mutation spectra of M13 vectors containing site-specific cis-syn, trans-syn-I, (6-4), and Dewar pyrimidone photoproducts of thymidylyl-(3'fwdarw 5')-thymidine in *Escherichia coli* under SOS conditions. *Biochemistry* **35**, 4146–4154.

Spiegel, S., Foster, D., and Kolesnick, R. (1996). Signal transduction through lipid second messengers. *Curr. Opin. Cell Biol.* **8**, 159–167.

Takács, Z., Csistalan, Zs., Sass, L., Laitat, E., Vass, I., and Tuba, Z. (1999). UV-B tolerance of bryophyte species with different degrees of desiccation tolerance. *J. Photochem. Photobiol. B: Biology* **48**, 210–215.

Tanada, T., and Hendricks, S.B. (1953). Photoreversal of ultraviolet effects in soybean leaves. *Am. J. Bot.* **40**, 634–637.

UNEP (United Nations Environment Programme) (1995). Environmental effects of ozone depletion: 1994 assessment. *Ambio* **24**, 137–196.

UNEP (United Nations Environment Programme) (1998). Environmental effects of ozone depletion: 1998 assessment. *J. Photochem. Photobiol. B: Biology* **46**, 1–108.

UNEP (United Nations Environment Programme) (2002). Environmental effects of ozone depletion and its interactions with climate change: 2002 assessment. *Photochem. Photobiol. Sci.* **2** (1, UNEP Special Issue), 1–72.

Wang, L., Xu, D., Dai, W., and Lu, L. (1999). An ultraviolet-activated K^+ channel mediates apoptosis of myeloblastic leukemia cells. *J. Biol. Chem.* **274**, 3678–3685.

Wu, J.H. (1971). Retardation f ultraviolet light accelerated leaf senescence by a cytokinin: N^6 benzyladenine. *Photochem. Photobiol.* **13**, 179–181.

Wu, J.H., Skokut, T., and Hartman, M. (1973). Ultraviolet-radiation-accelerated leaf chlorosis: Prevention of chlorosis by removal of epidermis or by floating leaf discs on water. *Photochem. Photobiol.* **18**, 71–77.

Yoshida, S., and Uemura, M. (1986). Lipid composition of plasma membranes and tonoplasts isolated from etiolated seedlings of mung bean (*Vigna radiata* L.). *Plant Physiol.* **82**, 807–812.

Yu, S.G., and Björn, L.O. (1999). Ultraviolet-B stimulates grand formation in chloroplasts in the African desert plant *Dimorphotheca pluvialis*. *J. Photochem. Photobiol. B.* **49**, 65–70.

20

Effects of Airborne Pollutants

Eva J. Pell, Jennifer D. Miller and
Douglas G. Bielenberg

I. Introduction

When leaves of plants are subjected to stress, a frequent response is the acceleration of the senescence process. Flückiger *et al.* (1979) observed that foliage of many trees and shrubs exhibited accelerated leaf abscission when the plants were grown in containers along a motorway. While the causal agent(s) inducing accelerated senescence was not defined in this study, the response is an indication of the role the atmosphere plays in altering leaf longevity. Occasional reports suggest that sulfur dioxide exposure can lead to more rapid aging of foliage (Kargiolaki *et al.*, 1991) and there is an extensive literature describing ozone (O_3) induction of premature or accelerated leaf senescence (Mikkelsen and Heide-Jørgensen, 1996; Ojanperä *et al.*, 1998; Pell *et al.*, 1997; Reich and Lassoie, 1985), as will be discussed below.

Ozone was first discovered as an air pollutant capable of causing injury to foliage in the 1950s (Richards *et al.*, 1958). While higher concentrations of the gas readily lead to regions of leaf necrosis, chronic exposures to relatively low concentrations of the gas have become associated with signs of accelerated leaf senescence (Pell *et al.*, 1997). In this chapter we will explore the evidence that O_3 induces accelerated leaf senescence, the mechanism by which this induction might occur, and the implications to the whole plant.

Ozone enters the leaf almost exclusively through stomates (Laisk *et al.*, 1989) and is either scavenged in the apoplastic fluid or begins to penetrate the cells. Since O_3 is so

Plant Cell Death Processes

reactive, it is thought that virtually no gas penetrates the cell membrane (Laisk *et al.*, 1989). Rather, reactions either within the wall or membrane are likely to generate reactive oxygen species (ROS) with high oxidizing potential, e.g. OH°, RO°, O^{2-}, or H_2O_2 (Kangasjärvi *et al.*, 1994; Ranieri *et al.*, 1999). While most of the ROS are highly reactive and will not persist, H_2O_2 is a more stable byproduct capable of movement throughout the cell (Levine *et al.*, 1994). The ability of cells to withstand elevated levels of O_3-generated oxidative stress will depend upon the capacity of the cell to scavenge ROS (Kangasjärvi *et al.*, 1994; Pell *et al.*, 1999).

II. Physiology and Biochemistry of Ozone-induced Accelerated Leaf Senescence

Normal senescence is associated with a change in leaf color and foliar abscission. It is widely accepted that O_3 accelerates the rate of leaf abscission in a wide range of species including tree species such as *Populus*, apple (*Malus domestica* Brokh.), beech (*Fagus sylvatica* L.), loblolly pine (*Pinus taeda* L.) (Braun and Flückiger, 1995; Nyman, 1986; Pell *et al.*, 1995; Stow *et al.*, 1992; Wiltshire *et al.*, 1993) and numerous agronomic species including soybean (*Glycine max* L.), chick pea (*Cicer arietinum*) and black gram (*Trigonella foenumgraecum*) (Heagle and Booker, 1998; Kasana, 1991). The response is illustrated by 3-year-old seedlings of beech that exhibited earlier shifts in foliar autumnal coloration after chronic O_3 exposure (Mikkelsen and Heide-Jørgensen, 1996).

The process of senescence, and evidence of O_3 as an agent accelerating this series of events, can be described at ultrastructural and biochemical levels. As the flag leaves of wheat (*Triticum aestivum* L.) senesce, the plasma membrane loosens from the cell wall, the vacuole increases in size, chloroplasts become smaller, while vesicles positioned between granal thylakoids and plastoglobuli increase in prominence (Ojanperä *et al.*, 1992). These responses occur earlier in O_3-treated plants, but the pattern follows that of normal senescence. Similar responses have been observed in other species including birch (*Betula pendula* Roth.) and beech (Mikkelsen and Heide-Jørgensen, 1996; Pääkkönen *et al.*, 1995). During normal senescence, cells show a reduction in chlorophyll content and an ordered loss of galactolipids and then phospholipids (Gut and Matile, 1988). Ozone induces a similar but more rapid change in concentration of these compounds in spring wheat (*Triticum aestivum* L. cv. Drabant) (Sandelius *et al.*, 1995).

Gas exchange capacity changes as leaves age normally. Typically net photosynthesis (A_{sat}) peaks when leaves reach full expansion and declines thereafter. Several groups have looked at the effects of O_3 on changes in the profile of A_{sat} over time (Clark *et al.*, 1996; Kull *et al.*, 1996; Mulholland *et al.*, 1997; Pell *et al.*, 1992) and found that the decline proceeded more rapidly as the leaf aged. While there is evidence that O_3 can directly affect stomatal conductance (Pearson and Mansfield, 1993), reduction in uptake of CO_2 does not explain the reduction in A_{sat} during chronic exposures that lead to accelerated senescence (Pell *et al.*, 1992). Analysis of A/C_i response curves suggests that the decline in A_{sat} is more likely a reflection of altered carboxylation (Clark *et al.*, 1996).

Changes in activity and quantity of ribulose-1,5-bisphosphate carboxylase/oxygenase (Rubisco) during leaf aging parallel changes in rates of A_{sat}. Ozone-induced accelerated senescence is associated with a more rapid loss of both activity and quantity of this key enzyme of photosynthesis (Pell *et al.*, 1994a). Dalling (1987) proposed that the degradation of Rubisco normally occurs when free radicals in the chloroplast structurally modify the protein, rendering it more vulnerable to proteolysis. Since the synthesis of Rubisco is limited after full leaf expansion, any process that accelerates degradation of the protein would lead to more rapid leaf senescence. Rubisco is vulnerable to O_3-induced structural modification and an associated increase in rate of proteolysis (Eckardt and Pell, 1995; Landry and Pell, 1993). While O_3 probably never reaches the chloroplast, the ROS formed by this air pollutant may raise the oxidizing potential of the cell.

III. Mechanisms of Ozone-induced Accelerated Leaf Senescence

During normal senescence, photosynthesis-associated genes (PAGs) decline. Ozone caused a rapid loss in *rbc*S transcript levels in potato (*Solanum tubersosum* L.) leaves immediately after acute exposures to the pollutant (Reddy *et al.*, 1993), and this response was also observed in *Arabidopsis* and tobacco (Bahl and Kahl, 1995; Conklin and Last, 1995). A prematuare decline in *rbc*S was observed after more prolonged exposures to lower doses of the gas (Glick *et al.*, 1995; Miller *et al.*, 1999). A drop in *rbc*L was also observed, but this occurred after the initial drop in *rbc*S. In addition, O_3 has been associated with a reduction in the transcript level of chlorophyll a/b binding protein (*cab*) and two senescence-down regulated genes of unidentified function (Conklin and Last, 1995; Miller *et al.*, 1999).

Several laboratories have been pursuing the identification of senescence-associated genes (SAGs) in a number of botanical systems; this subject is covered in detail elsewhere in this volume (Chapters 4 and 5). We have examined the ability of O_3 to induce a subset of SAGs isolated from *Arabidopsis*. When Landsberg *erecta* plants were treated with 0.15 $\mu L\ L^{-1}$ O_3 for 6 h day^{-1} for two weeks, lower leaves in the rosette showed signs of yellowing in the absence of any foliar injury while companion control plants remained green (Miller *et al.*, 1999). When we probed comparable aged leaves from plants in O_3-treated or control environments, we found that seven of 11 SAGs were induced (Fig. 20-1).

The gene products of two SAGs induced by O_3 have in common metal binding functions, namely copper chaperone (CCH) and blue copper-binding protein (BCB). These proteins may be important in the recycling of essential metals once the leaf is targeted for senescence, or may act as protectants against metal-catalyzed oxidation that leads to generation of free radicals (Weaver *et al.*, 1997; Himelblau *et al.*, 1998). Himelblau *et al.* (1998) have reported that 0.80 $\mu L\ L^{-1}$ O_3 increased mRNA levels of *CCH* by 30% in 30 min. We found that *CCH* was induced after 6 days of the chronic exposure described above (Miller *et al.*, 1999). *CCH* is a functional homologue of the Anti-oxidant 1 (*ATX1*) yeast gene, a gene identified for its ability to protect against oxygen toxicity in yeast lacking superoxide dismutase (Himelblau *et al.*, 1998).

BCB expression was induced within 2 days of the chronic O_3 exposure (Miller *et al.*, 1999). In another report, Richards *et al.* (1998) treated *Arabidopsis* with 0.30 $\mu L\ L^{-1}$ O_3 for 3 and 6 h and were able to demonstrate a 2–3-fold increase in transcript level of *BCB*. BCB is membrane bound with extensive similarity to blue Cu^{2+}-binding proteins plastocyanin and stellacyanin, and likely functions as an electron carrier (Van Gysel *et al.*, 1993). Retention of electron transport capabilities may be important during senescence for completion of degradative processes and maximum remobilization of ions (Weaver *et al.*, 1997).

ERD1, a protease regulator, encodes a ClpC-like protein that may interact with the ATP-dependent ClpP protease to facilitate proteolysis in the chloroplast (Weaver *et al.*, 1997) and was induced within 2 days of O_3 treatment (Miller *et al.*, 1999). More rapid decline in Rubisco levels during O_3-induced accelerated leaf senescence is attributed to proteolysis

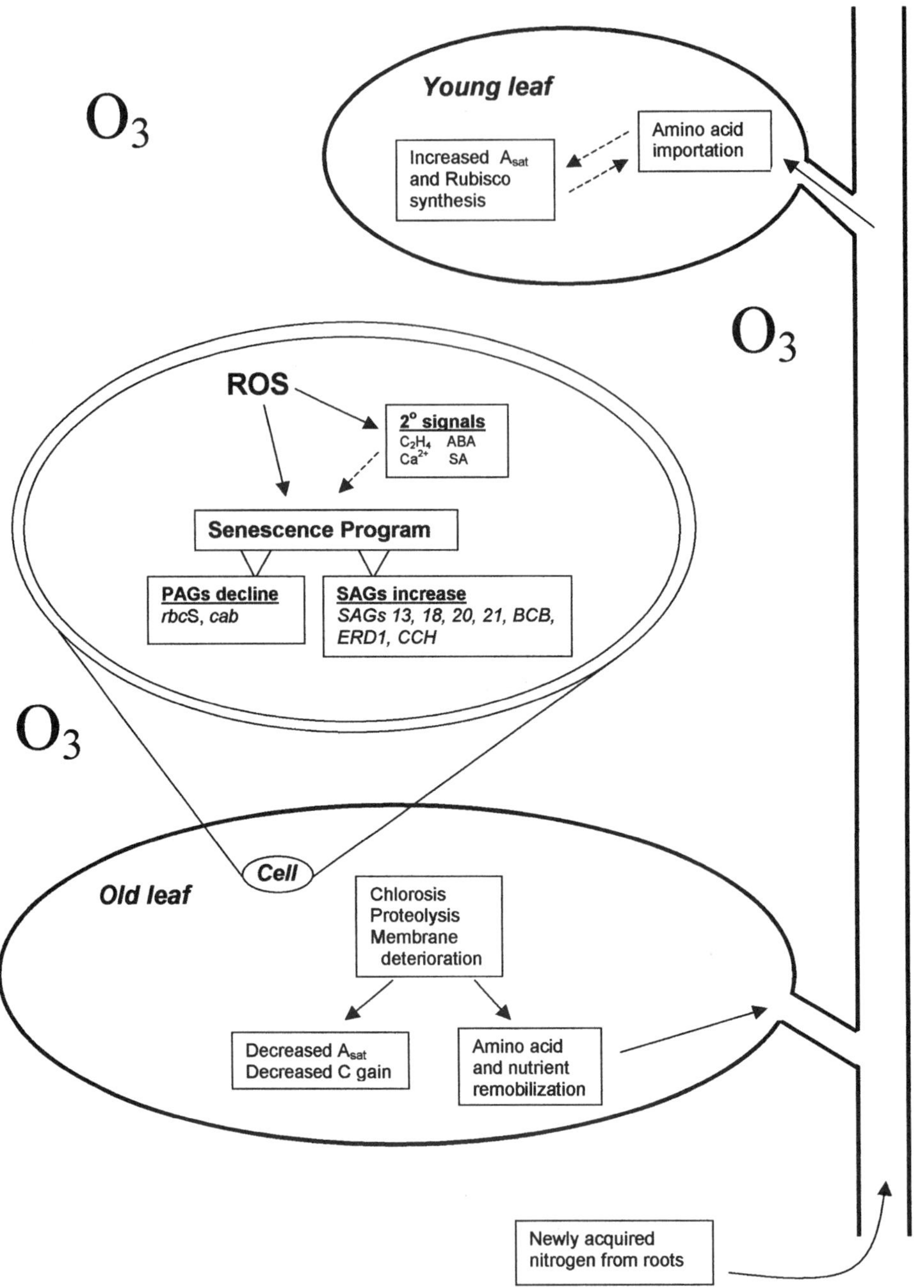

Figure 20-1. Ozone (O_3) induction of leaf senescence in older tissue. Reactive oxygen species (ROS), either directly or through induction of secondary signals, e.g. ethylene, abscisic acid (ABA), calcium, salicylic acid (SA), initiate a senescence program characterized by repression of photosynthesis-associated genes (PAGs) and induction of senescence-associated genes (SAGs) (Weaver *et al.*, 1997). Resultant physiological and biochemical effects are associated with impacts on younger foliage in the plant.

(Brendley and Pell, 1998; Eckardt and Pell, 1994). It is not known whether *ERD1* is involved in this response since the protein targets of ClpP are not known (Desimone *et al.*, 1998). *SAG12*, a gene that codes for a protease, was not induced by O_3 (Miller *et al.*, 1999). All proteases may not participate in O_3-induced accelerated senescence or may be induced at different times during the process.

SAG21 is similar to members of the late embryogenesis abundant (Lea) gene family (Weaver *et al.*, 1998), a family of proteins that protect developing seeds from dehydration. In our experiments, *SAG21* was strongly induced within 2 days of O_3 treatment. If *SAG21* has a similar protective function as Lea proteins, it might protect cells from damage during senescence (Weaver *et al.*, 1997), including O_3-induced senescence.

Two other genes, *SAGs 18* and *20*, were induced by O_3 after 6 days of exposure (Miller *et al.*, 1999). SAGs *18* and *20* show no strong similarity to other sequences in the database, although *SAG20* did show weak similarity to a potato wound-induced gene (Weaver *et al.*, 1998).

In the studies of Miller *et al.* (1999), *SAG13* was one of the first genes detected after 2 days of O_3 exposure. There is currently no known function of *SAG13* in senescence, although the gene does have sequence similarity to short-chain alcohol dehydrogenases (Weaver *et al.*, 1997). Gan and Amasino (personal communication) have transformed *Arabidopsis* with a *SAG13* promoter–GUS fusion construct. When these plants were treated with O_3, GUS expression was detected within the same time period as was the detection of the *SAG13* transcript (Miller *et al.*, 1999). The O_3-induced expression of *SAG13*-driven GUS activity was observed throughout the leaf. In contrast, *SAG13*-driven GUS activity in control tissue was first observed only along the leaf margins, in the same location where senescence-associated chlorosis initially was observed. The spatial difference between O_3-induced and normal expression of *SAG13* suggests that the O_3 response, at least in part, differs from that of natural senescence.

While transcripts for many SAGs were expressed early in response to O_3, this was not a universal response. As discussed above, *SAG12* was not detected in either O_3-treated or control samples during the 14 day exposure; a similar observation was made for *SAG19* (Miller *et al.*, 1999). *Atgsr2*, a cytosolic isoform of glutamine synthetase, is important for the assimilation of nitrogen and increases in senescing and dark-treated *Arabidopsis* leaves (Bernhard and Matile, 1994); however, *Atgsr2* was not induced by O_3 treatment (Miller *et al.*, 1999). *MT1* is a SAG able to restore copper resistance in metallothionein-deficient yeast (Weaver *et al.*, 1997). Transcript levels for *MT1* increased in abundance in both O_3-treated and non-treated leaves. Metallothioneins are thought to protect cells from the toxic effects of metals by binding and sequestering or removing the metals from undesirable locations, such as in the nucleus where sensitive DNA is located (Chubatsu and Meneghini, 1993). There are several explanations for the apparent absence of O_3-induced expression of some SAGs. It is possible that the genes in question, which are induced later during normal senescence, might have exhibited an O_3-accelerated response had sampling occurred later. Alternatively, plants at the age studied may have produced sufficient gene product to cope with the added O_3 stress or in cases where a SAG was part of a multi-gene family, other members of the gene family may have been induced by O_3.

It is likely that O_3 elicits a signal(s) that is necessary for the subsequent induction of SAGs. It is well known that O_3 treatment is associated with the emission of ethylene from leaves (Tingey *et al.*, 1976). More recently there have been reports that O_3 can induce genes involved in ethylene biosynthesis including the rapid induction of ACC oxidase in tomato (*Lycopersicon esculentum* Mill.) (Tuomainen *et al.*, 1997) and the induction of

ACC synthase in potato (*ACS4* and *ACS5*), tomato (*LE-ACS2*) and *Arabidopsis* (*ACS6*) (Schlagnhaufer *et al.*, 1995, 1997; Tuomainen *et al.*, 1997; Vahala *et al.*, 1998). In experiments described above, O_3 did induce *ACS6*, but the transcript appeared later than did the transcript for some of the SAGs, namely, *SAG13*, *ERD1*, *BCB* and *SAG21*, and concurrent with *SAG18*, *SAG20* and *CCH* (Miller *et al.*, 1999). Ethylene may play a role as a secondary signal in the O_3 response since this hormone can induce the appearance of *SAG13*, *SAG20*, *SAG21*, *BCB* and *ERD1* transcripts (Weaver *et al.*, 1998).

There are many mutants of *Arabidopsis* that lack ethylene perception, due to mutations in receptors or in down stream signaling components, e.g. *etr1-3* (Guzman and Ecker, 1990). No difference in SAG expression was found between the Columbia ecotype of *Arabidopsis* and *etr1-3*, following O_3 treatment (Miller, Arteca and Pell unpublished data). This work suggests that ethylene perception by the etr1-3 receptor is not required for induction of the O_3-responsive SAGs. The role of ethylene perception cannot be completely ruled out, due to the redundancy of ethylene perception in *Arabidopsis*. Other ethylene insensitive mutants are being studied to better define the role of this hormone in O_3-induced SAG expression (Miller, Arteca and Pell, unpublished data).

The potential for other hormones to act as signals in the induction of O_3-responsive SAGs remains to be examined. Abscisic acid (ABA), for example, is known to induce *ERD1* and *SAG13* (Weaver *et al.*, 1998). Since O_3 can induce stomatal closure (Pearson and Mansfield, 1993) and accelerate the rate of foliar abscission (Pell *et al.*, 1995), involvement of ABA is possible. Salicylic acid is induced during exposures to high O_3 concentrations and is involved in the induction of antioxidant gene expression and cell death (Rao and Davis, 1999; Sharma *et al.*, 1996; Yalpani *et al.*, 1994). However, the importance of salicylic acid in regulating the response to O_3 at concentrations that induce accelerated senescence is not known.

Calcium serves as a secondary messenger in many signal transduction pathways and changes in cytosolic calcium have been associated with senescence (Huang *et al.*, 1997). There are several reports that O_3 can induce changes in cytosolic calcium levels (Castillo and Heath, 1990; Clayton *et al.*, 1999). Clayton *et al.* (1999) report an increase in transcript accumulation of glutathione-S-transferase (GST), in conjunction with the second of two O_3-induced calcium peaks, suggesting a potential role of this cation as a signal. Whether changes in cytosolic calcium are necessary for induction of SAGs remains to be determined.

Oxidative stress may serve a signaling function in the induction of SAGs through generation of ROS, as discussed earlier in this chapter. Aluminum, which can also induce oxidative stress, increased *BCB* transcript levels (Richards *et al.*, 1998). The ROS generated during an O_3 exposure could directly regulate gene expression. Schubert *et al.* (1997) demonstrated that a specific region in the promoter of stilbene synthase, an enzyme responsible for synthesis of the secondary grape metabolite resveratrol, was required for induction by O_3 but not by a biological pathogen.

In recent years, sugars, including glucose and sucrose, have been identified as regulatory signals in a variety of metabolic processes including photosynthesis (Lalonde *et al.*, 1999). Ono and Watanabe (1997) have suggested that when sugar levels increase in the leaf, photosynthetic proteins may decline more rapidly. This may be related to a sugar-induced repression of protein synthesis. There is evidence that when plants are treated with O_3, foliar carbohydrates can accumulate (Lux *et al.*, 1997). Fialho and Bücker (1996) found transient increases in the levels of several sugars including glucose, fructose and sucrose in leaves of black poplar (*Populus nigra* cv. Loenen) treated with O_3 and SO_2 when compared with control tissue. The mechanism by which carbohydrates accumulate in the leaf in response to

O_3 is not understood, although it has been proposed that phloem loading may be inhibited (Darrall, 1989). However, a mechanistic linkage between O_3-induced accumulation of sugars, down regulation of photosynthesis-associated genes, and accelerated senescence is worthy of further consideration.

The hallmarks of programmed cell death (PCD) including DNA laddering and TUNEL positive cells, were found in naturally senescing leaves of five different plant species (Yen and Yang, 1998). There is evidence that relatively high concentrations of O_3 leading to hypersensitive-like lesions can induce PCD in hybrid poplar based upon the appearance of TUNEL positive cells (Koch and Davis, personal communication). Markers of PCD have not been studied during the response to low doses of O_3 in the induction of accelerated leaf senescence.

IV. Implications of Senescence to the Whole Plant

The detailed study of O_3 effects on older foliage has been described above. Taken in isolation, acceleration of the senescence process of a leaf translates into a shorter time period for carbon acquisition and accumulation of photosynthate. The decline in older foliage and the resulting remobilization of nutrients will have an impact on other growing tissues within the plant. Thus, while accelerated leaf senescence could be viewed as inherently adverse there are examples in which older leaves exhibit this O_3-induced response, but the whole plant is seemingly unaffected. For example, Mulholland *et al.* (1997) reported that in spring wheat, O_3 induced premature leaf senescence of the first seven leaves with no adverse effects in the flag leaf or on mean grain yield. When apple trees were treated with O_3, the oldest leaves exhibited accelerated senescence and abscission while younger leaves were apparently unaffected relative to control tissue (Wiltshire *et al.*, 1993). In contrast, other investigations have shown that there is a correlation between a more rapid rate of senescence of older leaves in O_3-treated plants, and greater performance of the respective younger tissue as measured by variables like carbon assimilation (Pell *et al.*, 1994b). The causes and connections between these events are still relatively uncertain; resources recaptured during leaf senescence may be translocated elsewhere in the plant, thus providing compensation to the stress (Manderscheid *et al.*, 1992; Temple and Riechers, 1995).

New leaf (or organ) growth in plants is often supported by reallocation of nutrients such as nitrogen from internal stores or from older, less productive foliage. It has been shown in loblolly and ponderosa pine (*Pinus ponderosa* Laws), and in trembling aspen (*Populus tremuloides*) and hybrid poplar (*Populus maximowizi x trichocarpa*) that accelerated loss of nitrogen and/or soluble leaf protein in older foliage, resulting from O_3 exposure, are associated with increases in nitrogen in younger foliar tissues (Pell *et al.*, 1994b; Temple and Riechers, 1995). In a recent study, Brendley and Pell (1998) showed that the loss of Rubisco in the oldest leaves of hybrid poplar could be attributed to accelerated degradation of the protein. For a transient period of time, younger leaves in the canopy exhibited higher concentrations and a higher rate of biosynthesis of Rubisco. Others have shown that amino acid concentrations increased in foliage of loblolly pine after an O_3 exposure (Manderscheid *et al.*, 1992).

The importance of internally recycled nutrients to the plant may relate to the availability of exogenous nitrogen. As nitrogen becomes less available to the roots, internal sources of nitrogen may become more important for new growth. To test this hypothesis we grew hybrid poplar plants in sand culture with a complete nutrient regime supplied daily through a fertigation system. Midway through the growing season, nitrogen levels in half of the treatments were reduced by 80 percent. Only O_3-treated plants (protocol as per Brendley and Pell, 1998) exhibited significant leaf abscission and the degree of abscission was significantly higher in those plants subjected to nitrogen withdrawal (Bielenberg *et al.*, 2001). Pääkkönen and Holopainen (1995) demonstrated that when birch seedlings were exposed to O_3, the magnitude of nitrogen supply was inversely proportional to the rate of autumnal senescence. When trembling aspen seedlings were grown with six different levels of nitrogen, plants in all treatments exhibited O_3-induced accelerated leaf senescence (Pell *et al.*, 1995). This response differed from that of Pääkkönen and Holopainen (1995) probably because nitrogen was delivered once through a time release fertilizer, resulting in a decline in the nutrient over the growing season, while the latter group maintained nitrogen levels with weekly doses of fertilizer.

It is significant that O_3-induced accelerated leaf senescence is influenced by nutrient availability. Typically plants receive the highest nitrogen dose early in the growing season, ironically when they are experiencing the least demand. Agricultural plants are fertilized at or near the planting date; these plants might receive a "top dressing" of nitrogen once plants have emerged, but thereafter nitrogen is allowed to deplete. Similarly, in temperate ecosystems nitrogen levels are highest in the spring when conditions support release of nutrients from litter that may have been deposited the previous autumn (Haynes, 1986).

While availability of nitrogen may play a role in influencing the rate of O_3-induced accelerated leaf senescence, other factors are also possible players in the process. A well-established effect of O_3 upon plants is an increase in the shoot to root ratio (Cooley and Manning, 1987). This is thought to occur by several mechanisms that act to cause reduced translocation of soluble carbohydrates to the roots (Cooley and Manning, 1987). Accelerated senescence may impact this process by removing the older foliage, the likely source of photosynthate for the roots (Dickson and Isebrands, 1991). A reduction in the transport of soluble carbohydrates to the roots may also impair energy dependent processes, e.g. nitrogen fixation (Pausch *et al.*, 1996). In addition, reduced root mass may affect nutrient acquisition via reduced uptake sites. However, Bielenberg *et al.* (2002) observed no effect of O_3 upon whole plant nitrogen uptake as measured by stable isotope uptake or growth analysis except as mediated by treatment-induced changes in whole plant biomass.

Earlier in this chapter we spoke of the relationship between senescence in the oldest leaves and increased performance in younger foliage, and suggested that the driving force for this shift derived from the aging foliage. It is possible that younger foliage of O_3-stressed plants was actually initiated earlier than in non-stressed plants, and the increase in net photosynthesis etc. could be a reflection of a developmental difference rather than an actual increase in performance. While we could not observe any difference in the rate of leaf initiation between hybrid poplar plants stressed by O_3 for 12 weeks and those not stressed (Bielenberg *et al.*, 2001), Held *et al.* (1991) did detect such an increase in the rate of leaf initiation when radish (*Raphanus sativus* L.) plants were stressed by O_3.

Multi-year chronic O_3 stress may also lead to whole plant senescence; this is of particular concern when tree species are the target. Under continued stress, plants can eventually reach a stage of carbohydrate exhaustion as a result of O_3-mediated decline in the carbon balance. Accelerated foliar senescence, and the associated loss of assimilatory potential along with increased respiration, may lead to reduced capacity for repair and increased vulnerability to disease. Pearce (1996) has shown that after prolonged O_3 stress, Sitka spruce (*Picea sitchensis* (Bong.) Carr.) infected with the root and butt-rot fungus *Heterobasidion annosum* (Fr.) exhibited impaired capacity to induce lignification in the bark tissue when compared with non-stressed trees. A well documented, O_3-associated decline in ponderosa pine in Southern California is thought to have occurred, in part, because these trees were unable to combat secondary ingress of root rotting fungi and infestation by pine bark beetles (Miller, 1982).

In conclusion, O_3 enters the leaf, breaks down into ROS, and rapidly initiates a senescence program (Fig. 20-1). As Mach and Greenberg discuss elsewhere in this book (Chapter 13) oxidation plays a central role in the aging process. Whether O_3 has some unique feature as a trigger or merely provides a significant increase in oxidizing potential remains to be determined. While the leaf is the primary receptor of this toxic air pollutant, the implications of O_3-induced accelerated leaf senescence radiate throughout the plant.

References

Bahl, A., and Kahl, G. (1995). Air pollutant stress changes the steady-state transcript levels of three photosynthesis genes. *Environ. Pollut.* **88**, 57–65.

Bernhard, W.R., and Matile, P. (1994). Differential expression of glutamine synthetase genes during the senescence of *Arabidopsis thaliana* rosette leaves. *Plant Science* **98**, 7–14.

Bielenberg, D.G., Lynch, J.P., and Pell, E.J. (2001). A decline in nitrogen availability affects plant responses to ozone. *New Phytol.* **151**, 413–425.

Bielenberg, D.G., Lynch, J.P., and Pell, E.J. (2002). Nitrogen dynamics during O_3-induced accelerated senescence and related compensation in hybrid poplar. *Plant Cell Environ.* **25**, 501–512.

Braun, S., and Flückiger, W. (1995). Effects of ambient ozone on seedlings of *Fagus sylvatica* L. and *Picea abies* (L.) Karst. *New Phytol.* **129**, 33–44.

Brendley, B.W., and Pell, E.J. (1998). Ozone-induced changes in biosynthesis of Rubisco and associated compensation to stress in foliage of hybrid poplar. *Tree Physiol.* **18**, 81–90.

Castillo, F.J., and Heath, R.L. (1990). Ca^{2+} transport in membrane vesicles from pinto bean leaves and its alteration after ozone exposure. *Plant Physiol.* **94**, 788–795.

Chubatsu, L.S., and Meneghini, R. (1993). Metallothionein protects DNA from oxidative damage. *Biochem. J.* **291**, 193–198.

Clark, C.S., Weber, J.A., Lee, E.H., and Hogsett, W.E. (1996). Reductions in gas exchange of *Populus tremuloides* caused by leaf aging and ozone exposure. *Can. J. For. Res.* **26**, 1384–1391.

Clayton, H., Knight, M.R., Knight, H., McAinsh, M.R., and Hetherington, A.M. (1999). Dissection of the ozone-induced calcium signature. *Plant Journal* **17**, 575–579.

Conklin, P.L., and Last, R.L. (1995). Differential accumulation of antioxidant mRNAs in *Arabidopsis thaliana* exposed to ozone. *Plant Physiol.* **109**, 203–212.

Cooley, D.R., and Manning, W.J. (1987). The impact of ozone on assimilate partitioning in plants: a review. *Environ. Pollut.* **47**, 95–113.

Dalling, M.J. (1987). Proteolytic enzymes and leaf senescence. In *Plant Senescence: Its Biochemistry and Physiology* (W.W. Thomson, E. A. Nothnagel, and R.C. Huffaker, Eds.), pp. 54–70. American Society of Plant Physiologists, Rockville, MD.

Darrall, N.M. (1989). The effect of air pollutants on physiological processes in plants. *Plant Cell Environ.* **12**, 1–30.

Desimone, M., Wagner, E., and Johanningmeier, U. (1998). Degradation of active-oxygen-modified ribulose-1,5-bisphosphate carboxylase/oxygenase by chloroplastic proteases requires ATP-hydrolysis. *Planta* **205**, 459–466.

Dickson, R.E., and Isebrands, J.G. (1991). Leaves as regulators of stress response. In *Response of Plants to Multiple Stresses* (H.A. Mooney, W.E. Winner, and E.J. Pell, Eds.), pp. 3–34. Academic Press, San Diego.

Eckardt, N.A., and Pell, E.J. (1994). O_3-induced degradation of Rubisco protein and loss of Rubisco mRNA in relation to leaf age in *Solanum tuberosum* L. *New Phytol.* **127**, 741–748.

Eckardt, N.A., and Pell, E.J. (1995). Oxidative modification of Rubisco from potato foliage in response to ozone. *Plant Physiol. Biochem.* **33**, 273–282.

Fialho, R.C., and Bücker, J. (1996). Changes in levels of foliar carbohydrates and *myo*-inositol before premature leaf senescence of *Populus nigra* induced by a mixture of O_3 and SO_2. *Can. J. Bot.* **74**, 965–970.

Flückiger, W., Oertli, J.J., Flückiger-Keller, H., and Braun, S. (1979). Premature senescence in plants along a motorway. *Environ. Pollut.* **19**, 171–176.

Glick, R.E., Schlagnhaufer, C.D., Arteca, R.N., and Pell, E.J. (1995). Ozone-induced ethylene emission accelerates the loss of Rubisco and nuclear-encoded mRNAs in senescing potato leaves. *Plant Physiol.* **109**, 891–898.

Gut, H., and Matile, P. (1988). Breakdown of galactolipids in senescent barley leaves. *Botanica Acta* **102**, 31–36.

Guzman, P., and Ecker, J.R. (1990). Exploiting the triple response of *Arabidopsis* to identify ethylene-related mutants. *Plant Cell* **2**, 513–523.

Haynes, R.J. (1986). *Mineral Nitrogen in the Plant-Soil System*. Academic Press, Orlando, FL.

Heagle, A.S., and Booker, F.L. (1998). Influence of ozone stress on soybean response to carbon dioxide enrichment: I. Foliar Properties. *Crop Science* **38**, 113–121.

Held, A.A., Mooney, H.A., and Gorham, H.N. (1991). Acclimation to ozone stress in radish: leaf demography and photosynthesis. *New Phytol.* **118**, 417–423.

Himelblau, E., Mira, H., Lin, S.-J., Culotta, V.C., Peñarrubia, L., and Amasino, R.M. (1998). Identification of a functional homolog of the yeast copper homeostasis gene ATX1 from Arabidopsis. *Plant Physiol.* **117**, 1227–1234.

Huang, F.-Y., Phiosoph-Hadas, S., Meier, S., Callaham, D.A., Sabato, R., Zelcer, A., and Hepler, P.K. (1997). Increases in cytosolic Ca^{2+} in parsley mesophyll cells correlate with leaf senescence. *Plant Physiol.* **115**, 51–60.

Kangasjärvi, J., Talvinen, J., Utriainen, M., and Karjalainen, R. (1994). Plant defence systems induced by ozone. *Plant Cell Environ.* **17**, 783–794.

Kargiolaki, H., Osborne, D.J., and Thompson, F.B. (1991). Leaf abscission and stem lesions (intumescences) on poplar clones after SO_2 and O_3 fumigation: a link with ethylene release? *J. Exp. Bot.* **42**, 1189–1198.

Kasana, M.A. (1991). Sensitivity of three leguminous crops to ozone as influenced by different stages of growth and development. *Environ. Pollut.* **69**, 131–149.

Kull, O., Sober, A., Coleman, M.D., Dickson, R.E., Isebrands, J.G., Gagnon, Z., and Karnosky, D.F. (1996). Photosynthetic responses of aspen clones to simultaneous exposures of ozone and CO_2. *Can. J. For. Res.* **26**, 639–648.

Laisk, A., Kull, O., and Moldau, H. (1989). Ozone concentration in leaf intercellular air spaces is close to zero. *Plant Physiol.* **90**, 1163–1167.

Lalonde, S., Boles, E., Hellmann, H., Barker, L., Patrick, J.W., Frommer, W.B., and Ward, J.M. (1999). The dual function of sugar carriers: transport and sugar sensing. *Plant Cell* **11**, 707–726.

Landry, L.G., and Pell, E.J. (1993). Modification of rubisco and altered proteolytic activity in O_3-stressed hybrid poplar (*Populus maximowizii x trichocarpa*). *Plant Physiol.* **101**, 1355–1362.

Levine, A., Tenhaken, R., Dixon, R., and Lamb, C. (1994). H_2O_2 from the oxidative burst orchestrates the plant hypersensitive disease resistance response. *Cell* **79**, 583–593.

Lux, D., Leonardi, S., Müller, J., Wiemken, A., and Flückiger, W. (1997). Effects of ambient ozone concentrations on contents of non-structural carbohydrates in young *Picea abies* and *Fagus sylvatica*. *New Phytol.* **137**, 399–409.

Manderscheid, R., Jäger, H.-J., and Kress, L.W. (1992). Effects of ozone on foliar nitrogen metabolism of *Pinus taeda* L. and implications for carbohydrate metabolism. *New Phytol.* **121**, 623–633.

Mikkelsen, T.N., and Heide-Jørgensen, H.S. (1996). Acceleration of leaf senescence in *Fagus sylvatica* L. by low levels of tropospheric ozone demonstrated by leaf colour, chlorophyll fluorescence and chloroplast ultrastructure. *Trees* **10**, 145–156.

Miller, J.D., Arteca, R.N., and Pell, E.J. (1999). Senescence-associated gene expression during ozone-induced leaf senescence in *Arabidopsis thaliana*. *Plant Physiol.* **120**, 1015–1023.

Miller, P.R. (1982). Ozone effects in the San Bernardino national forest. In *Symposium on Air Pollution and the Productivity of the Forest*, Izaak Walton League, Arlington, VA, pp. 161–197.

Mulholland, B.J., Craigon, J., Black, C.R., Colls, J.J., Atherton, J., and Landon, G. (1997). Impact of elevated atmospheric CO_2 and O_3 on gas exchange and chlorophyll content in spring wheat (*Triticum aestivum* L.). *J. Exp. Bot.* **48**, 1853–1863.

Nyman, B.F. (1986). Industrial air pollution and peroxidase activity in scots pine needles—two case studies. *Eur. J. For. Path.* **16**, 139–147.

Ojanperä, K., Sutinen, S., Pleijel, H., and Selldén, G. (1992). Exposure of spring wheat, *Triticum aesivum* L., cv. Drabant, to different concentrations of ozone in open-top chambers: effects on the ultrastructure of flag leaf cells. *New Phytol.* **120**, 39–48.

Ojanperä, K., Pätsikkä, E., and Yläranta, T. (1998). Effects of low ozone exposure of spring wheat on net CO_2 uptake, Rubisco, leaf senescence and grain filling. *New Phytol.* **138**, 451–460.

Ono, K., and Watanabe, A. (1997). Levels of endogenous sugars, transcripts of *rbcS* and *rbcL*, and of RuBisCO protein in senescing sunflower leaves. *Plant Cell Physiol.* **38**, 1032–1038.

Pääkkönen, E., and Holopainen, T. (1995). Influence of nitrogen supply on the response of clones of birch (*Betula pendula* Roth.) to ozone. *New Phytol.* **129**, 595–603.

Pääkkönen, E., Holopainen, T., and Kärenlampi, L. (1995). Ageing-related anatomical and ultrastructural changes in leaves of birch (*Betula pendula* Roth.) clones as affected by low ozone exposure. *Annals Bot.* **75**, 285–294.

Pausch, R.C., Mulchi, C.L., Lee, E.H., Forseth, I.N., and Slaughter, L.H. (1996). Use of ^{13}C and ^{15}N isotopes to investigate O_3 effects on C and N metabolism in soybeans. Part I. C fixation and translocation. *Agricul. Ecosystems Environ.* **59**, 69–80.

Pearce, R.B. (1996). Effects of exposure to high ozone concentrations on stilbenes in Sitka spruce (*Picea sitchensis* (Bong.) Carr.) bark and on its lignification response to infection with *Heterobasidion annosum* (Fr.) Bref. *Physiol. Mol. Plant Pathol.* **48**, 117–129.

Pearson, M., and Mansfield, T.A. (1993). Interacting effects of ozone and water stress on the stomatal resistance of beech (*Fagus sylvatica* L.). *New Phytol.* **123**, 351–358.

Pell, E.J., Eckardt, N., and Enyedi, A.J. (1992). Timing of ozone stress and resulting status of ribulose bisphoshate carboxylase/oxygenase and associated net photosynthesis. *New Phytol.* **120**, 397–405.

Pell, E.J., Eckardt, N.A., and Glick, R.E. (1994a). Biochemical and molecular basis for impairment of photosynthetic potential. *Photosyn. Res.* **39**, 453–462.

Pell, E.J., Temple, P.J., Friend, A.L., Mooney, H.A., and Winner, W.E. (1994b). Compensation as a plant response to ozone and associated stresses: An analysis of ROPIS experiments. *J. Environ. Qual.* **23**, 429–436.

Pell, E.J., Sinn, J.P., and Vinten-Johansen, C. (1995). Nitrogen supply as a limiting factor determining sensitivity of *Populus tremuloides* Michx. to ozone. *New Phytol.* **130**, 436–446.

Pell, E.J., Schlagnhaufer, C.D., and Arteca, R.N. (1997). Ozone-induced oxidative stress: mechanisms of action and reaction. *Physiol. Plant.* **100**, 264–273.

Pell, E.J., Sinn, J.P., Brendley, B.W., Samuelson, L., Vinten-Johansen, C., Tien, M., and Skillman, J. (1999). Differential response of four tree species to ozone-induced acceleration of foliar senescence. *Plant Cell Environ.* **22**, 779–790.

Ranieri, A., Castagna, A., Padu, E., Moldau, H., Rahi, M., and Soldatini, G.F. (1999). The decay of O_3 through direct reaction with cell wall ascorbate is not sufficient to explain the different degrees of O_3-sensitivity in two poplar clones. *J. Plant Physiol.* **154**, 250–255.

Rao, M.V., and Davis, K.R. (1999). Ozone-induced cell death occurs via two distinct mechanisms in *Arabidopsis*: the role of salicylic acid. *Plant Journal* **17**, 603–614.

Reddy, G.N., Arteca, R.N., Dai, Y.-R., Flores, H.E., Negm, F.B., and Pell, E.J. (1993). Changes in ethylene and polyamines in relation to mRNA levels of the large and small subunits of ribulose bisphosphate carboxylase/oxygenase in ozone-stressed potato foliage. *Plant Cell Environ.* **16**, 819–826.

Reich, P.B., and Lassoie, J.P. (1985). Influence of low concentration of ozone on growth, biomass partitioning and leaf senescence in young hybrid poplar plants. *Environ. Pollut. Ser. A* **39**, 39–51.

Richards, B.L., Middleton, J.T., and Hewitt, W.B. (1958). Air pollution with relation to agronomic crops. V. Oxidant stipple of grape. *Agron. J.* **50**, 559–561.

Richards, K.D., Schott, E.J., Sharma, Y.K., Davis, K.R., and Gardner, R.C. (1998). Aluminum induces oxidative stress genes in *Arabidopsis thaliana*. *Plant Physiol.* **116**, 409–418.

Sandelius, A.S., Näslund, K., Carlsson, A.S., Pleijel, H., and Selldén, G. (1995). Exposure of spring wheat (*Triticum aestivum*) to ozone in open-top chambers. Effects on acyl lipid composition and chlorophyll content of flag leaves. *New Phytol.* **131**, 231–239.

Schlagnhaufer, C.D., Glick, R.E., Arteca, R.N., and Pell, E.J. (1995). Molecular cloning of an ozone-induced 1-aminocyclopropane-1-carboxylate synthase cDNA and its relationship with a loss of *rbc*S in potato (*Solanum tuberosum* L.) plants. *Plant Mol. Biol.* **28**, 93–103.

Schlagnhaufer, C.D., Arteca, R.N., and Pell, E.J. (1997). Sequential expression of two 1-aminocyclopropane-1-carboxylate synthase genes in response to biotic and abiotic stresses in potato (*Solanum tuberosum* L.) leaves. *Plant Mol. Biol.* **35**, 683–688.

Schubert, R., Fischer, R., Hain, R., Schreier, P.H., Bahnweg, G., Ernst, D., and Sandermann Jr., H. (1997). An ozone-responsive region of the grapevine resveratrol synthase promoter differs from the basal pathogen-responsive sequence. *Plant Mol. Biol.* **34**, 417–426.

Sharma, Y.K., León, J., Raskin, I., and Davis, K.R. (1996). Ozone-induced responses in *Arabidopsis thaliana*: the role of salicylic acid in the accumulation of defense-related transcripts and induced resistance. *Proc. Natl. Acad. Sci. USA* **93**, 5099–5104.

Stow, T.K., Allen, H.L., and Kress, L.W. (1992). Ozone impacts on seasonal foliage dynamics of young loblolly pine. *Forest Science* **38**, 102–119.

Temple, P.J., and Riechers, G.H. (1995). Nitrogen allocation in ponderosa pine seedlings exposed to interacting ozone and drought stresses. *New Phytol.* **130**, 97–104.

Tingey, D.T., Standley, C., and Field, R.W. (1976). Stress ethylene evolution: a measure of ozone effects on plants. *Atmospheric Environ.* **10**, 969–974.

Tuomainen, J., Betz, C., Kangasjärvi, J., Ernst, D., Yin, Z.-H., Langebartels, C., and Sandermann Jr., H. (1997). Ozone induction of ethylene emission in tomato plants: regulation by differential accumulation of transcripts for the biosynthetic enzymes. *Plant Journal* **12**, 1151–1162.

Vahala, J., Schlagnhaufer, C.D., and Pell, E.J. (1998). Induction of an ACC synthase cDNA (ACS6) by ozone in light grown *Arabidopsis thaliana* leaves. *Physiol. Plant.* **103**, 45–50.

Van Gysel, A., Van Montagu, M., and Inzé, D. (1993). A negatively light-regulated gene from *Arabidopsis thaliana* encodes a protein showing high similarity to blue copper-binding proteins. *Gene* **136**, 79–85.

Weaver, L.M., Himelblau, E., and Amasino, R.M. (1997). Leaf senescence: Gene expression and regulation. *Genetic Engineering* **19**, 215–234.

Weaver, L.M., Gan, S., Quirino, B., and Amasino, R.M. (1998). A comparison of the expression patterns of several senescence-associated genes in response to stress and hormone treatment. *Plant Mol. Biol.* **37**, 455–469.

Wiltshire, J.J.J., Wright, C.J., Unsworth, M.H., and Craigon, J. (1993). The effects of ozone episodes on autumn leaf fall in apple. *New Phytol.* **124**, 433–437.

Yalpani, N., Enyedi, A.J., León, J., and Raskin, I. (1994). Ultraviolet light and ozone stimulate accumulation of salicylic acid, pathogenesis-related proteins and virus resistance in tobacco. *Planta* **193**, 372–376.

Yen, C.H., and Yang, C.H. (1998). Evidence for programmed cell death during leaf senescence in plants. *Plant Cell Physiol.* **39**, 922–927.

21

Physiology of Flower Senescence

Donald A. Hunter, Nathan E. Lange and
Michael S. Reid

I. Introduction

The evolution of the flowering habit was accompanied by the evolution of a programmed and often remarkably rapid sequence of opening, development, and senescence of the reproductive tissues and their associated organs. This is most clearly observed in the rapid senescence of petals. Unlike leaves, to which they are ontogenetically related and which may remain active for years, petals typically senesce within a few days of flower opening, and sometimes within a few hours. This has made the comparatively simple tissue of the petals a popular choice for studying the biology of senescence in plants (for reviews see Halevy and Mayak, 1979, 1981; Borochov and Woodson, 1989; Rubinstein, 2000).

II. Model Systems

In their studies on floral senescence, researchers have focused predominantly on selected flowers for reasons including the longevity of the flowers, the ease at which the flowers are produced, their ability to be transformed and their importance in the floral industry. Researchers concerned with the commercial aspects of flower senescence have focused predominantly on carnations and roses. Those interested primarily in studying senescence mechanisms have typically used flowers such as the morning glory (*Pharbitis*), tradescantia,

Plant Cell Death Processes

carnation (*Dianthus*), petunia, and daylily (*Hemerocallis*), but there also have been many studies examining the senescence of a wide variety of cut flowers, including recent work on iris, *Phalaenopsis* and sandersonia. Results from these studies have indicated that although the initiation of the senescence program in flowers differs between plant species, the biochemistry of the program appears to be remarkedly conserved.

III. Structural and Metabolic Changes Occurring during Senescence

Many flowers show dramatic changes in form during senescence. The stigma and anthers mature and senesce in a variety of manners typically designed to maximize outcrossing. Once pollination of the stigma is accomplished, petals often change color, wilt and/or abscise (Van Doorn, 1997).

A. Morphology

1. Color changes

Certain flowers show age-related changes in color, including the bright blue flowers of chicory that fade over the course of the day that they are open (Proctor and Creasey, 1969). Many flowers, particularly those in the Leguminosae and Verbenaceae (Van Doorn, 1997) change color in response to pollination. In the Arizona lupin (*Lupinus albifrons*), for example, pollination is followed by a rapid and dramatic change in color, in which the bright yellow spot on the "banner" petal of this blue flower changes in color to a reddish-purple (Stead and Reid, 1990). In *Cymbidium* orchids, pollination or even emasculation results in a rapid "blushing" of the lip.

B. Metabolic Changes

The senescence of floral organs is associated with dramatic changes in their physiology and metabolism, and in many flowers these changes are associated with changes in the rate and/or pathways of respiration.

1. Respiration

Maxie *et al.* (1973) demonstrated the close coincidence between the onset of wilting, respiratory climacteric, and burst of ethylene synthesis in cut carnations. A similar relationship has been shown in studies of ephemeral flowers such as those of *Ipomoea* (Matile and Winkenbach, 1971), tradescantia (Suttle and Kende, 1980), and *Hibiscus* (Woodson *et al.*, 1985).

i. Climacteric and non-climacteric flowers

The respiration of senescing flowers can be classified as climacteric or non-climacteric. Climacteric flowers, such as carnation and petunia, produce increased amounts of carbon dioxide during senescence (Bufler *et al.*, 1980; Whitehead *et al.*, 1984a,b). Often, climacteric flowers have also been found to be ethylene sensitive (See Section IV) and produce elevated levels of ethylene in parallel with the rise in respiration. However, there is not always a close association between increased respiration and elevated ethylene production. For example, in *Sandersonia aurantiaca* flowers the 6-fold increase in respiration that occurs during senescence is not accompanied with a corresponding rise in ethylene production (Eason and De Vre, 1995). Classification of a flower as climacteric can also depend upon whether respiration is calculated on a per floret or fresh weight basis (Serek *et al.*, 1994).

2. Other metabolic changes

Dramatic changes in petal metabolism presumably accompany the striking morphological and physiological changes occurring during petal senescence. Matile and Winkenbach (1971) reported increased activity of deoxyribonuclease, ribonuclease and β-glucosidase in senescing corollas of morning glory. Hobson and Nichols (1977) demonstrated in senescing carnation petals that activity of acid phosphatase, ribonuclease and ATPase increased, esterase changed little and diaphorase and some other dehydrogenases decreased. Some researchers have examined changes in wall-based enzymes since senescence involves quite striking changes in wall composition, cell size, shape and degree of attachment. Wiemken *et al.* (1974) reported increased activity of β-glucosidase, β-galactosidase, laminarinase and cellobiase in senescing corolla tissue of *Convolvulus tricolor* and, more recently, Panavas *et al.* (1998a) showed senescence-associated increases in specific activity of cellulase, polygalacturonase and β-galactosidase in daylily flowers.

Recent molecular studies point to changes in a plethora of key metabolic pathways, including glycolysis, lipid metabolism, the glyoxylate cycle, protein turnover, and cell wall metabolism (Panavas *et al.*, 1999; Hunter and Reid, 2001).

i. Carbohydrate metabolism

During senescence, daylily flowers lose up to 95% of their sugars and ^{14}C-sucrose applied to attached senescing flowers is rapidly translocated to other parts of the plant, particularly to the developing flower buds (Bieleski, 1993). In gladiolus, removal of this carbohydrate source by removal of the mature flowers, caused a substantial reduction in the opening of subsequent flowers (Waithaka *et al.*, 2000), which is consistent with the demonstrated export of radioactively labeled sugars from wilting gladiolus florets to younger buds (Yamane *et al.*, 1995).

ii. Phosphate ester metabolism

Researchers have used changes in the proportion of phosphate in the various phosphate ester pools of tissue to indicate changes in metabolism of senescing tissue. In daylily, Bieleski and Reid (1992) found that the uptake of phosphate by petal slices from a 100 μM solution increased during the onset of senescence from 6 to 10 nmoles g FW^{-1} h^{-1}. Half of the uptaken phosphate was esterified and of this 14% was in ATP. The cellular energy charge remained high at 0.86 during senescence, indicating that the tissue was actively metabolizing and that the respiration remained tightly coupled until near the end of senescence.

iii. Phospholipid metabolism

Petal senescence is accompanied by a loss of differential permeability of membranes (Thompson, 1988). In daylily petals, Bieleski and Reid (1992) demonstrated a 14-fold increase in sugar efflux, and a 5-fold increase in ion efflux coinciding with the senescence-associated respiratory increase. Increased membrane permeability has been attributed to a reduction in membrane fluidity resulting from changes in composition of the membrane. Arrhenius plots of the moments of ^{2}H-NMR spectra and fluorescence depolarization values measured from 1,6-diphenylhexatriene-labeled rose petal membrane lipids indicated that membrane lipid order increased with increasing age of the petals (Itzhaki *et al.*, 1995). This is consistent with previously observed increases in fatty acid saturation and increases in the sterol:phospholipid ratio in senescing petals of roses (Borochov *et al.*, 1982) and carnations (Thompson *et al.*, 1982).

The increase in the sterol:phospholipid ratio is in part due to a decline in polar membrane lipids. This decline in phospholipids has been shown to occur in petals prior to visual symptoms of senescence, e.g., in morning glory (Beutelmann and Kende, 1977), tradescantia (Suttle and Kende, 1980) and rose (Borochov *et al.*, 1982) and has been attributed to reduced biosynthesis (Itzhaki *et al.*, 1998). Borochov *et al.* (1994) studied the specific activities and products of a number of kinases and lipases involved in the synthesis and catabolism of membrane lipid using plasma membranes isolated from senescing petunia petals. After harvest, but prior to the onset of senescence, levels of the polar lipid headgroups, phosphatidic acid (PA) and phosphatidylinositol monophosphate (PIP) increased. With petal wilting, there was a significant decline in PA and PIP synthesis, but synthesis of phosphatidylinositol-4,5-bisphosphate remained high. The catabolic activities of type A and C phospholipases increased as the petals senesced. Travnicek *et al.* (1999) proposed that lipolytic activity is important late in senescence during breakdown of subcellular compartmentation where it is responsible for rapid degradation of residual cellular constituents. These workers found very high constitutive activity of phospholipases in enzyme preparations from corollas of morning glory, which resulted in almost instantaneous deacylation of all phospholipids upon homogenization of their tissue extracts. In fresh corollas, the enzyme(s) were found to be present but functionally latent, presumably due to localization in the vacuole.

C. Flower Wilting

Wilting of the corolla commonly marks the end of life for cut flowers. This is particularly true of roses, where failure to open, bending of the pedicel ("bent neck") and petal wilting are symptoms that often occur long before the flower would have senesced had it remained on the plant. In cut carnation flowers, Solomos and Gross (1997) demonstrated that petal wilting was accompanied by a decrease in water uptake. They found no evidence of obstruction of the vascular bundles and suggested that the petals lost their ability to absorb water.

IV. Regulation of Senescence

Flowers can be classified as either ethylene-sensitive or -insensitive based upon their responsiveness to exogenously supplied ethylene (Woltering and van Doorn, 1988). In many of the ethylene-sensitive flowers ethylene appears to be the natural regulator of senescence or floret abscission (carnation, petunia, morning glory, snapdragon). In these flowers, ethylene production rises during natural senescence or abscission, and pretreatment with inhibitors of ethylene production or action extends flower longevity. In another important group, the "ethylene-insensitive" flowers (iris, daylily, tulip, sandersonia), there is no significant increase in ethylene production during natural senescence, and inhibitors of ethylene production and action have no effect on flower longevity.

Flowers and other plant organs are able to respond to ethylene because of the presence of an ethylene transduction signaling pathway. For a review of the components of this pathway and their interaction see Chapter 24.

A. Ethylene-induced Senescence

For a number of flowers, especially in the Caryophyllaceae, Leguminaceae, and Orchidaceae, the onset of natural or pollination-induced senescence is associated with a climacteric rise in respiration triggered by an increase in ethylene production. For many others, ethylene is involved in the abscission of flowers, petals, or florets (Woltering and van Doorn, 1988). The rapidity of the ethylene response has made flowers whose senescence is coordinated by ethylene a favored system for exploring the action of ethylene and the genes involved.

1. Ethylene production during flower senescence

More than 20 years ago researchers showed that the rapid senescence of carnation flowers was associated with increased biosynthesis of ethylene (Maxie *et al.*, 1973) from ACC (Bufler *et al.*, 1980). The importance of ethylene in the life of carnation flowers was confirmed by examining cultivars that exhibited extended life. For example, flowers of the "Sandra" cultivar lasted twice as long in the vase as "White Sim" flowers and showed neither the normal increase in ethylene production nor a marked respiratory climacteric during their eventual senescence (Reid and Wu, 1992). The extended vase life of "Sandra" was attributed to repressed ethylene biosynthesis, since the application of exogenous ethylene hastened flower senescence.

2. Inhibitor studies

The most important inhibitors of ethylene-mediated flower senescence are inhibitors of ACC synthase [e.g. aminoethyoxyvinylglycine (AVG) and aminooxyacetic acid (AOA)], and two potent means of inhibiting ethylene action [silver ion, formulated as the anionic silver thiosulfate complex (STS) and a range of cyclopropenes].

i. AOA/AVG

Pyridoxal phosphate is a cofactor for ACC synthase, and compounds that inhibit pyridoxal phosphate-mediated reactions are effective inhibitors of ethylene synthesis. Two such compounds, AOA and AVG, have been shown to extend the vase life of flowers whose natural senescence is coordinated by ethylene (Baker *et al.*, 1977).

ii. STS

Beyer (1976) first demonstrated the dramatic inhibition of ethylene action by Ag^+, using the response of *Cattleya* blossoms to ethylene. However, because of the phytotoxicity of silver salts, this effect was a matter of academic interest until the report by Veen and Van der Geijn (1978) that the very stable anionic silver thiosulfate complex (STS) moved readily in the stems of cut flowers, and could be used, at low concentrations, to extend their life.

iii. Cyclopropenes

Cyclic olefins based on the cyclopropene ring have proved to be extremely effective and irreversible inhibitors of ethylene action and the most active of them, 1-methylcyclopropene (1-MCP) has been patented, licensed, and registered for use in preventing the effects of ethylene in ornamentals. We have demonstrated the action of 1-MCP in preventing flower senescence in a range of commercial crops (Serek *et al.*, 1995).

3. Gene expression studies

Jones (Chapter 4) describes the remarkable progress that has been made in the past decade in analyzing changes in gene expression during flower senescence. The genes controlling ethylene biosynthesis have been cloned from a number of species including carnation, petunia and roses, and antisense and co-suppressed constructs have been successfully engineered into some of these species, resulting in the predicted phenotypes. For example, Savin *et al.* (1995) transformed carnations with an antisense ACC oxidase gene under the control of the constitutive MAC promoter and demonstrated extended vase life of the transgenic flowers. The *Arabidopsis* gene (*ETR1-1*) that encodes a faulty ethylene receptor has now also been cloned (Chang *et al.*, 1993) allowing production of transgenic plants with reduced sensitivity to ethylene. Heterologous expression of the *ETR1-1* gene in petunia and tomato greatly extended the life of their flowers (Wilkinson *et al.*, 1997), and a similar effect has recently been demonstrated in transgenic carnations (Bovy *et al.*, 1999).

4. Effects of other plant hormones

Studies over the past two decades have demonstrated that senescence control probably depends on the interaction of different growth regulators. Treatment of carnations with cytokinins (Eisinger, 1977) or gibberellins (Saks and van Staden, 1992) has been shown to extend flower life, and treatment with high concentrations of auxins to reduce it (Sacalis and Nichols, 1980), although this latter effect may simply be due to auxin-stimulated ethylene production. A similar mechanism seems to explain the stimulation of orchid flower senescence by methyl jasmonate (Porat *et al.*, 1993). Exogenously applied abscisic acid (ABA) has been shown to accelerate flower senescence in standard roses (Halevy and Mayak, 1972), miniature roses (Muller *et al.*, 1999) and carnations (Mayak and Dilley, 1976).

B. Ethylene-insensitive Senescence

Petal senescence in many flowers is not responsive to ethylene (Woltering and van Doorn, 1988). In a number of monocotyledonous geophytes (e.g. daylily, iris, sandersonia), senescence occurs quite rapidly, but is not associated with increased ethylene production, nor is the flower life extended by pretreatment with STS or 1-MCP. For example, recent studies on *Sandersonia aurantiaca* (Eason and De Vrë, 1995) showed that propylene treatment (0.5% for 24 h) of the flowers did not alter the patterns of color change, fresh weight and respiration that accompany normal flower senescence. Likewise, STS failed to extend the vase life of the flowers and postharvest ethylene production by flowers was negligible (< 0.01 nl g^{-1} FW h^{-1}).

1. Physiology

The senescence events in daylily flowers are associated with dramatic changes in the content of carbohydrates in the corolla (Bieleski, 1995), loss of proteins (Lay-Yee *et al.*, 1992;

Stephenson and Rubenstein, 1998), increases in reactive oxygen species (Panavas and Rubenstein, 1998), changes in phosphate ester metabolism and leakage of vacuolar contents (Bieleski and Reid, 1992). Bieleski (1995) showed that the phloem remains highly active until late in senescence, presumably exporting the sugars, amino acids, and other products of cellular catabolism from the senescing tepals to developing buds further up the scape.

In *Iris x hollandica*, another very short-lived flower, tepals dramatically lose proteins shortly after opening so that by 2 days, when the first symptoms of senescence are apparent, the measured protein content is only 20% of that at harvest (Celikel and van Doorn, 1995).

2. Gene expression studies

In daylily, senescence has been shown to be associated with increases in activity of a range of enzymes (Panavas *et al.*, 1998a; Stephenson and Rubenstein, 1998; Guerrero *et al.*, 1998) and up-regulation of catabolic genes (Valpuesta *et al.*, 1995; Panavas *et al.*, 1999).

3. Effects of other plant hormones

Abscisic acid may have a role in controlling the senescence of ethylene-insensitive flowers. In daylily, the petal ABA content increases prior to the occurrence of visual symptoms of senescence and application of this hormone to the petals has been shown to advance physiological, biochemical and molecular events that would normally occur later in the flower as it naturally senesces (Panavas *et al.*, 1998b).

C. Other Regulatory Patterns

Not all ethylene-sensitive flowers are climacteric and produce elevated levels of ethylene during senescence. For example, the daffodil flower is ethylene-sensitive, non-climacteric, and produces only minimal amounts of ethylene throughout its maturation and senescence. Furthermore, repeated treatments of the flower with 1-MCP result in only a very modest increase in longevity which suggests that ethylene is not the primary regulator of its senescence. It appears therefore, that the absence of climacteric ethylene production or respiration is no signature for ethylene insensitivity or vice versa. It seems likely that the relationships among flower senescence, respiration, ethylene production and ethylene sensitivity need to be defined for each new species studied.

D. Lipids and the Regulation of Senescence

Changes in lipid composition of cut flowers have long been associated with the onset of senescence (Section IIIB2iii). Some workers have suggested that the onset of flower senescence is directed by changes in membrane structure, composition, or function. In a study of senescence in petunia, Borochov *et al.* (1997) demonstrated an age-related transient increase in the content of diacylglycerol (DAG), presumably derived from membrane breakdown, in petunia plasma membranes. The increase in ethylene associated with petal wilting appeared in petunia flowers well after phospholipid degradation and the increase in DAG had commenced. Senescence was accelerated by treatment with the DAG analogue, phorbol 12-myristate 13-acetate (PMA), and this acceleration was inhibited by STS. These workers suggested an active role for lipid metabolites like DAG in enhancing flower senescence, through regulation of ethylene production and action, or activation of kinases.

V. Pollination and Senescence

Van Doorn (1997) in an exhaustive survey of the literature found that pollination rapidly reduces floral attraction in numerous orchids, but among the other plant families studied, only about 60 genera were found to show pollination-induced shortening of floral attraction. In some instances, pollination was even found to increase the longevity of the flower. Where examined, the pollination-mediated decrease in floral longevity appeared to be due to the action of ethylene as inhibitors of ethylene synthesis or perception invariably blocked the effect of pollination.

A. The Pollination Signal

A number of studies have demonstrated that pollination results in increased sensitivity of floral tissues to exogenous ethylene (Halevy *et al.*, 1984; Porat *et al.*, 1994). Researchers have suggested various causes for the increased ethylene production and/or sensitivity of pollinated flowers, including physical stimulation, auxin, ethylene, pollen ACC, short-chain fatty acids, and a range of other molecules (O'Neill, 1997).

1. Physical stimulation

Gilissen (1976, 1977) noted that mechanical wounding of the stigma and style, like pollination, induced premature wilting of petunia flowers and speculated that the pollination-induced wilting was due to the wounding resulting from growth of the pollen tube. However, the postpollination syndrome cannot be solely the result of wounding inherent in the growth of the pollen tube, as the response to pollination can be rapid and has been shown to not require germination of the pollen tube (Zhang and O'Neill, 1993). Moreover, application of pollen to the stigma of petunia causes stimulation of ethylene synthesis in as little as five minutes (Pech *et al.*, 1987).

2. Auxins

Burg and Dijkman (1967) proposed that auxin was transferred to orchid stigmas from the pollen and then diffused to the column and labellum where it promoted the onset of autocatalytic ethylene production. However, studies with labeled IAA have shown that its movement is very slow in orchids (Strauss and Arditti, 1982). O'Neill and Nadeau (1997) have also noted that the levels of free auxin in orchid pollen are appreciably lower than those required to induce rates of ethylene production comparable to those found in pollinated flowers.

3. Pollen ACC

The very high concentrations of ACC present in petunia pollen (Whitehead *et al.*, 1984a), and the presence of a very high activity of ACC oxidase in the stigmas (Pech *et al.*, 1987) seemed to implicate ACC as the signal for the rapid onset of ethylene synthesis that follows pollination. However, Hoekstra and Weges (1986) showed that if they treated stigmas with AVG to prevent endogenous ACC synthesis, pollination-dependent ethylene synthesis was eliminated, indicating that pollen ACC was not the source of the early burst of ethylene synthesis in the stigmas. Recently, it has been shown that wild-type petunia flowers pollinated with transgenic pollen containing lower ACC concentrations produce just as much ethylene and senesce just as quickly as petunia flowers pollinated with pollen containing higher ACC concentrations (Lei *et al.*, 1996). The function of the high concentrations of ACC found in many types of pollen is not known, but it does not appear to be related to the early induction of ethylene biosynthesis in pollinated flowers.

4. Ethylene

Ethylene signaling is discussed in Chapter 8. We interpreted the evolution of radioactively labeled ethylene from petals of carnation after labeled ACC was applied to the stigma as evidence for movement of ACC from the stigma to the petals (Reid *et al.*, 1984). In contrast, Woltering *et al.* (1995) suggested that in *Cymbidium* orchids, it was radiolabeled ethylene produced in the rostellum (stigma), not the radiolabeled ACC that moved to the petals. These researchers showed that preventing conversion of exogenously supplied ACC to ethylene in the rostellum with cobalt chloride caused a reduction in the evolution of ethylene from the petals. They further demonstrated that incisions in the columnar tissue, allowing dissipation of ethylene but not affecting transfer of soluble substances, delayed coloration of the lip in ACC-treated flowers. In a later study, Woltering *et al.* (1999) demonstrated, in *Cymbidium* and petunia flowers treated with AVG and Co^{++} so that they could neither synthesize nor oxidize ACC, that radiolabeled ACC applied to the stigma was immobile, with less than 1% recovery in other floral tissues. In contrast, when ethylene or propylene was applied (at physiologically relevant levels) just to the central column, these gases were released from the petal tissues within half an hour. These results provide convincing evidence that it is interstitial ethylene gas, not ACC, which moves from the stigma to the petals following pollination.

5. Short-chain fatty acids

Whitehead and Halevy (1989) demonstrated a pollination-associated increase in the concentration of short-chain fatty acids in stylar exudates of petunia and found that application of stylar exudates or pure short-chain fatty acids to stigmas increased the ethylene sensitivity of the flowers. They subsequently demonstrated that these chemicals increased ethylene sensitivity and reduced vase life of carnation flowers (Whitehead and Vasiljevic, 1993), and have recently demonstrated changes in the activities of Acyl Co-A carboxylase (Whitehead *et al.*, 2000) that are consistent with the hypothesis that short-chain fatty acids are the mobile sensitivity factor in pollination. However, Woltering *et al.* (1993) were unable to confirm the suggested involvement of short-chain saturated fatty acids in senescence of carnation, *Cymbidium* and petunia.

6. Other signal molecules

Over the years researchers have proposed a diversity of other possible signals, including flavonoids, methyl jasmonate, and jasmonic acid (See Chapter 9). O'Neill (1997) and her associates investigated the role of some of these materials in *Phaleonopsis* orchid flowers and eliminated methyl jasmonate, pollen-derived lipids and proteins, flavonoids, and systemin because they failed to elicit either ethylene production and/or stigma closure within a time frame consistent with a role in early pollination signaling. In a subsequent study, Porat *et al.* (1998) were able to readily extract in water what they termed the "primary pollination signal" from pollinia of *Phalaenopsis* flowers. They found the extracted materials to be low molecular weight (<3000 Da), non-proteinaceous compounds that could be separated into several active fractions by ion exchange chromatography. Two of the seven HPLC active fractions stimulated the complete postpollination response of perianth senescence and ovary growth, one fraction co-chromatographed with authentic ACC, but none of the active fractions were found to co-elute with IAA. A particularly intriguing observation was that one of the other peaks induced perianth senescence, but not ovary growth. The nature and function of these different active fractions remains to be determined.

References

Baker, J.E., Wang, C.Y., Lieberman, M., and Hardenburg, R. (1977). Delay of senescence in carnations by a rhizobitoxine analog and sodium benzoate. *HortScience* **12**, 38–39.

Beutelmann, P., and Kende, H. (1977). Membrane lipids in senescing flower tissue of *Ipomoea tricolor*. *Plant Physiol.* **59**, 888–893.

Beyer, E.M.J. (1976). A potent inhibitor of ethylene action in plants. *Plant Physiol.* **58**, 268–271.

Bieleski, R.L. (1993). Fructan hydrolysis drives petal expansion in the ephemeral daylily flower. *Plant Physiol.* **103**, 213–219.

Bieleski, R.L. (1995). Onset of phloem export from senescent petals of daylily. *Plant Physiol.* **109**, 557–565.

Bieleski, R.L., and Reid, M.S. (1992). Physiological changes accompanying senescence in the ephemeral daylily flower. *Plant Physiol.* **98**, 1042–1049.

Borochov, A., and Woodson, W.R. (1989). Physiology and biochemistry of flower petal senescence. *Hortic. Rev.* **11**, 15–43.

Borochov, A., Halevy, A.H., and Shinitzky, M. (1982). Senescence and the fluidity of rose petal membranes. *Plant Physiol.* **69**, 296–299.

Borochov, A., Cho, M.H., and Boss, W.F. (1994). Plasma membrane lipid metabolism of petunia petals during senescence. *Physiol. Plant.* **90**, 279–284.

Borochov, A., Spiegelstein, H., and Philosoph, H.S. (1997). Ethylene and flower petal senescence: interrelationship with membrane lipid catabolism. *Physiol. Plant.* **100**, 606–612.

Bovy, A.G., Angenent, G.C., Dons, H.J.M., and van Altvorst, A.C. (1999). Heterologous expression of the *Arabidopsis* etr1-1 allele inhibits the senescence of carnation flowers. *Mol. Breed.* **5**, 301–308.

Bufler, G., Mor, Y., Reid, M.S., and Yang, S.F. (1980). Changes in 1-aminocyclopropane-1-carboxylic-acid content of cut carnation flowers in relation to their senescence. *Planta* **150**, 439–442.

Burg, S.P., and Dijkman, M.J. (1967). Ethylene and auxin participation in pollen induced fading of Vanda orchid blossoms. *Plant Physiol.* **42**, 1648–1650.

Celikel, F.G., and van Doorn, W. (1995). Solute leakage, lipid peroxidation and protein degradation during the senescence of iris tepals. *Physiol. Plant.* **94**, 515–521.

Chang, C., Kwok, S.F., Bleecker, A.B., and Meyerowitz, E.M. (1993). Arabidopsis ethylene-response gene Etr1: similarity of product to two-component regulators. *Science* **262**, 539–544.

Eason, J.R., De Vre, L. (1995). Ethylene-insensitive floral senescence in *Sandersonia aurantiaca* (Hook.). *N.Z. J. Crop Hort. Sci.* **23**, 447–454.

Eisinger, W. (1977). Role of cytokinins in carnation flower senescence. *Plant Physiol.* **59**, 707–709.

Gilissen, L.J.W. (1976). The role of the style as a sense organ in relation to wilting of the flower. *Planta* **131**, 201–202.

Gilissen, L.J.W. (1977). Style-controlled wilting of the flower. *Planta* **133**, 275–280.

Guerrero, C., Calle, M.M., Reid, M.S., Valpuesta, V., and De-la-Calle, M. (1998). Analysis of the expression of two thiolprotease genes from daylily (*Hemerocallis spp.*) during flower senescence. *Plant Mol. Biol.* **36**, 565–571.

Halevy, A.H., and Mayak, S. (1972). Senescence control of flowers. *Is. J. Bot.* **21**, 121.

Halevy, A.H., and Mayak, S. (1979). Senescence and postharvest physiology of cut flowers Part 1. *Hort. Rev.* **1**, 204–236.

Halevy, A.H., and Mayak, S. (1981). Senescence and postharvest physiology of cut flowers Part 2. *Hort. Rev.* **3**, 59–143.

Halevy, A.H., Whitehead, C.S., and Kofranek, A.M. (1984). Does pollination induce corolla abscission of cyclamen flowers by promoting ethylene production? *Plant Physiol.* **75**, 1090–1093.

Hobson, G.E., and Nichols, R. (1977). Enzyme changes during petal senescence in the carnation. *Ann. Appl. Biol.* **85**, 445–447.

Hoekstra, F.A., and Weges, R. (1986). Lack of control by early pistillate ethylene of the accelerated wilting of *Petunia hybrida* flowers. *Plant Physiol.* **80**, 403–408.

Hunter, D.A., and Reid, M.S. (2001). Senescence-associated gene expression in *Narcissus* "Dutch Master". *Acta Hortic.* **553**, 341–343.

Itzhaki, H., Mayak, S., and Borochov, A. (1998). Phosphatidylcholine turnover during senescence of rose petals. *Plant Physiol. Biochem.* **36**, 457–462.

Itzhaki, H., Davis, J.H., Borochov, A., Mayak, S., and Pauls, K.P. (1995). Deuterium magnetic resonance studies of senescence-related changes in the physical properties of rose petal membrane lipids. *Plant Physiol.* **108**, 1029–1033.

Lay-Yee, M., Stead, A.D., and Reid, M.S. (1992). Flower senescence in day-lily *Hemerocallis*. *Physiol. Plant.* **86**, 308–314.

Lei, C.H., Lindstrom, J.T., and Woodson, W.R. (1996). Reduction of 1-aminocyclopropane-1-carboxylic acid (ACC) in pollen by expression of ACC deaminase in transgenic Petunias. *Plant Physiol. supplement* **111**, 149.

Matile, P., and Winkenbach, F. (1971). Function of lysosomes and lysosomal enzymes in the senescing corolla of the morning glory (*Ipomoea purpurea*). *J. Exp. Bot.* **22**, 759–771.

Maxie, E.C., Farnham, D.S., Mitchell, F.G., Sommer, N.F., Parsons, R.A., Snyde, R.G., and Rae, H.L. (1973). Temperature and ethylene effects on cut flowers of carnation (*Dianthus carophyllus* L.). *J. Am. Soc. Hort. Sci.* **98**(6), 568–572.

Mayak, S., and Dilley, D.R. (1976). Regulation of senescence in carnation (*Dianthus caryophyllus*). Effect of abscisic acid and carbon dioxide on ethylene production. *Plant Physiol.* **58**, 663–665.

Muller, R., Stummann, B.M., Andersen, A.S., and Serek, M. (1999). Involvement of ABA in postharvest life of miniature potted roses. *Plant Growth Regul.* **29**, 143–150.

O'Neill, S.D. (1997). Pollination regulation of flower development. *Annu. Rev. Plant Physiol.* **48**, 547–574.

O'Neill, S.D., and Nadeau, J.A. (1997). Postpollination flower development. *Hort. Rev.* **19**, 1–58.

Panavas, T., and Rubenstein, B. (1998). Oxidative events during programmed cell death of daylily (*Hemerocallis* hybrid) petals. *Plant Sci.* **133**, 125–138.

Panavas, T., Reid, P.D., and Rubinstein, B. (1998a). Programmed cell death of daylily petals: Activities of wall-based enzymes and effects of heat shock. *Plant Physiol. Biochem.* **36**, 379–388.

Panavas, T., Walker, E.L., and Rubinstein, B. (1998b). Possible involvement of abscisic acid in senescence of daylily petals. *J. Exp. Bot.* **49**, 1987–1997.

Panavas, T., Pikula, A., Reid, P.D., Rubinstein, B., and Walker, E.L. (1999). Identification of senescence-associated genes from daylily petals. *Plant Mol. Biol.* **40**, 237–248.

Pech, J.C., Latche, A., Larrigaudiere, C., and Reid, M.S. (1987). Control of early ethylene synthesis in pollinated petunia flowers. *Plant Physiol. Biochem.* **25**, 431–437.

Porat, R., Borochov, A., and Halevy, A.H. (1993). Enhancement of *Petunia* and *Dendrobium* flower senescence by jasmonic acid methyl ester is via the promotion of ethylene production. *Plant Growth Regul.* **13**, 297–301.

Porat, R., Borochov, A., Halevy, A.H., and O'Neill, S.D. (1994). Pollination-induced senescence of *Phalaenopsis* petals: The wilting process, ethylene production and sensitivity to ethylene. *Plant Growth Regul.* **15**, 129–136.

Porat, R., Nadeau, J.A., Kirby, J.A., Sutter, E.G., and O'Neill, S.D. (1998). Characterization of the primary pollen signal in the postpollination syndrome of *Phalaenopsis* flowers. *Plant Growth Regul.* **24**, 109–117.

Proctor, J.T.A., and Creasy, L.L. (1969). An anthocyanin-decolorizing system in florets of *Cichoium intybus*. *Phytochem.* **8**, 1401–1403.

Reid, M.S., and Wu, M.J. (1992). Ethylene and flower senescence. *Plant Growth Regul.* **11**, 37–43.

Reid, M.S., Fujino, D.W., Hoffman, N.E., and Whitehead, C.S. (1984). 1-Aminocyclopropane-1-carboxylic acid (ACC)—the transmitted stimulus in pollinated flowers? *J. Plant Growth Regul.* **3**, 189–196.

Rubinstein, B. (2000). Regulation of cell death in flower petals. *Plant Mol. Biol.* **44**, 303–318.

Sacalis, J.N., and Nichols, R. (1980). Effects of 2,4-D uptake on petal senescence in cut carnation flowers. *HortScience* **15**, 499–500.

Saks, Y., and van Staden, J. (1992). Effect of gibberellic acid on carnation flower senescence: evidence that the delay of carnation flower senescence by gibberellic acid depends on the stage of flower development. *Plant Growth Regul.* **11**, 45–51.

Savin, K.W., Baudinette, S.C., Graham, M.W., Michael, M.Z., Nugent, G.D., Lu, C.Y., Chandler, S.F., and Cornish, E.C. (1995). Antisense ACC oxidase RNA delays carnation petal senescence. *HortScience* **30**, 970–972.

Serek, M., Jones, R.B., and Reid, M.S. (1994). Role of ethylene in opening and senescence of Gladiolus sp. flowers. *J. Am. Soc. Hortic. Sci.* **119**, 1014–1019.

Serek, M., Sisler, E.C., and Reid, M.S. (1995). Effects of 1-MCP on the vase life and ethylene response of cut flowers. *Plant Growth Regul.* **16**, 93–97.

Solomos, T., and Gross, K.C. (1997). Effects of hypoxia on respiration and the onset of senescence in cut carnation flowers (*Dianthus caryophyllus* L.). *Postharvest Biol. Tech.* **10**, 145–153.

Stead, A.D., and Reid, M.S. (1990). The effect of pollination and ethylene on the colour change of the banner spot of Lupinus albifrons (Bentham) flowers. *Ann. Bot.* **66**, 655–663.

Stephenson, P., and Rubenstein, B. (1998). Characterization of proteolytic activity during senescence in daylilies. *Physiol. Plant.* **104**, 463–473.

Strauss, M., and Arditti, J. (1982). Postpollination phenomena in orchid flowers. X. Transport and fate of auxin. *Bot. Gaz.* **143**, 286–293.

Suttle, J.C., and Kende, H. (1980). Ethylene action and loss of membrane integrity during petal senescence in *Tradescantia*. *Plant Physiol.* **65**, 1067–1072.

Thompson, J.E. (1988). The molecular basis for membrane deterioration during senescence. In *Senescence and Aging in Plants* (L.D. Noodén and A.C. Leopold, Eds.), pp. xx–xx. Academic Press, San Diego.

Thompson, J.E., Mayak, S., Shinitzky, M., and Halevy, A.H. (1982). Acceleration of membrane senescence in cut carnation flowers by treatment with ethylene. *Plant Physiol.* **69**, 859–863.

Travnicek, I., Rodoni, S., Schellenberg, M., and Matile, P. (1999). A thio-based phospholipase assay employed for the study of lipolytic acyl hydrolase in the ephemeral corolla of *Ipomoea tricolor*. *J. Plant Physiol.* **155**, 220–225.

Valpuesta, V., Lange, N.E., Guerrero, C., and Reid, M.S. (1995). Up-regulation of a cysteine protease accompanies the ethylene-insensitive senescence of daylily (*Hemerocallis*) flowers. *Plant Mol. Biol.* **28**, 575–582.

Van Doorn, W.G. (1997). Effects of pollination on floral attraction and longevity. *J. Exp. Bot.* **48**, 1615–1622.

Veen, H., and Van der Geijn, S.C. (1978). Mobility and ionic form of silver as related to longevity of cut carnations. *Planta* **140**, 93–96.

Waithaka, K., Dodge, L.L., and Reid, M.S. (2000). Carbohydrate traffic during the opening of Gladiolus flowers. *Acta Hortic.* (in press).

Whitehead, C.S., and Halevy, A.H. (1989). Ethylene sensitivity: the role of short-chain saturated fatty acids in pollination induced senescence of Petunia hybrida. *Plant Growth Regul.* **8**, 41–54.

Whitehead, C.S., and Vasiljevic, D. (1993). Role of short-chain saturated fatty acids in control of ethylene sensitivity in senescing carnation flowers. *Physiol. Plant.* **88**, 342–250.

Whitehead, C.S., Halevy, A.H., and Reid, M.S. (1984a). Roles of ethylene and 1-aminocyclopropane-1-carboxylic acid in pollination and wound-induced senescence of *Petunia hybrida*. *Physiol. Plant.* **61**, 643–648.

Whitehead, C.S., Halevy, A.H., and Reid, M.S. (1984b). Control of ethylene synthesis during development and senescence of carnation petals. *J. Am . Soc. Hortic. Sci.* **109**, 473–475.

Whitehead, C.S., Botha, L., and Niemann, N. (2000). The role of acetyl-CoA carboxylase in the control of ethylene sensitivity in senescing flowers. *Acta Hortic.* (in press).

Wiemken, G.V., Wiemken, A., and Matile, P. (1974). Cell wall breakdown in wilting flowers of *Ipomoea tricolor*. *Planta* **115**, 297–307.

Wilkinson, J.Q., Lanahan, M.B., Clark, D.G., Bleecker, A.B., Chang, C., Meyerowitz, E.M., and Klee, H.J. (1997). A dominant mutant receptor from *Arabidopsis* confers ethylene insensitivity in heterologous plants. *Nat. Biotech.* **15**, 444–447.

Woltering, E.J., and van Doorn, W.G. (1988). Role of ethylene in senescence of petals: Morphological and taxonomical relationships. *J. Exp. Bot.* **39**, 1605–1616.

Woltering, E.J., Hout, Mv., Somhorst, D., Harren, F., and Van Hout, M. (1993). Roles of pollination and short-chain saturated fatty acids in flower senescence. *Plant Growth Regul.* **12**, 1–10.

Woltering, E.J., Somhorst, D., and van der Veer, P.P. (1995). The role of ethylene in interorgan signaling during flower senescence. *Plant Physiol.* **109**, 1219–1225.

Woltering, E.J., Van der Bent, A., de Vrije, G.J., and Van Amerongen, A. (1999). Ethylene: interorgan signaling and modeling of binding site structure. In *Biology and Biotechnology of the Plant Hormone Ethylene* (A.K. Kanellis, C. Chang, H. Kende and D. Grierson, Eds.), pp. 163–174. Kluwer, Dordrecht.

Woodson, W.R., Hanchey, S.H., and Chisholm, D.N. (1985). Role of ethylene in the senescence of isolated Hibiscus petals. *Plant Physiol.* **79**, 679–683.

Yamane, K., Kotake, Y., Okada, T., Ogata, R., Fjeld, T., and Stromme, E. (1995). Export of ^{14}C-sucrose, ^{3}H-water, and fluorescent tracers from gladiolus florets to other plant parts associated with senescence. *Acta Hortic.* **405**, 269–276.

Zhang, X.S., and O'Neill, S.D. (1993). Ovary and gametophyte development are coordinately regulated by auxin and ethylene following pollination. *Plant Cell* **5**, 403–418.

22

Postharvest Senescence of Vegetables and Its Regulation

Barry J. Pogson and Stephen C. Morris

I. Definitions and Theories of Vegetable Senescence

The term vegetable is used to describe a harvested edible plant product that is a member of the following categories: tubers, bulbs, roots, leaves, flowers, stem and fruit vegetables. Since senescence in fruits is discussed elsewhere in this book, we will confine discussion to the other categories. There are several distinct patterns of natural plant senescence: 1. Overall senescence, 2. Top senescence, 3. Deciduous senescence, 4. Progressive senescence of leaves, and 5. Bottom senescence of storage organs during sprouting. While this scheme is not always applicable to harvested vegetables, there are many similarities, such as tubers resulting from type 2 and then progressing into type 5 senescence. Also, the mechanisms of harvest-induced senescence in asparagus and broccoli are similar to natural deciduous and flower senescence respectively (King and O'Donoghue, 1995; Pogson *et al.*, 1995).

When considering vegetable senescence, researchers often define the onset of substantial senescence as being when the limits of the marketable quality (or appearance) of the vegetable are reached; this is also referred to as storage life. Consequently, in different types of vegetables the criteria for senescence will differ. In leafy vegetables senescence is indicated by substantial chlorophyll loss and wilting. Whereas in roots and storage organs it is indicated by commencement of sprouting and increased pathogen infection (Schouten, 1987). Chlorophyll loss is an indicator of green tissue senescence, since chloroplast pigments, lipids and proteins are typically the first compounds to be degraded (Noodén *et al.*, 1997). This loss becomes significant 24–48 h at 20°C in the dark after detachment for leaves, spinach and broccoli.

Senescence is defined as a series of active degenerative processes of a cell, organ or organism that are under genetic control, whereas aging is considered a passive process of accumulated defects (Noodén, 1988a,b). Aging may advance the induction of senescence, for example leaves of *Arabidopsis* mutants impaired in photoprotection, appear to be more susceptible to free radical damage and senesce more rapidly (Björkman and Niyogi, 1998; Pogson *et al.*, 1998). Senescence requires active synthesis of cytoplasmic proteins and RNA, with over 30 senescence-associated genes (SAGs) being up regulated (Gan and Amasino, 1997). Even when chlorophyll (Chl) loss is 80%, further loss can be halted by inhibitors of cytoplasmic protein synthesis, or at times even reversed, as shown when pods are removed from senescing soybeans (Brady, 1988).

The membranes, enzymes and other cellular components of a typical plant cell are being continually synthesized (metabolites) and degraded (catabolites) (Brady, 1988). In immature tissue, the balance is in favor of synthesis, while in senescing tissues the balance is in favor of degradation. The shift to catabolism enables nutrient reallocation between organs (Emmerling, 1880). However, the nutrient deficiency of a sink organ is generally not the initiator of the senescence response (Noodén *et al.*, 1997). In general it is a combination of environmental or developmental cues initiating changes in hormone fluxes that promote the active degradation of the targeted tissue (Fig. 22-1; Noodén, 1988a). Although in specific instances like asparagus spear senescence, sucrose content may be an early signal for the subsequent degradative pathways (King and O'Donoghue, 1995). Cytokinins, gibberellins and to some extent auxins typically retard senescence, and other phytohormones such as ethylene, ABA and the jasmonates typically increase the rate of senescence (Fig. 22-1);

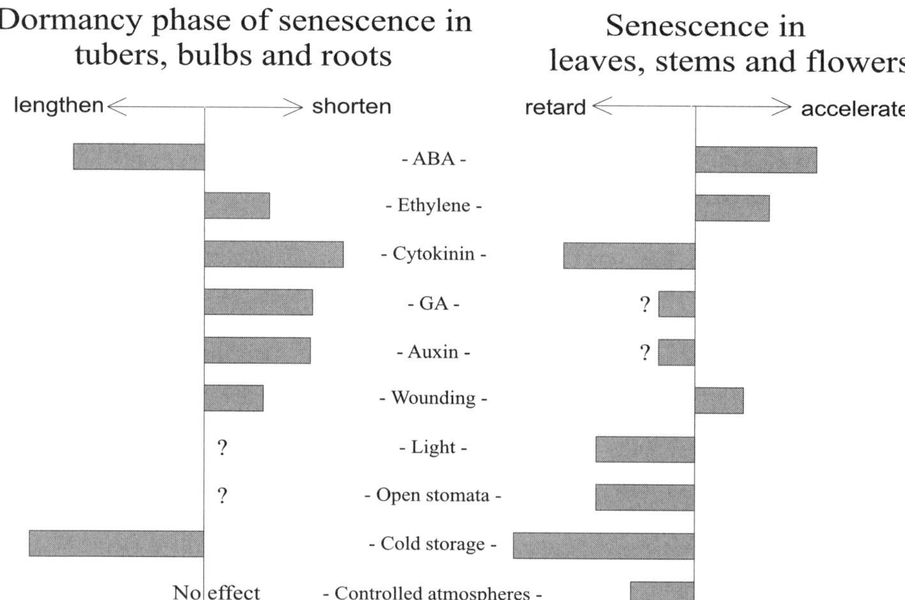

Figure 22-1. Effects of hormones and other factors on the senescence of detached green vegetables and on the dormant phase of senescence in functionally self-sufficient tubers, bulbs and roots.

however, one needs to consider how hormones influence the levels of each other and how they act synergistically and antagonistically.

During natural senescence correlative control and metabolite movement between tissues are involved (Noodén *et al.*, 1997). However, in detached plant parts many of these important interactions are not possible, except to a limited degree in vegetables with multiple tissue types (King and Morris, 1994; Pogson *et al.*, 1995). Similarities between detached leaves and certain rapidly growing leaf, flower and stem vegetables enable the insight from some studies on the former to be applied to the latter (King and O'Donoghue, 1995; Gan and Amasino, 1997; Noodén *et al.*, 1997; Bleecker, 1998). Yet, there are several important differences. First, vegetable tissue is generally much more mature than seedling leaves, which is significant since the stage of maturity affects the senescence process and thereby the way it can be controlled (Lipton, 1987; Ludford, 1987). Second, vegetables are represented by a wide range of tissue types, both within a given vegetable and between different vegetables. Thus, a given vegetable such as broccoli can include leaves, stems and flowers, and this may enable limited relocation of metabolites from mature sinks to immature tissues with high metabolic activity. Finally, many tubers, bulbs and root crops are capable of being "functionally self-sufficient" when detached. That is, the organ has the nutrient reserves, low respiration, effective transpiration barriers, active wound repair mechanisms and the necessary hormonal control system for sustained viability until sprouting, which produces the next generation. This phase is defined as dormancy, during which degradation and sprouting are both minimized and inhibited (Fig. 22-1).

The two major disturbances caused by detachment are dehydration and wounding. The water stress can be so severe that unless amelioration by storage in high humidity and cool temperatures is provided rapid death may occur (Lipton, 1987). Dehydration causes stomatal closure, which accelerates senescence and overrides the senescence-delaying effect of light (Thimann *et al.*, 1977). The senescence-delaying effect of light is ascribed to its effects on opening stomata and maintenance of the photosystems. Given these advantages it is surprising that use of light during storage is not standard commercial practice. However, if light were used during storage then the light intensity/temperature regimes would have to be chosen to limit photooxidative damage (Gray and Huner, 2000).

Wounding also tends to accelerate senescence and increase susceptibility to pathogen attack (Noodén, 1988a; Becker and Apel, 1993). While there are some genes induced by both natural leaf and detachment-induced senescence, other genes are only induced by wounding or wound-induced changes in ABA, ethylene or jasmonic acid (Becker and Apel, 1993; King and O'Donoghue, 1995). Additionally, wounding initiates a series of protective mechanisms and if metabolite reserves are limiting, this could induce an imbalance in normal metabolic maintenance activities. Under certain conditions normal development can be suspended as metabolic activity shifts towards producing protective structures and phytoalexins at the site of the wound (Morris *et al.*, 1989).

II. Specific Patterns of Senescence and its Regulation in Vegetables

An overview of vegetable senescence for each of the major categories is given below. Specific details on the senescence, respiration and handling of individual types of vegetables are readily available on several World Wide Web sites (see Morris, 1999; Moyer, 1999).

A. Tubers, Bulbs and Root Vegetables

Tubers, bulbs and some root vegetables are essentially functionally self-sufficient asexual reproductive structures. Examples of each category are potatoes (tuber), onions (bulb), carrots, horseradish and turnip (root vegetables). In general, senescence can be viewed in two phases: a dormant phase when little degradation occurs and sprouting is inhibited; followed by the active promotion of shoot sprouting and degradation of the storage organ, even under optimal storage conditions (Isenberg *et al.*, 1987; Ludford, 1987).

Interestingly, the roles of hormones regulating dormancy and degradation in storage organs are quite different to that for rapidly degrading leaves and flowers (Fig. 22-1). Dormancy is to a large part controlled by the balance between growth inhibitors (ABA) and growth promoting hormones, such as cytokinins, gibberellin and auxins (Fig. 22-1; Isenberg *et al.*, 1987; Ludford, 1987). It should be noted that the effects of ABA, cytokinin and gibberellin on senescence in dormant tubers and bulbs are opposite to their effects on green tissues (Fig. 22-1), whereas ethylene seems to consistently accelerate senescence. This paradox makes sense when we consider other tissues with a dormancy phase such as buds and seeds, where ABA promotes seed dormancy and GA and cytokinins shorten dormancy and stimulate germination and shoot growth (Khan and Tao, 1978). In general, levels of ABA and other less well-defined inhibitors are high for 2–8 weeks after harvest, then over the following months they decline as the growth promoting hormones increase, with the cytokinin levels often being the first to rise (Schouten, 1987).

Transgenically increasing endogenous cytokinins shortened the dormancy phase in potato (Ooms *et al.*, 1991). Likewise, exogenous application of cytokinins, ethylene and gibberellins and wounding shortens dormancy in bulbs and tubers, whereas ABA lengthens it (Fig. 22-1; Isenberg *et al.*, 1987). As dormancy ends, the various metabolite reserves, such as starch, in the organs are converted to forms more suited to transport and rapid metabolism in the sprouts, such as sucrose (Stoll and Weichmann, 1987).

The rates of respiration in bulbs, tubers and most root vegetables during dormancy are very low (Fig. 22-2); a feature with obvious advantages with respect to conserving reserves. Although, it should be noted that some root crops have higher respiration rates and a shorter storage life (Fig. 22-2). Controlled atmospheres (CAs) that reduce oxygen and increase carbon dioxide levels and also slow respiration, purportedly prolong storage life for many crops (Stoll and Weichman, 1987). However, CAs have little effect on the senescence of tubers, bulbs or root vegetables with already low respiration rates (Burton, 1978; Schouten, 1987; Kubo *et al.*, 1990). The structures around tubers and bulbs that are designed to reduce water loss also limit gas diffusion and internally would mimic CA conditions, which may lower respiration (Burton, 1978).

Water loss via transpiration of tubers and bulbs is very low, typically 0.0003% water loss day^{-1} Pa vpd^{-1} during dormancy (Ben-Yehoshua, 1987). However, transpiration rates of roots are markedly higher, typically 0.006–0.024% water loss day^{-1} Pa vpd^{-1} (Ben-Yehoshua, 1987; Stoll and Weichmann, 1987). This is a major difference between roots and tubers and it reflects the root's thin, easily damaged periderm.

In order to maximize viability during dormancy, bulbs and tubers contain high levels of phytoalexins to resist pathogen attack and have effective wound repair mechanisms to limit pathogen access (Morris and Lee, 1984). The types of phytoalexins include phenols, steroidal alkaloids and proteinase inhibitors. The wound repair mechanisms in potatoes include the deposition of a suberin barrier (lipids and lignin) in the cell walls immediately beneath the wound to limit pathogen access and water loss (Burton, 1978; Ludford, 1987;

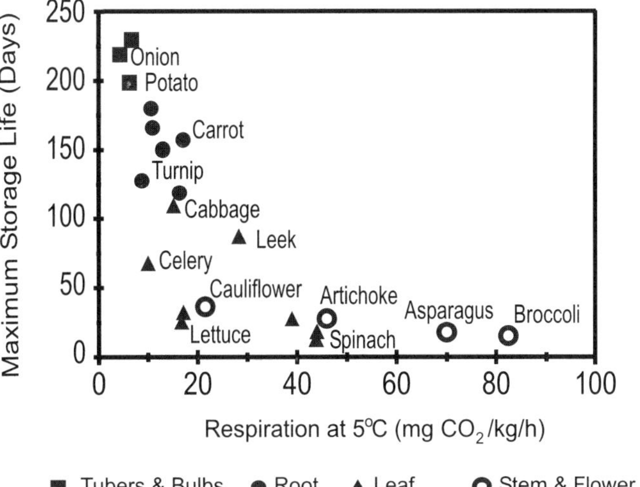

Figure 22-2. Respiration rate at 5°C and maximum storage life at optimal storage temperature for various vegetables. Respiration rates were obtained by converting heat output (W tonne^{-1}) to mgCO_2 kg^{-1} h^{-1} using the conversion factor of 0.34. Data are an average of values from experimental data, Morris (1999) and Moyer (1999).

Schouten, 1987; Morris *et al.*, 1989). Whereas, onions produce a barrier by rapidly desiccating the surrounding tissue. Problematically, defense mechanisms decline during dormancy and are optimal at storage temperatures higher than ideal for prolonging dormancy (Schouten, 1987; Morris *et al.*, 1989). This is a major feature affecting tuber and bulb senescence. The defense mechanisms can be enhanced by a pre-storage treatment at higher temperatures. Encouragingly, transgenic experiments have shown it is also possible to genetically increase the resistance of potatoes to disease and viruses during prolonged storage (Feher *et al.*, 1992).

B. Leafy Vegetables

Many crucifers, including brussel sprouts, cabbage, and crops such as lettuce, spinach, cress, and chives are included in this category. This diversity of crops is reflected in the very large range in refrigerated commercial storage life—from less than two weeks (lettuce and spinach), two to eight weeks (brussels sprouts and chinese cabbage), and even up to six months (cabbage, Fig. 22-2). For those vegetables with short storage, senescence to a large degree parallels that for detached dark-induced leaves, and this is reviewed elsewhere (Lipton, 1987; Gan and Amasino, 1997; Bleecker, 1998). In this section we will focus on one example that shows differences to classic leaf senescence and it is from the long storage end of the spectrum, namely cabbage.

Cabbage differs from seedling leaves in that both the tissue maturity and the arrangement of the leaves delay senescence. Like tubers, there are two phases: dormancy followed by a sprouting phase. In cold storage the dormancy may last 10–20 weeks depending upon the cultivar (Isenberg *et al.*, 1987). During this time, ABA-like inhibitors initially increase and then decline around the time that dormancy ends and GA and auxin levels rise (Isenberg *et al.*, 1987). Cytokinins remain low until the apical meristem starts to develop.

Variations in cytokinin-like substances between cultivars may correlate with the length of dormancy (Lipton, 1987). Therefore, although not so clearly defined, it seems that hormonal regulation in cabbages is in some ways similar to that in tubers and bulbs (Ludford, 1987). Production of ethylene by cabbages is low (0.6 nL g^{-1} h^{-1} at 20°C). However, exogenous application of ethylene to stored cabbage exacerbates the problem of "pepper spot" (black specks on the leaves) and has a dose-response effect in accelerating degreening, abscission, sugar loss and changes in organic acids (Ludford, 1987).

Cabbages have a low respiration rate (Fig. 22-2), as expected for a vegetable harvested at maturity. High concentrations of CO_2 (60%) have no effect on the respiration rate of cabbage (Kubo *et al.*, 1990). However, CA storage of 5% CO_2 and 2.5% O_2 will extend the storage life for several months (Lipton, 1987; Ludford, 1987), presumably through the effect on respiration of reducing oxygen.

The compact nature of a cabbage head reduces the transpiration by two orders of magnitude compared to spinach (Ben-Yehoshua, 1987), demonstrating the importance of tissue arrangement in regulating water loss. While wound repair is not a prominent feature, some alleochemical defense may be provided by glucosinolates, since they correlate positively with storage life (Chong and Bérard, 1983). Glucosinolates (thioglucosides) are sulfurous secondary metabolites that contribute to crucifer odor.

C. Stem Vegetables

The most important horticultural stem vegetable is asparagus. The asparagus spear is growing very rapidly at harvest (as much as 70 mm per day). Thus its senescence is rapid, becoming unmarketable within 4–5 days at 20°C (King and O'Donoghue, 1995). Changes involved include lignification, ammonia production and rapid loss of protein, lipids, sugars, Chl and disease resistance (resulting in tip rot) and changes in flavor, texture and aroma (King and O'Donoghue, 1995).

Ethylene is produced at fairly constant levels of 2–4 nL g^{-1} h^{-1} until day 4 at 15°C, and it increases in parallel with advanced deterioration (Haard *et al.*, 1974). Exogenous ethylene increases lignification (Haard *et al.*, 1974); however, its effects on other parameters of senescence are unclear.

Harvested asparagus spears have a very high rate of respiration (reflecting the rapid metabolism and pre-harvest growth), which declines after harvest (Fig. 22-2). This differentiates the pattern of harvest-induced senescence from that of more mature crops and provides some commonality with flower vegetables. Depletion of sucrose by 6 h correlates with a shift in the respiratory quotient from 1.05 at harvest, to 0.66 within 12 h at 20°C. This is indicative of a shift from carbohydrate to protein and lipid based catabolism (King and O'Donoghue, 1995). There is a coincident increase in β-galactosidase mRNA by 12 h, suggesting that galactose is also mobilized in response to depleted sucrose (O'Donoghue *et al.*, 1998). Fifty percent of the lipids are lost over 5 days and likewise depleted are proteins with an associated increase in asparagine, other free amino acids and ammonia (King and O'Donoghue, 1995).

There are changes in the abundance of specific mRNAs in postharvest asparagus, many within the first 12 h (King and O'Donoghue, 1995). Of the mRNAs that decrease, several encode proteins for growth related processes. Of those that increase, many are senescence-associated genes (SAGs), including asparagine synthetase (AS) mRNA. Many of the SAG mRNAs like AS can also be manipulated by altering sucrose concentration in asparagus tissue culture (Davies *et al.*, 1996). Thus, it may be that sucrose depletion stimulates an increase in asparagine, which is a translocateable form of available nitrogen (King and O'Donoghue, 1995). Harvest-induced senescence resulted in the up- and down-regulation of some of the same genes in naturally senescing asparagus ferns as was observed for barley (Becker and Apel, 1993; King and O'Donoghue, 1995). However, some of these mRNA and physiological changes are unique to spear senescence. In fact, treatment of asparagus spears with protein synthesis inhibitors actually accelerates senescence—an atypical result (Noodén, 1988b). This possibly reflects the atypical key role of sugar levels in asparagus senescence (Hurst *et al.*, 1996).

Cool storage reduces the respiration rate and consequently prolongs viability as there is a close relationship between storage life and accumulated CO_2 production (Brash *et al.*, 1995). That is, when asparagus are stored at different temperatures for different times until each reaches the same degree of senescence, the total CO_2 produced is the same (Brash *et al.*, 1995), demonstrating that regulating respiration rates is critical in delaying senescence of rapidly respiring organs.

D. Flower Vegetables

These types of vegetables include broccoli or calabrese (compact inflorescence of immature flower buds with stem), cauliflower (aborted floral meristem and stem tissue) and artichokes (a mixture of immature and mature tissue with large sepals and stem). As a group, the rate of senescence is generally rapid. The most prominent effect of senescence in broccoli is the readily observable loss of chlorophyll from the sepals of the immature floral buds after 2–3 days at 20°C, whereas there is no Chl loss from the stem (King and Morris, 1994).

The pattern of ethylene production in broccoli is frequently a 5–10-fold peak of wound-induced ethylene after harvest. However, this wound peak is not always observed and can be ameliorated by cold storage (Tian *et al.*, 1994; Pogson and Morris, 1997). After the peak, levels drop to 0.5–1.5 nL g^{-1} h^{-1}, where they remain for 2–3 days at 20°C, often slowly rising after that, sometimes in association with pathogen infection (King and Morris, 1994). Chlorophyll loss in broccoli florets can be delayed by ethylene synthesis inhibitors and accelerated by the application of exogenous ethylene, its analogues and ACC (Aharoni *et al.*, 1985; Tian *et al.*, 1994). The long-established effects of exogenous cytokinins on delaying senescence in broccoli are dramatic, postponing substantial chlorophyll, protein and turgor loss from 2–3 days to 6–8 days at 20°C, even while increasing ethylene production 2–5-fold (Rushing, 1990; Downs *et al.*, 1997). There are few reports about other hormones, except that GA delays chlorophyll loss (Aharoni *et al.*, 1985).

As flower vegetables range in their rate of senescence and growth at harvest, so their respiration rates range from moderate (cauliflower) to very high (broccoli, Fig. 22-2). Indeed, at harvest broccoli florets respire almost 1% of the dry weight per hour (King and Morris, 1994). This respiration rate declines by more than 65% within 12 h of harvest at 20°C, then stabilizes (Rushing, 1990; King and Morris, 1994). The respiratory quotient (RQ) shifts from above 1 to below 0.9, indicating a shift from catabolism of organic acids and sugars to proteins and lipids (King and Morris, 1994). Cold storage reduces respiration and prolongs storage life (Pogson and Morris, 1997). High CO_2 and low O_2 controlled atmospheres had marked effects on reducing respiration rates, but did not delay chlorophyll loss (Aharoni *et al.*, 1985; Kubo *et al.*, 1990). Thus, the general observation of CA reducing respiration and thereby delaying senescence is not apparent in broccoli and perhaps this reflects a regulatory function by its floral tissue.

Broccoli is an immature floral inflorescence, and it appears that the mechanism of detachment-induced senescence actually parallels the observed postpollination-induced senescence of mature flowers such as orchids (O'Neill *et al.*, 1993; Pogson *et al.*, 1995). That is, ACC oxidase activity and mRNA rise first and to 10–15-fold greater levels in the immature reproductive tissues than the vegetative sepals and petals of harvested broccoli (Tian *et al.*, 1994; Pogson *et al.*, 1995). ACC oxidase mRNA is induced within 2 h of harvest by a gene, ACC oxidase2, whose expression is reproductive tissue-specific. Subsequently, there is induction of another gene, ACC oxidase1, whose expression is essentially specific to sepals and senescing leaves (Pogson *et al.*, 1995). Furthermore, removing the reproductive organs or inhibiting ethylene biosynthesis by heat-treatment delays sepal senescence (Tian *et al.*, 1994, 1997). The nature of this signal, or whether sepal metabolites are transferred to the reproductive tissues is not known. Indirect evidence suggests relocation of metabolites from mature stem tissue to florets and treating florets with sucrose delays yellowing (King and Morris, 1994; Irving and Joyce, 1995). It may be that targeted degradation of vegetative tissue is an attempt to compensate for impending metabolite exhaustion and this could involve regulation by the reproductive structures in an attempt to maintain the reproductive system at the expense of the surrounding tissue (Pogson *et al.*, 1995).

Transpiration rates for broccoli are very high, $\sim 0.07\%$ loss day^{-1} Pa vpd^{-1}, a rate only exceeded by vegetables with loose or isolated leaves (Ben-Yehoshua, 1987; Isenberg *et al.*, 1987). Rates for cauliflower are moderate, 0.01–0.02% loss day^{-1} Pa vpd^{-1}. There seem to be major genotypic differences between broccoli cultivars with regard to susceptibility to postharvest pathogens, particularly during cold storage (Morris and Pogson, unpublished data), but no active wound repair like tubers.

III. Conclusions and Prospects

A key feature of senescence in vegetables is the two major categories: functionally and non-functionally self-sufficient, with the former generally having a dormant phase (Fig. 22-1). The short postharvest life of the second group is to a large part due to an inability to maintain homoeostasis after detachment. Since vegetables in this category are so poorly adapted to being detached from the plant it is difficult to separate the wounding effect of harvesting from the deprivation of nutrients, water and hormones. The two key handling practices are low temperature to reduce respiration, and high humidity to reduce transpiration. When one considers the respiration rates between vegetables and within ecotypes of a given species, it is clear that there is a strong relationship between respiration rate and the rate of senescence (Fig. 22-2). However, senescence is still strongly influenced by hormones, external stimuli and correlative controls from other tissues in complex vegetables. In a sense respiration imposes a load on the system, but a range of other controls determine the ability of the system to cope with the load.

The role of hormones in vegetable senescence varies depending on whether one is considering senescence, or the senescence-delaying stage of dormancy (Fig. 22-1). Ethylene, cold storage and wounding all perform similarly during both phases, while ABA, cytokinins, GA and auxins have opposite effects on each phase. The interrelation of the major hormones is understood to a degree in dormancy, but is not well understood in the degenerative phase of tubers. For other vegetable groups, such as stem and flower vegetables, little is known about changes of any endogenous hormones, except for ethylene. The paucity of information about ABA is surprising considering the important role this hormone plays in detached leaf systems. Another issue is that while hormones often promote or delay senescence, this is not an all or nothing response and the sensitivity of tissue to that hormone may regulate the response.

Our understanding of the mechanisms of senescence and programmed cell death has advanced dramatically in recent years and this will be of relevance to many crop species. A classic example of this is the expression of a cytokinin biosynthetic gene under the regulation of a senescence-induced promoter to markedly delay green leaf senescence (Gan and Amasino, 1997). This brings the discussion to a key point. Certainly, ethylene and cytokinins are targets for some, but definitely not all crops. So as our knowledge of the key signal(s) advances, the second generation of transgenics should be more specific and effective for a given organ. For instance, since broccoli senescence is to a degree regulated by reproductive tissues, targeting the key signal there may provide an even better solution than manipulating overall cytokinin production. This is because while cytokinins delay chlorophyll loss, they do not affect the decline in sugars or change the respiration rate on the first day after harvest (Downs *et al.*, 1997). Alternatively, senescence may be delayed if a greater movement of metabolites from stem tissues could be encouraged, or even simply if broccoli were harvested at sunset when starch levels are higher. In asparagus, it seems that respiration rates and metabolite depletion are critical. In this instance, the targets may be mobilization and relocation of alternative metabolites to the tip, combined with a transgene that represses the respiration that is rapidly induced by harvest.

The tubers and bulbs present their own unique targets. Improvements in pathogen resistance that enable sustained resistance at temperatures optimal for storage will be advantageous and initial work on this is encouraging (Feher *et al.*, 1992). Hormone manipulation may be of benefit, however, it needs to be under the regulation of appropriate organ-specific promoters, particularly when it involves differentiation hormones, like the cytokinins (Ooms *et al.*, 1991).

Progress in defining signal transduction pathways and identifying the regulatory class of senescence-induced genes (Bleecker, 1998) will further help unravel the initial steps in senescence. With the tools of molecular biology and the insight being gained into programmed cell death, the horticultural scientist can now start to delineate the order of events in the long-reported pleiotropic correlative changes for a given organ in a given crop. Thereafter, the key signal(s) can be targeted to achieve the maximal physiological effect on the broadest range of changes, whether by genetic manipulation, handling practices, or more likely, a combination of the two.

Acknowledgments

We would like to dedicate this to the memory of Colin Brady, who made a lifetime of significant contributions to plant senescence. We acknowledge the welcomed input over the years from our colleagues, including Colin Brady (CSIRO Horticulture), Graeme King and Chris Downs (N.Z. Crop & Food CRI) and Colin Turnbull (Univ. Queensland).

References

Aharoni, N., Philosoph-Hadas, S., and Barkai-Golan, R. (1985). Modified atmospheres to delay senescence and decay of Broccoli. In *Controlled Atmospheres for Storage and Transport of Perishable Agricultural Commodities* (S.M. Blankenship, Ed.), pp. 169–177. Horticulture Report, Vol. 126, Raleigh, NC.

Becker, W., and Apel, K. (1993). Differences in gene expression between natural and artificially induced leaf senescence. *Planta* **189**, 74–79.

Ben-Yehoshua, S. (1987). Transpiration, water stress and gas exchange. In *Postharvest Physiology of Vegetables* (J. Weichmann, Ed.), pp. 113–170. Marcel Dekker, New York.

Björkman, O., and Niyogi, K.K. (1998). Xanthophylls and excess energy dissipation: A genetic dissection in Arabidopsis. In *Photosynthesis: Mechanisms and Effects* (G. Garab, Ed.), Vol. 3, pp. 2085–2090. Kluwer Academic, Dordrecht.

Bleecker, A.B. (1998). The evolutionary basis of leaf senescence: method to the madness? *Curr. Opin. Plant Biol.* **1**, 73–78.

Brady, C.J. (1988). Nucleic acid and protein synthesis. In *Senescence and Aging in Plants* (L.D. Noodén and A.C. Leopold, Eds.), pp. 147–179. Academic Press, San Diego.

Brash, D.W., Charles, C.M., Wright, S., and Bycroft, B.L. (1995). Shelf-life of stored asparagus is strongly related to postharvest respiratory activity. *Postharvest Biol. Technol.* **5**, 77–81.

Burton, W.G. (1978). The physics and physiology of storage. In *The Potato Crop: The Scientific Basis for Improvement* (P.M. Morris, Ed.), pp. 545–606. Chapman & Hall, London.

Chong, C., and Bérard, L.S. (1983). Changes in glucosinolates during refrigerated storage of cabbage. *J. Amer. Soc. Hortic. Sci.* **108**, 688–691.

Davies, K.M., Seelye, J.F., Irving, D.E., Borst, W.M., Hurst, P.L., and King, G.A. (1996). Sugar regulation of harvest-related genes in asparagus. *Plant Physiol.* **111**, 877–883.

Downs, C.G., Somerfield, S.D., and Davey, M.C. (1997). Cytokinin treatment delays senescence but not sucrose loss in harvested broccoli. *Postharvest Biol. Technol.* **11**, 93–100.

Emmerling, A. (1880). Landwirtz. *Versuchs-stat.* **24**, 113.

Feher, A., Skyrabin, K.G., Balazs, E., Preiszner, J., Shulga, O.A., Zakharyev, V.M., and Dudits, D. (1992). Expansion of PVX coat protein gene under the control of extensin-gene promoter confers virus resistance on transgenic potato plants. *Plant Cell Reports* **11**, 48–52.

Gan, S.S., and Amasino, R.M. (1997). Making sense of senescence—Molecular genetic regulation and manipulation of leaf senescence. *Plant Physiol.* **113**, 313–319.

Gray, G.R., and Huner, N.P. (2000). Low temperature effects. In *Programmed Cell Death and Related Processes in Plants* (L.D. Noodén, Ed.), pp. . Academic Press, San Diego.

Haard, N.F., Sharma, S.C., Wolfe, R., and Frenkel, C. (1974). Ethylene induced isoperoxidase changes during fibre formation in postharvest asparagus. *J. Food Sci.* **39**, 452.

Hurst, P.L., Borst, W.M., and Sinclair, B.K. (1996). Protein synthesis inhibitors accelerate the postharvest senescence of asparagus and induce tiprot. *N. Z. J. Crop Hortic. Sci.* **24**, 191–197.

Irving, D.E., and Joyce, D.C. (1995). Sucrose supply can increase longevity of broccoli (brassica-oleracea) branchlets kept at 22-degrees-C. *Plant Growth Regul.* **17**, 251–256.

Isenberg, F.M.R., Ludford, P.M., and Thomas, T.H. (1987). Hormonal alterations during the postharvest period. In *Postharvest Physiology of Vegetables* (J. Weichmann, Ed.), pp. 45–49. Marcel Dekker, New York.

Khan, A.A., and Tao, K.L. (1978). Phytohormones, seed dormancy and germination. In *Phytohormones and Related Compounds—A Comprehensive Treatise* (D.S. Letham, P.B. Goodwin, and T.J. Higgins, Eds.),Vol. 2, pp. 371–422. Elsevier, New York.

King, G.A., and Morris, S.C. (1994). Early compositional changes during postharvest senescence of broccoli. *J. Am. Soc. Hortic. Sci.* **119**, 1000–1005.

King, G.A., and O'Donoghue, E.M. (1995). Unravelling senescence: New opportunities for delaying the inevitable in harvested fruit and vegetables. *Trends Food Sci. Technol.* **6**, 385–389.

Kubo, Y., Inaba, A., and Nakumura, R. (1990). Respiration and C_2H_4 production in various harvested crops held in CO_2-enriched atmospheres. *J. Am. Soc. Hortic. Sci.* **115**, 975–978.

Lipton, W.J. (1987). Senescence of leafy vegetables. *Hortic. Sci.* **22**, 854–859.

Lipton, W.J. (1990). Postharvest biology of fresh asparagus. *Hortic. Rev.* **12**, 69–155.

Ludford, P.M. (1987). Postharvest hormone changes in vegetables and fruit. In *Plant Hormones and their Role in Plant Growth and Development* (P.J. Darvies, Ed.), pp. 574–592. Marinus Nijhoff, Dordrecht.

Morris, S.C. (1999). Fruit and vegetable storage information. Sydney Postharvest Laboratory, URL: http://www.postharvest.com.au/Produce_Information.htm.

Morris, S.C., and Lee, T.H. (1984). The toxicity and teratogenicity of *Solanaceae* glycoalkaloids, particularly those of the potato (*Solanum tuberosum*): a review. *Food Technol.* **36**, 118–124.

Morris, S.C., Forbes-Smith, M.R., and Scriven, F.M. (1989). Determination of optimum conditions for suberization, wound periderm formation, cellular desiccation and pathogen resistance in wounded *Solanum tuberosum* tubers. *Physiol. Mol. Plant Pathol.* **35**, 177–190.

Moyer, P. (1999). Produce facts. Postharvest Technology Research and Information Center, University of California, Davis, URL: http://postharvest.ucdavis.edu/Produce.

Noodén, L.D. (1988a). The phenomena of senescence and aging. In *Senescence and Aging in Plants* (L.D. Noodén and A.C. Leopold, Eds.), pp. 1–50. Academic Press, London.

Noodén, L.D. (1988b). Postude and prospects. In *Senescence and Aging in Plants* (L.D. Noodén and A.C. Leopold, Eds.), pp. 499–517. Academic Press, London.

Noodén, L.D., Guiamet, J.J., and John, I. (1997). Senescence mechanisms. *Physiol. Plant.* **101**, 746–753.

O'Donoghue, E.M., Somerfield, S.D., Sinclair, B.K., and King, G.A. (1998). Characterization of the harvest-induced expression of beta-galactosidase in Asparagus officinalis. *Plant Physiol. Biochem.* **36**, 721–729.

O'Neill, S.D., Nadeau, J.A., Zhang, X.S., Bui, A.Q., and Halevy, A.H. (1993). Interorgan regulation of ethylene biosynthetic genes by pollination. *Plant Cell* **5**, 419–432.

Ooms, G., Risiott, R., Kendall, A., Keys, A., Lawlor, D., Smith, S., Turner, J., and Young, A. (1991). Phenotypic changes in T-*cyt*-transformed potato plants are consistent with enhanced sensitivity of specific cell types to normal regulation by root-derived cytokinin. *Plant Mol. Biol.* **17**, 727–743.

Pogson, B.J., and Morris, S.C. (1997). Consequences of cool storage of broccoli on physiological and biochemical changes and subsequent senescence at 20°C. *J. Am. Soc. Hortic. Sci.* **122**, 553–558.

Pogson, B.J., Downs, C.G., and Davies, K.M. (1995). Differential expression of two 1-aminocyclopropane-1-carboxylic acid oxidase genes in broccoli after harvest. *Plant Physiol.* **108**, 651–657.

Pogson, B.J., Niyogi, K.K., Björkman, O., and DellaPenna, D. (1998). Altered xanthophyll compositions adversely affect chlorophyll accumulation and nonphotochemical quenching in Arabidopsis mutants. *Proc. Natl. Acad. Sci. USA* **95**, 13324–13329.

Rushing, J.W. (1990). Cytokinins affect respiration, ethylene production, and chlorophyll retention of packaged broccoli florets. *Hortic. Sci.* **25**, 88–90.

Schouten, S.P. (1987). Bulbs and tubers. In *Postharvest Physiology of Vegetables* (J. Weichmann, Ed.), pp. 555–581. Marcel Dekker, New York.

Stoll, K., and Weichmann, J. (1987). Root vegetables. In *Postharvest Physiology of Vegetables* (J. Weichmann, Ed.), pp. 541–553. Marcel Dekker, New York.

Thimann, K.V., Tetley, R.M., and Krivak, B.M. (1977). Metabolism of oat leaves during senescence. V Senescence in light. *Plant Physiol.* **59**, 448–454.

Tian, M.S., Downs, C.G., Lill, R.E., and King, G.A. (1994). A role for ethylene in the yellowing of broccoli after harvest. *J. Am. Soc. Hortic. Sci.* **119**, 276–281.

Tian, M.S., Islam, T., Stevenson, D.G., and Irving, D.E. (1997). Color, ethylene production, respiration, and compositional changes in broccoli dipped in hot water. *J. Am. Soc. Hortic. Sci.* **122**, 112–116.

23

Evolutionary and Demographic Approaches to the Study of Whole Plant Senescence

Deborah Ann Roach

I. Introduction

There have been two distinct definitions of "plant senescence" which have developed within the literature. First, physiologists and cell biologists use the term senescence to describe the continual turnover of cells and plant parts that occurs within an individual as part of an internally controlled program of development. In cases of monocarpy (semelparity), this program can be responsible for the death of the whole organism. The details of this program of "physiological senescence" within individuals are addressed in the other chapters of this book. The second, alternative approach to senescence is termed "evolutionary senescence" and it addresses theories and experimental evidence explaining variation in mortality patterns among individuals within populations and between species. Senescence, as viewed by most animal and evolutionary biologists (see Chapter 1), is a decline in age-specific survival and reproduction with advancing age. The evolutionary theories of senescence are designed to explain *why* this senescence occurs in most species, and to explain the variation in the rates of evolutionary senescence between different species. It is this, evolutionary, population-level, approach to senescence that will be considered in this chapter. This chapter will apply a demographic approach to senescence that has traditionally been used exclusively by animal biologists and theoreticians, to the study of plants and the determination of mortality patterns and longevity.

From an evolutionary perspective, the phenomenon of senescence presents a paradox: Why should a trait that causes an individual to have an increased probability of dying with age persist in a population? The theory of evolution by natural selection suggests that heritable traits that improve the survival and reproduction of individuals should spread through a species because of their higher rates of transmission. We know that there is genetic variation in mortality patterns and longevity in populations. In classic studies with the fruit fly, *Drosophila melanogaster*, for example, researchers have successfully used artificial selection experiments to extend the life span and reduce rates of senescence (cf. Rose, 1984). This large degree of genetic variation in life span is expected in most species, and these results suggest that natural selection could potentially act on this genetic variation to change the time of onset and rate of senescence. Moreover, we also know that there are large differences between species in the rates of senescence. Birds, for example, generally live longer and have a lower rate of increase in mortality with age, in other words a lower rate of senescence, than mammals of comparative size. Some species are thus more effective at preventing or repairing damage than others are. The presence of genetic variation for senescence both within and between species needs to be explained from an evolutionary perspective. The objective of this chapter is to present the study of whole plant senescence within an evolutionary and demographic context. In the first part of this chapter the theories which have been proposed to explain the evolution and persistence of senescence will be discussed, and experimental tests of the theories will be evaluated. To study senescence at the level of the whole plant, demographic evidence for a decline in mortality and reproduction with age is essential. In the second part of this chapter, demographic evidence for senescence in plants will be evaluated and the techniques and problems that are unique to demographic studies of whole plant senescence will be discussed.

II. Evolutionary Approaches

A. Theories of Senescence

Both physiological and evolutionary senescence show genetic control, yet the details of our understanding of senescence at these two levels is very different (Bleecker, 1998). Physiological studies have demonstrated that within an individual plant, there is a genetic program which results in an orderly degenerative process leading either to the reabsorption of nutrients and the abscission of plant parts, or in the case of semelparous species, to the death of the whole organism (Noodén, 1988). Physiological senescence of plant parts is an active process that requires energy and protein synthesis and often involves the redistribution of nutrients and photosynthates. It is a beneficial process which increases an individual's chance for survival, for example by preparing a plant for harsh environmental conditions (e.g., winter), or by allowing a plant to shed unnecessary and inefficient structures.

Genetic control of evolutionary senescence is clear from the fact that there are species-specific differences in maximum life span and from the inheritance of longevity traits within populations. An evolutionary approach to the study of senescence seeks to understand the ultimate reasons why senescence occurs, and how we can explain variation in the rates of senescence between species. Given that there is genetic control of evolutionary senescence, it is interesting to ask why there has not been selection for increased longevity and perhaps even infinite life span. Instead, this phenomenon of senescence, which increases the probability of dying with increasing age, persists in populations and because of this, it is often considered to reflect a failure of natural selection to act against a deleterious trait.

The paradox of senescence from an evolutionary point of view is that, at the level of the individual, there is no scenario in which an increased rate of dying with age can be considered an adaptive trait, in other words a trait directly favored by natural selection. Historically, the one direct "adaptive" theory of senescence in the animal literature suggested that senescence was a mechanism to eliminate older, weaker, individuals from a population in order to conserve resources for younger more vigorous individuals (Weismann, 1889). Within the plant literature, it has been suggested that mortality will be selectively advantageous for an individual if it would result in a better chance for the immediate colonization of its offspring (Wilson, 1997). These hypotheses rely on the principles of group selection, and they can be shown to be theoretically improbable given that selection acts primarily at the level of the individual (Charlesworth, 1980).

Evolutionary theorists have alternatively suggested that senescence may be a non-adaptive trait which has evolved indirectly as a consequence of a selective premium on genes with favorable effects on survival or fecundity early in the life history. The key element to allow the evolution of senescence is the fact that the force of natural selection on survival and fertility necessarily decreases with age because organisms have a nonzero chance of dying from external causes such as predation and accidents. Medawar (1952) provided the foundation for the major theoretical progress on the evolution of senescence. He suggested that, even in the hypothetical complete absence of senescence, if constant fertility is assumed, the reproductive output of each age class declines with age, because survivorship from birth is a decreasing function of age. As a result, the relative importance of traits expressed at late ages, to the lifetime fitness of an individual, is less than the importance of traits expressed at earlier ages. This implies that the intensity of selection will be stronger on age-specific genes acting early in life, than it will on genes that affect traits later in life. Some genes, which have their effect at very late ages, after an organism is usually dead, can completely escape the influences of selection. The first direct experimental test of Medawar's hypothesis is presented later in this chapter (Section IIB).

With this major assumption, that natural selection is weaker at later ages, two major evolutionary theories, mutation accumulation and pleiotropy, have been proposed to explain the evolution and persistence of senescence in populations. The evolutionary theory of mutation accumulation suggests that due to the decline in the strength of selection on age-specific characters, mutations with a deleterious effect at the end of the life cycle will accumulate, resulting in a decrease in age-specific mortality and fecundity with increasing age (Medawar, 1952). In other words, because they will be effectively neutral, genes with deleterious effects will spread in the population because they affect only late-life traits. This type of mutation accumulation is distinct from the somatic mutations that may accumulate during the life of an individual. It refers instead to the large number of deleterious genes that are expressed at late ages and which persist in the population as a result of a change in the mutation–selection balance. The accumulation of mutations is thus due either to increasingly ineffective selection against recurrent deleterious mutations, or from genetic drift.

The second major evolutionary theory, antagonistic pleiotropy, suggests that genes affecting traits negatively late in life may get established in a population if those genes also have positive effects at an earlier age (Williams, 1957; Williams, 1957; for plant example see Roach, 1986). This theory is based on an optimality approach in which late-life performance is sacrificed for early survival or reproduction (Partridge and Barton, 1993). A closely related theory, the disposable soma theory, is a variation of the pleiotropy theory and incorporates a trade-off between reproduction, on the one hand, and maintenance and repair, on the other (Kirkwood and Holliday, 1979). This theory suggests that if there are a limited amount of resources available to the organism, then the allocation of energy to early reproduction will be favored at the expense of somatic maintenance and repair. Repair will consequently occur at a sub-optimal level leading to senescence. The lack of repair leads to an accumulation of damage to somatic cells and an increased probability of death at later ages. One of the primary assumptions of the disposable soma theory is that there is an early separation of germ line from the soma. This separation does not occur until just prior to reproduction in plants, yet as will be discussed later in this chapter, a violation of this assumption does not appear to be an important factor in the evolution of senescence.

The evolutionary theories of senescence suggest that the decline with age in reproduction and survival is the result of the failure of natural selection to act against genes with late age-specific effects (Charlesworth, 1980). The common thread of the evolutionary theories is that senescence is a consequence of a selective premium on genes with favorable effects on survival and fecundity early in the life history. The theories suggest that senescence is a trait that has evolved because of the absence of selection against degenerative changes in old age, or because of positive selection for mutations that increase early success even when they have later deleterious effects. There are several lines of evidence that can be used to evaluate the theories. First, the assumptions and predictions of the theories can be directly tested through experimental manipulation. Secondly, evidence needs to be obtained from demographic studies to determine whether there is, in fact, evidence for a change in mortality with increasing age, and to determine how species differ with respect to their mortality patterns. Both of these approaches will be used in the remainder of this chapter.

B. Tests of the Theories

1. Medawar's hypothesis

At the foundation of the evolutionary theories of senescence are the theoretical predictions of Medawar (1952). He suggested that in a natural environment a population would experience a high level of random, age-independent, mortality due to harsh environmental conditions. As a result, there is an ever-decreasing proportion of individuals that survive to reproduce at older ages. Harmful genes whose time of expression is beyond the natural range of life span for a particular species, can accumulate in a population with little or no check because the effect of mortality from external causes will be to reduce the force of selection in each successive portion of the life span. As a result, natural selection is ineffective at preventing an accumulation of deleterious genes that affect late-ages because they have a minimal impact on the population. Medawar concluded that aging is the result of the cumulative expression of deleterious genes in individuals that live longer than the average life span of the species in its natural environment. He expanded this further to suggest that in populations removed from their natural environment, a larger percentage of individuals will live to relatively late ages, and it is only under these circumstances that the expression of late deleterious genes would be evident and senescence could be observed. A direct test of Medawar's hypothesis would be to compare the demography of a single species under natural conditions and under artificial conditions in which life span was artificially extended. The previously unpublished study described here is the first direct test of this hypothesis.

The objective of this experiment was to evaluate the mortality of a single species in its natural environment and under idealized conditions. The experiment was done with *Rumex hastatulus*, a dioecious, wind-pollinated, weedy colonizer of disturbed sites, which is known to show substantial variation in individual life span from annual, to biennial, to short-lived perennial (Radford *et al.*, 1968). For the field experiment, the objective was to grow the plants under the same local conditions under which selection had been acting and had shaped the life history. The observed mortality pattern would then be an accurate assessment of natural field mortality. For this experiment, 646 three-week old seedlings were planted into the field, 20 cm apart, and except for an initial mowing, immediately prior to planting, the natural vegetation was undisturbed. For the experiment in the greenhouse, the objective was to grow plants under idealized conditions and to increase the proportion of individuals in the population living to late ages. To do this, there were two treatments. In the first, undisturbed census treatment, 637 plants were grown in individual pots, and watered and fertilized on a regular basis. In the second greenhouse treatment, 50 plants were inhibited from flowering by removing any flower buds from the base of the plant as soon as they appeared. This latter treatment was used to assess the effects of reproduction on mortality under these idealized conditions. The field and greenhouse populations were all censused monthly until every individual had died, and the mortality patterns of the different populations were contrasted.

The results of these experiments show that survivorship in the field and in the greenhouse was very different (Fig. 23-1). The mean life span in the field was 146 ± 80 days. Only 4 of 646 individuals survived until the second growing season and only one of these individuals flowered. Mortality was high, constant, and age-independent, in this experimental field population. This pattern is consistent with Medawar's conjecture about species in their natural populations. A similar high rate of mortality was observed for a population of natural seedlings of this species, which were marked in the field following emergence (Roach, unpublished).

Survivorship in the greenhouse was very high: 84% of the population survived to the day of first flowering during the second growing season (day 350). Yet, post-flowering survival

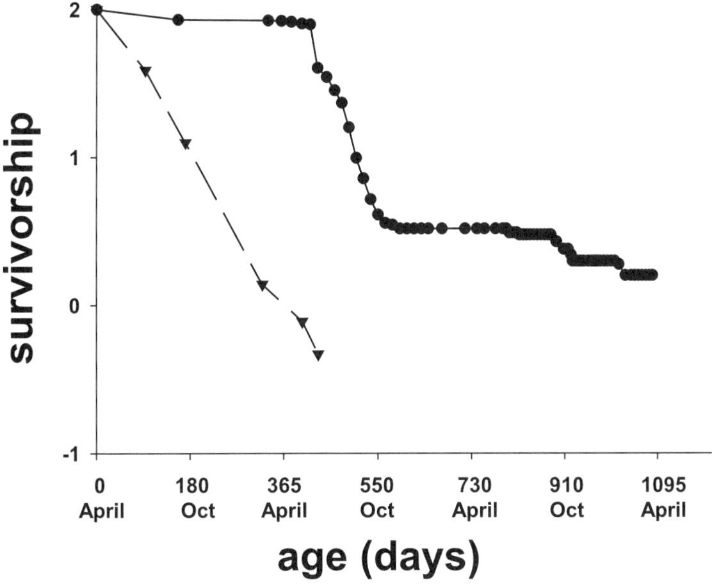

Figure 23-1. Survivorship curves (log percent) for *R. hastatulus* field (triangles) and undisturbed greenhouse (circles) populations.

was very low; only 9% of the undisturbed treatment plants were still alive post-flowering (day 525). Clearly, the developmental shift from vegetative to reproductive growth results in an increase in mortality. This may be an example of a pleiotropic trade-off between reproduction and survival, but the genetic correlations between these traits could not be measured with this experimental design. The mean life span in the undisturbed treatment in the greenhouse was 444 ± 259 days. There were no differences in mean life span for male and female plants that survived to flower.

The survival curves (Fig. 23-1) clearly show that, as predicted by Medawar, a larger proportion of the population in the greenhouse survived to later ages than in the field. Despite the fact that survivorship curves have traditionally been used in plant demography studies, it is in fact difficult to discern the pattern of mortality by looking at changes in the slope of a survival curve (Carey *et al.*, 1992). Thus, in order to evaluate mortality patterns for the two greenhouse populations a plot of the age-specific mortality was calculated (Fig. 23-2). The mortality curve for the undisturbed treatment shows a mortality increase following reproduction and then a decline. There is a second, lower, increase in mortality following reproduction during the third growing season. For the 50 deflowered plants, 90% of the individuals were still alive at the post-flowering period during the second growing season. The mean life span for the non-flowering plants was 1064 ± 589 days. At the end of the third growing season, both populations showed a steady increase in mortality with increasing age of the plants, and there were no differences in mortality patterns for the undisturbed and non-flowering treatments.

The evidence from this study suggests that there may be multiple mechanisms that cause senescence in *R. hastatulus*. In a natural population, if random environmental factors do not cause a constant rate of mortality, then senescence may be manifest following the developmental shift to reproduction. If, on the other hand, flowering is inhibited, or plants

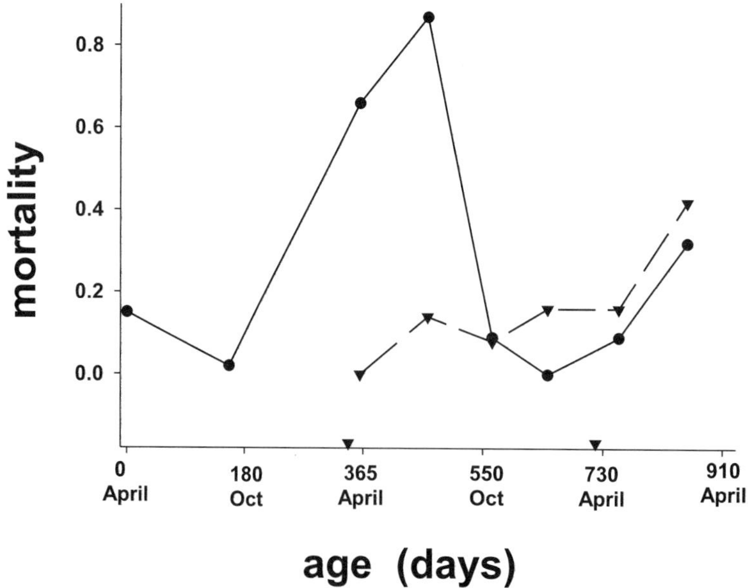

Figure 23-2. Mortality curves for *R. hastatulus* undisturbed (circles) and de-flowered (triangles) greenhouse populations. Triangles on the age-axis designate time of reproduction.

survive flowering and live to extremely late ages, another type of senescence may still occur but for different reasons. The experimental manipulations show that the late-age increase in mortality after three years of growing in the greenhouse was not linked to reproduction, because there were no differences in the mortality rates between the reproducing and deflowered plants. There are two possible explanations for the late-age mortality increase for these greenhouse populations. First, it is possible that there may have been an environmental change caused by a cumulative deterioration of the soil environment. The plants were grown in the same small pots for the entire experiment. There was no visual evidence that the plants had become pot-bound (pers. obs.) but there may have been a deterioration of the soil environment, caused possibly by an accumulation of root pathogens or other soil microbes, which may have caused an increase in mortality over time and age. Long-lived plants grown in the field may also experience a deterioration of their immediate soil environment. Experimental work with *Krigia dandelion*, for example, showed that there can be a negative feedback between species and their soil communities, which can result in reduced survival (Bever, 1994). Changes in mortality caused strictly by a progressive deterioration of the environment are not considered evidence for senescence. However, alternatively, the increase in mortality at the later ages may be due to deleterious genetic traits expressed at ages not normally attained by this species. The increases in mortality in the greenhouse plants occurred when the individuals were between 500–1100 days old. The mortality rates from the field study clearly show that few, if any, individuals could be expected to live under natural conditions to these extreme ages. In this case the increase in late-age mortality would be due to the expression of genes with deleterious effects at late ages, which have persisted in the population because they have not been exposed to strong selection. In order to test this hypothesis more vigorously, one would need to measure the changes in additive genetic variance in life history traits with age (Charlesworth, 1990).

2. Evidence for selection on patterns of senescence

The best test of the predictions of senescence would be an experimental change in late-age mortality and reproduction in response to a change in age-specific selection. Whereas these types of selection experiments have never been directly done with plants, there is indirect evidence from comparative intra- and inter-specific studies of plants growing under contrasting environmental conditions. These studies support the hypothesis that patterns of senescence have been, and can be, molded by natural selection. The theoretical prediction is that senescence rates change in a predictable fashion when there are long-term external environmental conditions that increase or decrease survivorship. Furthermore, it suggests that the evolution of species life span differences is explained by the evolution of senescence. The intensity of selection on early performance varies with ecological circumstances, and this leads to variation in the rate of senescent decline in different species. An externally imposed increase in survivorship will favor the evolution of postponed senescence. Conversely, when the level of externally imposed mortality is high, species that invest in rapid growth and reproduction have a selective advantage over species or individuals that invest in prolonged growth and perhaps never have an opportunity to reproduce. A high level of environmental mortality should be associated with short life span and a rapid rate of senescence.

Testing the effect of exogenous mortality on rates of senescence is difficult in a natural environment, because the variation among populations in exogenous mortality must be separated from any other possible evolutionary differences in the life histories among the study populations (Partridge and Barton, 1996). To date, the best examples of the influence of external environmentally imposed mortality patterns on life span and mortality rates in natural plant populations are two studies, one with *Poa annua*, and one comparative study with two species of *Lobelia*. In what has now become a classic example of the comparative effect of exogenous mortality under natural conditions, populations of annual meadow grass (*P. annua*), from frequently disturbed meadows and less disturbed meadows, were compared in a common garden experiment (Law *et al.*, 1977; Law, 1979). The results of this experiment showed that seeds derived from plants from the frequently disturbed sites showed higher early reproduction but shorter life spans and higher rates of mortality. Whereas the design of this experiment did not eliminate the possibility that there may have been correlated environmental variables which selected for different life histories in the disturbed and undisturbed meadows, the experiment was sufficiently large to suggest that the results describe underlying genetic differences between the populations found in these two environments.

In a second study with two species of *Lobelia*, one semelparous and one iteroparous, Young (1990) found that the different life histories of these species could be attributed to demographic variation between sites caused by different environmental conditions. In the drier sites, where there was a higher risk of mortality, selection favored an annual, semelparous life history. Under these extreme environmental conditions, individuals flowered so infrequently, and suffered such high mortality between reproductive episodes, that the probability of longer life span and future reproduction was outweighed by the greater fecundity associated with semelparity. Both of these comparative studies show that, as predicted by the evolutionary theories, there can be differential evolution of mortality patterns, either within species or among closely related species, in response to different levels of exogenous mortality. In other words, genetic differences between populations in patterns of senescence can evolve in response to different levels of age-specific selection caused by external environmental factors.

III. Demographic Approaches

A. Distinguishing Age and Environment

In order to demonstrate a change in mortality patterns with age, studies on the evolution of senescence require, first and foremost, an accurate accounting of the age-specific mortality of individuals. This requires that marked individuals be followed for their entire life span. In one of the few studies that has documented the age-specific mortality from emergence to death for a perennial plant in its natural environment, Canfield (1957) found evidence for increased mortality with increasing age in several tussock range grasses (Fig. 23-3). This study provides excellent data on a long-lived iteroparous species, and it was the first study to document a senescence-like increase in mortality with age (Harper, 1977; Watkinson, 1992; Roach, 1993). It should be noted however, that one of the major difficulties associated with studying mortality patterns in natural populations is that there may be large season- or environment-dependent increases in mortality, and it may be difficult to distinguish endogenous causes of mortality due to senescence from these exogenous changes over time. In Canfield's study, the increase in mortality over time may have been due to the fact that the external environmental conditions became less favorable for survival. It is not

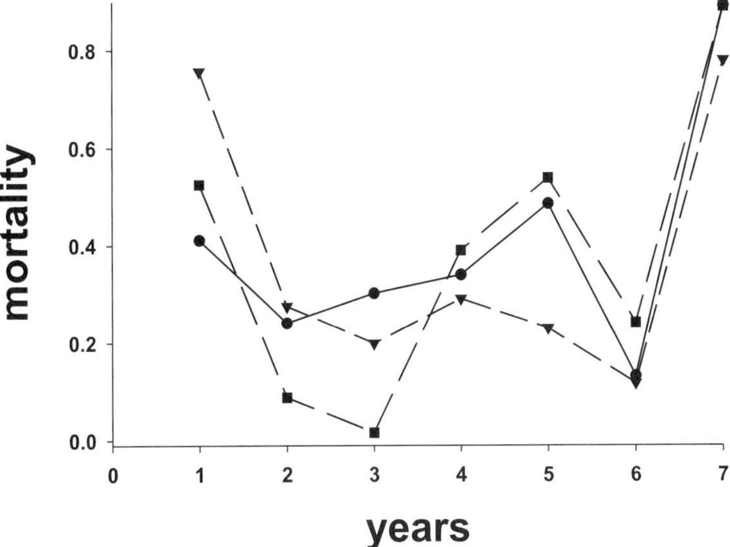

Figure 23-3. Mortality curves for three perennial range grasses: *Bouteloua hirsuta* (circles), *B. chondrosiodes* (squares), and *B. eriopoda* (triangles). Data derived from Canfield (1957), ungrazed treatments.

clear whether the increase in mortality was due to the increasing age of individuals in the population, or to a deterioration of the extrinsic environment. In order to definitively show that senescence is occurring in natural plant populations, experiments need to be designed with the proper controls to distinguish these different causes of mortality.

In order to differentiate between age- and environment-dependent changes in mortality, an additional cohort, of a different age, but experiencing the same environment, is needed. In the first study of this kind, an experiment was done with *Ipomea purpurea* (common morning glory) an indeterminate annual, which continues to grow, flower, and set seed, until an exogenous climatic event, i.e., frost, causes the death of individuals at the end of the growing season. This study was designed to determine: (1) whether there was a change in mortality patterns within the growing season prior to the abrupt change in mortality due to environmental deterioration; (2) whether there was a change in reproduction within the growing season; and (3) whether changes in mortality and reproduction were due to increasing age of the individuals in the population or to changes in the environment. In order to distinguish age- and environment-dependent changes, two plantings were made six weeks apart. In this species, nearly all seeds that are planted at the same time germinate synchronously; thus the two plantings created two cohorts that differed in age by six weeks. Given that the plants from the two cohorts were grown together in the same field, any differences in cohort performance could be attributed to age-specific variation, rather than to changes in the environment over time. Plants were censused regularly for mortality, and mature seeds were collected every three days. All plants that were still alive in mid-November were killed by a hard frost.

The comparative analysis of the mortality patterns of the two cohorts demonstrates a senescent decline in the oldest cohort during the growing season (Fig. 23-4). As seedlings, each cohort showed an increase in mortality after which mortality decreased. As the growing season progressed, the older cohort showed an increase in mortality, much higher than the

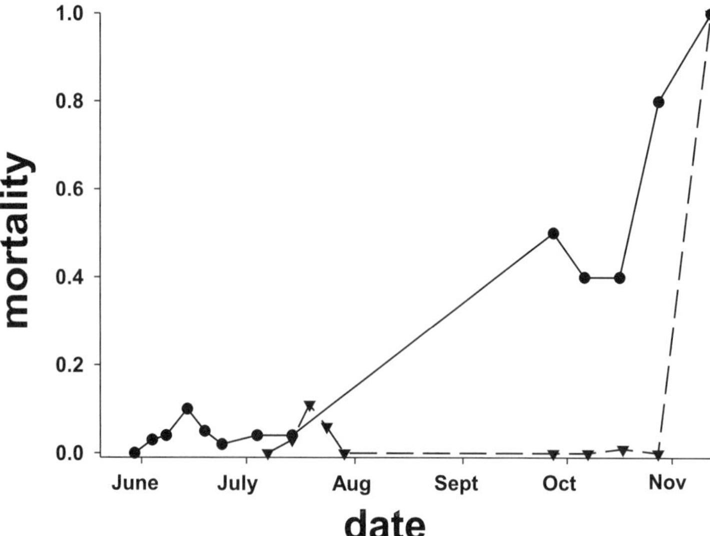

Figure 23-4. Mortality curves for two different-aged cohorts of *Ipomea purpurea*. Circles represent cohort 1; triangles (with dashed line) represent the later planted, younger, cohort 2. All remaining plants died in mid-November following a hard frost.

younger cohort, suggesting that it was age rather than changes in environmental conditions that were causing these mortality patterns. The younger cohort may have shown a similar increase at a later time if the populations had not been killed by the frost at the end of the growing season.

Senescence may also be demonstrated through a decrease in the number or quality of offspring produced with age. In plants, seed weight is a major determinant of offspring quality (cf. Wulff, 1986). Changes in seed mass over age and time were compared for the two morning glory cohorts. At the end of the growing season, when both cohorts were producing seeds, there was a significant decline in mean seed mass for both cohorts (Fig. 23-5). This decline in offspring quality was due to a decline in the quality of the environment as the growing season progressed and concurrently as the plants aged. However, when a comparison is made across the two cohorts, there was a greater decline in seed mass in the older cohort. This difference is manifest as a significant difference between the two cohorts in the slope of the regression of individual seed mass on date (cohort 1: slope $= -0.11$; cohort 2: slope $= -0.03$). During this same period, there was no evidence for a trade-off between seed mass and seed number. The younger cohort showed an increase in the total mass of all seeds produced per day (slope $= +0.83$, $p < 0.0001$), and a positive correlation between seed mass and seed number ($r = 0.11$, $p < 0.02$). The older cohort showed no significant change in total seed mass per day during this same period (slope $= -0.10$, $p > 0.10$) and no significant correlation between these two traits ($r = 0.02$, $p > 0.5$). This analysis shows clear evidence for a senescent decline in reproduction. The results of this study with an annual morning glory show that even within short-lived semelparous species, there may be age-dependent variation in mortality and reproduction. Yet, it is only with the proper experimental controls, in this case a set of different-aged cohorts, that we can distinguish age-dependent and environment-dependent patterns.

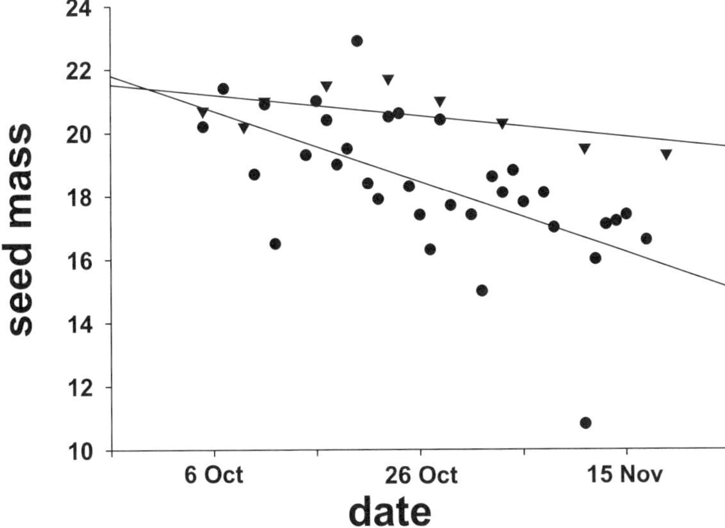

Figure 23-5. Changes in mean seed mass over time for two different-aged cohorts of *Ipomea purpurea*. Cohort 1 (circles) regression: $y = -0.11x + 51.9$ ($r^2 = 0.37$). Cohort 2 (triangles) regression: $y = -0.03x + 30.4$ ($r^2 = 0.38$).

B. Unique Features of Plants

1. Growth forms and theoretical expectations

A unique feature, distinguishing plants and animals is the indeterminate growth that takes place in the meristems of plants. Meristem organization may be an important factor determining variation in life span and mortality patterns of different species. Variation within the plant kingdom ranges from unitary organisms to clonal species. A plant species with a unitary organization and determinate growth is expected to follow demographic patterns similar to model animal systems such as *Drosophila*. On the other hand, plants with unitary organization and indeterminate growth, or clonal species with more modular construction, will have multiple meristems, and this can have important implications for population dynamics. Production of new meristems may allow a clonal species to escape determinacy and may facilitate an escape from whole organism senescence.

Historically, the demography of plants was considered to be relatively intractable because of this variation in growth forms, specifically because an "individual" is sometimes difficult to delineate. For unitary species, identification of an individual is not difficult and age- or size-specific mortality and reproduction are relatively easy to follow. For species with clonal growth, Harper (1977) introduced the term "genet" to refer to all individuals that were derived from the same zygote, and "ramet" as the modules of a genet, which may in some cases become severed from the parent plant and grow as independent individuals. As long as the birth rate of ramets exceeds the death rate, a genet will survive in the population. Ramet demography may, in some cases, act as a buffer for the dynamics of a genet. In a computer simulation based on several years of data from populations of *Ranunculus repens*, it was found that the decay rate of the genet can be buffered by the decay rate of the ramets, which has the effect of conserving genetic variability within the population (Soane and Watkinson, 1979).

To date, there have been no comparative studies of the demography of ramets and genets, thus it is not known whether the demography of a population of genetically identical ramets reflects the vigor of the genet. With respect to studies on evolutionary processes such as senescence, it is critical that demographic studies be made at the level of the genet (Silander, 1985). There is limited evidence showing variation in age-specific demography of genets. In a study of the pasture grasses *Agrostis stolonifera* and *Lolium perenne*, Bullock *et al.* (1996) showed tiller production for different-aged genets changed in response to grazing. The existence of extremely long-lived clones is often cited as evidence that some clonal species lack whole organism senescence (cf. Watkinson and White, 1985; Noodén and Guiamét, 1996; Gardner and Mangel, 1997). Unfortunately, given their extreme longevity, it may be impossible to get a true understanding of late-age dynamics. There is clearly a need for more comparative studies of the demography of genets and ramets.

Variation in growth forms within the plant kingdom may necessitate a reevaluation of some of the evolutionary theories of senescence. In particular, the assumption of the theories that a species shows age-specific gene expression may be violated. A genet which grows to consist of a collection of ramets of different ages, cannot be expected to show age-specific gene action, particularly if the ramets are detached from the parent plant and grow independently. There is evidence, however, from bamboo that independent ramets show synchronized genetic control of flowering time, but the nature of this internal genetic clock and its interaction with the environment is not known (John and Nadguada, 1999). Moreover, even without independent growth, the turnover of tissues within an individual may allow an escape from senescence. For example, in the forest herb *Arisaema*, old tissue is sloughed off from the bottom of corms as new tissue is added to the top so that no part of any plant, even one 20 years old, is ever really over 4 years old. If there is complete turnover of tissue within an individual, in other words if no tissue ever achieves a chronologically old age, then there may be no reason to expect evolutionary senescence or any increasing mortality with age (Bierzychudek, 1982).

A second assumption of the evolutionary theories of senescence, which requires further evaluation with respect to clonal species, is the assumption that the force of selection declines with increasing age. This issue has been addressed in several recent theoretical papers. An analysis by Gardner and Mangel (1997) suggests that it is the rate of sexual reproduction relative to vegetative growth that is the important determinate of the strength of selection. Higher rates of sexual reproduction and lower rates of clonal reproduction result in a more rapid drop in the strength of selection with clonal age following an initial peak [see also, Orive (1995) and Pedersen (1995)]. Experimental results by Bell (1984) with six asexual freshwater invertebrates support this generalization, but there have been no comparative studies with plants of the demography of closely related species with different rates of sexual reproduction and clonal growth.

2. Somatic mutations and age

Another important distinction between plant and animal development is the timing of the differentiation between germ and soma plasma. In mammals, insects, and many other animals, the germ cells are segregated early in development, whereas in plants, the germ line does not segregate from the soma until just before reproduction. It was theoretically suggested that senescence should be found to be inevitable for any organism for which there is an early segregation between germ and soma (Williams, 1957). This has led several researchers to predict that patterns of evolutionary senescence may not be the same in animals and plants (cf. Kirkwood and Holliday, 1979; Rose, 1991). Yet, the reason why this segregation is necessary for the evolution of senescence has never been explicitly stated (Roach, 1993). Two scenarios are presented below, in which the presence of multiple meristems and the late separation of germ and soma may influence patterns of plant senescence.

Individual longevity may be affected if somatic mutations result in intraorganismal selection. In long-lived plants with multiple meristems, a genetic mosaic may develop within individual plants, and this may have one of two consequences. On the one hand, intraorganismal selection has primarily been seen as a mechanism for eliminating deleterious somatic mutations (cf. Klekowski and Kazarinova-Fukshansky, 1984; Otto and Orive, 1995). Natural selection can act among cells within an individual to eliminate deleterious mutations that appear in a ramet or module, with little or no impact on the fitness of the genet. Conversely, selection may allow the spread of a favorable mutation within a clonal plant, which may result in an evolutionary change if an advantageous mutation goes to fixation (Pineda-Krch and Fagerström, 1999). If somatic mutations occur at a level high enough to allow selective changes within the life span of an individual, then this could allow for a closer tracking of an individual with its environment thus potentially resulting in increased longevity. A recent stochastic model demonstrates that there is a high probability of an advantageous somatic mutation going to fixation through a mitotic cell lineage in the presence of intraorganismal selection (Pineda-Krch and Fagerström, 1999). Genetically different individuals may thus originate through a succession of chimeric ramet generations without sexual reproduction (Fagerström *et al.*, 1998). The genetic mosaic within an individual may be evolutionarily important, particularly as a defense against rapidly evolving pathogens, parasites, and herbivores (Whitham and Slobodchikoff, 1981). Recent work on genetic polymorphism within the chloroplast genome in *Senecio vulgaris* has shown that there can be within-plant selection between different DNA chloroplast types (Frey, 1999). Additionally, work with *Chenopodium album* (Darmency, 1994) suggests that within-individual selection may be an important mechanism for the development of within-generation resistance to herbicides. With respect to the evolution of senescence, the important factor is whether or not there are some species which can escape senescence because intraorganismal selection allows individual genotypes to increase their longevity by becoming more "fine-tuned" to their environment over time. The evolutionary importance of intraorganismal selection in plants cannot be assessed until further studies are conducted (Pineda-Krch and Fagerström, 1999; Sutherland and Watkinson, 1986; Otto and Orive, 1995). Moreover, the importance of this to senescence across the plant kingdom is not known, but it will only potentially be relevant to extremely long-lived species with multiple meristems.

The lack of separation between germ line and soma may also be important in the evaluation of the age-specific quality of offspring, particularly if somatic mutations become incorporated into the germ line. If there is a decline in the quality of offspring produced at later ages, then the intensity of selection on late-life traits will decline and will result in the evolution of more rapid senescence. The chance of incorporating a somatic mutation into a gamete depends on how many cell generations there are before gametes are formed and how many cell lines produce germ cells (Jerling, 1985). It is clear that individuals in some species can live to extremely old ages, for example, clones of creosote bush are estimated to be 9,000–10,000 years old (Vasek, 1980). Furthermore, some species cover extensive areas, and thus can potentially have large numbers of meristems. Over time, and with a large number of meristems, there is thus the theoretical potential for mutations to become incorporated into the germ line. In a study of the effects of tree age on pollen, seed, and seedling characteristics in Great Basin bristlecone pine trees (*Pinus longaeva*), Connnor and

Lanner (1991) evaluated trees ranging in age from 23 to 4,257 years at both high and low elevation sites. They had hypothesized that there should be an increase in somatic mutations with age, particularly at high elevation sites since radiation can severely damage pollen. Their results showed no relationship between tree age and any reproductive variables and no differences between the two sites. To date, evidence suggests that vertical inheritance of somatic mutations is rare (reviewed in Schmid, 1990). Thus, with the possible exception of variation arising from intraorganismal selection, the late separation of germ line and soma does not appear to be an important factor in the evolution of senescence.

IV. Conclusions

Studies of evolutionary plant senescence use both demographic and evolutionary approaches to understand *why* the deleterious phenomenon of senescence persists in populations. Evaluating plant senescence within a demographic and evolutionary context is a relatively new approach (Watkinson, 1992; Roach, 1993; Pedersen, 1999), and there are many studies that will need to be done before we will have a clear understanding of the processes that have shaped life histories at the latest ages. Some guidelines and issues that should be considered in future studies are discussed below.

There is currently a paucity of high quality data on age-specific demography from plant populations. Despite the fact that the demography of plants is comparatively easy, few studies have followed a population of individuals for their entire life cycle, particularly for species which live more than a few years (reviewed by Watkinson, 1992; Roach, 1993). Geographically and taxonomically, studies that have been done, to date, have barely begun to sample the diversity of the plant kingdom (Franco and Silvertown, 1990). The studies which have been done have used small sample sizes, and there have been very few attempts to estimate the variance in age-specific life history traits.

One of the limitations of gerontological studies, in all species, has been sample size (Finch, 1990). Survivorship curves based on small sample sizes can provide estimates of life expectancy at birth, but it is not possible to estimate mortality rates late in life because so few individuals remain alive at older ages. Moreover, in field experiments there is a high rate of random mortality affecting survival at all ages. With a large sample size, it will be possible to attain a more accurate understanding of the inherent changes in mortality at different ages.

It is important to emphasize that data on age-specific mortality, not just life span data, are needed to understand the evolution of senescence. Plant species show a wide range of variation in their life spans, ranging from a few weeks, for some ephemeral annuals, to over 1,000 years for many conifer species. The presence of senescence cannot be inferred from life span measures for several reasons. First, evidence for senescence is derived from a change in the shape of the mortality curve for the population. Data on the maximum longevity of a species give the endpoints for the curve, but tell us nothing about a change in the shape of the curve with age (Bell, 1992). Secondly, two species can show different longevities due to differences in their annual mortality rates irrespective of any differences in senescence (Partridge and Barton, 1996). Thus, more complete demographic data on age-specific mortality are critical.

As it has been defined here, senescence refers to an increase in adult mortality with age. Theoretical studies have shown that factors other than senescence may cause mortality to increase with age (Abrams, 1993; Blarer *et al.*, 1995, McNamara and Houston, 1996).

Specifically, the theoretical expectation is that an optimized life history may show an increase in mortality due to an increased reproductive effort late in life. For plant species with determinate growth, in other words a fixed size at sexual maturity, reproductive output is expected to decline with age, but for some plant species, which increase in size with age, an increase in late reproduction may be observed. The change in age-specific reproductive effort, and its consequences for mortality rates has not been experimentally evaluated. Furthermore, selection acts jointly on survival and reproduction. Consequently, future studies with complete data on age-specific mortality and age-specific fecundity may demonstrate that some combination of mortality and fecundity will be a better measure, to detect the presence of aging in a population and to compare rates of aging, than mortality measures alone (Partridge and Barton, 1993, 1996). Plants are good experimental organisms to test these ideas, because age-specific reproduction is relatively easy to quantify.

The comparative biology of different species is one of the most useful approaches in evolutionary biology. There are a wide variety of growth forms, life spans, and mortality patterns within the plant kingdom and species have traditionally been classified either by their longevity, annual, biennial, perennial, or by their number of reproductive episodes, monocarpic, polycarpic. With more complete data on a wider range of species, perhaps we can look forward to the time when a thorough phylogenetic analysis of the evolution of whole plant senescence can be considered.

Finally, it is hoped that future research will begin to make a link between physiological and evolutionary approaches to plant senescence. Perhaps then, we will be able to make a bridge between our understanding of the physiological processes which occur within an individual, how the processes of physiological senescence change with age, and how they affect life history traits and the evolution of senescence.

References

Abrams, P.A. (1993). Does increased mortality favor the evolution of more rapid senescence? *Evolution* **47**, 877–887.

Bell, G. (1984). Evolutionary and nonevolutionary theories of senescence. *American Naturalist* **124**, 600–603.

Bell, G. (1992). Mid-life crisis. *Evolution* **46**, 854–856.

Bever, J.D. (1994). Feedback between plants and their soil communities in an old field community. *Ecology* **75**, 1965–1977.

Bierzychudek, P. (1982). Life histories and demography of shade-tolerant temperate forest herbs: A review. *New Phytologist* **90**, 757–776.

Blarer, A., Doebeli, M., and Stearns, S.C. (1995). Diagnosing senescence: Inferring evolutionary causes from phenotypic patterns can be misleading. *Proceedings of the Royal Society of London, Series B* **262**, 305–312.

Bleecker, A.B. (1998). The evolutionary basis of leaf senescence: method to the madness? *Current Opinion in Plant Biology* **1**, 73–78.

Bullock, J., Silvertown, J., and Clear Hill, B. (1996). Plant demographic responses to environmental variation: Distinguishing between effects on age structure and effects on age-specific vital rates. *Journal of Ecology* **84**, 733–743.

Canfield, R.H. (1957). Reproduction and life span of some perennial grasses of southern Arizona. *Journal of Range Management* **10**, 199–203.

Carey, J.R., Liedo, P., Orozco, D., and Vaupel, J.W. (1992). Slowing mortality rates at older ages in large Medfly cohorts. *Science* **258**, 457–460.

Charlesworth, B. (1980). *Evolution in Age-Structured Populations.* Cambridge University Press, Cambridge, UK.

Charlesworth, B. (1990). Optimization models, quantitative genetics, and mutation. *Evolution* **44**, 520–538.

Connor, K.F., and Lanner, R.M. (1991). Effects of tree age on pollen, seed, and seedling characteristics in Great Basin Bristlecone Pine. *Botanical Gazette* **152**, 107–113.

Darmency, H. (1994). Genetics of herbicide resistance in weeds and crops. In *Herbicide Resistance in Plants* (S.B. Powels and J.A.M. Holtum, Eds.), pp. 263–297. Lewis, London.

Fagerström, T., Briscoe, D., and Sunnucks, P. (1998). Evolution of mitotic cell-lineages in multicellular organisms. *Trends in Ecology and Evolution* **13**, 117–120.

Finch, C.E. (1990). *Longevity, Senescence, and the Genome.* University of Chicago Press, Chicago.

Franco, M., and Silvertown, J. (1990). Plant demography: What do we know? *Evolutionary Trends in Plants* **4**, 74–76.

Frey, J.E. (1999). Genetic flexibility of plant chloroplasts. *Nature* **398**, 115–116.

Gardner, S., and Mangel, M. (1997). When can a clonal organism escape senescence? *American Naturalist* **150**, 462–490.

Harper, J.L. (1977). *Population Biology of Plants.* Academic Press, New York.

Jerling, L. (1985). Are plants and animals alike? A note on evolutionary plant population ecology. *Oikos* **45**, 150–153.

John, C.K., and Nadgauda, R.S. (1999). In vitro-induced flowering in bamboos. *In Vitro Cellular and Developmental Biology-Plant* **35**, 309–315.

Kirkwood, T.B.L., and Holliday, F.R.S. (1979). The evolution of ageing and longevity. *Proceedings of the Royal Society of London, Series B* **205**, 531–546.

Klekowski, E.J., Jr., and Kazarinova-Fukshansky, N. (1984). Shoot-apical meristems and mutation: selective loss of disadvantageous cell genotypes. *American Journal of Botany* **71**, 28–34.

Law, R. (1979). The cost of reproduction in annual meadow grass. *American Naturalist* **113**, 3–16.

Law, R., Bradshaw, A.D., and Putwain, P.D. (1977). Life history variation in *Poa annua. Evolution* **31**, 233–246.

McNamara, J.M., and Houston, A.I. (1996). State-dependent life histories. *Nature* **380**, 215–221.

Medawar, P.B. (1952). *An Unsolved Problem of Biology.* H.K. Lewis, London.

Noodén, L.D. (1988). The phenomena of senescence and aging. In *Senescence and Aging in Plants* (L.D. Noodén and A.C. Leopold, Eds.), pp. 2–50. Academic Press, San Diego.

Noodén, L.D., and Guiamét, J.J. (1996). Genetic control of senescence and aging in plants. In *Handbook of the Biology of Aging* (E.L. Schneider and J.W. Rowe, Eds.), pp. 94–118. Academic Press, San Diego.

Orive, M.E. (1995). Senescence in organisms with clonal reproduction and complex life histories. *American Naturalist* **145**, 90–108.

Otto, S.P., and Orive, M.E. (1995). Evolutionary consequences of mutation and selection within an individual. *Genetics* **141**, 1173–1187.

Partridge, L., and Barton, N.H. (1993). Optimality, mutation and the evolution of ageing. *Nature* **362**, 305–311.

Partridge, L., and Barton, N.H. (1996). On measuring the rate of ageing. *Proceedings of the Royal Society of London, Series B* **263**, 1365–1371.

Pedersen, B. (1995). An evolutionary theory of clonal senescence. *Theoretical Population Biology* **47**, 292–320.

Pedersen, B. (1999). Senescence in plants. In *Life History Evolution in Plants* (T.O. Vuorisalo and P.K. Mutikainen, Eds.), pp. 239–274. Kluwer, Boston.

Pineda-Krch, M., and Fagerström, T. (1999). On the potential for evolutionary change in meristematic cell lineages through intraorganismal selection. *Journal of Evolutionary Biology* **12**, 681–688.

Radford, A.E., Ahles, H.E., and Bell, C.R. (1968). *Manual of the Vascular Flora of the Carolinas.* University of North Carolina Press, Chapel Hill.

Roach, D.A. (1986). Life history variation in *Geranium carolianum*. I. Covariation between characters at different stages of the life cycle. *American Naturalist* **128**, 47–57.

Roach, D.A. (1993). Evolutionary senescence in plants. *Genetica* **91**, 53–64.

Rose, M.R. (1984). Laboratory evolution of postponed senescence in *Drosophilia melanogaster. Evolution* **38**, 1004–1010.

Rose, M.R. (1991). *Evolutionary Biology of Aging.* Oxford University Press, New York.

Schmid, B. (1990). Some ecological and evolutionary consequences of modular organization and clonal growth in plants. *Evolutionary Trends in Plants* **4**, 25–34.

Silander, J.A. (1985). Microevolution in clonal plants. In *Population Biology and Evolution of Clonal Organisms* (J.B. Jackson, L.W. Buss, and R.E. Cook, Eds.), pp. 107–152. Yale University Press, New Haven.

Soane, I.D., and Watkinson, A.R. (1979). Clonal variation in populations of *Ranunculus repens. New Phytologist* **82**, 557–573.

Sutherland, W.J., and Watkinson, A.R. (1986). Do plants evolve differently? *Nature* **320**, 305.

Vasek, F.C. (1980). Creosote Bush: Long-lived clones in the Mojave Desert. *American Journal of Botany* **67**, 246–255.

Watkinson, A. (1992). Plant senescence. *Trends in Ecology and Evolution* **7**, 417–420.

Watkinson, A.R., and White, J. (1985). Some life-history consequences of modular construction in plants. *Philosophical Transactions of the Royal Society of London, Series B* **313**, 31–51.

Weismann, A. (1889). The duration of life (a paper presented in 1881). In *Essays upon Heredity and Kindred Biological Problems* (E.B. Poulton, S. Schonland, and A.E. Shipley, Eds.), pp. 1–66. Clarendon Press, Oxford. Reprint: Dabor Science, Oceanside, NY.

Whitham, T.G., and Slobodchikoff, C.N. (1981). Evolution by individuals, plant-herbivore interactions, and mosaics of genetic variability: the adaptive significance of somatic mutations in plants. *Oecologia* **49**, 287–292.

Williams, G.C. (1957). Pleiotropy, natural selection, and the evolution of senescence. *Evolution* **11**, 398–411.

Wilson, J.B. (1997). An evolutionary perspective on the 'death hormone' hypothesis in plants. *Physiologia Plantarum* **99**, 5111–516.

Wulff, R.D. (1986). Seed size variation in *Desmodium paniculatum*. III. Effects on reproductive yield and competitive ability. *Journal of Ecology* **74**, 115–121.

Young, T.P. (1990). Evolution of semelparity in Mount Kenya lobelias. *Evolutionary Ecology* **4**, 157–171.

24

Flower Longevity

Tia-Lynn Ashman

I. Introduction

Floral longevity (the length of time from anthesis to flower death) plays an important role in the reproductive ecology of plants. The length of time a flower is open can directly influence the total number of pollinator visits, which, in turn, can affect the amount and diversity of pollen the flower receives, and the amount of pollen it disseminates. Additionally, floral longevity contributes to determining the number of flowers open at any given time (floral display size), the duration of floral display, and the total number of flowers per plant (Primack, 1985). Thus, floral longevity can indirectly influence the attractiveness of the entire plant to pollinators, as well as to flower/seed predators or parasites. Moreover, acting through floral display size, floral longevity can influence the potential for geitonogamous self-pollination (Primack, 1985). Ultimately floral longevity influences many factors which determine the quantity and quality of progeny a plant produces. Over the period of time that a flower functions and contributes to plant fitness, it receives resources to remain alive and attractive to pollinators. Such floral maintenance expenditures may compete with future flower production or other plant functions if plant resources are limited (Ashman and Schoen, 1996b, 1997). Thus, the optimal lifespan of a flower reflects the balance between fitness benefits and maintenance (Ashman and Schoen, 1994, 1995) and/or ecological costs (e.g., risk of geitonogamy or predation) of continued floral longevity. In this chapter, floral longevity is viewed as a component of a plant's life history strategy, and thus, a character shaped by natural selection to fit a plant's ecological context.

II. Patterns of Flower Longevity in Nature

A. Variation

Kerner von Marilaun (1895) first noted the vast range of flower longevities that exist in Angiosperms: flower lifetimes can vary from the very short span of a few hours or single day [e.g., morning glories (*Ipomoea* spp.), evening primroses (*Oenothera* spp.)], to many days [e.g., columbines (*Aquilegia* spp.), buttercups (*Rannunculus* spp.)], to as long as a few months [e.g., orchids (*Cypripedium* spp., *Odontoglossum* spp.)]. This dramatic diversity in floral lifetimes (shown in Fig. 24-1) suggests that floral longevity is a character that reflects adaptation to a variety of ecological conditions.

B. Ecological and Phylogenetic Correlates of Flower Longevity

In an effort to understand the mechanisms underlying this vast variation in floral longevity several researchers have used a comparative approach to characterize ecological and phylogenetic associations with floral longevity. These studies identified several factors that may influence floral longevity, most notable among them being habitat/climate, pollinator type and availability, breeding system and taxonomic association. Kerner von Marilaun (1895) was the first to suggest that variation in floral longevity was related to level of pollinator visitation, among other things. He suggested that habitats with unpredictable weather would select for plants with few, long-lived flowers, whereas habitats with weather that was predictably favorable for pollination would select for plants with many, short-lived flowers. Continuing Kerner von Marilaun's line of reasoning, recent workers (Primack, 1985; Stratton, 1989; Seres and Ramirez, 1995; Kunin and Shmida, 1997) have predicted that habitats or climates unfavorable or unpredictable for pollinator visitation would select for longer-lived flowers, while habitats where metabolic costs of floral longevity (e.g., transpiration) are likely to be high (e.g., dry climates) would select for short-lived flowers. Primack's (1985) survey revealed that flowers living only a single day predominate in tropical forests, whereas flowers of temperate forests generally last from three to nine days. Seres and Ramirez (1995) also found that species in the dry forests of Venezuela had floral longevities of about one day, while those of cloud forests had longer lifetimes.

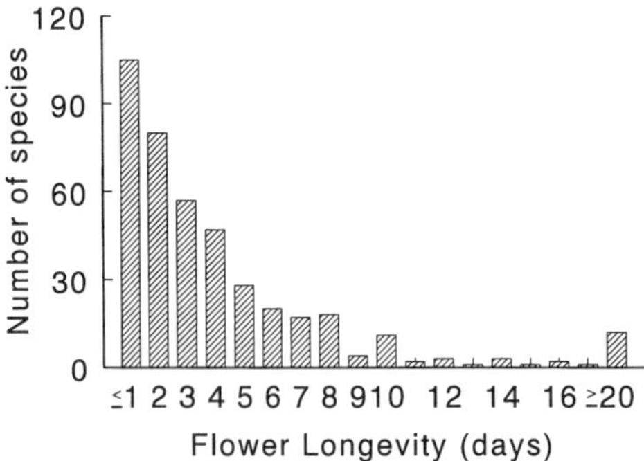

Figure 24-1. Frequency distribution of flower longevities in 419 species. When longevities were available for more than one species within a genus, the mean for the genus is represented. Data from Stratton, 1989; Seres and Ramirez, 1995; Ashman and Schoen, 1996a; Kunin and Shmida, 1998.

Stratton (1989), however, did not find a significant difference between flower longevities of cloud and mesic forests of Costa Rica. Primack (1985) also uncovered associations of floral longevity with elevation and time of flowering season: flowers at higher elevations lived longer than those at lower elevations, and those blooming in the spring lived longer than those blooming in the summer. While these predictions regarding climate and pollinator availability are generally upheld, it is impossible to discern the underlying causes of these patterns. That is, do flowers live longer in temperate forests, in the spring, or at high elevations because these habitats/climates support less frequent pollinator service, or because the costs of floral maintenance are lower, or both? Detailed knowledge of pollinator activity levels, and schedules of actual pollen deposition and removal, as well as metabolic costs of flower maintenance for plants in different habitats are required to evaluate the causal mechanisms underlying these associations of floral longevity with habitat/climate.

Pollinator guild is also hypothesized to influence flower longevity, because pollinators can differ in their tendency to forage under inclement weather, and/or differ in their pollen transferring efficiencies. However, there is little concordance among comparative studies in the effects of pollinator class on floral longevity. Across habitat types, Primack (1985) did not find consistent differences in flower longevity among pollinator classes, although there were some within-habitat trends. Likewise, Stratton (1989) found flowers pollinated by hummingbirds lived longer than bee- or lepidopteran-pollinated species, but the significance of this result depended on habitat type (i.e., a significant effect of pollinator type was found in cloud forests, but not in moist forests). On the other hand, Seres and Ramirez (1995) found the opposite trend (bee-pollinated species tended to live longer than bird-pollinated species) in Venezuelan forests. Taken together, these data suggest that an interaction between pollinator class and habitat type exists, perhaps owing to different cost/benefit ratios (cost of maintaining flowers versus pollinator availability/efficiency) produced in each environment.

A more direct test of the association between pollinators, their foraging rates or abundances, and floral longevity is provided by a correlation between actual pollinator visitation rates and floral longevity. Using data on 39 species gleaned from the literature, Ashman and Schoen (1996b) found a strong negative correlation between flower longevity and pollinator visitation rate ($r = -0.57$; $P < 0.0001$), that is, species visited frequently had shorter flower longevities than those visited infrequently. Moreover, because pollinator abundance may depend on the size of the plant population, Kunin and Shmida (1997) argued that rare plant species may be faced with unusually high selective pressures to attract pollinators and thus, should have longer-lived flowers than common species. Testing this hypothesis with data on 52 species of annual crucifers native to Mediterranean habitats of Israel, Kunin and Shmida (1997) found not only that rare species had longer floral lifetimes than common ones, but also that the difference between them was especially pronounced in self-incompatible species, that is, those that rely entirely on pollinator visits to fulfil reproductive function.

Breeding system, in general, is expected to impact flower longevity. Flowers of self-incompatible or out-crossing species are likely to have been selected to remain open longer than self-pollinating species in order to achieve full reproductive function. Primack (1985) found striking associations between flower lifetime and breeding system within several taxa. Likewise, out-crossing crucifer species had flower longevities 25% longer than inbreeding species [analysis of data in Kunin and Shmida's (1997) appendix; $t = 2.69$; $P = 0.01$; $n = 52$]. Moreover, dramatic differences in floral longevity have been seen in plants where male and female reproductive functions are separated into different flowers (e.g., monoecy and dioecy). Female flowers live longer than male flowers in just about every species surveyed (Primack, 1985; Seres and Ramirez, 1995). It may be that sexual dimorphism in flower longevity reflects selection for different optimal lifetimes in different-sexed flowers. Flowers with male function often experience higher pollinator visitation rates than female flowers (e.g., Ashman and Stanton, 1991; Delph and Lively, 1992) suggesting that male fitness may accrue faster, and thus, be satisfied more easily than female fitness. So, while strong associations between floral longevity and pollinator classes may be habitat dependent, the associations of floral longevity with pollinator availability and breeding system appear strong and ubiquitous.

Given that floral longevity is a trait that is expected to evolve, closely related species may have more similar floral longevities than distantly related species. In the absence of phylogenetic data, taxonomic affiliation has been used to test this hypothesis. Both Primack (1985) and Stratton (1989) discovered a strong association between floral longevity and family membership. Stratton (1989) concluded that taxonomic affiliation was the most important factor influencing floral longevity in his data set. When he controlled for taxonomy, all of the significant ecological patterns in his data were reduced to non-significant trends. In contrast, Primack (1985) concluded that flower longevity was more related to overall floral ecology than to phylogeny in his data set that spanned several plant communities. It seems likely that on some level phylogeny may constrain the evolution of floral longevity; however, larger data sets and data on phylogenetic relationships will be needed to fully resolve the issue.

III. Evolutionary Perspective

The comparative studies reviewed above have identified several ecological, genetic and physiological factors that contribute to the variation in flower longevities. In an effort to evaluate how these key features interact, Ashman and Schoen (1994, 1995, 1996a, 1997) have incorporated several of them into an evolutionary model of floral longevity. This model provides a framework for quantitatively evaluating variation in floral longevity. In particular, it has been used to make specific evolutionary predictions regarding the optimal lifetime of a flower, given a specified set of ecological conditions.

A. Model of Optimal Flower Longevity

By viewing floral longevity as an allocation strategy subject to natural selection, Ashman and Schoen (1994, 1995, 1996a, 1997) have extended the work of Primack (1985) (and Lloyd) to incorporate the evolution of floral longevity into the well developed theoretical framework provided by evolutionary stable strategy models that address variation in life history (e.g., Charnov, 1982; Stearns, 1992). In its basic form, this model focuses on two fundamental factors that determine the fitness consequences of a flower longevity strategy. The first factor involves the resource constraints that are imposed on plant activities during flowering, i.e., that resources available during flowering are allocated among the competing functions (flower construction, maintenance, and production). The second factor relates to how quickly the flower's reproductive function is fulfilled. As pollen is removed and disseminated over time, the flower's potential contribution to reproductive fitness through male function increases, and likewise, as the amount of pollen deposited on the stigma increases and ovules become fertilized, the plant gains increased fitness through female function. Taken together, these two factors determine the fitness consequences of a flower longevity strategy and allow one to evaluate floral longevities in a resource allocation framework.

1. Assumptions

Ashman and Schoen's (1994, 1995, 1996b) floral longevity model rests on three main assumptions. First, it assumes that variation in floral longevity is heritable and can be optimized by natural selection. The close correlation between floral longevity in the absence of pollinators ("maximum floral longevity") and in the presence of pollinators ("realized floral longevity") among species (MFL = 0.88RFL + 1.72; $P < 0.0001$; $n = 23$) strongly suggests that genetic variation in floral longevity exists among wild species (Ashman and Schoen, 1996b). However, no study has determined the degree of genetic variation in floral longevity within wild plant species. In contrast, substantial evidence for genetic variation in floral longevity within plant populations has come from studies of horticulturally important species. For example, Scott *et al.* (1994) found significant genetic variation in maximum longevity of intact flowers of two species of cacti. Several other studies of floricultural crops report significant heritabilities for flower longevity in cut flowers (Van Eijk and Eikelboom, 1976; Harding *et al.*, 1981; van der Meulen-Muisers *et al.*, 1998). The assumption of genetic determination of floral longevity does not ignore the influence of post-pollination physiological changes in flowers seen in some species (reviewed in Gori, 1983; Van Doorn, 1997). Indeed, floral senescence caused by pollination may, in some cases, interact with genetic variation in floral longevity and thereby introduce additional variation in floral longevity of a population. It is worth noting, however, that pollination does not induce floral senescence in many species (e.g., Gori, 1983; Motten, 1986; Edwards and Jordan, 1992; Aizen, 1993; Kunin and Shmida, 1997; Van Doorn, 1997; Bell and Cresswell, 1998). Thus, pollination or lack of pollination is not the only factor determining floral longevity. The role of pollination induced plasticity in floral longevity is discussed in greater detail below.

Second, the model assumes that there is a fixed pool of resources for flowering [or flowering and fruiting (Ashman and Schoen, 1997)] and, hence, a trade-off between the number of flowers (or fruits) made, floral construction, and floral maintenance. The idea is that floral maintenance (nectar production, floral respiration and transpiration) represents a cost that must be met by the plant and thus, is likely to compete for resources with other reproductive functions. Not only can nectar represent a considerable investment for animal-pollinated plants (Southwick, 1984; Zimmerman and Pyke, 1986; Pyke, 1991; but see Harder and Barrett, 1992), but daily respiration can involve a carbon investment of similar magnitude (Ashman and Schoen, 1996b). For example, carbon costs of daily floral maintenance can range from <1% to 25% of flower construction costs (Ashman and Schoen, 1996b). Furthermore, water transpired by flowers can be a major component of a plant's flowering water budget (Nobel, 1977; Galen *et al.*, 1999). Resource trade-offs among reproductive functions are common, such as trade-offs between flower and fruit production, or between flower number and their size (reviewed in Ashman and Schoen, 1996b). Only one study has addressed the involvement of flower maintenance in such trade-offs, but firmly established that flower maintenance can be involved in resource trade-offs. Specifically, Ashman and Schoen (1997) manipulated flower lifetime in *Clarkia tembloriensis* and revealed that an increase in longevity led to a decrease in fruit weight and seed number, but not in flower number.

Third, the model assumes that each flower has the same potential to contribute to the overall reproductive success of the plant, and that male (pollen export) and female (pollen receipt) components of fitness accrue at constant daily rates. In this way, a flower's remaining contribution to plant fitness diminishes with time, i.e., as the proportion of pollen that remains undispersed and the proportion of ovules that are unfertilized approach zero.

2. Model formulation

In Ashman and Schoen's model, optimal floral longevity is determined by the interaction of three major factors: (1) the daily cost of maintaining a flower (m) relative to the cost of constructing a new flower (c), such that the total maintenance cost per flower over its lifetime (t) is given by tmc; (2) the rate at which pollen is received to fertilize ovules, "female fitness accrual rate" [$g(t)$]; (3) the rate at which pollen is disseminated and enters the pool of pollen that competes to fertilize ovules, "male fitness accrual rate" [$p(t)$]. These fitness accrual functions take the form $1 - e^{-rt}$ where r is a rate constant for fitness accrual which may differ for female and male fitness. The number of flowers produced in a single flowering season [$F(t)$] having flowers that each remain open for t days is expressed as $R/(c + tmc)$. Optimal floral longevities are found by solving the function $w = \frac{1}{2}\{F(t)/F(t*)\}\{P(t)/P(t*) + G(t)/G(t*)\}$, where t denotes the floral longevity of a rare mutant and t^* denotes the evolutionarily stable strategy of floral longevity for the t that yields maximal fitness. Different combinations of the above parameters result in different optimal floral longevities which can be displayed in two dimensions as "isoclines", that is, combinations of daily pollen and seed fitness-accrual rates and floral maintenance cost for which values of $t^* = 1, 2, \ldots, T$ days yields maximum fitness (Fig. 24-2).

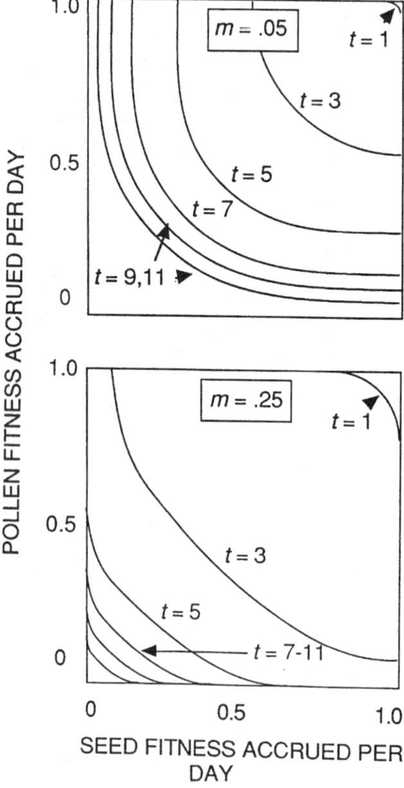

Figure 24-2. The effects of low ($m = 0.05$; top) and high ($m = 0.25$; bottom) maintenance costs and various rates (slow to high: 0 to 1) of male and female fitness accrual on optimal floral longevity (t). A given optimal floral longevity is represented by a line of equal fitness (isocline).

3. Predictions

It can be seen from Fig. 24-2 that long-lived flowers are favored when low daily floral maintenance cost and low daily rate of male and female fitness accrual are combined. When maintenance costs are relatively low (e.g., $m = 0.05$), however, small changes in the daily pollen and seed fitness-accrual rates have relatively large effects on the optimal floral longevity. Small decreases in the speed of fitness accrual select for longer-lived flowers. On the other hand, when maintenance costs are high (e.g., $m = 0.25$), most combinations of daily pollen and seed fitness-accrual rates select for reduced floral longevities, and only very slow fitness-accrual rates select for long-lived flowers.

B. Test of the Model

Two data sets have been used to test the basic model (Ashman and Schoen, 1994, 1996b) and have done so with much success. In an analysis of the combined data sets (Fig. 24-3), it can be seen that, on the whole, observed floral longevities are strongly correlated with those predicted by the optimal floral longevity model ($r = 0.67$; $P = 0.002$; $n = 19$). However, the fit between observed and predicted flower lifetimes for some individual species is poor. It is unknown whether these discrepancies reflect limitations of the model or in the collection of empirical data. Possible limitations are discussed with reference to the species with the most discordant predicted and optimal longevities (e.g., *Erythronium americanum*,

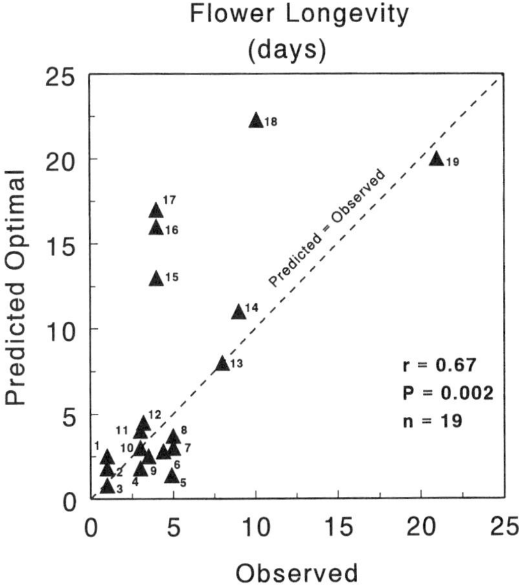

Figure 24-3. The correlation between observed and predicted floral longevities. Predicted floral longevity is based on the model of Ashman and Schoen (1994). Data are from Ashman and Schoen (1994) and (1996a). Data points refer to species as follows: 1. *Oenothera flava*, 2. *Lewisia pygmaea*, 3. *Pontederia cordata*, 4. *Raphanus sativus*, 5. *Sidalcea oregana* ssp. *spicata*, 6. *Campsis radicans*, 7. *Mertensia ciliata*, 8. *Ascelpias syriaca*, 9. *Impatiens capensis*, 10. *Geranium caespitosum*, 11. *Ipomopsis aggregata*, 12. *Polemonium delicatum*, 13. *Zigadenus elegans*, 14. *Aquilegia caerulea*, 15. *Cerastium arvense*, 16. *Pedicularis sudetica*, 17. *Sedum laceolatum*, 18. *Erythronium americanum*, 19. *Trillium grandiflorum*.

Cerastium arvense, *Sedum laceolatum*, *Pedicularis sudetica*). These four species had longer predicted flower longevities than observed ones, and at least two empirical factors can lead to exceptionally long predicted longevities. First, underestimating maintenance costs or choosing an inappropriate currency will result in inflated optimal longevities, especially when maintenance costs are low. This may have been the case for at least three of the four species with very low estimates of maintenance in terms of carbon (*Erythronium americanum*, *Cerastium arvense*, *Sedum laceolatum*). Measuring maintenance cost in terms of water rather than carbon might have been more appropriate currency for species living in dry environments such as *Sedum* and *Cerastium*. Second, if fitness accrual rates are underestimated, then predicted flower longevities will be overestimates. This might be the case for *Erythronium americanum*, where male and female fitness accrual were assumed to be equal (Ashman and Schoen, 1996b): if male fitness actually accrues faster than female, then the predicted floral longevity would be an overestimate. Lastly, it should be pointed out that these data are estimates from a single flowering season, and any atypical conditions (e.g., non-native pollinators, unusual weather conditions) during data collection could result in estimates of fitness parameters that do not represent the conditions under which the longevity phenotype evolved, and thus are likely to reduce the fit between the data and the model predictions. Features that were not included in the basic model formulation that would reduce predicted optimal floral longevities include flowers combining to form the attractive unit, plasticity in floral longevity in response to pollination, and selection for reduced longevity imposed by seed predators or geitonogamous self-pollination. Several of these have been explored in extended versions of the model and are discussed below.

C. Extensions of the Model

Extensions of the basic model can add greater realism and inform on how other floral traits or ecological factors influence floral longevity. A number of extensions are possible, and those that have been considered thus far include, temporal separation in the maturation of stigma and anthers (dichogamy), the inflorescence as the unit of pollinator attraction, reduction in maintenance costs with flower age, and resource trade-offs that include fruit production. Staggered maturation of the sexual functions can constrain optimal floral longevity, that is, a flower must be maintained at least as long as required for the delayed sexual function to commence. When dichogamy [either female structures maturing before male (protogyny) or the reverse (protandry)] is incorporated into the model, optimal floral longevity shifts in the direction of longer-lived flowers. However, the extent of the shift depends specifically on fitness accrual rates. For example, when pollen accrual is slow, protogyny shifts the optimal flower toward longer-lived flowers to a greater degree than when pollen accrual is fast (Schoen and Ashman, 1995). In many species, individual flowers combine to increase attractiveness of the plant to pollinators (e.g., Gori, 1983; Snow *et al.*, 1996). Incorporating the number of open flowers (floral display) into the model of floral longevity, shows that selection would favor shorter optimal flower lifetimes relative to those selected for when single flowers are the unit of attraction (Schoen and Ashman, 1995). The basic model assumes constant floral maintenance cost over time; however, such costs can decline as a flower ages (e.g., Coorts, 1975; Weiss, 1991). Incorporating reduced maintenance leads to an increase in the optimal floral longevity (Schoen and Ashman, 1995). Lastly, altering the basic model to allow an overlap in the resources used for fruit production with flowering allows application of the model to plants in which flowers are produced sequentially over an extended period of time. When flower maintenance is allowed to trade-off with fruits, the qualitative predictions of the basic model are upheld: high maintenance costs select for shorter-lived flowers (Ashman and Schoen, 1997). In addition, Charnov (1996) has simplified the basic floral longevity model, and shown that floral longevity can be cast as a life history invariant.

IV. Other Issues

A. The Role of Plasticity in Floral Longevity

Many authors have noted that in addition to species-specific flower longevities, some flowers have shorter lifetimes in the presence of pollinators than in their absence (Motten, 1986; reviewed in Ashman and Schoen, 1996b; Van Doorn, 1997). A plastic response to pollination environment (i.e., removal and/or deposition of pollen) would effectively minimize flower maintenance costs, and if these savings outweigh the cost of maintaining enzyme systems responsible for plastic response, then plasticity would be adaptive. Furthermore, if such a response is heritable, then selection acting on it could bring floral longevity closer to optimal (Ashman and Schoen, 1996b).

Experiments directly testing the effects of pollen deposition and pollen removal on flower lifetime have shown that both can shorten it by promoting early senescence of corollas (Table 24-1, and references above); however, not all species show this response (Table 24-1; also reviewed in Van Doorn, 1997). In fact, pollination-induced plasticity in flower longevity is not homogeneously distributed among flowering plants, rather it appears to be associated with other floral characters. In particular, pollination-induced senescence may be concentrated in plant families with relatively long-lived flowers, but absent from groups with very short-lived flowers (Stead, 1992; Van Doorn, 1997). Orchids fall into the former category, while many tropical families (e.g., Rubiaceae) may fall into the latter (Stratton, 1989; Stead, 1992; Van Doorn, 1997). However, more data are needed for a rigorous test of this hypothesis, because some exceptions clearly exist [e.g., long-lived flowers that do not respond to pollination, and short-lived ones that do (Table 24-1)]. Lack of pollination-induced senescence in short-lived flowers may be related to their insensitivity to ethylene, a known regulator of senescence in flowers that respond to pollination (Stead, 1992). Moreover, flowers with short lifetimes may not respond to pollination because their senescence "program" has already been initiated by the time their flowers receive pollen (see Chapter 22). For example, Rubinstein and co-workers (Panavas *et al.*, 1998a, 1998b) found that the "one-day" flowers of daylily begin the physiological and genetic processes of senescence even before the flowers open. This might indicate that optimal floral lifetime in this species is shorter than the time required to complete the senescence process, and the only way to achieve the optimal lifetime is to initiate senescence before anthesis.

The benefits of pollination-induced senescence are expected to differ for flowers that have simultaneous maturation of male and female organs (homogamous) versus those with staggered maturation (dichogamous). Pollination-induced senescence would be favored in homogamous flowers when both male and female function are satisfied at relatively equivalent rates, that is, in flowers where a senescence response to pollen deposition would not lead to loss of further fitness opportunities via pollen dispersal (Ashman and Schoen, 1996b). Alternatively, homogamous plants might simply respond to fulfilment of the function that is least easily satisfied (Proctor and Harder, 1995). Proctor and Harder (1995) reasoned that in deceit-pollinated orchids male function would be more easily satisfied than female

Table 24-1. Studies of the Effects of Pollen Removal and Deposition on Flower Longevity

	Floral longevity		Floral longevity reduced by		
Species	without pollination (days)	with pollination (days)	pollen removal	pollen deposition	References
Brassica napus (Brassicaceae)	2.7	1.5	yes	no	Bell and Cresswell, 1998
Calypso bulbosa (Orchidaceae)	8-11	4	no	yes	Proctor and Harder, 1995
Campanula rapunculoides (Campanulaceae)	—	3	yes	yes	Richardson and Stephenson, 1989
Chlorea alpina (Orchidaceae)	8–10	3.7	no/yes[a]	yes	Clayton and Aizen, 1996
Lilium philadelphium (Liliaceae)	9	9.5	—	no	Edwards and Jordan, 1992
Lobelia cardinalis (Lobeliaceae)	12. 3	8	yes	yes	Devlin and Stephenson, 1984
Portulaca umbraticola (Portulaceae)	0.15	0.16	—	no/yes[b]	Aizen, 1993

[a]Pollen removal did not affect flower longevity if pollen deposition had taken place, but if it had not, then pollen removal significantly decreased floral longevity.
[b]Deposition of out-cross pollen had no effect on floral longevity, but self-pollen reduced floral longevity.

function because the latter requires a pollinator to be fooled twice. To support this hypothesis they provide evidence that pollen deposition is less frequent than removal, and that floral longevity is shortened by pollen deposition (pollinia insertion) but not removal in deceit-pollinated *Calypso bulbosa*. Other evidence supporting Proctor and Harder's hypothesis comes from a study in homogamous *Brassica napus* where male function accrues more slowly than female, and here, flower senescence is hastened by pollen removal but not pollen deposition (Bell and Cresswell, 1998). It follows then, that senescence might be triggered by the fulfilment of different sex functions in plants with temporal separation in maturation of the sex organs [i.e., female organs mature first then male organs (protogyny), or the reverse (protandry)]. Flower senescence should be triggered by pollen removal, but not pollen deposition, in protogynous flowers, and conversely, senescence should be triggered by pollen deposition, but not removal, in protandrous flowers (Webb and Littleton, 1987; Stead, 1992). Although no studies address these hypotheses directly, there is some evidence that floral longevity is reduced by pollen deposition in protandrous flowers, but not in protogynous flowers (Motten, 1986; Webb and Littleton, 1987). It is interesting to note, however, that while pollen removal does not trigger flower senescence *per se* in protandrous *Lobelia cardinalis* and *Campanula rapunculoides*, it does hasten the onset of the female phase thereby shortening the entire flower lifetime (Devlin and Stephenson, 1984; Richardson and Stephenson, 1989). This phenomenon suggests that plasticity in floral longevity is not necessarily limited directly to senescence-related cues, but may also indirectly involve other developmental cues. The above-mentioned handful of studies has generated some interesting patterns of plastic responses in floral longevity; however, more work is needed to rigorously explore the relationship between these responses and maximal flower lifetime (in the absence of pollinators), dichogamy, and synchrony of male and female fitness accrual rates in the presence of pollinators.

B. Other Selective Forces on Floral Longevity

1. Flower/seed predators/parasites

In addition to increasing attractiveness to pollinators, flower longevity (either directly or indirectly through floral display size and/or duration) may also increase the plant's apparency to flower or seed predators. Such an effect would select for reduced floral longevity, and thus, counter selection acting through pollination success to extend floral longevity (Ashman and Schoen, 1996a). However, few studies have addressed the role of floral longevity in predation or disease risk directly. Stiles (1975) observed that nectar-rich *Heliconia* flowers were attacked a few hours after anthesis and their ovaries damaged by various non-pollinating nectivores, leading him to suggest that the very short (half-day) lifetime of these tropical flowers might have evolved in response to predation. Another study (Shykoff *et al.*, 1996) reported that the longer-lived female flowers of dioecious *Silene latifolia* and *S. dioica* are more prone to a pollinator-transmitted anther smut disease than conspecific male flowers. Plants with larger floral displays have also been seen to be subject to greater flower and/or seed herbivory (English-Loeb and Karban, 1992; Cunningham, 1995), although the relationship between floral display and floral longevity in these species is unknown. Studies which experimentally manipulate floral longevity and observe predation and/or disease responses, as well as the net fitness consequences are needed to firmly evaluate the strength of selection imposed on floral longevity by predators. Consideration of predator and parasite interactions with floral longevity will provide a more holistic view of the forces shaping floral longevity (Shykoff *et al.*, 1996; Ashman and Schoen, 1996b).

2. Geitonogamous inbreeding

Self-compatible plants with greater than one flower open per day may be subject to self-pollination among flowers (geitonogamy). As a result the longer individual flowers live the greater the number of flowers open at a single time for a given rate of opening, and thus, the greater the risk of geitonogamy. Plants with large floral displays often receive more pollinator visits per plant, but they also incur greater proportions of selfed progeny (reviewed in Snow *et al.*, 1996). Primack (1985) suggested that such geitonogamous self-pollination and subsequent inbreeding depression would select for shorter-lived flowers, thereby countering selection for increased flower longevity through pollination success. Likewise, Dobkin (1987) speculated that the presentation of a single "one-day" flower per plant by hummingbird-pollinated tropical plants was a means to prevent self-pollination. Empirical studies are needed that determine if the costs of geitonogamous self-pollination outweigh the benefits of extended flower longevity with respect to pollen dispersal and deposition. Comparisons between self-compatible and self-incompatible species may allow dissection of the benefits of extended floral longevity through increased attraction versus the costs through reduced progeny fitness.

V. Conclusions and Future Directions

The adaptive nature of flower lifespan has long been appreciated, and the recent treatment of floral longevity as a resource allocation problem has firmly seated its study in the context of life history and resource allocation theory. Empirical studies have elucidated patterns underlying variation in floral longevity, and theoretical models have provided a valuable framework for evaluating these patterns, illuminating key causal features and, perhaps most importantly, generating testable predictions. In particular, it has been established that the rate of pollination success (pollen deposition and removal) and cost of maintaining flowers combine to determine an optimal flower longevity strategy. However, more information is needed on the roles that plasticity, geitonogamous self-pollination, and flower/seed predators play in the evolution of floral longevity strategies. In addition, a concentrated effort to understand the constraints on the evolution of floral longevity is needed. Specifically, studies that determine if the evolution of flower longevity is constrained by the time required to complete the physiological process of flower senescence, by a lack of genetic variation within species, or by phylogeny are needed.

References

Aizen, M.O. (1993). Self-pollination shortens flower lifetime in *Portulaca umbraticola* H.B.K. (Portulacaceae). *International Journal of Plant Sciences* **154**, 412–415.

Ashman, T.-L., and Schoen, D.J. (1994). How long should flowers live? *Nature* 788–791.

Ashman, T.-L., and Schoen, D.J. (1996a). Ashman and Schoen reply (to 'Flower life span and disease risk', Shykoff *et al.*, van Doorn). *Nature* **379**, 780.

Ashman, T.-L., and Schoen, D.J. (1996b). Floral longevity: fitness consequences and resource costs. In *Floral Biology: Studies on Floral Evolution in Animal-Pollinated Plants* (D.G. Lloyd and S.C.H. Barrett, Eds.), pp. 112–139. Chapman and Hall, New York.

Ashman, T.-L., and Schoen, D.J. (1997). The cost of floral longevity in *Clarkia tembloriensis*—an experimental investigation. *Evolutionary Ecology* **11**, 289–300.

Ashman, T.-L., and Stanton, M.L. (1991). Seasonal variation in pollination dynamics of the sexually dimorphic species, *Sidalcea oregana* ssp. *spicata* (Malvaceae). *Ecology* **72**, 993–1003.

Bell, S.A., and Cresswell, J.E. (1998). The phenology of gender in homogamous flowers—temporal change in the residual sex function of flowers of oil-seed rape (*Brassica napus*). *Functional Ecology* **12**, 298–306.

Charnov, E.L. (1982). *The Theory of Sex Allocation*. Princeton University Press, Princeton, NJ.

Charnov, E.L. (1996). Optimal flower lifetimes. *Evolutionary Ecology* **10**, 245–248.

Clayton, S., and Aizen, M.O. (1996). Effects of pollinia removal and insertion on flower longevity in *Choraea alpina* (Orchidaceae). *Evolutionary Ecology* **10**, 653–660.

Coorts, G.D. (1975). Internal metabolic changes in cut flowers. *HortScience* **8**, 195–198.

Cunningham, S.A. (1995). Ecological constraints on fruit initiation by *Calyptrogyne ghiesbreghtiana* (Arecaceae): Floral herbivory, pollen availability, and visitation by pollinating bats. *American Journal of Botany* **82**, 1527–1536.

Delph, L.F., and Lively, C.M. (1992). Pollinator visitation, floral display, and nectar production of the sexual morphs of a gynodioecious shrub. *Oikos* **63**, 161–170.

Devlin, B., and Stephenson, A.G. (1984). Factors that influence the duration of the staminate and pistillate phases of *Lobelia cardinalis* flowers. *Botanical Gazette* **145**, 323–328.

Dobkin, D.S. (1987). Synchronous flower abscission in plants pollinated by hermit hummingbirds and the evolution of one-day flowers. *Biotropica* **19**, 90–93.

Edwards, J., and Jordan, R.J. (1992). Reversible anther opening in *Lilium philidelphicum* (Liliaceae): a possible means for enhancing male fitness. *American Journal of Botany* **79**, 144–148.

English-Loeb, G.M., and Karban, R. (1992). Consequences of variation in flowering phenology for seed head herbivory and reproductive success in *Erigeron glaucus* (Compositse). *Oecologia* **89**, 588–595.

Galen, C., Sherry, R.A., and Carroll, A.B. (1999). Are flowers physiological sinks or faucets? Costs and correlates of water use by flowers of *Polemonium viscosum*. *Oecologia* **118**, 461–470.

Gori, D.F. (1983). Post-pollination phenomena and adaptive floral changes. In *Handbook of Experimental Pollination Biology* (C.E. Jones and R.J. Little, Eds.), pp. 31–49. Van Nostrand Reinhold, New York.

Harder, L.D., and Barrett, S.C.H. (1992). The energy cost of bee pollination for *Pontederia cordata* (Pontederiaceae). *Functional Ecology* **6**, 226–233.

Harding, J., Byrne, T., and Nelson, R.L. (1981). Heritability of cut-flower vase longevity in gerbera. *Euphytica* **30**, 653–657.

Kerner von Marilaun, A. (1895). *The Natural History of Plants, Their Forms, Growth and Reproduction, and Distribution*. Henry Holt, New York.

Kunin, W.E., and Shmida, A. (1997). Plant reproductive traits as a function of local, regional, and global abundance. *Conservation Biology* **11**, 183–192.

Motten, A.F. (1986). Pollination ecology of the spring wildflower community of a temperate deciduous forest. *Ecological Monographs* **56**, 21–42.

Nobel, P.S. (1977). Water relations of flowering in *Agave deserti*. *Botanical Gazette* **138**, 1–6.

Panavas, T., Reid, P.D., and Rubinstein, B. (1998a). Programmed cell death of daylily petal: activities of wall-based enzymes and effects of heat shock. *Plant Physiology and Biochemistry* **36**, 379–388.

Panavas, T., Walker, E.L., and Rubinstein, B. (1998b). Possible involvement of abscisic acid in senescence of daylily petals. *Journal of Experimental Botany* **49**, 1987–1997.

Primack, R.B. (1985). Longevity of individual flowers. *Annual Review of Systematics and Ecology* **16**, 15–37.

Proctor, H.C., and Harder, L.D. (1995). Effect of pollination success on floral longevity in the orchid *Calypso bulbosa* (Orchidaceae). *American Journal of Botany* **82**, 1131–1136.

Pyke, G.H. (1991). What does it cost a plant to produce floral nectar? *Nature* **350**, 58–59.

Richardson, T.E., and Stephenson, A.G. (1989). Pollen removal and pollen deposition affect the duration of the staminate and pistillate phases in *Campanula rapunculoides*. *American Journal of Botany* **76**, 532–538.

Schoen, D.J., and Ashman, T.-L. (1995). The evolution of floral longevity: a resource allocation to maintenance versus construction of repeated structures in modular organisms. *Evolution* **49**, 131–139.

Scott, D., Boyle, T.H., and Han, S.S. (1994). Floral development and flower longevity in *Rhipsalidopsis* and *Schlumbergera* (Cactaceae). *HorScience* **29**, 898–900.

Seres, A., and Ramirez, N. (1995). Biologia floral y polinizacion de algunas monocotiledoneas de un bosque nublado Venezolano. *Annals of the Missouri Botanical Garden* **82**, 61–81.

Shykoff, J.A., Buchell, E., and Kaitz, O. (1996). Flower lifespan and disease risk. *Nature* **379**, 779.

Snow, A.A., Spira, T.P., Simpson, R., and Klips, R.A. (1996). The ecology of geitonotgamous pollination. In *Floral Biology: Studies on Floral Evolution in Animal-Pollinated Plants* (D.G. Lloyd and S.C.H. Barrett, Eds.), pp. 191–216. Chapman and Hall, New York.

Southwick, E.E. (1984). Photosynthate allocation to floral nectar: a neglected energy investment. *Ecology* **65**, 1775–1779.

Stead, A.D. (1992). Pollination-induced flower senescence: a review. *Plant Growth Regulation* **11**, 13–20.

Stearns, S.C. (1992). *The Evolution of Life Histories*. Oxford University Press, Oxford.

Stiles, F.G. (1975). Ecology, flowering phenology, and humming bird pollination of some Costa Rican *Heliconia* species. *Ecology* **56**, 285–301.

Stratton, D.A. (1989). Longevity of individual flowers in a Costa Rican cloud forest: Ecological correlates and phylogenetic constraints. *Biotropica* **21**, 308–318.

van der Meulen-Muisers, J.J.M., van Oeveren, J.C., and van Tuyl, J.M. (1998). Genotypic variation in postharvest flower longevity of Asiatic hybrid lilies. *Journal of the American Horticulture Society* **123**, 283–287.

Van Doorn, W.G. (1997). Effects of pollination on floral attraction and longevity. *Journal of Experimental Botany* **48**, 1615–1622.

Van Eijk, J.P., and Eikelboom,W. (1976). Possibilities of selection for keeping quality in tulip breeding. *Euphytica* **25**, 353–359.

Webb, C.J., and Littleton, J. (1987). Flower longevity and protandry in two species of *Gentiana* (Gentianaceae). *Annals of the Missouri Botanical Garden* **74**, 51–57.

Weiss, M.R. (1991). Floral colour changes as cues to pollinators. *Nature* **354**, 221–223.

Zimmerman, M., and Pyke, G.H. (1986). Reproduction in *Polemonium*: patterns and implications of floral nectar production and standing crops. *American Journal of Botany* **73**, 1405–1415.

25

Ecology of Leaf Senescence

Kihachiro Kikuzawa

I. Introduction

Leaf senescence is a genetically coded process of the programmed death of a leaf (Chapter 1). Usually, it is measured as a decrease in photosynthetic rate or as other changes in chloroplasts (Thomas and Stoddart, 1980; Noodén, 1988). Concomitantly, senescence is influenced by micro-environmental conditions, particularly light availability, and these change as plants grow (Hikosaka *et al.*, 1994). The available light often decreases due to shading by surrounding plants or self-shading. Limiting nutrients, especially nitrogen, are redistributed from old, shaded leaves at the bottom of a plant to new leaves at the top which can receive full light (Chapter 14). Here, leaf senescence and nutrient reallocation are considered to be an adaptive behavior in which a plant replaces leaves to adapt to environmental changes (Hikosaka, 1996).

The timing of leaf death differs among plant species. Some species exhibit short leaf longevities of less than 30 days, and others have leaf longevities greater than 10 years (Chabot and Hicks, 1982). Leaf longevity also differs among individual plants within a species depending on the environmental conditions. It also changes within an individual plant or with the development of a plant. The wide range of leaf life spans suggests that leaf longevity is an adaptive feature (Kikuzawa and Ackerly, 1999). In this chapter, I will review longevity of leaves of higher plants by considering the replacement of leaves to be an adaptive strategy of plants.

II. Photosynthesis

Since leaves are photosynthetic organs, leaf senescence and leaf replacement must be considered from the point of view of carbon economy of plants (Kikuzawa, 1991). Some plant species have leaves with a higher photosynthetic rate per unit leaf area, while other species have lower rates (Larcher, 1975; Ceulemans and Saugier, 1991). Since up to 75% of leaf organic nitrogen is present in the chloroplasts (Poorter and Evans, 1998), nitrogen contents are associated with rates of photosynthesis (Reich *et al.*, 1995; Ninemets *et al.*, 2002). The dependence of photosynthetic capacity on nitrogen content varies strikingly among species when both are expressed on a leaf area basis. In particular, leaves of evergreen trees and shrubs show very low photosynthetic capacities relative to those of crop and herbaceous plants (Hikosaka *et al.*, 1998). Of the ten dicotyledonous species (four woody and six herbaceous species) compared by Poorter and Evans (1998), the herbaceous group had higher specific leaf areas (SLA: leaf area per unit leaf mass) and lower carbon concentration per unit leaf mass. This group allocated more nitrogen to thylakoids and to a key photosynthetic enzyme (Rubisco) while the woody group did the reverse (Poorter and Evans, 1998; Hikosaka *et al.*, 1998; Evans and Poorter, 2001). In brief, the slow-growing species invest less in their photosynthetic apparatus and more in their non-photosynthetic functions such as defensive and supporting tissues (Chabot and Hicks, 1982). Such systematic differences in thickness of the photosynthetic apparatus (SLA), nitrogen investment patterns and photosynthetic rates are observed between slow-growing and fast-growing species within herbs (Poorter and Evans, 1998; Rothstein and Zak, 2001), within trees (Poorter and Evans, 1998; Mooney *et al.*, 1984; Atkin *et al.*, 1998; Bauer *et al.*, 2001) and even within a wider range of photosynthetic organisms including cyanobacteria. Photosynthesis rates (mg C g C^{-1} h^{-1}) declined with increasing thickness of the photosynthetic tissue and were positively scaled to chlorophyll a (Enriquez *et al.*, 1996).

The general course of the photosynthetic rate per unit leaf area during leaf ontogeny is a rather steep increase to a maximum value, often achieved before leaves complete their expansion and reach the greatest amount of chlorophyll. This peak usually is followed by a slower decline in photosynthetic rate (Sestak, 1981). Some tree species, in particular understorey trees in tropical rain forests (Kursar and Coley, 1992a,b) and temperate evergreen broad-leaved trees (Miyazawa *et al.*, 1998), showed delayed greening. Expanding leaves sometimes show a white or red color with no visible chlorophyll. The nitrogen contents and photosynthetic rates also remained low for 10–30 days after leaf expansion. In some evergreen coniferous trees, the time when the newly emerged leaves reach the maximum photosynthetic rate is also delayed (Kajimoto, 1990; Matsumoto, 1984). After reaching the maximum photosynthetic rate, photosynthesis usually declines with increasing leaf age (Matsumoto, 1984; Oren *et al.*, 1986; Kajimoto, 1990). The decline in photosynthesis has been approximated by a straight line (Oren *et al.*, 1986). Net CO_2 uptake also decreased linearly with increasing leaf age in some tropical tree species (Zots and Winter, 1994; Kitajima *et al.*, 1997) and temperate deciduous tree species (Koike, 1990). In some deciduous trees, however, the photosynthetic rate measured at light-saturation increased to a maximum near the completion of leaf expansion, was constant until autumn, and then rapidly declined until leaf death (Jurik, 1986).

Decline in photosynthetic rate with leaf age is associated with decrease in nitrogen content in the leaf (Field and Mooney, 1983). Some changes in chloroplast ultrastructure were observed such as swelling of the thylakoids (Noodén, 1988) and appearance of plastoglobuli (Nilsen *et al.*, 1988; Noodén, 1988; Gepstein, 1988). In *Rhododendron maximum*, the shortest leaf longevity is associated with the most rapid reduction in photosynthetic capacity and the most rapid development of chloroplast plastoglobuli (Nilsen *et al.*, 1988). The disappearance of chlorophyll is also a prominent feature of leaf senescence (Gepstein, 1988). In many cases, however, the decrease in chlorophyll is slower than that of some Calvin-cycle enzymes, soluble proteins, electron carriers and electron-transport capacity (Hikosaka, 1996). In erect herbaceous plants, the irradiance received by a leaf decreases as the plant grows because of shading by upper new leaves (Hikosaka, 1996). There is a gradient of photon flux density within a canopy of plant population. With the degradation of micro-environmental conditions, in particular light conditions, it is adaptive to remove nitrogen from older, shaded leaves and to reinvest it in newly emerged leaves (Mooney and Gulmon, 1982; Hirose and Werger, 1987). The whole canopy carbon gain will be maximized when nitrogen is distributed greatest to the leaves which receive the highest photon flux density (Field, 1983). In fact, photon flux density, leaf photosynthetic capacity and leaf nitrogen content all decrease with depth into a canopy (Kull and Kruijt, 1999). Actual leaf nitrogen distribution within a canopy of *Solidago altissima* was similar to that predicted from a model for optimal canopy-photosynthesis (Hirose and Werger, 1987).

Decline in photosynthetic rate with increasing leaf age is often attributed either to endogenously controlled senescence or to environmentally induced senescence, or both. As an illustration of these phenomena, the environmental effect can be distinguished from endogenously initiated senescence, by a unique experimental system in which a vine is grown horizontally to avoid mutual shading (Hikosaka *et al.*, 1994; Hikosaka, 1996). When only a limited amount of nitrogen was supplied, the nitrogen content of leaves dropped drastically with decrease of photon flux density and also decreased with advance of leaf age and position along the vine even when the vine is horizontal, i.e., without shading. The result strongly suggests that this decline in photosynthetic rate can be attributed both to environmental factors and to endogenously controlled senescence.

III. Herbivory

Herbivory, mostly by insect larvae, is an important factor affecting leaf life span. Longevity of leaves of seedlings and saplings of a tropical pioneer tree, *Heliocarpus appendiculatus*, is determined by herbivory (Nunez-Farfan and Dirzo, 1989). Leaves damaged by gall flies or aphids are shed early in the season and, therefore, they have shortened longevity. Early shedding of affected leaves is considered to be part of a defensive strategy by plants. By shedding leaves, plants can reduce insect density, which would act as density-dependent mortality factor of herbivorous populations. For example, at least 98% of the aphids in the galls of abscised leaves died (Williams and Whitham, 1986).

Various defensive systems have evolved to protect leaves from herbivore damage. Leaf toughness is a proximate mechanical factor discouraging feeding by insects. The toughness increases with leaf age and renders mature leaves a poor source of food for Lepidoptera larvae (Feeny, 1970). In contrast, nitrogen and water contents of young leaves are usually higher than those in mature leaves—an apparently unavoidable consequence of cell growth. In the tropics, almost 70% of a leaf's lifetime damage occurs while it is expanding, suggesting that selection for young leaf defenses should be intense (Coley and Barone, 1996). Differential defense may be deployed age-specifically, with younger, more valuable leaves (Harper, 1989) or tissues being more heavily defended than older, less valuable ones (Woodman and Fernandes, 1991).

The timing of leafing also acts as protection or escape from herbivore damage. Oak trees that produced leaves earlier in the spring were attacked lightly, while trees that leafed out later were attacked heavily by a leaf-mining moth. The difference among trees in leaf production phenology explained 61% of the variation in leaf miner densities (Mopper and Simberloff, 1995). Delayed greening in tropical trees is also considered to be a strategy to avoid herbivory. Here, chloroplast development is postponed until after the leaf has reached full size, has toughened, and is better protected from herbivores (Kursar and Coley, 1992a,b; Coley and Barone, 1996). Leaves can also be flushed synchronously, saturating herbivores with an abundance of leaves to increase the number of leaves that escape damage before they develop resistance (Aide, 1988; Coley and Barone, 1996).

IV. Leaf Longevity

As mentioned above, a wide range of longevity of leaves of various plant species has been observed. The life span of floating leaves of aquatic floating plants ranged from 13 to 40 days, while that of emergent leaves of the same species was 35–57 days (Tsuchiya, 1991). The short life span of floating leaves seems to be closely related to the fact that old leaves are submerged, lose their photosynthetic function quickly and are replaced by new leaves early. Leaf longevity in a soybean plant increases with increased height of the leaf on the stem, longevity of leaves at the lowest position was about 20 days and that at the higher position was about 50 days (Miyaji and Tagawa, 1979). Leaf longevity in an annual herb, *Ambrosia trifida*, was from about 20 to 90 days (Abul-Fatih and Bazzaz, 1980). The average leaf life spans of 29 perennial herbaceous plant species of central Europe ranged from 41 to 95 days (Diemer *et al.*, 1992). Leaf longevities for 41 deciduous tree species in Hokkaido, northern Japan, were from 90 to 200 days (Kikuzawa, 1983), while those of evergreen shrubs in the understorey of the same forest were from 2 to 3 years (Kikuzawa, 1984). Longevity of leaves of coniferous trees in temperate and subarctic forests usually reached 5 years (Kayama *et al.*, 2002) and was sometimes more than 20 years (Chabot and Hicks, 1982; Schulze *et al.*, 1986). Leaf longevity of seedlings of a tropical pioneer tree ranged from 27 to 37 days (Ackerly and Bazzaz, 1995). Leaf life spans of some trees at the understory of tropical forests were usually more than 2 years, and some were more than 5 years (Bentley, 1979).

Some patterns and trends were observed concerning leaf longevities of various plant species in relation to their physiological traits, architectures and habitat conditions. Differences in leaf longevities among different life forms or architectures were attributed to the differences in the costs of supporting tissues (Kikuzawa and Ackerly, 1999). Supporting structures are greater in tall trees than in seedlings of the same species. Leaf longevities of tree seedlings (Seiwa and Kikuzawa, 1991; Seiwa, 1998) were shorter than those of tall trees of the same species (Kikuzawa, 1983), reflecting the difference in the amount of supporting tissues (Kikuzawa and Ackerly, 1999).

There is a relationship between leaf longevity and photosynthetic capacity. That is, the longer the life of a leaf in general, the lower the leaf's intrinsic photosynthetic capacity. For example, deciduous leaves generally have higher photosynthetic capacities than evergreen leaves (Mooney and Gulmon, 1982). Plants occurring in habitats with limited water, light, or nutrients should have lower carboxylating enzyme contents and lower intrinsic photosynthetic capacities (Mooney and Gulmon, 1982; Casper *et al.*, 2001), because they cannot conduct higher photosynthesis in the limited environmental conditions even if they have higher capacity. In order to maintain positive carbon gain in such environments, plants must retain their leaves for longer (Kikuzawa, 1995b).

V. Theories of Leaf Longevity

Often, plants in resource-limited habitats retain leaves for longer to compensate for low photosynthetic rates (Small, 1972; Moore, 1980; Kikuzawa, 1995b). Hence, there is a trade-off between high photosynthetic rate and long leaf longevity (Reich *et al.*, 1992). Williams *et al.* (1989) considered that the ratio of daily carbon gain to leaf construction costs must be related to leaf longevity. This idea was supported by the comparison of leaf longevities of several tree species (Sobrado, 1991). Moreover, the photosynthetic rate of a leaf decreases with increase in time by nutrient reallocation or by leaf aging. Thus, the decline in leaf photosynthetic rate with increase in time must be incorporated into the model.

Carbon gain by a leaf is given as:

$$G = \int_0^t p(t)\,\mathrm{d}t - C \tag{25-1}$$

where $p(t)$ is the photosynthetic rate and is a decreasing function of leaf age t. The optimum leaf longevity (t^*) to maximize gain per unit time (G/t) is obtained by the following equation (Kikuzawa, 1991).

$$t^* = \left(\frac{2bC}{a}\right)^{1/2} \tag{25-2}$$

where, a is the maximum photosynthetic rate and is measured as the actual photosynthetic rate (carbon · leaf area^{-1} time^{-1}) at the time of the highest rate. The rate a/b is the decreasing rate of photosynthesis with time (carbon · leaf area^{-1} time^{-2}), b being the intercept of the x axis or potential leaf longevity (time). C is the construction cost of a leaf (glucose · leaf area^{-1}).

From equation (2), it is supposed that higher photosynthetic rate (a), lower construction cost (C) and shorter potential leaf longevity (b) will shorten leaf longevity, while the reverse will promote leaf longevity. Increased nutrient supply will promote increased photosynthetic rate and thus decrease leaf longevity. Experimental addition of nutrients or water shortened leaf longevity (Shaver, 1981; Lajtha and Whitford, 1989; Cordel *et al.*, 2001). Greater leaf mass per unit leaf area (LMA; the inverse of SLA) is closely related to parameter C (Kikuzawa, 1995b) and positively correlated with leaf longevity (Wright and Westoby, 2002). Actual leaf longevity correlated to potential leaf longevity (Kitajima *et al.*, 1997; Ackerly, 1999). Reich *et al.* (1991, 1992) have summarized some empirical relationships for some traits in trees or in leaves with leaf longevity.

Two patterns of shoot elongation of trees were found in a tropical rain forest: ever-growing and intermittent (Koriba, 1948, 1958). Similarly, two leaf emergence patterns were known in deciduous broad-leaved trees—leaves appearing successively, one after another, or leaves appearing simultaneously, all at once (Kikuzawa, 1983, 1995a)—as well as in tropical forests (Lowman, 1992). In temperate forests, successive leafing trees are usually found in open habitats such as flood plains and large gaps where light, water and nutrients are available (Kikuzawa, 1988, 1995a), while simultaneous flush trees are usually found in forest canopies and the forest understorey. Linkages were found among high photosynthetic rate, short leaf longevity and successive leafing of pioneer trees and among low photosynthetic rate, long leaf longevity and simultaneous leafing (Kikuzawa, 1988, 1995a; Koike, 1988). The autumnal leaf coloring pattern is also linked with the leafing pattern. Successive leafing trees first turned color in the lower branches, while simultaneous-leafing trees started in the peripheral zone of a crown (Koike, 1990).

Plants with several leaf cohorts, which appear at different times, can differently utilize the light resource, which fluctuates during the growing season. The winter green shrub, *Daphne kamtchatica* var. *jezoensis*, in a deciduous broad-leaved forest leafs out in autumn as well as in spring (Kikuzawa, 1984; Lei and Koike, 1998), both are timed to appear when the canopy trees are leafless (Seiwa and Kikuzawa, 1996). Similarly, an evergreen herbaceous species, *Heuchera americana*, produces leaves twice in a year; in spring and in autumn (Skillman *et al.*, 1996). The cotton sedge *Eriophorum vaginatum* can make use of opportunities for early- and late-season photosynthesis by successive leafing (Defoliart *et al.*, 1988). This species produced leaves successively, and nutrients are redistributed from older leaves to actively growing leaves. This efficient use of resources may be important in enabling this species to dominate nutrient-poor sites (Jonasson and Chapin, 1985).

VI. Leaf Habit and Geographical Pattern

Extension of equation (1) to seasonal environments explains the leaf habit or evergreenness and deciduousness (Kikuzawa, 1991, 1996):

$$G = \int_0^f p(t)\,\mathrm{d}t + \int_1^{1+f} p(t)\,\mathrm{d}t + \cdots + \int_{[t]}^t p(t)\,\mathrm{d}t - \int_0^t m(t)\,\mathrm{d}t - C \qquad (25\text{-}3)$$

where f is the fractional length of the favorable season for photosynthesis within a year, and m is the maintenance cost of a leaf. When the favorable period is 1.0, a plant can conduct photosynthesis throughout a year, and thus all plants are evergreen. When favorable and unfavorable seasons alternate within a year, the deciduous habit where all leaves are shed before the unfavorable season will appear. However, when the length of the unfavorable season is short, many plants will show an evergreen habit, because the decrease in carbon gain during the unfavorable season is small relative to the cost of making a new leaf. When the length of the unfavorable period increases beyond this point, the strategy of discarding old leaves before the unfavorable period and carrying out photosynthesis by using new leaves at the beginning of the next favorable period is selected instead of maintaining old leaves during the unfavorable period. Thus, with increasing the length of the unfavorable period, the percentage of deciduous habit will increase. When the length of the unfavorable period increases further, a leaf growing during a single season would have little time for maximum photosynthesis (Bell and Bliss, 1977). In this situation, leaf longevity will be extended to more than one year and evergreenness will be selected again (Kikuzawa, 1991).

The changes in developmental patterns in the world biogeographic regions verify the above prediction (Landsberg and Gower, 1997). Tropical evergreen trees become deciduous when planted in seasonal environments (Koriba, 1948). In Greenland, many evergreen and wintergreen species were recorded at a northern site (77° 45′ N), but the percentage of deciduous species increases at a southern site (72° 50′) (Bell and Bliss, 1977). Leaf life span of perennial herbaceous plants declined significantly with increasing seasonality related to latitude and duration of the annual growth period (Diemer, 1998). Leaf longevities in species at the northernmost latitudes were greater than those at less northerly latitudes. Although these observations apply to changes in growing season with latitude, a similar bimodal trend in the percentage of evergreenness parallel occurs with altitude (Kikuzawa, 1996). For example, leaf longevity of mid-latitude, low-altitude European plants was shorter than in the Alps (Karlsson, 1992).

Leaf longevities of three dwarf-shrub species were compared in a mountain area of Hokkaido, northern Japan, where the length of the favorable period differs naturally with the difference in the length of snow-free period caused by the difference in the depth of snow cover during winter (Kikuzawa and Kudo, 1995). Leaf longevity of two evergreen species decreased with longer favorable periods, while a deciduous plant showed increased leaf longevity as the length of the favorable period increased. These contradictory patterns in the changes of leaf longevity with the length of the favorable period were well explained by a simulation using equation (3) (Kikuzawa and Kudo, 1995). Since photosynthetic rates of evergreen species are low and leaf costs are high, evergreen species in the sites of shorter favorable periods compensate for their low photosynthetic rates by prolonging their leaf longevity. The photosynthetic rate of deciduous species is high but decreases rapidly with time, so deciduous species prolong their leaf longevity in sites with long favorable periods to the limit in order to use the season fully. Hence, the plastic changes in leaf longevity with varying favorable periods seem to be adaptive traits for maximizing photosynthetic gain for the plants (Kikuzawa and Kudo, 1995).

VII. Leaf Longevity in Different Ecosystems

Mechanical structures and chemicals which act as defenses against herbivores during a leaf's life span may also act as defenses after leaf fall against attack by soil animals and microorganisms and thus delay leaf decay (Takeda *et al.*, 1987). Hence, we can predict that leaves with short life span decay faster on the forest floor than leaves with long longevity. For example, *Dendrocnid excelsa* has soft mesomorphic leaves with short life span of about 7 months in a tropical rain forest in Australia and consequently its leaves decay rapidly, requiring just over 4 months. In contrast, the leaf litter of *Nothofagus moorei*, whose leaf is tough and toxic with a life span of about 2 years, decays slowly with a half-life of 16 months (Lowman, 1992). The existence of a linkage between leaf turnover, which is the inverse of leaf longevity, and the turnover of fallen litter suggests the linkage between ecosystem-level matter flow and leaf longevity.

An important difference among ecosystems, e.g., forest, grassland and aquatic ecosystems, is the turnover rates of organic materials (Cyr and Pace, 1993). Differences in turnover rate of organic materials may be attributable to the differences in leaf longevity of the "structural organisms" i.e., those organisms that create or provide the physical structure of the ecosystem such as trees in a forest (Huston, 1994). Kikuzawa and Ackerly (1999) incorporated the costs for supporting tissues into equation (2) and obtained a more comprehensive equation. A great difference in leaf longevity among plant life forms may be attributed to the difference in supporting costs, which are greatest in trees and smallest in aquatic plants (Kikuzawa and Ackerly, 1999).

Matter flow in ecosystems is characterized by the size of supporting tissues of photosynthetic organs in the structural organisms (Huston, 1994). The large size in forests results in greater longevity of leaves. Slow turnover or longer life span of leaves needs mechanical structure which results in slow turnover of fallen leaves in the detritus layer.

Leaf longevity is determined by a balance between costs and benefits of a leaf. The benefit of the leaf is the photosynthetic gain which declines with increase in leaf age and with degradation of micro-environmental conditions around the leaf. Costs are construction and maintenance of the leaf and its supporting tissues. The balance produces short or long leaf longevities, depending on the resource availability of environments. Leaf longevities are usually linked with the patterns of leaf emergence and defense against herbivores. Costs for supporting tissues of structural organisms are different among ecosystems. Longevities of living and fallen leaves are interrelated with each other and affect the rate of matter flow within an ecosystem. Thus, leaf longevity is a key factor in ecosystems. Moreover, these considerations are also central in designing more efficient and productive crops.

References

Abul-Fatih, H.A., and Bazzaz, F.A. (1980). The biology of Ambrosia trifida L. IV. Demography of plants and leaves. *New Phytologist* **84**, 107–111.

Ackerly, D.D., (1999). Self-shading, carbon gain and leaf dynamics: a test of alternative optimality models. *Oecologia* **119**, 300–310.

Ackerly, D.D., and Bazzaz, F.A. (1995). Leaf dynamics, self-shading and carbon gain in seedlings of a tropical pioneer tree. *Oecologia* **101**, 289–298.

Aide, T.M. (1988). Herbivory as a selective agent on the timing of leaf production in a tropical understory community. *Nature* **336**, 574–575.

Atkin, O.K., Schortemeyer, M., McFarlane, N., and Evans, J.R. (1998). Variation in the components of relative growth rate in ten Acacia species from contrasting environments. *Plant, Cell and Environment* **21**, 1007–1017.

Bauer, G.A., Berntson, G.M., and Bazzaz, F.A. (2001). Regenerating temperate forests under elevated CO_2 and nitrogen deposition: comparing biochemical and stomatal limitation of photosynthesis. *New Phytologist* **152**, 249–266.

Bell, K.L., and Bliss, L.C. (1977). Overwinter phenology of plants in a polar semi-desert. *Arctic* **30**, 118–121.

Bentley, B.L. (1979). Longevity of individual leaves in a tropical rainforest under-story. *Annals of Botany* **43**, 119–121.

Casper, B.B., Forseth, I.N., Kempenich, H. Seltzer, S., and Xavier, K. (2001). Drought prolongs leaf life span in the herbaceous desert perennial *Cryptantha flava*. *Functional Ecology* **15**, 740–747.

Ceulemans, R.J., and Saugier, B. (1991). Photosynthesis. In *Physiology of Trees* (A.S. Raghavendra, Ed.), pp. 21–50. John Wiley and Sons, New York.

Chabot, B.F., and Hicks, D.J. (1982). The ecology of leaf life spans. *Annual Review of Ecology and Systematics* **13**, 229–259.

Coley, P.D., and Barone, J.A. (1996). Herbivory and plant defenses in tropical forests. *Annual Review of Ecology and Systematics* **27**, 305–335.

Cordell, S., Goldstein, G., Meinzer, F.C., and Vitousek, P.M. (2001). Regulation of leaf life-span and nutrient-use efficiency of *Metrosideros polymorpha* trees at two extremes of a long chronosequence in Hawaii. *Oecologia* **127**, 198–206.

Cyr, H., and Pace, M.L. (1993). Magnitude and patterns of herbivory in aquatic and terrestrial ecosystems. *Nature* **361**, 148–150.

Defoliart, L.S., Griffith, M., Chapin, F.S., III, and Jonasson, S. (1988). Seasonal patterns of photosynthesis and nutrient storage in *Eriophorum vaginatum* L., an arctic sedge. *Functional Ecology* **2**, 185–194.

Diemer, M. (1998). Life span and dynamics of leaves of herbaceous perennials in high-elevation environments: 'news from the elephant's leg'. *Functional Ecology* **12**, 413–425.

Diemer, M., Korner, Ch., and Prock, S. (1992). Leaf life spans in wild perennial herbaceous plants: a survey and attempts at a functional interpretation. *Oecologia* **89**, 10–16.

Enriquez, S., Duarte, C.M., Sand-Jensen, K., and Nielsen, S.L. (1996). Broad-scale comparison of photosynthetic rates across phototrophic organisms. *Oecologia* **108**, 197–206.

Evans, J.R., and Poorter, H. (2001). Photosynthetic acclimation of plants to growth irradiance: the relative importance of specific leaf area and nitrogen partitioning in maximizing carbon gain. *Plant, Cell and Environment* **24**, 755–767.

Feeny, P. (1970). Seasonal changes in oak leaf tannins and nutrients as a cause of spring feeding by winter moth caterpillars. *Ecology* **51**, 565–581.

Field, C. (1983). Allocating leaf nitrogen for the maximization of carbon gain: leaf age as a control on the allocation program. *Oecologia* **56**, 341–347.

Field, C., and Mooney, H.A. (1983). Leaf age and seasonal effects on light, water, and nitrogen use efficiency in a California shrub. *Oecologia* **56**, 348–355.

Gepstein, S. (1988). Photosynthesis. In *Senescence and Aging in Plants* (L.D. Noodén and A.C. Leopold, Eds.), pp. 85–109. Academic Press, San Diego.

Harper, J.L. (1989). The value of a leaf. *Oecologia* **80**, 53–58.

Hikosaka, K. (1996). Effects of leaf age, nitrogen nutrition and photon flux density on the organization of the photosynthetic apparatus in leaves of a vine (Ipomoea tricolor Cav.) grown horizontally to avoid mutual shading of leaves. *Planta* **198**, 144–150.

Hikosaka, K., Terashima, I., and Katoh, S. (1994). Effects of leaf age, nitrogen nutrition and photon flux density on the distribution of nitrogen among leaves of a vine (Ipomoea tricolor Cav.) grown horizontally to avoid mutual shading of leaves. *Oecologia* **97**, 451–457.

Hikosaka, K., Hanba, Y.T., Hirose, T., and Terashima, I. (1998). Photosynthetic nitrogen-use efficiency in leaves of woody and herbaceous species. *Functional Ecology* **12**, 896–905.

Hirose, T., and Werger, M.J.A. (1987). Maximizing daily canopy photosynthesis with respect to the leaf nitrogen allocation pattern in the canopy. *Oecologia* **72**, 520–526.

Huston, M.A. (1994). *Biological Diversity*. Cambridge University Press, Cambridge, UK.

Jonasson, S., and Chapin, F.S., III (1985). Significance of sequential leaf development for nutrient balance of the cotton sedge, *Eriophorum vaginatum* L. *Oecologia* **67**, 511–518.

Jurik, T.W. (1986). Seasonal patterns of leaf photosynthetic capacity in successional northern hardwood tree species. *American Journal of Botany* **73**, 131–138.

Kajimoto, T. (1990). Photosynthesis and respiration of *Pinus pumila* needles in relation to needle age and season. *Ecological Research* **5**, 333–340.

Karlsson, P.S. (1992). Leaf longevity in evergreen shrub: variation within and among European species. *Oecologia* **91**, 346–349.

Kayama, S., Sasa, K., and Koike, T. (2002). Needle life span, photosynthetic rate and nutrient concentration of *Picea glehnii*, *P. jezoensis* and *P. abies* planted on serpentine soil in northern Japan. *Tree Physiology* **22**, 707–716.

Kikuzawa, K. (1983). Leaf survivals of woody plants in deciduous broad-leaved forests. *Canadian Journal of Botany* **61**, 2133 –2139.

Kikuzawa, K. (1984). Leaf survivals of woody plants in deciduous broad-leaved forests. 2. Small trees and shrubs. *Canadian Journal of Botany* **62**, 2551–2556.

Kikuzawa, K. (1988). Leaf survivals of tree species in deciduous broad-leaved forests. *Plant Species Biology* **3**, 67–76.

Kikuzawa, K. (1991). A cost-benefit analysis of leaf habit and leaf longevity of trees and their geographical pattern. *American Naturalist* **138**, 1250–1263.

Kikuzawa, K. (1995a). Leaf phenology as an optimal strategy for carbon gain in plants. *Canadian Journal of Botany* **73**, 158–163.

Kikuzawa, K. (1995b). The basis for variation in leaf longevity of plants. *Vegetatio* **121**, 89–100.

Kikuzawa, K. (1996). Geographical distribution of leaf life span and species diversity of trees simulated by a leaf-longevity model. *Vegetatio* **122**, 61–67.

Kikuzawa, K., and Ackerly, D. (1999). Significance of leaf longevity in plants. *Plant Species Biology* **14**, 39–46.

Kikuzawa, K., and Kudo, G. (1995). Effects of favorable period length on the leaf lifespan of several alpine shrubs—Implication by the cost-benefit model. *Oikos* **73**, 214–220.

Kitajima, K., Mulkey, S.S., and Wright, S.J. (1997). Decline of photosynthetic capacity with leaf age in relation to leaf longevities for five tropical canopy tree species. *American Journal of Botany* **84**, 702–708.

Koike, T. (1988). Leaf structure and photosynthetic performance as related to the forest succession of deciduous broad-leaved trees. *Plant Species Biology* **3**, 778–799.

Koike, T. (1990). Autumn coloring, photosynthetic performance and leaf development of deciduous broad-leaved trees in relation to forest succession. *Tree Physiology* **7**, 21–32.

Koriba, K. (1948). On the origin and meaning of deciduousness viewed from the seasonal habit of trees in the tropics. *Ecology and Physiology (Kyoto)* **2**, 85–93.

Koriba, K. (1958). On the periodicity of tree-growth in the tropics, with reference to the mode of branching, the leaf-fall, and the formation of the resting bud. *Gardens Bulletin* **17**, 11–81.

Kull, O., and Kruijt, B. (1999). Acclimation of photosynthesis to light: a mechanistic approach. *Functional Ecology* **13**, 24–36.

Kursar, T.A., and Coley, P.D. (1992a). Delayed development of the photosynthetic apparatus in tropical rain forest species. *Functional Ecology* **6**, 411–422.

Kursar, T.A., and Coley, P.D. (1992b). Delayed greening in tropical leaves: an antiherbivore defense? *Biotropica* **24**, 256–262.

Lajtha, K., and Whitford, W.G. (1989). The effect of water and nitrogen amendments on photosynthesis, leaf demography, and resource-use efficiency in *Larrea tridentata*, a desert evergreen shrub. *Oecologia* **80**, 341–348.

Landsberg, J.J., and Gower, S.T. (1997). *Applications of Physiological Ecology to Forest Management*. Academic Press, New York.

Larcher, W. (1975). *Physiological Plant Ecology*. Springer, Berlin.

Lei, T.T., and Koike, T. (1998). Some observations of phenology and ecophysiology of *Daphene kamtscatica* Maxi., var. *jezoensis* (Maxim.) Ohwi, a shade deciduous shrub, in the forest of northern Japan. *Journal of Plant Research* **111**, 207–212.

Lowman, M.D. (1992). Leaf growth dynamics and herbivory in five species of Australian rain-forest canopy trees. *Journal of Ecology* **80**, 433–447.

Matsumoto, Y. (1984). Photosynthetic production in *Abies veitchii* advance growths growing under different light environmental conditions (II) photosynthesis and respiration. *Bulletin of the Tokyo University Forests* **73**, 229–252.

Miyaji, K., and Tagawa, H. (1979). Longevity and productivity of leaves of a cultivated annual *Glycine mas* Merrill. I. Longevity of leaves in relation to density and sowing time. *New Phytologist* **82**, 233–244.

Miyazawa, S., Satomi, S., and Terashima, I. (1998). Slow leaf development of evergreen broad-leaved tree species in Japanese warm temperate forests. *Annals of Botany* **82**, 859–869.

Mooney, H.A., and Gulmon, S.L. (1982). Constraints on leaf structure and function in reference to herbivory. *BioScience* **32**, 198–206.

Mooney, H.A., Field, C., and Vazques-Yanes, C. (1984). Photosynthetic characteristics of wet tropical forest plants. In *Physiological Ecology of Plants of the Wet Tropics* (E. Medina, H.A. Mooney and C. Vazques-Yanes, Eds.), pp. 113–128. Junk, The Hague.

Moore, P. (1980). The advantages of being evergreen. *Nature* **285**, 535.

Mopper, S., and Simberloff, D. (1995). Differential herbivory in an oak population: the role of plant phenology and insect performance. *Ecology* **76**, 1233–1241.

Nilsen, E.T., Stetler, D.A., and Gassman, C.A. (1988). Influence of age and microclimate on the photochemistry of *Rhododendron maximum* leaves II. Chloroplast structure and photosynthetic light response. *American Journal of Botany* **75**, 1526–1534.

Ninemets, U., Portsmuth, A., and Truus, L. (2002). Leaf structural and photosynthetic characteristics, and biomass allocation to foliage in relation to foliar nitrogen content and tree size in three Betula species. *Annals of Botany* **89**, 191–204.

Noodén, L.D. (1988). The phenomena of senescence and aging. In *Senescence and Aging in Plants* (L.D. Noodén and A.C. Leopold, Eds.), pp. 1–50, Academic Press, San Diego.

Nunez-Farfan, J., and Dirzo, R. (1989). Leaf survival in relation to herbivory in two tropical pioneer species. *Oikos* **54**, 71–74.

Oren, R., Schulze, E.-D., Matyssek, R., and Zimmermann, R. (1986). Estimating photosynthetic rate and annual carbon gain in conifers from specific leaf weight and leaf biomass. *Oecologia* **70**, 187–193.

Poorter, H., and Evans, J.R. (1998). Photosynthetic nitrogen-use efficiency of species that differ inherently in specific leaf area. *Oecologia* **116**, 26–37.

Reich, P.B., Uhl, C., Walters, M.B., and Ellsworth, D.S. (1991). Leaf lifespan as a determinant of leaf structure and function among 23 tree species in Amazonian forest communities. *Oecologia* **86**, 16–24.

Reich, P.B., Walters, M.B., and Ellsworth, D.S. (1992). Leaf life-span in relation to leaf, plant and stand characteristics among diverse ecosystems. *Ecological Monograph* **62**, 365–3921.

Reich, P.B., Kloeppel, B.D., Ellsworth, D.S., and Walters, M.B. (1995). Different photosynthesis-nitrogen relations in deciduous hardwood and evergreen coniferous tree species. *Oecologia* **104**, 24–30.

Rothstein, D.E., and Zak, D.R. (2001). Photosynthetic adaptation and acclimation to exploit seasonal periods of direct irradiance in three temperate, deciduous-forest herbs. *Functional Ecology* **15**, 722–731.

Schulze, E.D., Kuppers, M., and Matyssek, R. (1986). The roles of carbon balance and branching pattern in the growth of woody species. In *On the Economy of Plant Form and Function* (T.J. Givnish, Ed.), pp. 585–602. Cambridge University Press, Cambridge, UK.

Seiwa, K. (1998). Advantages of early germination for growth and survival of seedlings of *Acer mono* under different overstorey phenologies in deciduous broad-leaved forests. *Journal of Ecology* **86**, 219–228.

Seiwa, K., and Kikuzawa, K. (1991). Phenology of tree seedlings in relation to seed size. *Canadian Journal of Botany* **69**, 532–538.

Seiwa, K., and Kikuzawa, K. (1996). Importance of seed size for the establishment of seedlings of five deciduous broad-leaved tree species. *Vegetatio* **123**, 51–64.

Sestak, Z. (1981). Leaf ontogeny and photosynthesis. In *Physiological Processes Limiting Plant Growth* (C.B. Johnson, Ed.), pp. 147–158. Butterworths, London.

Shaver, G.R. (1981). Mineral nutrition and leaf longevity in an evergreen shrub, *Ledum palustre* ssp. *decumbens*. *Oecologia* **49**, 362–365.

Skillman, J.B., Strain, B.R., and Osmond, C.B. (1996). Contrasting patterns of photosynthetic acclimation and photoinhibition in two evergreen herbs from a winter deciduous forest. *Oecologia* **107**, 446–455.

Small, E. (1972). Photosynthetic rates in relation to nitrogen recycling as an adaptation to nutrient deficiency in peat bog plants. *Canadian Journal of Botany* **50**, 2227–2233.

Sobrado, M.A. (1991). Cost-benefit relationships in deciduous and evergreen leaves of tropical dry forest species. *Functional Ecology* **5**, 608–616.

Takeda, H., Ishida, Y., and Tsutsumi, T. (1987). Decomposition of leaf litter in relation to litter quality and site conditions. *Memoirs of the College of Agriculture, Kyoto University* **180**, 17–38.

Thomas, H., and Stoddart, J.L. (1980). Leaf senescence. *Annual Review of Plant Physiology* **31**, 83–111.

Tsuchiya, T. (1991). Leaf life span of floating-leaved plants. *Vegetatio* **7**, 149–160.

Williams, A.G., and Whitham, T.G. (1986). Premature leaf abscission: an induced plant defense against gall aphids. *Ecology* **67**, 1619–1627.

Williams, K., Field, C.B., and Mooney, H.A. (1989). Relationships among leaf construction cost, leaf longevity, and light environment in rain-forest plants of the genus Piper. *American Naturalist* **133**, 198–211.

Woodman, R.L., and Fernandes, G.W. (1991). Differential mechanical defense: herbivory, evapotranspiration, and leaf-hairs. *Oikos* **60**, 11–19.

Wright, I.J., and Westoby, M. (2002). Leaves at low versus high rainfall: coordination of structure, lifespan and physiology. *New Phytologist* **155**, 403–416.

Zots, G., and Winter, K. (1994). Photosynthesis of a tropical canopy tree, *Ceiba pentandra*, in a lowland forest in Panama. *Tree Physiology* **14**, 1291–1301.

26

Light Control of Senescence

Larry D. Noodén and Michael J. Schneider

I. Introduction

For a long time, it has been recognized that leaf senescence is influenced by light, but this is mostly acceleration of senescence by darkness or dim light, which tend to be unnatural treatments (Meyer, 1918; Simon, 1967). The role of light in actually controlling senescence under natural conditions is not as clear. Nonetheless, considerable evidence indicates that light is an important environmental regulator of leaf senescence, and it supplies important environmental signals.

Traditionally, the role of light in leaf senescence has been viewed mainly in terms of its ability to provide the energy for photosynthesis thereby sustaining life, but clearly, light does more. Like the light control of stem elongation (Quail *et al.*, 1995; Smith, 1995; Christie and Briggs, 2001), the light control of leaf senescence is turning out to be complex. In fact, it may be even more complex than for growth processes, because light also controls the formation of the photosynthetic apparatus whose components are common measures of senescence, and it may cause phototoxicity. The relevant literature, and therefore also this chapter, focuses primarily on leaves, although the senescence of other organs may also respond to light. Light has three kinds of information that is useful in guiding/signaling plant development including senescence: (1) quantity, e.g., light dosage as well as darkness; (2) quality, e.g., spectral composition, especially the ratio of red/far-red (R/FR) light;

and (3) timing, e.g., duration or photoperiod. In this chapter, we will summarize the diverse literature dealing with light control of leaf senescence, with the admonition that this story is closer to its beginning than its end. Nonetheless, it can be seen that all three types of light information are used in regulating senescence.

The light treatments used to study light effects on leaf senescence fall into the following groups: darkness, light quantity or intensity, light quality, photoperiod and high (excess) light. To some extent, these groupings also reflect the natural light controls, and this chapter will, therefore, be organized around these groupings except that phototoxicity will be covered in Chapter 18.

Before getting into the effects of these different light treatments on leaf senescence, it should be noted that the parameters used to measure leaf senescence are primarily chloroplastic, e.g., chlorophyll and also total protein (mainly in chloroplasts, Chapter 1), and these may be influenced by light independently of senescence. Ideally, longevity should also be measured; however, that is rare.

As in other aspects of development, correlative controls (the influence of one part on another) are important in light regulation of senescence. In this way, light may act on one organ to produce changes at others, but this complication is usually not taken into account.

Not only does the light control of leaf senescence reflect important environmental controls of plant development, but it has many practical applications as well. Of course, light is required for the photosynthesis that sustains plant growth and long-term survival, but light controls are much more complex than that (Quail *et al.*, 1995; Smith, 1995; Christie and Briggs, 2001). Some problems in the culture of plants under artificial lighting, especially at lower intensities, may be due in part to miscuing by the unnatural spectral quality of most artificial light sources. Even the postharvest longevity of green vegetables and leaf-bearing cut flowers may be influenced by these light signals.

II. Effects of Darkness

An old and extensive literature shows that light deprivation, particularly darkness, accelerates leaf senescence in many species (Meyer, 1918; Simon, 1967; Goldthwaite and Laetsch, 1967). Quite reasonably, it was assumed for many years that the primary action of light on leaf senescence was through photosynthesis, i.e., lack of light for photosynthesis caused starvation and thereby senescence and death. Haber *et al.* (1969) were among the first to challenge this idea, and one of their lines of evidence was that 3-(3,4-dichlorophenyl)-1,1-dimethylurea (DMCU), an inhibitor of photosynthesis, did not block the light retardation of leaf senescence. Although this experiment is open to the limitation that DCMU does not inhibit cyclic photophorylation (De Greef *et al.*, 1971), it seems unlikely that this illumination could sustain life just through cyclic photophosphorylation. Thimann, Tetley and Krivak (1977) also observed that DCMU does not block the senescence-delaying effect of light even as it blocks the light-induced increase in sugars. However, some (Goldthwaite and Laetsch, 1967; Cuello *et al.*, 1987) have reported that DCMU does prevent the light effect in some tissues, and it is not clear why these differences exist.

The starvation idea seemed to be supported by the observation that supplying sugar to leaves subjected to darkness could delay senescence (Goldthwaite and Laetsch, 1967), but many other reports show that sugars may actually accelerate Chl loss in some leaves, even in darkness (e.g., sugar repression, see Section III below and Table 15-1 in Chapter 15 on whole plant senescence). Later in this chapter, it will be shown that quantities of light too small to contribute significantly to photosynthesis can retard leaf senescence very effectively, which also argues against the idea that light prevents senescence simply by providing energy for photosynthesis. In addition, it should be noted here that senescence induced by darkness differs from natural senescence in some significant ways, e.g., changes in ultrastructure (Butler and Simon, 1971; Noodén, 1988; Weaver *et al.*, 1998), and therefore, darkness may induce more changes than just those in natural senescence. Furthermore, the effects of darkness are under correlative control, i.e., darkness has a more rapid effect when applied to a single leaf than when applied to a whole plant (Weaver and Amasino, 2001), which also indicates that more than simple deprivation of photosynthesis is involved.

III. Effects of Light Quantity

Although the induction of senescence by darkness could be viewed as an extreme and unnatural case, photosynthesis and light excitation pressure may sometimes be involved in the senescence-delaying effects of light (Ono *et al.*, 2001).

Light dosage effects on leaf senescence are best viewed relative to the CO_2 compensation point, that light intensity where photosynthetic CO_2 fixation equals or compensates for the release of CO_2 by respiration (Veierskov, 1987). In these experiments, white light intensities below the compensation point do not retard senescence (Chl loss) over a long time in detached leaves from three different species (pea, oat and fuchsia). Only irradiance at or above the compensation point maintains the leaf Chl. It was suggested that depletion of stored resources in darkness or in light below the compensation point is important in initiating senescence. Interestingly, increasing the light intensity above 17 $\mu m\ m^{-2}\ s^{-1}$ actually decreases Chl retention suggesting phototoxicity in these leaves. Similarly, shading (decreasing the light intensity from 360 to 90 $\mu m\ m^{-2}\ s^{-1}$) delays the loss of photosynthetic components and extends rye grass leaf longevity (Mae *et al.*, 1993).

Another approach to the effects of light doses is to compare continuous white light (CWL) versus pulsed white light. In papaya leaf disks, CWL (*ca.* 55 $\mu m\ m^{-2}\ s^{-1}$) does not retard Chl loss relative to darkness, but 5-min pulses of white light (*ca.* 55 $\mu m\ m^{-2}\ s^{-1}$) given at 12-h intervals do (Biswal and Choudhury, 1986). Since the irradiance in the CWL treatment is fairly low, it is difficult to invoke phototoxicity; however, phototoxicity can occur at lower light intensities (Chapter 18; Tyystjärvi and Aro, 1996). Clearly, pulsed light is very effective in retarding Chl loss, so the light effects are generally not simply "dose"-related. Although excitation pressure is an important factor in the formation and maintenance of the photosynthetic machinery, it may not be important in senescence processes per se (Ono *et al.*, 2001). In the next section, it can be seen that light quality is important.

IV. Effects of Light Quality

The results of studies on the effects of different kinds of light, particularly red (R) and far-red (FR), on leaf senescence are quite complex, and they demonstrate some differences among species and among different conditions. Phytochrome is implicated as an important photoreceptor for light retardation of leaf senescence, but it does not seem to be the only photosensor, i.e., the blue-light receptors, cryptochrome and phototropin, may also be involved.

The treatments applied to alter senescence generally fall into the low fluence category, and they are usually given during or around a dark period. The treatments consist of continuous colored light in place of white light (WL), pulses of light given during an extended dark period or "end-of-the-day" treatments (which are actually given in lieu of daylight at the start of the dark period as opposed to being added to WL at the end of the day).

Several studies (Haber *et al.*, 1969; Cuello *et al.*, 1987, 1989; Van Doorn and Van Lieburg, 1993) have shown that continuous R light does inhibit Chl loss compared to darkness, but FR usually does not with some exceptions (Sugiura, 1963). The vast majority of the studies on the effects of different light types use short (usually 30-sec to 20-min) pulses of R and/or FR given repeatedly at intervals during a prolonged (3- to 10-day) dark period (Sugiura, 1963; De Greef *et al.*, 1971; Pfeiffer and Kleudgen, 1980; Steinitz *et al.*, 1980; Tucker, 1981; Biswal and Sharma, 1976; Sen *et al.*, 1984; Biswal and Choudhury, 1986; Cuello *et al.*, 1989; Kappers *et al.*, 1998). In many cases, these R pulses retard senescence as effectively as continuous WL (Biswal and Sharma, 1976; Steinitz *et al.*, 1980; Tucker, 1981; Kappers *et al.*, 1998). Even if the pulses of FR have no effect by themselves, they may reverse the R pulses. Not surprisingly, however, the FR pulses do not always reverse the R pulses; in some cases, the action of the R may have proceeded beyond reversibility by the time the FR pulse was given. Sometimes, FR alone will accelerate Chl loss (Sugiura, 1963; Clark *et al.*, 1991; Rousseaux *et al.*, 1997); however, FR may also retard Chl loss (Okada and Katoh, 1998). Only a few studies (Pfeiffer and Kleudgen, 1980; Clark *et al.*, 1991; Okada *et al.*, 1992) use "end-of-the-day" treatments. These also indicate that R retards senescence, while FR reverses the R effect. Clearly, phytochrome mediates the light retardation of leaf senescence in many species, but not all. In bean leaves and dogwood, phytochrome may not be involved (Goldthwaite, 1974; Tucker, 1981). Interestingly, this difference cannot be explained away as a difference between sun and shade plants, for bean is a sun plant and dogwood a shade plant. At this point, it is not possible to say which type of phytochrome mediates the low fluence responses.

In cases where phytochrome is not the photoreceptor mediating light retardation of senescence, then what is? Of course, those (see Section II above) invoking light action through photosynthesis would point to Chl and the accessory pigments. Although Haber *et al.* (1969) did report that the action spectrum for light retardation of Chl loss resembled Chl, they argued against mediation through photosynthesis. Others (Cuello *et al.*, 1989) have reported that continuous blue (B) light and even green (G) retard Chl loss, and the photosynthesis inhibitor DCMU blocked this effect. Interestingly, FR does produce some reversal of B. Several possible explanations exist for the B effects including action through Chl, through blue absorption by phytochrome or through the blue-light receptors.

Few of these light effects by R, B, or FR clearly qualify as HIRs (high irradiance responses), although the full-sunlight studies are HIRs and some of those cases with continuous irradiation might be. HIRs do not show R–FR reversibility and they are proportional to irradiance or fluence rates (Taiz and Zeiger, 1998). Phototoxicity is a different phenomenon, and that is discussed in Chapter 18. The promotion of the senescence of lower leaves in soybean and sunflower plants by continuous FR supplements to lower the R/FR ratio probably is a HIR (Guiamét *et al.*, 1989; Rousseaux *et al.*, 1996). These effects are of considerable ecological significance, because light passing through the upper leaves under field conditions would have reduced R relative to FR, and this change in R/FR might serve as a signal to trigger senescence of the lowest leaves, so that their resources can be reallocated in the upper shoot. Interestingly, the R/FR ratio seems to be more critical than light intensity in triggering the senescence of shaded leaves in many but not all cases (Smith, 1995; Rousseaux *et al.*, 1996; Ono *et al.*, 2001).

This senescence pattern observed in the growing plants, such as sunflower here, is termed progressive senescence and is under correlative control, i.e. induction by the growing apex (Noodén, 1980). This correlative control can be influenced by the R/FR given to the shoot apex as well (Guiamét *et al.*, 1989). Clearly, light controls leaf senescence through phytochrome under many circumstances, but these controls are complex. Blue-light receptors and possibly even chlorophyll may also be involved.

V. Photoperiod Effects

The early work (1930s) on photoperiod effects was directed mainly toward control of autumnal leaf senescence and shedding in woody plants and the culture of plants in controlled environments.

Autumnal senescence occurs at a time of decreasing day length, and therefore, short days (SDs) would be expected to cause leaf senescence/shedding in trees and herbaceous plants if they are photoperiodic. In many, but apparently not all species, autumnal senescence is induced by SDs (Matzke, 1936; Jester and Kramer, 1939; Ashby, 1950; Olmstead, 1951; Krizek *et al.*, 1966). Interestingly, street lights delay leaf senescence and shedding in many tree species (Matzke, 1936). The effect of this night interruption tends to be greater close to the light source and more pronounced in areas with milder winters (Larry Noodén, personal observations). Under continuous days in a glasshouse, some leaves on red maple saplings lasted almost 2 years, far beyond their natural longevity (Jester and Kramer, 1939). Since SDs with interrupted nights behave like long day (LD) photoperiods, the SD-photoperiod effects are not due simply to light dosage (Ashby, 1951; Krizek *et al.*, 1966).

LD photoperiods can also promote leaf senescence and plant death in some species (Schwabe, 1970; Khan and Padhy, 1978; Kang and Cleland, 1990; Trippi and Brulfert, 1973; Matta *et al.*, 1979; Kar, 1986). Longer day lengths may promote senescence proportionately more in excised rice leaves (Kar, 1986) and this suggests possible phototoxicity; however, in butterfly flower leaves, day lengths beyond 14 h promote less (Matta *et al.*, 1979). In Arabidopsis, LDs promote leaf senescence at 300 μm m^{-2} s^{-1}; however, when the light intensity is reduced to 180 μm m^{-2} s^{-1}, LDs do not promote leaf senescence (Noodén *et al.*, 1996). Here, what looked like a photoperiod effect turned out to be a light-dose effect.

Because the reproductive structures commonly trigger leaf senescence (correlative controls) and photoperiod often controls their development, the effect of photoperiod on leaf senescence could be indirect, via the reproductive structures. This does not seem to be the case in cocklebur (Krizek *et al.* , 1966), but it may apply in other monocarpic plants where the reproductive structures play such a dominant role (Noodén, 1980, Chapter 15 of this volume). Nonetheless, photoperiod can also alter senescence in detached leaves (Schwabe, 1970; Misra and Biswal, 1973; Kar, 1986; Kang and Cleland, 1990) and therefore can also act directly on the leaves.

Another important effect of flower-inducing photoperiods is the cessation of vegetative growth as the plants shift their resources to reproductive development (Noodén, 1980). For example, LDs induce apex senescence in certain pea varieties (Proebsting and Davies, 1978). The result is that "worn-out" leaves, etc. are not replaced, and eventually this could contribute to the death of the plant.

In early studies on the effects of controlled environments on plant development, it was noticed that some species grow poorly or even die under continuous days. For example, continuous illumination injures geranium, tobacco, coleus and potato plants, while it actually kills tomato plants (Arthur *et al.*, 1930; Wheeler and Tibbitts, 1986). Others such as winter wheat actually grow better under continuous illumination, and many species are not significantly affected. The detrimental effects of continuous illumination may be photoperiod effects or phototoxicity; however, in tomato plants, it has been ascribed to interference with the plants' endogenous rhythms (Highkin and Hanson, 1954). Generally, the photoperiod effects on leaf senescence have not been tested to determine if they are due to insufficient light (SDs) or excess light (LDs); however, most are probably true photoperiod effects as opposed to light-dose effects as in Section III above.

VI. Relationships between Light Effects on Senescence and Hormones

All of the major hormones can influence senescence of leaves and other parts (Noodén and Leopold, 1988). Cytokinin seems particularly important in retarding senescence, and ethylene may be an important promoter, even in leaves. Recently, brassinosteroids have also been implicated as senescence promoters (Section VII below; Chory and Li, 1997). Light has been shown to influence positively or negatively the levels of most of these hormones (Wareing and Thompson, 1976; Hart, 1988), but there are relatively few reports showing a direct linkage between the light effects on senescence and those on the hormone concentrations, i.e., that light influences senescence through an effect on the hormones. Simple correlations, e.g., light increasing the hormone along with retardation of senescence, are a good first step but do not themselves establish causality. Likewise, the ability of exogenous applications to substitute for or duplicate the action of light are good first steps, but again more is needed to establish causality.

A related question is: Do the hormones and light act through the same pathway? One simple approach is to determine whether or not light can exert effects above those produced by saturating doses of hormones or vice versa with hormones acting on saturating light treatments, but data of this type are very limited. For example, R or B pulses could retard Chl loss in papaya leaf disks already treated with saturating concentrations of cytokinin (Biswal and Choudhury, 1986). In alstroemeria cuttings, R is able to further delay Chl loss when applied to cuttings treated with doses of gibberellins that maximally delay Chl loss (Kappers *et al.*, 1998). These data suggest that hormones and light may act on different pathways when they delay senescence. On the other hand, sometimes light does not produce an additional delay in senescence when combined with a saturating dose of cytokinin (Sugiura, 1963), and cytokinin is sometimes effective in delaying senescence in darkness but not in light (Cuello *et al.*, 1987). These data suggest light and cytokinin act through the same pathway in some cases. Since light has been shown to act differently on different species or even on the same species under different conditions, the role of hormones may likewise differ between species and even with the circumstances.

Apex senescence in the G2 pea line has been probed through photoperiod manipulations, and it appears to be linked to gibberellins which retard apex senescence (Proebsting and Davies, 1978).

Although the data are still quite limited, they do suggest that light effects on senescence, particularly retardation of senescence, may sometimes be mediated by hormones with cytokinin and gibberellin being prime candidates.

VII. Genetic Alterations of Light Control of Senescence

The information available on genetic controls of the regulation of senescence by light is very incomplete. Nonetheless, it provides a useful glimpse at what can be achieved through genetic probes of these light controls.

Due to the technical ease of screening for mutations to inhibit dark-induced leaf yellowing, several of these have been reported (Chapter 5). All of the stay-green mutations tested delay dark-induced leaf yellowing and sometimes other senescence processes (Thomas, 1987; Canfield *et al.*, 1995). These observations do indicate that natural senescence and dark-induced senescence share some biochemistry even if they are not identical.

Given the usage of chloroplast parameters to measure leaf senescence, it is difficult to separate the effects of light on the amounts of the photosynthetic machinery from its effects on senescence per se. It is well known that light promotes chloroplast development including chlorophyll formation (Scheer, 1991), and therefore, all of these light effects would counter what we normally call senescence, but are they specifically antisenescence effects?

Some evidence (see Section VI above) shows that light may act through effects on the cytokinin, gibberellin and ethylene hormones. Genetic studies indicate that brassinosteroids may mediate some light effects (Chory and Li, 1997). Thus, it is significant that *det-2*, which blocks brassinosteroid synthesis, also inhibits leaf senescence. Exogenous applications indicate they do, in fact, promote senescence (Clouse and Sasse, 1998).

Engineering tobacco plants to overexpress PHYA inhibits leaf senescence (Cherry *et al.*, 1991; Jordan *et al.*, 1995), but overexpression of PHYB in potato (Thiele *et al.*, 1999) has similar effects. The excess PHYA also counteracts the senescence-promoting effects of continuous FR (Rousseaux *et al.*, 1997). Although these observations do not themselves produce a general explanation of how light controls senescence, it is clear that the use of mutants and genetically-engineered plants will enable the dissection of the roles of the different photoreceptors in the light control of senescence.

References

Arthur, J.W., Guthrie, J.D., and Newell, J.M. (1930). Some effects of artificial climates on the growth and chemical composition of plants. *American Journal of Botany* **17**, 416–482.

Ashby, E. (1950). Studies in the morphogenesis of leaves VI. Some effects of day upon leaf shape in *Ipomoea caerula. New Phytologist* **49**, 375–387.

Biswal, B., and Choudhury, N.K. (1986). Photocontrol of chlorophyll loss in papaya leaf discs. *Plant and Cell Physiology* **27**, 1439–1444.

Biswal, U.C., and Sharma, R. (1976). Phytochrome regulation of senescence in detached barley leaves. *Zeitschrift Für Pfanzenphysiologie* **80**, 71–76.

Butler, R.D., and Simon, E.W. (1971). Ultrastructural aspects of senescence in plants. *Advances in Gerontology Research* **3**, 73–129.

Canfield, M.R., Guiamét, J.J., and Noodén, L.D. (1995). Alteration of soybean seedling development in darkness and light by the stay-green mutation *cytG* and *Gd1d2*. *Annals of Botany* **75**, 143–150.

Cherry, J.R., Hershey, H.P., and Vierstra, R.D. (1991). Characterization of tobacco expressing functional oat phytochrome. Domains responsible for the rapid degradation of Pfr are conserved between monocots and dicots. *Plant Physiology* **96**, 775–785.

Chory, J., and Li, J. (1997). Gibberellins, brassinosteroids and light-regulated development. *Plant Cell and Environment* **20**, 801–806.

Christie, J.M., and Briggs, W.R. (2001). Blue light sensing in higher plants. *Journal of Biological Chemistry* **276**, 11457–11460.

Clark, D.G., Kelly, J.W., and Decoteau, D.R. (1991). Influence of end-of-day red and far-red light on potted roses. *Journal Environmental Horticulture* **9**, 127–130.

Clouse, S.D., and Sasse J.M. (1998). Brassinosteroids: Essential regulators of plant growth and development. *Annual Review of Plant Physiology and Plant Molecular Biology* **49**, 427–451.

Cuello, J., Quiles, M.J., and Sabater, B. (1987). Control by phytochrome of the synthesis of protein related to senescence in chloroplasts of barley (*Hordeum vulgare*). *Physiologia Plantarum* **71**, 341–344.

Cuello, J., Quiles, M.J., and Sabater, B. (1989). Differential effects of light and phytohormones in the senescence of apical and basal segments of barley leaves. *Phyton* **50**, 133–140.

De Greef, J., Butler, W.L., and Roth, T.F. (1971). Control of senescence in *Marchantia* by phytochrome. *Plant Physiology* **48**, 407–412.

Goldthwaite, J. (1974). Energy metabolism of *Rumex* leaf tissue in the presence of senescence-regulating hormones and sucrose. *Plant Physiology* **54**, 399–403.

Goldthwaite, J.J., and Laetsch, W.M. (1967). Regulation of senescence in bean leaf disks by light and chemical growth regulators. *Plant Physiology* **42**, 1757–1762.

Guiamét, J.J., Willemoes, J.G., and Montaldi, E.R. (1989). Modulation of progressive leaf senescence by the red:far-red ratio of incident light. *Botanical Gazette* **150**, 148–151.

Haber, A.H., Thompson, P.J., Walne, P.L., and Triplett, L.L. (1969). Nonphotosynthetic retardation of chloroplast senescence by light. *Plant Physiology* **44**, 1619–1628.

Hart, J.W. (1988). *Light and Plant Growth*. Unwin Hyman, London.

Highkin, H.R., and Hanson, J.B. (1954). Possible interaction between light-dark cycles and endogenous daily rhythms on the growth of tomato plants. *Plant Physiology* **29**, 301–302.

Jester, J.R., and Kramer, P.J. (1939). The effect of length of day on the height growth of certain forest tree seedlings. *Journal of Forestry* **37**, 796–803.

Jordan, E.T., Hatfield, P.M., Hondred, D., Talon, M., Zeevaart, J.A.D., and Vierstra, R.D. (1995). Phytochrome A overexpression in transgenic tobacco: correlation of dwarf phenotype with high concentrations of phytochrome in vascular tissue and attenuated gibberellin levels. *Plant Physiology* **107**, 797–805.

Kang, B.G., and Cleland, C.F. (1990). Characterization of senescence in *Lemna gibba* G3: A determinate growth system. *Plant and Cell Physiology* **31**, 661–666.

Kappers, I.F., Jordi, W., Tsesmetzis, N., Maas, F.M., and Van Der Plas, L.H.W. (1998). Ga(4) does not require conversion into Ga(1) to delay senescence of Alstroemeria hybrida leaves. *Journal of Plant Growth Regulation* **17**, 89–93.

Kar, M. (1986). The effect of photoperiod on chlorophyll loss and lipid peroxidation in excised senescing rice leaves. *Journal of Plant Physiology* **123**, 389–394.

Khan, P.A., and Padhy, B. (1978). Effects of photoperiod and certain chemicals on chlorophyll retention of excised leaves of *Eleusine corocana* during senescence. *Experimental Gerontology* **13**, 19–24.

Krizek, D.T., McIlrath, W.J., and Vergara, B.S. (1966). Photoperiodic induction of senescence in Xanthium plants. *Science* **151**, 95–96.

Mae, T., Thomas, H., Gay, A.P., Makino, A., and Hidema, J. (1993). Leaf development in *Lolium temulentum*: photosynthesis and photosynthetic proteins in leaves senescing under different irradiances. *Plant Cell Physiology* **34**, 391–399.

Matta, F.B., Thomason, R.C., and Pan, C.Y. (1979). The effects of photoperiod and ethylene inhibitors on flowering and vegetative senescence of *Schizanthus x wisetonensis* and *Nemesia strumosa*. *HortScience* **14**, 503–504.

Matzke, E.B. (1936). The effect of street lights in delaying leaf-fall in certain trees. *American Journal Botany* **23**, 446–452.

Meyer, A. (1918). Eiweissstoffwechsel und Vergilben der Laubblätter von Tropaeolum majus. *Flora* **111**, 85–127.

Misra, G., and Biswal, U.C. (1973). Factors concerned in leaf senescence. I. Effects of age, chemicals, petiole and photoperiod on senescence is detached leaves of *Hibiscus rosa-sinensis*. L. *Botanical Gazette* **134**, 5–11.

Noodén, L.D. (1980). Senescence in the whole plant. In *Senescence in Plants* (K.V. Thimann, Ed.). CRC Press, Boca Raton, FL.

Noodén, L.D. (1988). The phenomena of senescence and aging. In *Senescence and Aging in Plants* (L.D. Noodén and A.C. Leopold, Eds.), pp. 1–50. Academic Press, San Diego, CA.

Noodén, L.D., and Leopold, A.C. (Eds.) (1988). *Senescence and Aging in Plants*. Academic Press, San Diego, CA.

Noodén, L.D., Hillsberg, J.W., and Schneider, M.J. (1996). Induction of leaf senescence in *Arabidopsis thaliana* by long days through a light-dosage effect. *Physiologia Plantarum* **96**, 491–495.

Okada, K., and Katoh, S. (1998). Two long-term effects of light that control the stability of proteins related to photosynthesis during senescence of rice leaves. *Plant Cell Physiology* **39**, 394–404.

Okada, K., Inoue, Y., Satoh, K., and Katoh, S. (1992). Effects of light on degradation of chlorophyll and proteins during senescence of detached rice leaves. *Plant Cell Physiology* **33**, 1183–1191.

Olmstead, C.E. (1951). Experiments on photoperiodism, dormancy, and leaf age and abscission in sugar maple. *Botanical Gazette* **112**, 365–393.

Ono, K., Nishi, Y., Watanabe, A., and Terashima, I. (2001). Possible mechanisms of adaptive leaf senescence. *Plant Biology* **3**, 234–243.

Pfeiffer, H., and Kleudgen, H.K. (1980). Untersuchungen zur Phytochromsteuerung der Seneszenz im Photosyntheseapparat von *Hordeum vulgare* L. *Zeitschrift Für Pfanzenphysiologie* **100**, 437–445.

Proebsting, W.M., and Davies, P.J. (1978). Photoperiod-induced changes in gibberellin metabolism in relation to senescence in genetic lines of peas. *Plant Physiology* **61**, 112.

Quail, P.H., Boylan, M.T., Parks, B.M., Short, T.W., Xu, Y., and Wagner, D. (1995). Phytochromes: Photosensory perception and signal transduction. *Science* **268**, 675–680.

Rousseaux, M.C., Hall, A.J., and Sanchez, R.A. (1996). Far-red enrichment and photosynthetically active radiation level influence leaf senescence in field-grown sunflower. *Physiologia Plantarum* **96**, 217–224.

Rousseaux, M.C., Ballare, C.L., Jordan, E.T., and Vierstra, R.D. (1997). Directed overexpression of *PHYA* locally suppresses stem elongation and leaf senescence responses to far-red radiation. *Plant Cell and Environment* **20**, 1551–1558.

Scheer, H. (Ed.) (1991). *Chlorophylls*. CRC Press, Boca Raton, FL.

Schwabe, W.W. (1970). The control of leaf senescene in *Kleinia articulata* by photoperiod. *Annals of Botany* **34**, 43–55.

Sen, N.K., Patra, H.K., and Mishra, D. (1984). Phytochrome regulation of biochemical and enzymatic changes during senescence of excised rice leaves. *Zeitschrift Für Pflanzenphysiologie* **113**(2), 95–103.

Simon, E.W. (1967). Types of leaf senescence. *Symposium of the Society for Experimental Biology* **21**, 215–230.

Smith, H. (1995). Physiological and ecological function within the phytochrome family. *Annual Review of Plant Physiology and Plant Molecular Biology* **46**, 289–315.

Steinitz, B., Cohen, A., and Leshem, B. (1980). Factors controlling the retardation of chlorophyll degradation during senescence of detached statice (*Limonium sinuatum*) flower stalks. *Zeitschrift Für Pfanzenphysiologie* **100**, 343–349.

Sugiura, M. (1963). Effect of red and far-red light on protein and phosphate metabolism in tobacco leaf disks. *Botanical Magazine* **76**, 174–180.

Taiz, L., and Zeiger, E. (1998). *Plant Physiology* (2nd ed.). Sinauer Associates, Sunderland, MA.

Thiele, A., Herold, M., Lenk, I., Quail, P.H., and Gatz, C. (1999). Heterologous expression of Arabidopsis phytochrome B in transgenic potato influences photosynthetic performance and tuber development. *Plant Physiology* **120**, 73–81.

Thimann, K.V., Tetley, R.M., and Krivak, B.M. (1977). Metabolism of oat leaves during senescence. V. Senescence in light. *Plant Physiology* **59**, 448–454.

Thomas, H. (1987). *Sid*: a Mendelian locus controlling thylakoid membrane disassembly in senescing leaves of *Festuca pratensis*. *Theoretical and Applied Genetics* **73**, 551–555.

Trippi, V.S., and Brulfert, J. (1973). Organization of the morphophysiologic unit in *Anagallis arvensis* and its relation with the perpetuation mechanism and senescence. *American Journal Botany* **60**, 641–647.

Tucker, D.J. (1981). Phytochrome regulation of leaf senescence in cucumber and tomato. *Plant Science Letters* **23**, 103–108.

Tyystjärvi, E., and Aro, E.M. (1996). The rate constant of photoinhibition, measured in lincomycin-treated leaves, is directly proportional to light intensity. *Proceedings of the National Academy of Sciences of the United States of America* **93**, 2213–2218.

Van Doorn, W.G., and Van Lieburg, M.J. (1993). Interaction between the effects of phytochrome and gibberellic acid on the senescence of *Alstroemeria pelegrina* leaves. *Physiologia Plantarum* **89**, 182–186.

Veierskov, B. (1987). Irradiance-dependent senescence of isolated leaves. *Physiologia Plantarum* **71**, 316–320.

Wareing, P.F., and Thompson, A.G. (1976). Rapid effects of red light on hormone levels. In *Light and Plant Development* (H. Smith, Ed.), pp. 285–294. Butterworths, London.

Weaver, L.M., and Amasino, R.M. (2001). Senescence is induced in individually darkened Arabidopsis leaves but inhibited in whole darkened plants. *Plant Physiology* **127**, 876–886.

Weaver, M., Gan, S., Quirino, B., and Amasino, R.M. (1998). A comparison of the expression patterns of several senescence-associated genes in response to stress and hormone treatment. *Plant Molecular Biology* **37**, 455–469.

Wheeler, R.M., and Tibbitts, T.W. (1986). Growth and tuberization of potato (*Solanum tuberosum*) under continuous light. *Plant Physiology* **80**, 801–804.

EPILOGUE

In reading/editing these chapters, three general points seemed clear to me. First, the study of programmed cell death and senescence is entering a new phase that will provide great expansion of our understanding of these processes. Twenty years ago, when gene cloning was starting to be applied widely, many believed the biochemistry of senescence would soon be revealed. A lot of advances have been made, but we still do not know what senescence or programmed cell death are at the molecular level. Moreover, we cannot be absolutely confident about how to measure it except by measuring the end product, death. Now, it does look like new tools, some deriving from molecular biology, will at last help us to understand exactly what this (these) process(es) is (are).

Second, there have been some serious lapses in overlooking evidence. The widely accepted idea that senescence is caused by a sugar deficiency, discussed in Chapter 15, is an important case in point. A field cannot flourish and grow if significant contrary evidence is widely ignored. Presently, I believe there is a problem with determining changes in gene expression (mRNA levels) using electrophoretic gels loaded (normalized) on the basis of rRNA. This can produce a misleading picture of gene expression during senescence (Chapter 1) and will complicate the identification of the controls of genes whose expression increases during senescence. Hopefully, this can be rectified before defensive postures are solidified, or it will take a generation to get past the problem and an important field will be hindered.

Third, there needs to be more dialogue between the disparate disciplines covered in this volume. This will bring new ideas and innovations in both directions. As genomics, proteomics and other new approaches reveal more information about coordinate gene expression, there is more interest in how all the cell components and cells themselves work together as systems. Bringing in ideas from these different disciplines will facilitate and stimulate the development of systems analysis. Hopefully, this volume will help to form bridges between the various diciplines related to senescence and programmed cell death.

Larry D. Noodén

INDEX